ELECTRIC POWER TRANSFORMER ENGINEERING

The ELECTRIC POWER ENGINEERING Series
Series Editor Leo L. Grigsby

Published Titles

Computational Methods for Electric Power Systems
Mariesa Crow

Distribution System Modeling and Analysis
William H. Kersting

Electric Drives
Ion Boldea and Syed Nasar

Electrical Energy Systems
Mohamed E. El-Hawary

Electric Power Substations Engineering
John D. McDonald and Leo L. Grigsby

Electric Power Transformer Engineering
James H. Harlow

***Electromechanical Systems, Electric Machines,
and Applied Mechatronics***
Sergey E. Lyshevski

The Induction Machine Handbook
Ion Boldea and Syed Nasar

***Linear Synchronous Motors:
Transportation and Automation Systems***
Jacek Gieras and Jerry Piech

Power Quality
C. Sankaran

Power System Operations and Electricity Markets
Fred I. Denny and David E. Dismukes

Forthcoming Titles

The Electric Generators Handbook
Ion Boldea and Leo L. Grigsby

Electric Power Distribution Handbook
Tom Short

ELECTRIC POWER TRANSFORMER ENGINEERING

Edited by
James H. Harlow

CRC PRESS

Boca Raton   London   New York   Washington, D.C.

Library of Congress Cataloging-in-Publication Data

Electric power transformer engineering / edited by James H. Harlow.
    p. cm. — (The Electric Power Engineering Series ; 9)
    Includes bibliographical references and index.
    ISBN 0-8493-1704-5 (alk. paper)
    1. Electric transformers. I. Harlow, James H. II. title. III. Series.

TK2551.E65 2004
621.31′4—dc21   2003046134

Preface

Transformer engineering is one of the earliest sciences within the field of electric power engineering, and power is the earliest discipline within the field of electrical engineering. To some, this means that transformer technology is a fully mature and staid industry, with little opportunity for innovation or ingenuity by those practicing in the field.

Of course, we in the industry find that premise to be erroneous. One need only scan the technical literature to recognize that leading-edge suppliers, users, and academics involved with power transformers are continually reporting novelties and advancements that would have been totally insensible to engineers of even the recent past. I contend that there are three basic levels of understanding, any of which may be appropriate for persons engaged with transformers in the electric power industry. Depending on day-to-day involvement, the individual's posture in the field can be described as:

- Curious — those with only peripheral involvement with transformers, or a nonprofessional lacking relevant academic background or any particular need to delve into the intricacies of the science
- Professional — an engineer or senior-level technical person who has made a career around electric power transformers, probably including other heavy electric-power apparatus and the associated power-system transmission and distribution operations
- Expert — those highly trained in the field (either practically or analytically) to the extent that they are recognized in the industry as experts. These are the people who are studying and publishing the innovations that continue to prove that the field is nowhere near reaching a technological culmination.

So, to whom is this book directed? It will truly be of use to any of those described in the previous three categories.

The *curious* person will find the material needed to advance toward the level of professional. This reader can use the book to obtain a deeper understanding of many topics.

The *professional*, deeply involved with the overall subject matter of this book, may smugly grin with the self-satisfying attitude of, "I know all that!" This person, like myself, must recognize that there are many transformer topics. There is always room to learn. We believe that this book can also be a valuable resource to professionals.

The *expert* may be so immersed in one or a few very narrow specialties within the field that he also may benefit greatly from the knowledge imparted in the peripheral specialties.

The book is divided into three fundamental groupings: The first stand-alone chapter is devoted to *Theory and Principles*. The second chapter, *Equipment Types*, contains nine sections that individually treat major transformer types. The third chapter, which contains 14 sections, addresses *Ancillary Topics* associated with power transformers. Anyone with an interest in transformers will find a great deal of useful information.

I wish to recognize the interest of CRC Press and the personnel who have encouraged and supported the preparation of this book. Most notable in this regard are Nora Konopka, Helena Redshaw, and Gail Renard. I also want to acknowledge Professor Leo Grigsby of Auburn University for selecting me to edit the "Transformer" portion of his *The Electric Power Engineering Handbook* (CRC Press, 2001), which forms the basis of this handbook. Indeed, this handbook is derived from that earlier work, with the addition of four wholly new chapters and the very significant expansion and updating of much of the other earlier work. But most of all, appreciation is extended to each writer of the 24 sections that comprise this handbook. The authors' diligence, devotion, and expertise will be evident to the reader.

James H. Harlow
Editor

Editor

James H. Harlow has been self-employed as a principal of Harlow Engineering Associates, consulting to the electric power industry, since 1996. Before that, he had 34 years of industry experience with Siemens Energy and Automation (and its predecessor Allis-Chalmers Co.) and Beckwith Electric Co., where he was engaged in engineering design and management. While at these firms, he managed groundbreaking projects that blended electronics into power transformer applications. Two such projects (employing microprocessors) led to the introduction of the first intelligent-electronic-device control product used in quantity in utility substations and a power-thyristor application for load tap changing in a step-voltage regulator.

Harlow received the BSEE degree from Lafayette College, an MBA (statistics) from Jacksonville State University, and an MS (electric power) from Mississippi State University. He joined the PES Transformers Committee in 1982, serving as chair of a working group and a subcommittee before becoming an officer and assuming the chairmanship of the PES Transformers Committee for 1994–95. During this period, he served on the IEEE delegation to the ANSI C57 Main Committee (Transformers). His continued service to IEEE led to a position as chair of the PES Technical Council, the assemblage of leaders of the 17 technical committees that comprise the IEEE Power Engineering Society. He recently completed a 2-year term as PES vice president of technical activities.

Harlow has authored more than 30 technical articles and papers, most recently serving as editor of the transformer section of *The Electric Power Engineering Handbook*, CRC Press, 2001. His editorial contribution within this handbook includes the section on his specialty, *LTC Control and Transformer Paralleling*. A holder of five U.S. patents, Harlow is a registered professional engineer and a senior member of IEEE.

Contributors

Dennis Allan
MerlinDesign
Stafford, England

Hector J. Altuve
Schweitzer Engineering
    Laboratories, Ltd.
Monterrey, Mexico

Gabriel Benmouyal
Schweitzer Engineering
    Laboratories, Ltd.
Longueuil, Quebec, Canada

Behdad Biglar
Trench Ltd.
Scarborough, Ontario,
    Canada

Wallace Binder
WBBinder
Consultant
New Castle, Pennsylvania

Antonio Castanheira
Trench Brasil Ltda.
Contegem, Minas Gelais, Brazil

Craig A. Colopy
Cooper Power Systems
Waukesha, Wisconsin

Robert C. Degeneff
Rensselaer Polytechnic Institute
Troy, New York

Scott H. Digby
Waukesha Electric Systems
Goldsboro, North Carolina

Dieter Dohnal
Maschinenfabrik Reinhausen
    GmbH
Regensburg, Germany

Douglas Dorr
EPRI PEAC Corporation
Knoxville, Tennessee

Richard F. Dudley
Trench Ltd.
Scarborough, Ontario, Canada

Ralph Ferraro
Ferraro, Oliver & Associates, Inc.
Knoxville, Tennessee

Dudley L. Galloway
Galloway Transformer
    Technology LLC
Jefferson City, Missouri

Anish Gaikwad
EPRI PEAC Corporation
Knoxville, Tennessee

Armando Guzmán
Schweitzer Engineering
    Laboratories, Ltd.
Pullman, Washington

James H. Harlow
Harlow Engineering Associates
Mentone, Alabama

Ted Haupert
TJ/H2b Analytical Services
Sacramento, California

William R. Henning
Waukesha Electric Systems
Waukesha, Wisconsin

Philip J. Hopkinson
HVOLT, Inc.
Charlotte, North Carolina

Sheldon P. Kennedy
Niagara Transformer
    Corporation
Buffalo, New York

Andre Lux
KEMA T&D Consulting
Raleigh, North Carolina

Arindam Maitra
EPRI PEAC Corporation
Knoxville, Tennessee

Arshad Mansoor
EPRI PEAC Corporation
Knoxville, Tennessee

Shirish P. Mehta
Waukesha Electric Systems
Waukesha, Wisconsin

Harold Moore
H. Moore & Associates
Niceville, Florida

Dan Mulkey
Pacific Gas & Electric Co.
Petaluma, California

Randy Mullikin
Kuhlman Electric Corp.
Versailles, Kentucky

Alan Oswalt
Consultant
Big Bend, Wisconsin

Paulette A. Payne
Potomac Electric Power
  Company (PEPCO)
Washington, DC

Dan D. Perco
Perco Transformer Engineering
Stoney Creek, Ontario, Canada

Gustav Preininger
Consultant
Graz, Austria

Jeewan Puri
Transformer Solutions
Matthews, North Carolina

Leo J. Savio
ADAPT Corporation
Kennett Square, Pennsylvania

Michael Sharp
Trench Ltd.
Scarborough, Ontario, Canada

H. Jin Sim
Waukesha Electric Systems
Goldsboro, North Carolina

Robert F. Tillman, Jr.
Alabama Power Company
Birmingham, Alabama

Loren B. Wagenaar
America Electric Power
Pickerington, Ohio

Contents

1

Theory and Principles

Dennis Allan
MerlinDesign

Harold Moore
H. Moore and Associates

Transformers are devices that transfer energy from one circuit to another by means of a common magnetic field. In all cases except autotransformers, there is no direct electrical connection from one circuit to the other.

When an alternating current flows in a conductor, a magnetic field exists around the conductor, as illustrated in Figure 1.1. If another conductor is placed in the field created by the first conductor such that the flux lines link the second conductor, as shown in Figure 1.2, then a voltage is induced into the second conductor. The use of a magnetic field from one coil to induce a voltage into a second coil is the principle on which transformer theory and application is based.

1.1 Air Core Transformer

Some small transformers for low-power applications are constructed with air between the two coils. Such transformers are inefficient because the percentage of the flux from the first coil that links the second coil is small. The voltage induced in the second coil is determined as follows.

$$E = N \, d\phi/dt \, 10^8 \qquad (1.1)$$

where N is the number of turns in the coil, $d\phi/dt$ is the time rate of change of flux linking the coil, and ϕ is the flux in lines.

At a time when the applied voltage to the coil is E and the flux linking the coils is ϕ lines, the instantaneous voltage of the supply is:

$$e = \sqrt{2} \, E \cos \omega t = N \, d\phi/dt \, 10^8 \qquad (1.2)$$

$$d\phi/dt = (\sqrt{2} \cos \omega t \, 10^8)/N \qquad (1.3)$$

The maximum value of ϕ is given by:

$$\phi = (\sqrt{2} \, E \, 10^8)/(2 \, \pi \, f \, N) \qquad (1.4)$$

Using the MKS (metric) system, where ϕ is the flux in webers,

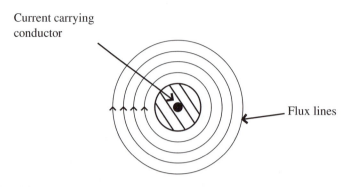

FIGURE 1.1 Magnetic field around conductor.

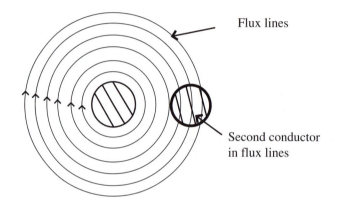

FIGURE 1.2 Magnetic field around conductor induces voltage in second conductor.

$$E = N \, d\phi/dt \qquad (1.5)$$

and

$$\phi = (\sqrt{2}E)/(2 \, \pi \, f \, N) \qquad (1.6)$$

Since the amount of flux ϕ linking the second coil is a small percentage of the flux from the first coil, the voltage induced into the second coil is small. The number of turns can be increased to increase the voltage output, but this will increase costs. The need then is to increase the amount of flux from the first coil that links the second coil.

1.2   Iron or Steel Core Transformer

The ability of iron or steel to carry magnetic flux is much greater than air. This ability to carry flux is called permeability. Modern electrical steels have permeabilities in the order of 1500 compared with 1.0 for air. This means that the ability of a steel core to carry magnetic flux is 1500 times that of air. Steel cores were used in power transformers when alternating current circuits for distribution of electrical energy were first introduced. When two coils are applied on a steel core, as illustrated in Figure 1.3, almost 100% of the flux from coil 1 circulates in the iron core so that the voltage induced into coil 2 is equal to the coil 1 voltage if the number of turns in the two coils are equal.

Continuing in the MKS system, the fundamental relationship between magnetic flux density (B) and magnetic field intensity (H) is:

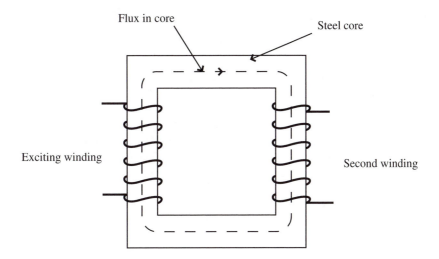

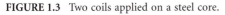

FIGURE 1.3 Two coils applied on a steel core.

$$B = \mu_0 H \tag{1.7}$$

where μ_0 is the permeability of free space $\equiv 4\pi \times 10^{-7}$ Wb A^{-1} m^{-1}.

Replacing B by ϕ/A and H by $(I\,N)/d$, where

ϕ = core flux in lines

N = number of turns in the coil

I = maximum current in amperes

A = core cross-section area

the relationship can be rewritten as:

$$\phi = (\mu\,N\,A\,I)/d \tag{1.8}$$

where

d = mean length of the coil in meters

A = area of the core in square meters

Then, the equation for the flux in the steel core is:

$$\phi = (\mu_0\,\mu_r\,N\,A\,I)/d \tag{1.9}$$

where μ_r = relative permeability of steel ≈ 1500.

Since the permeability of the steel is very high compared with air, all of the flux can be considered as flowing in the steel and is essentially of equal magnitude in all parts of the core. The equation for the flux in the core can be written as follows:

$$\phi = 0.225\,E/fN \tag{1.10}$$

where

E = applied alternating voltage

f = frequency in hertz

N = number of turns in the winding

In transformer design, it is useful to use flux density, and Equation 1.10 can be rewritten as:

$$B = \phi/A = 0.225\,E/(f\,A\,N) \tag{1.11}$$

where B = flux density in tesla (webers/square meter).

1.3   Equivalent Circuit of an Iron-Core Transformer

When voltage is applied to the exciting or primary winding of the transformer, a magnetizing current flows in the primary winding. This current produces the flux in the core. The flow of flux in magnetic circuits is analogous to the flow of current in electrical circuits.

When flux flows in the steel core, losses occur in the steel. There are two components of this loss, which are termed "eddy" and "hysteresis" losses. An explanation of these losses would require a full chapter. For the purpose of this text, it can be stated that the hysteresis loss is caused by the cyclic reversal of flux in the magnetic circuit and can be reduced by metallurgical control of the steel. Eddy loss is caused by eddy currents circulating within the steel induced by the flow of magnetic flux normal to the width of the core, and it can be controlled by reducing the thickness of the steel lamination or by applying a thin insulating coating.

Eddy loss can be expressed as follows:

$$W = K[w]^2[B]^2 \text{ watts} \qquad (1.12)$$

where

> K = constant
> w = width of the core lamination material normal to the flux
> B = flux density

If a solid core were used in a power transformer, the losses would be very high and the temperature would be excessive. For this reason, cores are laminated from very thin sheets, such as 0.23 mm and 0.28 mm, to reduce the thickness of the individual sheets of steel normal to the flux and thereby reducing the losses. Each sheet is coated with a very thin material to prevent shorts between the laminations. Improvements made in electrical steels over the past 50 years have been the major contributor to smaller and more efficient transformers. Some of the more dramatic improvements include:

- Development of cold-rolled grain-oriented (CGO) electrical steels in the mid 1940s
- Introduction of thin coatings with good mechanical properties
- Improved chemistry of the steels, e.g., Hi-B steels
- Further improvement in the orientation of the grains
- Introduction of laser-scribed and plasma-irradiated steels
- Continued reduction in the thickness of the laminations to reduce the eddy-loss component of the core loss
- Introduction of amorphous ribbon (with no crystalline structure) — manufactured using rapid-cooling technology — for use with distribution and small power transformers

The combination of these improvements has resulted in electrical steels having less than 40% of the no-load loss and 30% of the exciting (magnetizing) current that was possible in the late 1940s.

The effect of the cold-rolling process on the grain formation is to align magnetic domains in the direction of rolling so that the magnetic properties in the rolling direction are far superior to those in other directions. A heat-resistant insulation coating is applied by thermochemical treatment to both sides of the steel during the final stage of processing. The coating is approximately 1-μm thick and has only a marginal effect on the stacking factor. Traditionally, a thin coat of varnish had been applied by the transformer manufacturer after completion of cutting and punching operations. However, improvements in the quality and adherence of the steel manufacturers' coating and in the cutting tools available have eliminated the need for the second coating, and its use has been discontinued.

Guaranteed values of real power loss (in watts per kilogram) and apparent power loss (in volt-amperes per kilogram) apply to magnetization at 0° to the direction of rolling. Both real and apparent power loss increase significantly (by a factor of three or more) when CGO is magnetized at an angle to the direction of rolling. Under these circumstances, manufacturers' guarantees do not apply, and the transformer manufacturer must ensure that a minimum amount of core material is subject to cross-magnetization, i.e., where the flow of magnetic flux is normal to the rolling direction. The aim is to minimize the total core loss and (equally importantly) to ensure that the core temperature in the area is maintained within safe limits. CGO strip cores operate at nominal flux densities of 1.6 to 1.8 tesla (T). This value compares with 1.35 T used for hot-rolled steel, and it is the principal reason for the remarkable improvement achieved in the 1950s in transformer output per unit of active material. CGO steel is produced in two magnetic qualities (each having two subgrades) and up to four thicknesses (0.23, 0.27, 0.30, and 0.35 mm), giving a choice of eight different specific loss values. In addition, the designer can consider using domain-controlled Hi-B steel of higher quality, available in three thicknesses (0.23, 0.27, and 0.3 mm).

The different materials are identified by code names:

- CGO material with a thickness of 0.3 mm and a loss of 1.3 W/kg at 1.7 T and 50 Hz, or 1.72 W/kg at 1.7 T and 60 Hz, is known as M097–30N.
- Hi-B material with a thickness of 0.27 mm and a loss of 0.98 W/kg at 1.7T and 50 Hz, or 1.3 W/kg at 1.7 T and 60 Hz, is known as M103–27P.
- Domain-controlled Hi-B material with a thickness of 0.23 mm and a loss of 0.92 W/kg at 1.7T and 50 Hz, or 1.2 W/kg at 1.7 T and 60 Hz, is known as 23ZDKH.

The Japanese-grade ZDKH core steel is subjected to laser irradiation to refine the magnetic domains near to the surface. This process considerably reduces the anomalous eddy-current loss, but the laminations must not be annealed after cutting. An alternative route to domain control of the steel is to use plasma irradiation, whereby the laminations can be annealed after cutting.

The decision on which grade to use to meet a particular design requirement depends on the characteristics required in respect of impedance and losses and, particularly, on the cash value that the purchaser has assigned to core loss (the capitalized value of the iron loss). The higher labor cost involved in using the thinner materials is another factor to be considered.

No-load and load losses are often specified as target values by the user, or they may be evaluated by the "capitalization" of losses. A purchaser who receives tenders from prospective suppliers must evaluate the tenders to determine the "best" offer. The evaluation process is based on technical, strategic, and economic factors, but if losses are to be capitalized, the purchaser will always evaluate the "total cost of ownership," where:

Cost of ownership = capital cost (or initial cost) + cost of losses

Cost of losses = cost of no-load loss + cost of load loss + cost of stray loss

For loss-evaluation purposes, the load loss and stray loss are added together, as they are both current-dependent.

Cost of no-load loss = no-load loss (kW) × capitalization factor ($/kW)

Cost of load loss = load loss (kW) × capitalization factor ($/kW)

For generator transformers that are usually on continuous full load, the capitalization factors for no-load loss and load loss are usually equal. For transmission and distribution transformers, which normally operate at below their full-load rating, different capitalization factors are used depending on the planned load factor. Typical values for the capitalization rates used for transmission and distribution transformers are $5000/kW for no-load loss and $1200/kW for load loss. At these values, the total cost of ownership of the transformer, representing the capital cost plus the cost of power losses over 20 years, may be more than twice the capital cost. For this reason, modern designs of transformer are usually low-loss designs rather than low-cost designs.

Figure 1.4 shows the loss characteristics for a range of available electrical core-steel materials over a range of values of magnetic induction (core flux density).

The current that creates rated flux in the core is called the magnetizing current. The magnetizing circuit of the transformer can be represented by one branch in the equivalent circuit shown in Figure 1.5. The core losses are represented by Rm and the excitation characteristics by Xm. When the magnetizing current, which is about 0.5% of the load current, flows in the primary winding, there is a small voltage

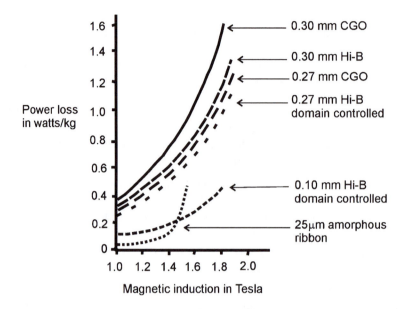

FIGURE 1.4 Loss characteristics for electrical core-steel materials over a range of magnetic induction (core flux density).

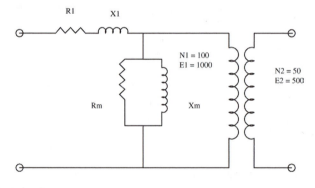

FIGURE 1.5 Equivalent circuit.

drop across the resistance of the winding and a small inductive drop across the inductance of the winding. We can represent these impedances as R1 and X1 in the equivalent circuit. However, these voltage drops are very small and can be neglected in the practical case.

Since the flux flowing in all parts of the core is essentially equal, the voltage induced in any turn placed around the core will be the same. This results in the unique characteristics of transformers with steel cores. Multiple secondary windings can be placed on the core to obtain different output voltages. Each turn in each winding will have the same voltage induced in it, as seen in Figure 1.6. The ratio of the voltages at the output to the input at no-load will be equal to the ratio of the turns. The voltage drops in the resistance and reactance at no-load are very small, with only magnetizing current flowing in the windings, so that the voltage appearing across the primary winding of the equivalent circuit in Figure 1.5 can be considered to be the input voltage. The relationship E1/N1 = E2/N2 is important in transformer design and application. The term E/N is called "volts per turn."

A steel core has a nonlinear magnetizing characteristic, as shown in Figure 1.7. As shown, greater ampere-turns are required as the flux density B is increased from zero. Above the knee of the curve, as the flux approaches saturation, a small increase in the flux density requires a large increase in the ampere-turns. When the core saturates, the circuit behaves much the same as an air core. As the flux

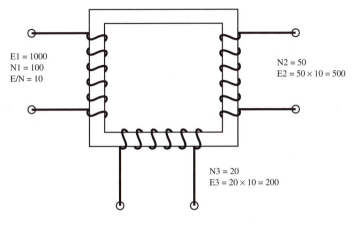

FIGURE 1.6 Steel core with windings.

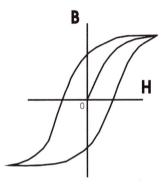

FIGURE 1.7 Hysteresis loop.

density decreases to zero, becomes negative, and increases in a negative direction, the same phenomenon of saturation occurs. As the flux reduces to zero and increases in a positive direction, it describes a loop known as the "hysteresis loop." The area of this loop represents power loss due to the hysteresis effect in the steel. Improvements in the grade of steel result in a smaller area of the hysteresis loop and a sharper knee point where the B-H characteristic becomes nonlinear and approaches the saturated state.

1.4 The Practical Transformer

1.4.1 Magnetic Circuit

In actual transformer design, the constants for the ideal circuit are determined from tests on materials and on transformers. For example, the resistance component of the core loss, usually called no-load loss, is determined from curves derived from tests on samples of electrical steel and measured transformer no-load losses. The designer will have curves similar to Figure 1.4 for the different electrical steel grades as a function of induction. Similarly, curves have been made available for the exciting current as a function of induction.

A very important relationship is derived from Equation 1.11. It can be written in the following form:

$$B = 0.225 \ (E/N)/(f \ A) \tag{1.13}$$

The term E/N is called "volts per turn": It determines the number of turns in the windings; the flux density in the core; and is a variable in the leakage reactance, which is discussed below. In fact, when the designer starts to make a design for an operating transformer, one of the first things selected is the volts per turn.

The no-load loss in the magnetic circuit is a guaranteed value in most designs. The designer must select an induction level that will allow him to meet the guarantee. The design curves or tables usually show the loss per unit weight as a function of the material and the magnetic induction.

The induction must also be selected so that the core will be below saturation under specified overvoltage conditions. Magnetic saturation occurs at about 2.0 T in magnetic steels but at about 1.4 T in amorphous ribbon.

1.4.2 Leakage Reactance

Additional concepts must be introduced when the practical transformer is considered,. For example, the flow of load current in the windings results in high magnetic fields around the windings. These fields are termed leakage flux fields. The term is believed to have started in the early days of transformer theory, when it was thought that this flux "leaked" out of the core. This flux exists in the spaces between windings and in the spaces occupied by the windings, as seen in Figure 1.8. These flux lines effectively result in an impedance between the windings, which is termed "leakage reactance" in the industry. The magnitude of this reactance is a function of the number of turns in the windings, the current in the windings, the leakage field, and the geometry of the core and windings. The magnitude of the leakage reactance is usually in the range of 4 to 20% at the base rating of power transformers.

The load current through this reactance results in a considerable voltage drop. Leakage reactance is termed "percent leakage reactance" or "percent reactance," i.e., the ratio of the reactance voltage drop to the winding voltage × 100. It is calculated by designers using the number of turns, the magnitudes of the current and the leakage field, and the geometry of the transformer. It is measured by short-circuiting one winding of the transformer and increasing the voltage on the other winding until rated current flows in the windings. This voltage divided by the rated winding voltage × 100 is the percent reactance voltage or percent reactance. The voltage drop across this reactance results in the voltage at the load being less than the value determined by the turns ratio. The percentage decrease in the voltage is termed "regulation," which is a function of the power factor of the load. The percent regulation can be determined using the following equation for inductive loads.

$$\%\mathrm{Reg} = \%R(\cos\,\phi) + \%X(\sin\,\phi) + \{[\%X(\cos\,\phi) - \%R(\sin\,\phi)]^2/200\} \qquad (1.14)$$

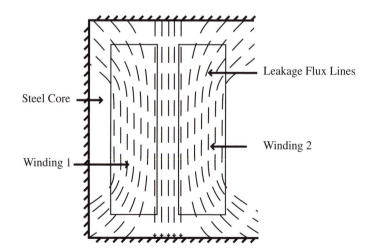

FIGURE 1.8 Leakage flux fields.

where

%Reg = percentage voltage drop across the resistance and the leakage reactance

%R = percentage resistance = (kW of load loss/kVA of transformer) × 100

%X = percentage leakage reactance

ϕ = angle corresponding to the power factor of the load = $\cos^{-1}$ pf

For capacitance loads, change the sign of the sine terms.

In order to compensate for these voltage drops, taps are usually added in the windings. The unique volts/turn feature of steel-core transformers makes it possible to add or subtract turns to change the voltage outputs of windings. A simple illustration of this concept is shown in Figure 1.9. The table in the figure shows that when tap 4 is connected to tap 5, there are 48 turns in the winding (maximum tap) and, at 10 volts/turn, the voltage E2 is 480 volts. When tap 2 is connected to tap 7, there are 40 turns in the winding (minimum tap), and the voltage E2 is 400 volts.

1.4.3 Load Losses

The term *load losses* represents the losses in the transformer that result from the flow of load current in the windings. Load losses are composed of the following elements.

- Resistance losses as the current flows through the resistance of the conductors and leads
- Eddy losses caused by the leakage field. These are a function of the second power of the leakage field density and the second power of the conductor dimensions normal to the field.
- Stray losses: The leakage field exists in parts of the core, steel structural members, and tank walls. Losses and heating result in these steel parts.

Again, the leakage field caused by flow of the load current in the windings is involved, and the eddy and stray losses can be appreciable in large transformers. In order to reduce load loss, it is not sufficient to reduce the winding resistance by increasing the cross-section of the conductor, as eddy losses in the conductor will increase faster than joule heating losses decrease. When the current is too great for a single conductor to be used for the winding without excessive eddy loss, a number of strands must be used in parallel. Because the parallel components are joined at the ends of the coil, steps must be taken to

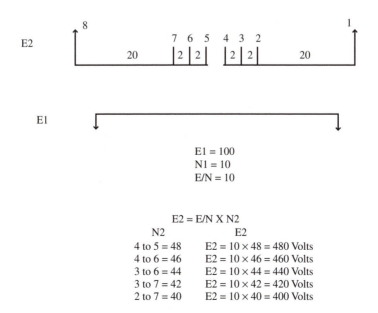

FIGURE 1.9 Illustration of how taps added in the windings can compensate for voltage drops.

circumvent the induction of different EMFs (electromotive force) in the strands due to different loops of strands linking with the leakage flux, which would involve circulating currents and further loss. Different forms of conductor transposition have been devised for this purpose.

Ideally, each conductor element should occupy every possible position in the array of strands such that all elements have the same resistance and the same induced EMF. Conductor transposition, however, involves some sacrifice of winding space. If the winding depth is small, one transposition halfway through the winding is sufficient; or in the case of a two-layer winding, the transposition can be located at the junction of the layers. Windings of greater depth need three or more transpositions. An example of a continuously transposed conductor (CTC) cable, shown in Figure 1.10, is widely used in the industry. CTC cables are manufactured using transposing machines and are usually paper-insulated as part of the transposing operation.

Stray losses can be a constraint on high-reactance designs. Losses can be controlled by using a combination of magnetic shunts and/or conducting shields to channel the flow of leakage flux external to the windings into low-loss paths.

1.4.4 Short-Circuit Forces

Forces exist between current-carrying conductors when they are in an alternating-current field. These forces are determined using Equation 1.15:

$$F = B\ I \sin \theta \tag{1.15}$$

where

> F = force on conductor
> B = local leakage flux density
> θ = angle between the leakage flux and the load current. In transformers, sin θ is almost always equal to 1

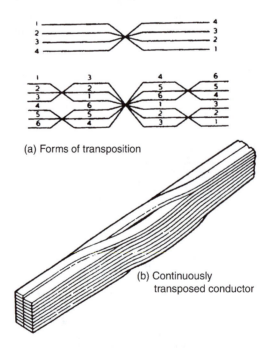

(a) Forms of transposition

(b) Continuously
    transposed conductor

FIGURE 1.10 Continuously transposed conductor cable.

Thus

$$B = \mu \ I \qquad (1.16)$$

and therefore

$$F \propto I^2 \qquad (1.17)$$

Since the leakage flux field is between windings and has a rather high density, the forces under short-circuit conditions can be quite high. This is a special area of transformer design. Complex computer programs are needed to obtain a reasonable representation of the field in different parts of the windings. Considerable research activity has been directed toward the study of mechanical stresses in the windings and the withstand criteria for different types of conductors and support systems.

Between any two windings in a transformer, there are three possible sets of forces:

- Radial repulsion forces due to currents flowing in opposition in the two windings
- Axial repulsion forces due to currents in opposition when the electromagnetic centers of the two windings are not aligned
- Axial compression forces in each winding due to currents flowing in the same direction in adjacent conductors

The most onerous forces are usually radial between windings. Outer windings rarely fail from hoop stress, but inner windings can suffer from one or the other of two failure modes:

- Forced buckling, where the conductor between support sticks collapses due to inward bending into the oil-duct space
- Free buckling, where the conductors bulge outwards as well as inwards at a few specific points on the circumference of the winding

Forced buckling can be prevented by ensuring that the winding is tightly wound and is adequately supported by packing it back to the core. Free buckling can be prevented by ensuring that the winding is of sufficient mechanical strength to be self-supporting, without relying on packing back to the core.

1.4.5 Thermal Considerations

The losses in the windings and the core cause temperature rises in the materials. This is another important area in which the temperatures must be limited to the long-term capability of the insulating materials. Refined paper is still used as the primary solid insulation in power transformers. Highly refined mineral oil is still used as the cooling and insulating medium in power transformers. Gases and vapors have been introduced in a limited number of special designs. The temperatures must be limited to the thermal capability of these materials. Again, this subject is quite broad and involved. It includes the calculation of the temperature rise of the cooling medium, the average and hottest-spot rise of the conductors and leads, and accurate specification of the heat-exchanger equipment.

1.4.6 Voltage Considerations

A transformer must withstand a number of different normal and abnormal voltage stresses over its expected life. These voltages include:

- Operating voltages at the rated frequency
- Rated-frequency overvoltages
- Natural lightning impulses that strike the transformer or transmission lines
- Switching surges that result from opening and closing of breakers and switches
- Combinations of the above voltages

- Transient voltages generated due to resonance between the transformer and the network
- Fast transient voltages generated by vacuum-switch operations or by the operation of disconnect switches in a gas-insulated bus-bar system

This is a very specialized field in which the resulting voltage stresses must be calculated in the windings, and withstand criteria must be established for the different voltages and combinations of voltages. The designer must design the insulation system to withstand all of these stresses.

References

Kan, H., Problems related to cores of transformers and reactors, *Electra*, 94, 15–33, 1984.

2
Equipment Types

H. Jin Sim
Scott H. Digby
Waukesha Electric Systems

Dudley L. Galloway
Galloway Transformer Technology LLC

Dan Mulkey
Pacific Gas & Electric Company

Gustav Preininger
Consultant

Sheldon P. Kennedy
Niagara Transformer Corporation

Paulette A. Payne
PEPCO

Randy Mullikin
Kuhlman Electric Corp.

Craig A. Colopy
Cooper Power Systems

Arindam Maitra
Anish Gaikwad
Arshad Mansoor
Douglas Dorr
EPRI PEAC Corporation

Ralph Ferraro
Ferraro, Oliver & Associates

Richard F. Dudley
Michael Sharp
Antonio Castanheira
Behdad Biglar
Trench Ltd.

2.1 Power Transformers

H. Jin Sim and Scott H. Digby

2.1.1   Introduction

ANSI/IEEE defines a transformer as a static electrical device, involving no continuously moving parts, used in electric power systems to transfer power between circuits through the use of electromagnetic induction. The term *power transformer* is used to refer to those transformers used between the generator and the distribution circuits, and these are usually rated at 500 kVA and above. Power systems typically consist of a large number of generation locations, distribution points, and interconnections within the system or with nearby systems, such as a neighboring utility. The complexity of the system leads to a variety of transmission and distribution voltages. Power transformers must be used at each of these points where there is a transition between voltage levels.

Power transformers are selected based on the application, with the emphasis toward custom design being more apparent the larger the unit. Power transformers are available for step-up operation, primarily used at the generator and referred to as generator step-up (GSU) transformers, and for step-down operation, mainly used to feed distribution circuits. Power transformers are available as single-phase or three-phase apparatus.

The construction of a transformer depends upon the application. Transformers intended for indoor use are primarily of the dry type but can also be liquid immersed. For outdoor use, transformers are usually liquid immersed. This section focuses on the outdoor, liquid-immersed transformers, such as those shown in Figure 2.1.1.

FIGURE 2.1.1   20 MVA, 161:26.4 × 13.2 kV with LTC, three phase transformers.

2.1.2 Rating and Classifications

2.1.2.1 Rating

In the U.S., transformers are rated based on the power output they are capable of delivering continuously at a specified rated voltage and frequency under "usual" operating conditions without exceeding pre-scribed internal temperature limitations. Insulation is known to deteriorate with increases in temperature, so the insulation chosen for use in transformers is based on how long it can be expected to last by limiting the operating temperature. The temperature that insulation is allowed to reach under operating condi-tions essentially determines the output rating of the transformer, called the kVA rating. Standardization has led to temperatures within a transformer being expressed in terms of the rise above ambient tem-perature, since the ambient temperature can vary under operating or test conditions. Transformers are designed to limit the temperature based on the desired load, including the average temperature rise of a winding, the hottest-spot temperature rise of a winding, and, in the case of liquid-filled units, the top liquid temperature rise. To obtain absolute temperatures from these values, simply add the ambient temperature. Standard temperature limits for liquid-immersed power transformers are listed in Table 2.1.1.

The normal life expectancy of a power transformer is generally assumed to be about 30 years of service when operated within its rating. However, under certain conditions, it may be overloaded and operated beyond its rating, with moderately predictable "loss of life." Situations that might involve operation beyond rating include emergency rerouting of load or through-faults prior to clearing of the fault condition.

Outside the U.S., the transformer rating may have a slightly different meaning. Based on some standards, the kVA rating can refer to the power that can be input to a transformer, the rated output being equal to the input minus the transformer losses.

Power transformers have been loosely grouped into three market segments based on size ranges. These three segments are:

1. Small power transformers: 500 to 7500 kVA
2. Medium power transformers: 7500 to 100 MVA
3. Large power transformers: 100 MVA and above

Note that the upper range of small power and the lower range of medium power can vary between 2,500 and 10,000 kVA throughout the industry.

It was noted that the transformer rating is based on "usual" service conditions, as prescribed by standards. Unusual service conditions may be identified by those specifying a transformer so that the desired performance will correspond to the actual operating conditions. Unusual service conditions include, but are not limited to, the following: high (above 40°C) or low (below −20°C) ambient temper-atures, altitudes above 1000 m above sea level, seismic conditions, and loads with total harmonic distor-tion above 0.05 per unit.

2.1.2.2 Insulation Classes

The insulation class of a transformer is determined based on the test levels that it is capable of with-standing. Transformer insulation is rated by the BIL, or basic impulse insulation level, in conjunction with the voltage rating. Internally, a transformer is considered to be a non-self-restoring insulation system, mostly consisting of porous, cellulose material impregnated by the liquid insulating medium. Externally,

TABLE 2.1.1 Standard limits for Temperature Rises Above Ambient

Average winding temperature rise	65°C[a]
Hot spot temperature rise	80°C
Top liquid temperature rise	65°C

[a]The base rating is frequently specified and tested as a 55°C rise.

the transformer's bushings and, more importantly, the surge-protection equipment must coordinate with the transformer rating to protect the transformer from transient overvoltages and surges. Standard insulation classes have been established by standards organizations stating the parameters by which tests are to be performed.

Wye-connected windings in a three-phase power transformer will typically have the common point brought out of the tank through a neutral bushing. (See Section 2.2, Distribution Transformers, for a discussion of wye connections.) Depending on the application — for example in the case of a solidly grounded neutral versus a neutral grounded through a resistor or reactor or even an ungrounded neutral — the neutral may have a lower insulation class than the line terminals. There are standard guidelines for rating the neutral based on the situation. It is important to note that the insulation class of the neutral may limit the test levels of the line terminals for certain tests, such as the applied-voltage or "hi-pot" test, where the entire circuit is brought up to the same voltage level. A reduced voltage rating for the neutral can significantly reduce the cost of larger units and autotransformers compared with a fully rated neutral.

2.1.2.3 Cooling Classes

Since no transformer is truly an "ideal" transformer, each will incur a certain amount of energy loss, mainly that which is converted to heat. Methods of removing this heat can depend on the application, the size of the unit, and the amount of heat that needs to be dissipated.

The insulating medium inside a transformer, usually oil, serves multiple purposes, first to act as an insulator, and second to provide a good medium through which to remove the heat.

The windings and core are the primary sources of heat, although internal metallic structures can act as a heat source as well. It is imperative to have proper cooling ducts and passages in the proximity of the heat sources through which the cooling medium can flow so that the heat can be effectively removed from the transformer. The natural circulation of oil through a transformer through convection has been referred to as a "thermosiphon" effect. The heat is carried by the insulating medium until it is transferred through the transformer tank wall to the external environment. Radiators, typically detachable, provide an increase in the surface area available for heat transfer by convection without increasing the size of the tank. In smaller transformers, integral tubular sides or fins are used to provide this increase in surface area. Fans can be installed to increase the volume of air moving across the cooling surfaces, thus increasing the rate of heat dissipation. Larger transformers that cannot be effectively cooled using radiators and fans rely on pumps that circulate oil through the transformer and through external heat exchangers, or coolers, which can use air or water as a secondary cooling medium.

Allowing liquid to flow through the transformer windings by natural convection is identified as "nondirected flow." In cases where pumps are used, and even some instances where only fans and radiators are being used, the liquid is often guided into and through some or all of the windings. This is called "directed flow" in that there is some degree of control of the flow of the liquid through the windings. The difference between directed and nondirected flow through the winding in regard to winding arrangement will be further discussed with the description of winding types (see Section 2.1.5.2).

The use of auxiliary equipment such as fans and pumps with coolers, called forced circulation, increases the cooling and thereby the rating of the transformer without increasing the unit's physical size. Ratings are determined based on the temperature of the unit as it coordinates with the cooling equipment that is operating. Usually, a transformer will have multiple ratings corresponding to multiple stages of cooling, as the supplemental cooling equipment can be set to run only at increased loads.

Methods of cooling for liquid-immersed transformers have been arranged into cooling classes identified by a four-letter designation as follows:

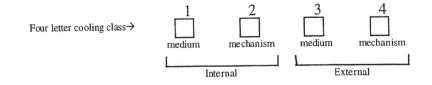

TABLE 2.1.2 Cooling Class Letter Description

		Code Letter	Description
Internal	First Letter	O	Liquid with flash point less than or equal to 300°C
	(Cooling medium)	K	Liquid with flash point greater than 300°C
		L	Liquid with no measurable flash point
	Second Letter	N	Natural convection through cooling equipment and windings
	(Cooling mechanism)	F	Forced circulation through cooling equipment, natural convection in windings
		D	Forced circulation through cooling equipment, directed flow in man windings
External	Third letter	A	Air
	(Cooling medium)	W	Water
	Fourth letter	N	Natural convection
	(Cooling medium)	F	Forced circulation

Table 2.1.2 lists the code letters that are used to make up the four-letter designation.

This system of identification has come about through standardization between different international standards organizations and represents a change from what has traditionally been used in the U.S. Where OA classified a transformer as liquid-immersed self-cooled in the past, it is now designated by the new system as ONAN. Similarly, the previous FA classification is now identified as ONAF. FOA could be OFAF or ODAF, depending on whether directed oil flow is employed or not. In some cases, there are transformers with directed flow in windings without forced circulation through cooling equipment.

An example of multiple ratings would be ONAN/ONAF/ONAF, where the transformer has a base rating where it is cooled by natural convection and two supplemental ratings where groups of fans are turned on to provide additional cooling so that the transformer will be capable of supplying additional kVA. This rating would have been designated OA/FA/FA per past standards.

2.1.3 Short-Circuit Duty

A transformer supplying a load current will have a complicated network of internal forces acting on and stressing the conductors, support structures, and insulation structures. These forces are fundamental to the interaction of current-carrying conductors within magnetic fields involving an alternating-current source. Increases in current result in increases in the magnitude of the forces proportional to the square of the current. Severe overloads, particularly through-fault currents resulting from external short-circuit events, involve significant increases in the current above rated current and can result in tremendous forces inside the transformer.

Since the fault current is a transient event, it will have the asymmetrical sinusoidal waveshape decaying with time based on the time constant of the equivalent circuit that is characteristic of switching events. The amplitude of the symmetrical component of the sine wave is determined from the formula,

$$I_{sc} = I_{rated}/(Z_{xfmr} + Z_{sys}) \qquad (2.1.1)$$

where Z_{xfmr} and Z_{sys} are the transformer and system impedances, respectively, expressed in terms of per unit on the transformer base, and I_{sc} and I_{rated} are the resulting short-circuit (through-fault) current and the transformer rated current, respectively. An offset factor, K, multiplied by I_{sc} determines the magnitude of the first peak of the transient asymmetrical current. This offset factor is derived from the equivalent transient circuit. However, standards give values that must be used based on the ratio of the effective ac (alternating current) reactance (x) and resistance (r), x/r. K typically varies in the range of 1.5 to 2.8.

As indicated by Equation 2.1.1, the short-circuit current is primarily limited by the internal impedance of the transformer, but it may be further reduced by impedances of adjacent equipment, such as current-limiting reactors or by system power-delivery limitations. Existing standards define the maximum magnitude and duration of the fault current based on the rating of the transformer.

The transformer must be capable of withstanding the maximum forces experienced at the first peak of the transient current as well as the repeated pulses at each of the subsequent peaks until the fault is cleared or the transformer is disconnected. The current will experience two peaks per cycle, so the forces will pulsate at 120 Hz, twice the power frequency, acting as a dynamic load. Magnitudes of forces during these situations can range from several hundred kilograms to hundreds of thousands of kilograms in large power transformers. For analysis, the forces acting on the windings are generally broken up into two subsets, radial and axial forces, based on their apparent effect on the windings. Figure 2.1.2 illustrates the difference between radial and axial forces in a pair of circular windings. Mismatches of ampere-turns between windings are unavoidable — caused by such occurrences as ampere-turn voids created by sections of a winding being tapped out, slight mismatches in the lengths of respective windings, or misalignment of the magnetic centers of the respective windings — and result in a net axial force. This net axial force will have the effect of trying to force one winding in the upward direction and the other in the downward direction, which must be resisted by the internal mechanical structures.

The high currents experienced during through-fault events will also cause elevated temperatures in the windings. Limitations are also placed on the calculated temperature the conductor may reach during fault conditions. These high temperatures are rarely a problem due to the short time span of these events, but the transformer may experience an associated increase in its "loss of life." This additional "loss of

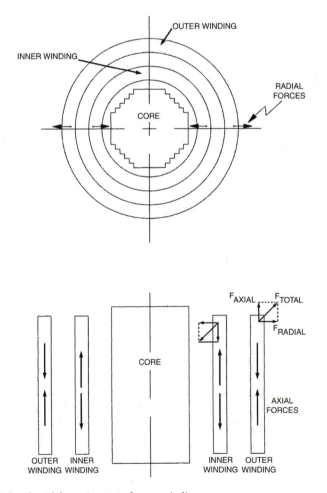

FIGURE 2.1.2 Radial and axial forces in a transformer winding.

life" can become more prevalent, even critical, based on the duration of the fault conditions and how often such events occur. It is also possible for the conductor to experience changes in mechanical strength due to the annealing that can occur at high temperatures. The temperature at which this can occur depends on the properties and composition of the conductor material, such as the hardness, which is sometimes increased through cold-working processes or the presence of silver in certain alloys.

2.1.4 Efficiency, Losses, and Regulation

2.1.4.1 Efficiency

Power transformers are very efficient, typically 99.5% or greater, i.e., real power losses are usually less than 0.5% of the kVA rating at full load. The efficiency is derived from the rated output and the losses incurred in the transformer. The basic relationship for efficiency is the output over the input, which according to U.S. standards translates to

$$\text{efficiency} = [\text{kVA rating}/(\text{kVA rating} + \text{total losses})] \times 100\% \qquad (2.1.2)$$

and generally decreases slightly with increases in load. Total losses are the sum of the no-load and load losses.

2.1.4.2 Losses

The no-load losses are essentially the power required to keep the core energized. These are commonly referred to as "core losses," and they exist whenever the unit is energized. No-load losses depend primarily upon the voltage and frequency, so under operational conditions they vary only slightly with system variations. Load losses, as the terminology might suggest, result from load currents flowing through the transformer. The two components of the load losses are the I^2R losses and the stray losses. I^2R losses are based on the measured dc (direct current) resistance, the bulk of which is due to the winding conductors and the current at a given load. The stray losses are a term given to the accumulation of the additional losses experienced by the transformer, which includes winding eddy losses and losses due to the effects of leakage flux entering internal metallic structures. Auxiliary losses refer to the power required to run auxiliary cooling equipment, such as fans and pumps, and are not typically included in the total losses as defined above.

2.1.4.3 Economic Evaluation of Losses

Transformer losses represent power that cannot be delivered to customers and therefore have an associated economic cost to the transformer user/owner. A reduction in transformer losses generally results in an increase in the transformer's cost. Depending on the application, there may be an economic benefit to a transformer with reduced losses and high price (initial cost), and vice versa. This process is typically dealt with through the use of "loss evaluations," which place a dollar value on the transformer losses to calculate a total owning cost that is a combination of the purchase price and the losses. Typically, each of the transformer's individual loss parameters — no-load losses, load losses, and auxiliary losses — are assigned a dollar value per kW ($/kW). Information obtained from such an analysis can be used to compare prices from different manufacturers or to decide on the optimum time to replace existing transformers. There are guides available, through standards organizations, for estimating the cost associated with transformers losses. Loss-evaluation values can range from about $500/kW to upwards of $12,000/kW for the no-load losses and from a few hundred dollars per kW to about $6,000 to $8,000/kW for load losses and auxiliary losses. Specific values depend upon the application.

2.1.4.4 Regulation

Regulation is defined as the change (increase) in the output voltage that occurs when the load on the transformer is reduced from rated load to no load while the input voltage is held constant. It is typically expressed as a percentage, or per unit, of the rated output voltage at rated load. A general expression for the regulation can be written as:

$$\% \text{ regulation} = [(V_{NL} - V_{FL})/V_{FL}] \times 100 \qquad (2.1.3)$$

where V_{NL} is the voltage at no load and V_{FL} is the voltage at full load. The regulation is dependent upon the impedance characteristics of the transformer, the resistance (r), and more significantly the ac reactance (x), as well as the power factor of the load. The regulation can be calculated based on the transformer impedance characteristics and the load power factor using the following formulas:

$$\% \text{ regulation} = pr + qx + [(px - qr)^2/200] \qquad (2.1.4)$$

$$q = \text{SQRT} (1 - p^2) \qquad (2.1.5)$$

where p is the power factor of the load and r and x are expressed in terms of per unit on the transformer base. The value of q is taken to be positive for a lagging (inductive) power factor and negative for a leading (capacitive) power factor.

It should be noted that lower impedance values, specifically ac reactance, result in lower regulation, which is generally desirable. However, this is at the expense of the fault current, which would in turn increase with a reduction in impedance, since it is primarily limited by the transformer impedance. Additionally, the regulation increases as the power factor of the load becomes more lagging (inductive).

2.1.5   Construction

The construction of a power transformer varies throughout the industry. The basic arrangement is essentially the same and has seen little significant change in recent years, so some of the variations can be discussed here.

2.1.5.1   Core

The core, which provides the magnetic path to channel the flux, consists of thin strips of high-grade steel, called laminations, which are electrically separated by a thin coating of insulating material. The strips can be stacked or wound, with the windings either built integrally around the core or built separately and assembled around the core sections. Core steel can be hot- or cold-rolled, grain-oriented or non-grain oriented, and even laser-scribed for additional performance. Thickness ranges from 0.23 mm to upwards of 0.36 mm. The core cross section can be circular or rectangular, with circular cores commonly referred to as cruciform construction. Rectangular cores are used for smaller ratings and as auxiliary transformers used within a power transformer. Rectangular cores use a single width of strip steel, while circular cores use a combination of different strip widths to approximate a circular cross-section, such as in Figure 2.1.2. The type of steel and arrangement depends on the transformer rating as related to cost factors such as labor and performance.

Just like other components in the transformer, the heat generated by the core must be adequately dissipated. While the steel and coating may be capable of withstanding higher temperatures, it will come in contact with insulating materials with limited temperature capabilities. In larger units, cooling ducts are used inside the core for additional convective surface area, and sections of laminations may be split to reduce localized losses.

The core is held together by, but insulated from, mechanical structures and is grounded to a single point in order to dissipate electrostatic buildup. The core ground location is usually some readily accessible point inside the tank, but it can also be brought through a bushing on the tank wall or top for external access. This grounding point should be removable for testing purposes, such as checking for unintentional core grounds. Multiple core grounds, such as a case whereby the core is inadvertently making contact with otherwise grounded internal metallic mechanical structures, can provide a path for circulating currents induced by the main flux as well as a leakage flux, thus creating concentrations of losses that can result in localized heating.

The maximum flux density of the core steel is normally designed as close to the knee of the saturation curve as practical, accounting for required overexcitations and tolerances that exist due to materials and manufacturing processes. (See Section 2.6, Instrument Transformers, for a discussion of saturation curves.) For power transformers the flux density is typically between 1.3 T and 1.8 T, with the saturation point for magnetic steel being around 2.03 T to 2.05 T.

There are two basic types of core construction used in power transformers: core form and shell form.

In core-form construction, there is a single path for the magnetic circuit. Figure 2.1.3 shows a schematic of a single-phase core, with the arrows showing the magnetic path. For single-phase applications, the windings are typically divided on both core legs as shown. In three-phase applications, the windings of a particular phase are typically on the same core leg, as illustrated in Figure 2.1.4. Windings are

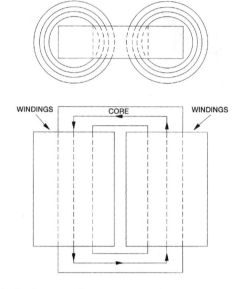

FIGURE 2.1.3 Schematic of single-phase core-form construction.

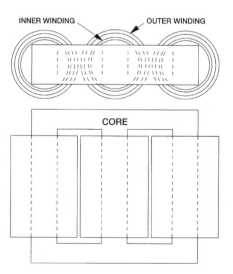

FIGURE 2.1.4 Schematic of three-phase core-form construction.

FIGURE 2.1.5 "E"-assembly, prior to addition of coils and insertion of top yoke.

constructed separate of the core and placed on their respective core legs during core assembly. Figure 2.1.5 shows what is referred to as the "E"-assembly of a three-phase core-form core during assembly.

In shell-form construction, the core provides multiple paths for the magnetic circuit. Figure 2.1.6 is a schematic of a single-phase shell-form core, with the two magnetic paths illustrated. The core is typically stacked directly around the windings, which are usually "pancake"-type windings, although some applications are such that the core and windings are assembled similar to core form. Due to advantages in short-circuit and transient-voltage performance, shell forms tend to be used more frequently in the largest transformers, where conditions can be more severe. Variations of three-phase shell-form construction include five- and seven-legged cores, depending on size and application.

2.1.5.2 Windings

The windings consist of the current-carrying conductors wound around the sections of the core, and these must be properly insulated, supported, and cooled to withstand operational and test conditions. The terms *winding* and *coil* are used interchangeably in this discussion.

Copper and aluminum are the primary materials used as conductors in power-transformer windings. While aluminum is lighter and generally less expensive than copper, a larger cross section of aluminum conductor must be used to carry a current with similar performance as copper. Copper has higher mechanical strength and is used almost exclusively in all but the smaller size ranges, where aluminum conductors may be perfectly acceptable. In cases where extreme forces are encountered, materials such as silver-bearing copper can be used for even greater strength. The conductors used in power transformers are typically stranded with a rectangular cross section, although some transformers at the lowest ratings may use sheet or foil conductors. Multiple strands can be wound in parallel and joined together at the ends of the winding, in which case it is necessary to transpose the strands at various points throughout the winding to prevent circulating currents around the loop(s) created by joining the strands at the ends. Individual strands may be subjected to differences in the flux field due to their respective positions within the winding, which create differences in voltages between the strands and drive circulating currents through the conductor loops. Proper transposition of the strands cancels out these voltage differences and eliminates or greatly reduces the circulating currents. A variation of this technique, involving many rectangular conductor strands combined into a cable, is called continuously transposed cable (CTC), as shown in Figure 2.1.7.

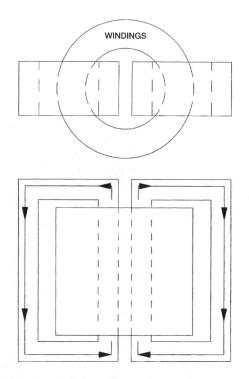

FIGURE 2.1.6 Schematic of single-phase shell-form construction.

FIGURE 2.1.7 Continuously transposed cable (CTC).

In core-form transformers, the windings are usually arranged concentrically around the core leg, as illustrated in Figure 2.1.8, which shows a winding being lowered over another winding already on the core leg of a three-phase transformer. A schematic of coils arranged in this three-phase application was also shown in Figure 2.1.4. Shell-form transformers use a similar concentric arrangement or an interleaved arrangement, as illustrated in the schematic Figure 2.1.9 and the photograph in Figure 2.1.13.

FIGURE 2.1.8 Concentric arrangement, outer coil being lowered onto core leg over top of inner coil.

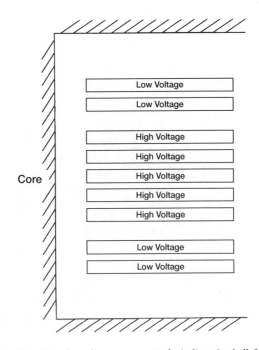

FIGURE 2.1.9 Example of stacking (interleaved) arrangement of windings in shell-form construction.

With an interleaved arrangement, individual coils are stacked, separated by insulating barriers and cooling ducts. The coils are typically connected with the inside of one coil connected to the inside of an adjacent coil and, similarly, the outside of one coil connected to the outside of an adjacent coil. Sets of coils are assembled into groups, which then form the primary or secondary winding.

When considering concentric windings, it is generally understood that circular windings have inherently higher mechanical strength than rectangular windings, whereas rectangular coils can have lower associated material and labor costs. Rectangular windings permit a more efficient use of space, but their use is limited to small power transformers and the lower range of medium-power transformers, where the internal forces are not extremely high. As the rating increases, the forces significantly increase, and there is need for added strength in the windings, so circular coils, or shell-form construction, are used. In some special cases, elliptically shaped windings are used.

Concentric coils are typically wound over cylinders with spacers attached so as to form a duct between the conductors and the cylinder. As previously mentioned, the flow of liquid through the windings can be based solely on natural convection, or the flow can be somewhat controlled through the use of strategically placed barriers within the winding. Figures 2.1.10 and 2.1.11 show winding arrangements comparing nondirected and directed flow. This concept is sometimes referred to as guided liquid flow.

A variety of different types of windings have been used in power transformers through the years. Coils can be wound in an upright, vertical orientation, as is necessary with larger, heavier coils; or they can be wound horizontally and placed upright upon completion. As mentioned previously, the type of winding depends on the transformer rating as well as the core construction. Several of the more common winding types are discussed here.

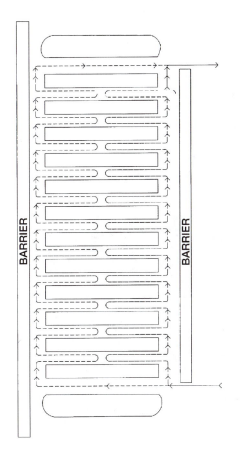

FIGURE 2.1.10 Nondirected flow.

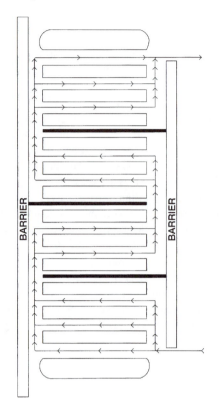

FIGURE 2.1.11   Directed flow.

2.1.5.2.1   Pancake Windings

Several types of windings are commonly referred to as "pancake" windings due to the arrangement of conductors into discs. However, the term most often refers to a coil type that is used almost exclusively in shell-form transformers. The conductors are wound around a rectangular form, with the widest face of the conductor oriented either horizontally or vertically. Figure 2.1.12 illustrates how these coils are typically wound. This type of winding lends itself to the interleaved arrangement previously discussed (Figure 2.1.13).

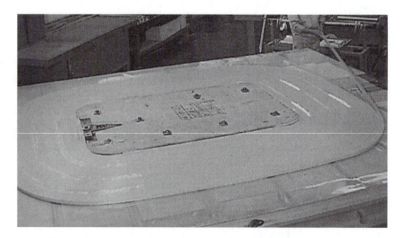

FIGURE 2.1.12   Pancake winding during winding process.

FIGURE 2.1.13 Stacked pancake windings.

2.1.5.2.2 *Layer (Barrel) Windings*

Layer (barrel) windings are among the simplest of windings in that the insulated conductors are wound directly next to each other around the cylinder and spacers. Several layers can be wound on top of one another, with the layers separated by solid insulation, ducts, or a combination. Several strands can be wound in parallel if the current magnitude so dictates. Variations of this winding are often used for applications such as tap windings used in load-tap-changing (LTC) transformers and for tertiary windings used for, among other things, third-harmonic suppression. Figure 2.1.14 shows a layer winding during assembly that will be used as a regulating winding in an LTC transformer.

2.1.5.2.3 *Helical Windings*

Helical windings are also referred to as screw or spiral windings, with each term accurately characterizing the coil's construction. A helical winding consists of a few to more than 100 insulated strands wound in parallel continuously along the length of the cylinder, with spacers inserted between adjacent turns or discs and suitable transpositions included to minimize circulating currents between parallel strands. The manner of construction is such that the coil resembles a corkscrew. Figure 2.1.15 shows a helical winding during the winding process. Helical windings are used for the higher-current applications frequently encountered in the lower-voltage classes.

2.1.5.2.4 *Disc Windings*

A disc winding can involve a single strand or several strands of insulated conductors wound in a series of parallel discs of horizontal orientation, with the discs connected at either the inside or outside as a crossover point. Each disc comprises multiple turns wound over other turns, with the crossovers alternating between inside and outside. Figure 2.1.16 outlines the basic concept, and Figure 2.1.17 shows typical crossovers during the winding process. Most windings of 25-kV class and above used in coreform transformers are disc type. Given the high voltages involved in test and operation, particular attention is required to avoid high stresses between discs and turns near the end of the winding when subjected to transient voltage surges. Numerous techniques have been developed to ensure an acceptable voltage distribution along the winding under these conditions.

FIGURE 2.1.14 Layer windings (single layer with two strands wound in parallel).

FIGURE 2.1.15 Helical winding during assembly.

2.1.5.3 Taps-Turns Ratio Adjustment

The ability to adjust the turns ratio of a transformer is often desirable to compensate for variations in voltage that occur due to the regulation of the transformer and loading cycles. This task can be accomplished by several means. There is a significant difference between a transformer that is capable of changing the ratio while the unit is on-line (a load tap changing [LTC] transformer) and one that must be taken off-line, or de-energized, to perform a tap change.

Most transformers are provided with a means of changing the number of turns in the high-voltage circuit, whereby a part of the winding is tapped out of the circuit. In many transformers, this is done using one of the main windings and tapping out a section or sections, as illustrated by the schematic in Figure 2.1.18.

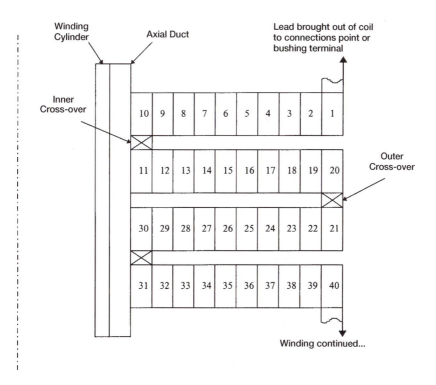

FIGURE 2.1.16 Basic disc winding layout

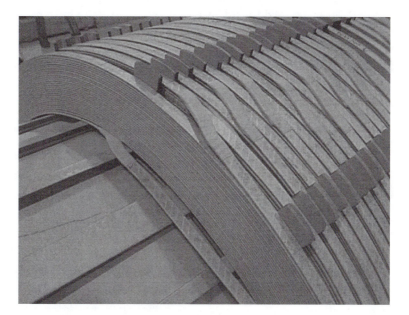

FIGURE 2.1.17 Disc winding inner and outer crossovers.

With larger units, a dedicated tap winding may be necessary to avoid the ampere-turn voids that occur along the length of the winding. Use and placement of tap windings vary with the application and among manufacturers. A manually operated switching mechanism, a DETC (de-energized tap changer), is normally provided for convenient access external to the transformer to change the tap position. When LTC capabilities are desired, additional windings and equipment are required, which significantly increase the size and cost

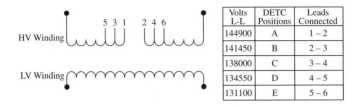

Volts L-L	DETC Positions	Leads Connected
144900	A	1 – 2
141450	B	2 – 3
138000	C	3 – 4
134550	D	4 – 5
131100	E	5 – 6

FIGURE 2.1.18 High-voltage winding schematic and connection diagram for 138-kV example.

of the transformer. This option is specified on about 60% of new medium and large power transformers. Figure 2.1.19 illustrates the basic operation by providing a sample schematic and connection chart for a transformer supplied with an LTC on the low-voltage (secondary) side. It should be recognized that there would be slight differences in this schematic based on the specific LTC being used. Figure 2.1.19 also shows a sample schematic where an auxiliary transformer is used between the main windings and the LTC to limit the current through the LTC mechanism.

It is also possible for a transformer to have dual voltage ratings, as is popular in spare and mobile transformers. While there is no physical limit to the ratio between the dual ratings, even ratios (for example 24.94 × 12.47 kV or 138 × 69 kV) are easier for manufacturers to accommodate.

2.1.6 Accessory Equipment

2.1.6.1 Accessories

There are many different accessories used to monitor and protect power transformers, some of which are considered standard features, and others of which are used based on miscellaneous requirements. A few of the basic accessories are briefly discussed here.

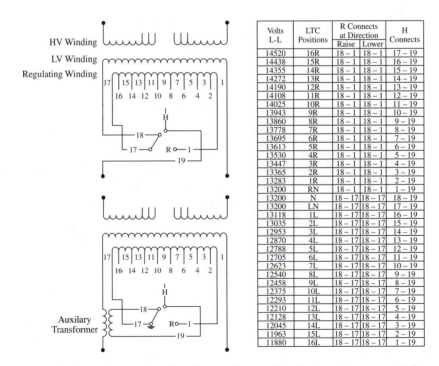

Volts L-L	LTC Positions	R Connects at Direction		H Connects
		Raise	Lower	
14520	16R	18 – 1	18 – 1	17 – 19
14438	15R	18 – 1	18 – 1	16 – 19
14355	14R	18 – 1	18 – 1	15 – 19
14272	13R	18 – 1	18 – 1	14 – 19
14190	12R	18 – 1	18 – 1	13 – 19
14108	11R	18 – 1	18 – 1	12 – 19
14025	10R	18 – 1	18 – 1	11 – 19
13943	9R	18 – 1	18 – 1	10 – 19
13860	8R	18 – 1	18 – 1	9 – 19
13778	7R	18 – 1	18 – 1	8 – 19
13695	6R	18 – 1	18 – 1	7 – 19
13613	5R	18 – 1	18 – 1	6 – 19
13530	4R	18 – 1	18 – 1	5 – 19
13447	3R	18 – 1	18 – 1	4 – 19
13365	2R	18 – 1	18 – 1	3 – 19
13283	1R	18 – 1	18 – 1	2 – 19
13200	RN	18 – 1	18 – 1	1 – 19
13200	N	18 – 17	18 – 17	18 – 19
13200	LN	18 – 17	18 – 17	17 – 19
13118	1L	18 – 17	18 – 17	16 – 19
13035	2L	18 – 17	18 – 17	15 – 19
12953	3L	18 – 17	18 – 17	14 – 19
12870	4L	18 – 17	18 – 17	13 – 19
12788	5L	18 – 17	18 – 17	12 – 19
12705	6L	18 – 17	18 – 17	11 – 19
12623	7L	18 – 17	18 – 17	10 – 19
12540	8L	18 – 17	18 – 17	9 – 19
12458	9L	18 – 17	18 – 17	8 – 19
12375	10L	18 – 17	18 – 17	7 – 19
12293	11L	18 – 17	18 – 17	6 – 19
12210	12L	18 – 17	18 – 17	5 – 19
12128	13L	18 – 17	18 – 17	4 – 19
12045	14L	18 – 17	18 – 17	3 – 19
11963	15L	18 – 17	18 – 17	2 – 19
11880	16L	18 – 17	18 – 17	1 – 19

FIGURE 2.1.19 Schematic and connection chart for transformer with a load tap changer supplied on a 13.2-kV low voltage.

2.1.6.1.1 *Liquid-Level Indicator*

A liquid-level indicator is a standard feature on liquid-filled transformer tanks, since the liquid medium is critical for cooling and insulation. This indicator is typically a round-faced gauge on the side of the tank, with a float and float arm that moves a dial pointer as the liquid level changes.

2.1.6.1.2 *Pressure-Relief Devices*

Pressure-relief devices are mounted on transformer tanks to relieve excess internal pressures that might build up during operating conditions. These devices are intended to avoid damage to the tank. On larger transformers, several pressure-relief devices may be required due to the large quantities of oil.

2.1.6.1.3 *Liquid-Temperature Indicator*

Liquid-temperature indicators measure the temperature of the internal liquid at a point near the top of the liquid using a probe inserted in a well and mounted through the side of the transformer tank.

2.1.6.1.4 *Winding-Temperature Indicator*

A winding-temperature simulation method is used to approximate the hottest spot in the winding. An approximation is needed because of the difficulties involved in directly measuring winding temperature. The method applied to power transformers involves a current transformer, which is located to incur a current proportional to the load current through the transformer. The current transformer feeds a circuit that essentially adds heat to the top liquid-temperature reading, which approximates a reading that models the winding temperature. This method relies on design or test data of the temperature differential between the liquid and the windings, called the winding gradient.

2.1.6.1.5 *Sudden-Pressure Relay*

A sudden- (or rapid-) pressure relay is intended to indicate a quick increase in internal pressure that can occur when there is an internal fault. These relays can be mounted on the top or side of the transformer, or they can operate in liquid or gas space.

2.1.6.1.6 *Desiccant (Dehydrating) Breathers*

Desiccant breathers use a material such as silica gel to allow air to enter and exit the tank, removing moisture as the air passes through. Most tanks are somewhat free breathing, and such a device, if properly maintained, allows a degree of control over the quality of air entering the transformer.

2.1.6.2 Liquid-Preservation Systems

There are several methods to preserve the properties of the transformer liquid and associated insulation structures that it penetrates. Preservation systems attempt to isolate the transformer's internal environment from the external environment (atmosphere) while understanding that a certain degree of inter-action, or "breathing," is required to accommodate variations in pressure that occur under operational conditions, such as expansion and contraction of liquid with temperature. Free-breathing systems, where the liquid is exposed to the atmosphere, are no longer used. The most commonly used methods are outlined as follows and illustrated in Figure 2.1.20.

- Sealed-tank systems have the tank interior sealed from the atmosphere and maintain a layer of gas — a gas space or cushion — that sits above the liquid. The gas-plus-liquid volume remains constant. Negative internal pressures can exist in sealed-tank systems at lower loads or temperatures with positive pressures as load and temperatures increase.

- Positive-pressure systems involve the use of inert gases to maintain a positive pressure in the gas space. An inert gas, typically from a bottle of compressed nitrogen, is incrementally injected into the gas space when the internal pressure falls out of range.

- Conservator (expansion tank) systems are used both with and without air bags, also called bladders or diaphragms, and involve the use of a separate auxiliary tank. The main transformer tank is completely filled with liquid; the auxiliary tank is partially filled; and the liquid expands and

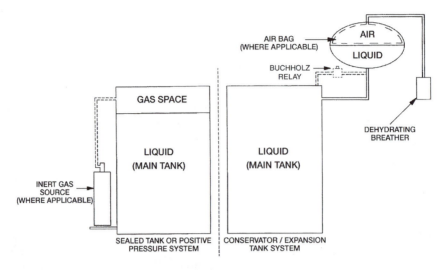

FIGURE 2.1.20 General arrangements of liquid preservation systems.

contracts within the auxiliary tank. The auxiliary tank is allowed to "breathe," usually through a dehydrating breather. The use of an air bag in the auxiliary tank can provide further separation from the atmosphere.

2.1.6.2.1. "Buchholz" Relay

On power transformers using a conservator liquid-preservation system, a "Buchholz" relay can be installed in the piping between the main transformer tank and the conservator. The purpose of the Buchholz relay is to detect faults that may occur in the transformer. One mode of operation is based on the generation of gases in the transformer during certain minor internal faults. Gases accumulate in the relay, displacing the liquid in the relay, until a specified volume is collected, at which time a float actuates a contact or switch. Another mode of operation involves sudden increases in pressure in the main transformer tank, a sign of a major fault in the transformer. Such an increase in pressure forces the liquid to surge through the piping between the main tank and the conservator, through the "Buchholz" relay, which actuates another contact or switch.

2.1.6.2.2. Gas-Accumulator Relay

Another gas-detection device uses a system of piping from the top of the transformer to a gas-accumulator relay. Gases generated in the transformer are routed to the gas-accumulator relay, where they accumulate until a specified volume is collected, actuating a contact or switch.

2.1.7 Inrush Current

When a transformer is taken off-line, a certain amount of residual flux remains in the core due to the properties of the magnetic core material. The residual flux can be as much as 50 to 90% of the maximum operating flux, depending the type of core steel. When voltage is reapplied to the transformer, the flux introduced by this source voltage builds upon that already existing in the core. In order to maintain this level of flux in the core, which can be well into the saturation range of the core steel, the transformer can draw current well in excess of the transformer's rated full-load current. Depending on the transformer design, the magnitude of this current inrush can be anywhere from 3.5 to 40 times the rated full-load current. The waveform of the inrush current is similar to a sine wave, but largely skewed to the positive or negative direction. This inrush current experiences a decay, partially due to losses that provide a dampening effect. However, the current can remain well above rated current for many cycles.

This inrush current can have an effect on the operation of relays and fuses located in the system near the transformer. Decent approximations of the inrush current require detailed information regarding the transformer design, which may be available from the manufacturer but is not typically available to the application engineer. Actual values for inrush current depend on where in the source-voltage wave the switching operations occur, the moment of opening affecting the residual flux magnitude, and the moment of closing affecting the new flux.

2.1.8 Transformers Connected Directly to Generators

Power transformers connected directly to generators can experience excitation and short-circuit conditions beyond the requirements defined by ANSI/IEEE standards. Special design considerations may be necessary to ensure that a power transformer is capable of withstanding the abnormal thermal and mechanical aspects that such conditions can create.

Typical generating plants are normally designed such that two independent sources are required to supply the auxiliary load of each generator. Figure 2.1.21 shows a typical one-line diagram of a generating station. The power transformers involved can be divided into three basic subgroups based on their specific application:

1. Unit transformers (UT) that are connected directly to the system
2. Station service transformers (SST) that connect the system directly to the generator auxiliary load
3. Unit auxiliary transformers (UAT) that connect the generator directly to the generator auxiliary load

In such a station, the UAT will typically be subjected to the most severe operational stresses. Abnormal conditions have been found to result from several occurrences in the operation of the station. Instances of faults occurring at point F in Figure 2.1.21 — between the UAT and the breaker connecting it to the auxiliary load — are fed by two sources, both through the UT from the system and from the generator itself. Once the fault is detected, it initiates a trip to disconnect the UT from the system and to remove the generator excitation. This loss of load on the generator can result in a higher voltage on the generator, resulting in an increased current contribution to the fault from the generator. This will continue to feed the fault for a time period dependent upon the generator's fault-current decrement characteristics. Alternatively, high generator-bus voltages can result from events such as generator-load rejection, resulting in overexcitation of a UAT connected to the generator bus. If a fault were to occur between the UAT and the breaker connecting it to the auxiliary load during this period of overexcitation, it could exceed the thermal and mechanical capabilities of the UAT. Additionally, nonsynchronous paralleling of the UAT and the SST, both connected to the generator auxiliary load, can create high circulating currents that can exceed the mechanical capability of these transformers.

Considerations can be made in the design of UAT transformers to account for these possible abnormal operating conditions. Such design considerations include lowering the core flux density at rated voltage to allow for operation at higher V/Hz without saturation of the core, as well as increasing the design

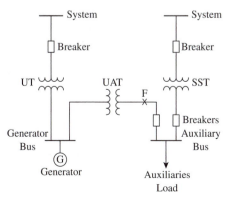

FIGURE 2.1.21 Typical simplified one-line diagram for the supply of a generating station's auxiliary power.

margin on the mechanical-withstand capability of the windings to account for the possibility of a fault occurring during a period of overexcitation. The thermal capacity of the transformer can also be increased to prevent overheating due to increased currents.

2.1.9 Modern and Future Developments

2.1.9.1 High-Voltage Generator

Because electricity is generated at voltage levels that are too low to be efficiently transmitted across the great distances that the power grid typically spans, step-up transformers are required at the generator. With developments in high-voltage-cable technology, a high-voltage generator, called the Powerformer™ (ABB Generation, Västerås, Sweden), has been developed that will eliminate the need for this GSU transformer and associated equipment. This Powerformer reportedly can be designed to generate power at voltage levels between 20 kV and 400 kV to feed the transmission network directly.

2.1.9.2 High-Temperature Superconducting (HTS) Transformer

Superconducting technologies are being applied to power transformers in the development of high-temperature superconducting (HTS) transformers. In HTS transformers, the copper and aluminum in the windings would be replaced by superconductors. In the field of superconductors, high temperatures are considered to be in the range of 116K to 144K, which represents a significant deviation in the operating temperatures of conventional transformers. At these temperatures, insulation of the type currently used in transformers would not degrade in the same manner. Using superconducting conductors in transformers requires advances in cooling, specifically refrigeration technology directed toward use in transformers. The predominant cooling medium in HTS development has been liquid nitrogen, but other media have been investigated as well. Transformers built using HTS technology would reportedly be smaller and lighter, and they would be capable of overloads without experiencing "loss of life" due to insulation degradation, using instead increasing amounts of the replaceable coolant. An additional benefit would be an increase in efficiency of HTS transformers over conventional transformers due to the fact that resistance in superconductors is virtually zero, thus eliminating the I^2R component of the load losses.

References

American National Standard for Transformers — 230 kV and below 833/958 through 8333/10417 kVA, Single-Phase; and 750/862 through 60000/80000/100000 kVA, Three-Phase without Load Tap Changing; and 3750/4687 through 60000/80000/100000 kVA with Load Tap Changing — Safety Requirements, ANSI C57.12.10-1997, National Electrical Manufacturers Association, Rosslyn, VA, 1998.

Bean, R.L., Chackan, N., Jr., Moore, H.R., and Wentz, E.C., *Transformers for the Electric Power Industry*, McGraw-Hill, New York, 1959.

Gebert, K.L. and Edwards, K.R., *Transformers*, 2nd ed., American Technical Publishers, Homewood, IL, 1974.

Goldman, A.W. and Pebler, C.G., *Power Transformers*, Vol. 2, Electrical Power Research Institute, Palo Alto, CA, 1987.

Hobson, J.E. and Witzke, R.L., *Power Transformers and Reactors, Electrical Transmission and Distribution Reference Book*, 4th ed., Central Station Engineers of the Westinghouse Electric Corporation, Westinghouse Electric, East Pittsburgh, PA, 1950, chap. 5.

IEEE, Guide for Transformers Directly Connected to Generators, IEEE C57.116-1989, Institute of Electrical and Electronics Engineers, Piscataway, NJ, 1989.

IEEE, Standard General Requirements for Liquid-Immersed Distribution, Power, and Regulating Transformers, IEEE C57.12.00-1999, Institute of Electrical and Electronics Engineers, Piscataway, NJ, 1999.

IEEE, Standard Terminology for Power and Distribution Transformers, ANSI/IEEE C57.12.80-1978, Institute of Electrical and Electronics Engineers, Piscataway, NJ, 1998.

Leijon, M., Owman, O., Sörqvist, T., Parkegren, C., Lindahl, S., and Karlsson, T., Powerformer™: A Giant Step in Power Plant Engineering, presented at IEEE International Electric Machines and Drives Conference, Seattle, WA, 1999.

Mehta, S.P., Aversa, N., and Walker, M.S., Transforming transformers, *IEEE Spectrum*, 34, 7, 43–49, July 1997.

Vargo, S.G., Transformer Design Considerations for Generator Auxiliary and Station Auxiliary Transformers, presented at 1976 Electric Utility Engineering Conference, 1976.

2.2 Distribution Transformers

Dudley L. Galloway and Dan Mulkey

2.2.1 Historical Background

2.2.1.1 Long-Distance Power

In 1886, George Westinghouse built the first long-distance alternating-current electric lighting system in Great Barrington, MA. The power source was a 25-hp steam engine driving an alternator with an output of 500 V and 12 A. In the middle of town, 4000 ft away, transformers were used to reduce the voltage to serve light bulbs located in nearby stores and offices (Powel, 1997).

2.2.1.2 The First Transformers

Westinghouse realized that electric power could only be delivered over distances by transmitting at a higher voltage and then reducing the voltage at the location of the load. He purchased U.S. patent rights to the transformer developed by Gaulard and Gibbs, shown in Figure 2.2.1(a). William Stanley, Westinghouse's electrical expert, designed and built the transformers to reduce the voltage from 500 to 100 V on the Great Barrington system. The Stanley transformer is shown in Figure 2.2.1(b).

2.2.1.3 What Is a Distribution Transformer?

Just like the transformers in the Great Barrington system, any transformer that takes voltage from a primary distribution circuit and "steps down" or reduces it to a secondary distribution circuit or a

FIGURE 2.2.1 (Left) Gaulard and Gibbs transformer; (right) William Stanley's early transformer. (By permission of ABB Inc., Raleigh, NC.)

consumer's service circuit is a distribution transformer. Although many industry standards tend to limit this definition by kVA rating (e.g., 5 to 500 kVA), distribution transformers can have lower ratings and can have ratings of 5000 kVA or even higher, so the use of kVA ratings to define transformer types is being discouraged (IEEE, 2002b).

2.2.2 Construction

2.2.2.1 Early Transformer Materials

From the pictures in Figure 2.2.1, the Gaulard-Gibbs transformer seems to have used a coil of many turns of iron wire to create a ferromagnetic loop. The Stanley model, however, appears to have used flat sheets of iron, stacked together and clamped with wooden blocks and steel bolts. Winding conductors were most likely made of copper from the very beginning. Several methods of insulating the conductor were used in the early days. Varnish dipping was often used and is still used for some applications today. Paper-tape wrapping of conductors has been used extensively, but this has now been almost completely replaced by other methods.

2.2.2.2 Oil Immersion

In 1887, the year after Stanley designed and built the first transformers in the U.S., Elihu Thompson patented the idea of using mineral oil as a transformer cooling and insulating medium (Myers et al., 1981). Although materials have improved dramatically, the basic concept of an oil-immersed cellulosic insulating system has changed very little in well over a century.

2.2.2.3 Core Improvements

The major improvement in core materials was the introduction of silicon steel in 1932. Over the years, the performance of electrical steels has been improved by grain orientation (1933) and continued improvement in the steel chemistry and insulating properties of surface coatings. The thinner and more effective the insulating coatings are, the more efficient a particular core material will be. The thinner the laminations of electrical steel, the lower the losses in the core due to circulating currents. Mass production of distribution transformers has made it feasible to replace stacked cores with wound cores. C-cores were first used in distribution transformers around 1940. A C-core is made from a continuous strip of steel, wrapped and formed into a rectangular shape, then annealed and bonded together. The core is then sawn in half to form two C-shaped sections that are machine-faced and reassembled around the coil. In the mid 1950s, various manufacturers developed wound cores that were die-formed into a rectangular shape and then annealed to relieve their mechanical stresses. The cores of most distribution transformers made today are made with wound cores. Typically, the individual layers are cut, with each turn slightly lapping over itself. This allows the core to be disassembled and put back together around the coil structures while allowing a minimum of energy loss in the completed core. Electrical steel manufacturers now produce stock for wound cores that is from 0.35 to 0.18 mm thick in various grades. In the early 1980s, rapid increases in the cost of energy prompted the introduction of amorphous core steel. Amorphous metal is cooled down from the liquid state so rapidly that there is no time to organize into a crystalline structure. Thus it forms the metal equivalent of glass and is often referred to as metal glass or "met-glass." Amorphous core steel is usually 0.025 mm thick and offers another choice in the marketplace for transformer users that have very high energy costs.

2.2.2.4 Winding Materials

Conductors for low-voltage windings were originally made from small rectangular copper bars, referred to as "strap." Higher ratings could require as many as 16 of these strap conductors in parallel to make one winding having the needed cross section. A substantial improvement was gained by using copper strip, which could be much thinner than strap but with the same width as the coil itself. In the early 1960s, instability in the copper market encouraged the use of aluminum strip conductor. The use of aluminum round wire in the primary windings followed in the early 1970s (Palmer, 1983). Today, both aluminum and copper conductors are used in distribution transformers, and the choice is largely dictated

FIGURE 2.2.2 Typical three-phase pad-mounted distribution transformer. (By permission of ABB Inc., Jefferson City, MO.)

by economics. Round wire separated by paper insulation between layers has several disadvantages. The wire tends to "gutter," that is, to fall into the troughs in the layer below. Also, the contact between the wire and paper occurs only along two lines on either side of the conductor. This is a significant disadvantage when an adhesive is used to bind the wire and paper together. To prevent these problems, manufacturers often flatten the wire into an oval or rectangular shape in the process of winding the coil. This allows more conductor to be wound into a given size of coil and improves the mechanical and electrical integrity of the coil (Figure 2.2.4).

2.2.2.5   Conductor Insulation

The most common insulation today for high-voltage windings is an enamel coating on the wire, with kraft paper used between layers. Low-voltage strip can be bare with paper insulation between layers. The use of paper wrapping on strap conductor is slowly being replaced by synthetic polymer coatings or wrapping with synthetic cloth. For special applications, synthetic paper such as DuPont's Nomex®[1] can be used in place of kraft paper to permit higher continuous operating temperatures within the transformer coils.

2.2.2.5.1   *Thermally Upgraded Paper*
In 1958, manufacturers introduced insulating paper that was chemically treated to resist breakdown due to thermal aging. At the same time, testing programs throughout the industry were showing that the estimates of transformer life being used at the time were extremely conservative. By the early 1960s, citing the functional-life testing results, the industry began to change the standard average winding-temperature rise for distribution transformers, first to a dual rating of 55/65°C and then to a single 65°C rating (IEEE, 1995). In some parts of the world, the distribution transformer standard remains at 55°C rise for devices using nonupgraded paper.

2.2.2.6   Conductor Joining

The introduction of aluminum wire, strap, and strip conductors and enamel coatings presented a number of challenges to distribution transformer manufacturers. Aluminum spontaneously forms an insulating

---

[1] Nomex® is a registered trademark of E.I. duPont de Nemours & Co., Wilmington, DE.

oxide coating when exposed to air. This oxide coating must be removed or avoided whenever an electrical connection is desired. Also, electrical-conductor grades of aluminum are quite soft and are subject to cold flow and differential expansion problems when mechanical clamping is attempted. Some methods of splicing aluminum wires include soldering or crimping with special crimps that penetrate enamel and oxide coatings and seal out oxygen at the contact areas. Aluminum strap or strip conductors can be TIG (tungsten inert gas)-welded. Aluminum strip can also be cold-welded or crimped to other copper or aluminum connectors. Bolted connections can be made to soft aluminum if the joint area is properly cleaned. "Belleville" spring washers and proper torquing are used to control the clamping forces and contain the metal that wants to flow out of the joint. Aluminum joining problems are sometimes mitigated by using hard alloy tabs with tin plating to make bolted joints using standard hardware.

2.2.2.7 Coolants

2.2.2.7.1 *Mineral Oil*

Mineral oil surrounding a transformer core-coil assembly enhances the dielectric strength of the winding and prevents oxidation of the core. Dielectric improvement occurs because oil has a greater electrical withstand than air and because the dielectric constant of oil (2.2) is closer to that of the insulation. As a result, the stress on the insulation is lessened when oil replaces air in a dielectric system. Oil also picks up heat while it is in contact with the conductors and carries the heat out to the tank surface by self-convection. Thus a transformer immersed in oil can have smaller electrical clearances and smaller conductors for the same voltage and kVA ratings.

2.2.2.7.2 *Askarels*

Beginning about 1932, a class of liquids called askarels or polychlorinated biphenyls (PCB) was used as a substitute for mineral oil where flammability was a major concern. Askarel-filled transformers could be placed inside or next to a building where only dry types were used previously. Although these coolants were considered nonflammable, as used in electrical equipment they could decompose when exposed to electric arcs or fires to form hydrochloric acid and toxic furans and dioxins. The compounds were further undesirable because of their persistence in the environment and their ability to accumulate in higher animals, including humans. Testing by the U.S. Environmental Protection Agency has shown that PCBs can cause cancer in animals and cause other noncancer health effects. Studies in humans provide supportive evidence for potential carcinogenic and noncarcinogenic effects of PCBs (http://www.epa.gov). The use of askarels in new transformers was outlawed in 1977 (Claiborne, 1999). Work still continues to retire and properly dispose of transformers containing askarels or askarel-contaminated mineral oil. Current ANSI/IEEE standards require transformer manufacturers to state on the nameplate that new equipment left the factory with less than 2 ppm PCBs in the oil (IEEE, 2000).

2.2.2.7.3 *High-Temperature Hydrocarbons*

Among the coolants used to take the place of askarels in distribution transformers are high-temperature hydrocarbons (HTHC), also called high-molecular-weight hydrocarbons. These coolants are classified by the National Electric Code as "less flammable" if they have a fire point above 300°C. The disadvantages of HTHCs include increased cost and a diminished cooling capacity from the higher viscosity that accompanies the higher molecular weight.

2.2.2.7.4 *Silicones*

Another coolant that meets the National Electric Code requirements for a less-flammable liquid is a silicone, chemically known as polydimethylsiloxane. Silicones are only occasionally used because they exhibit biological persistence if spilled and are more expensive than mineral oil or HTHCs.

2.2.2.7.5 *Halogenated Fluids*

Mixtures of tetrachloroethane and mineral oil were tried as an oil substitute for a few years. This and other chlorine-based compounds are no longer used because of a lack of biodegradability, the tendency to produce toxic by-products, and possible effects on the Earth's ozone layer.

2.2.2.7.6 *Esters*

Synthetic esters are being used in Europe, where high-temperature capability and biodegradability are most important and their high cost can be justified, for example, in traction (railroad) transformers. Transformer manufacturers in the U.S. are now investigating the use of natural esters obtained from vegetable seed oils. It is possible that agricultural esters will provide the best combination of high-temperature properties, stability, biodegradability, and cost as an alternative to mineral oil in distribution transformers (Oommen and Claiborne, 1996).

2.2.2.8 Tank and Cabinet Materials

A distribution transformer is expected to operate satisfactorily for a minimum of 30 years in an outdoor environment while extremes of loading work to weaken the insulation systems inside the transformer. This high expectation demands the best in state-of-the-art design, metal processing, and coating technologies. A typical three-phase pad-mounted transformer is illustrated in Figure 2.2.2.

2.2.2.8.1 *Mild Steel*

Almost all overhead and pad-mounted transformers have a tank and cabinet parts made from mild carbon steel. In recent years, major manufacturers have started using coatings applied by electrophoretic methods (aqueous deposition) and by powder coating. These new methods have largely replaced the traditional flow-coating and solvent-spray application methods.

2.2.2.8.2 *Stainless Steel*

Since the mid 1960s, single-phase submersibles have almost exclusively used AISI 400-series stainless steel. These grades of stainless were selected for their good welding properties and their tendency to resist pit-corrosion. Both 400-series and the more expensive 304L (low-carbon chromium-nickel) stainless steels have been used for pad mounts and pole types where severe environments justify the added cost. Transformer users with severe coastal environments have observed that pad mounts show the worst corrosion damage where the cabinet sill and lower areas of the tank contact the pad. This is easily explained by the tendency for moisture, leaves, grass clippings, lawn chemicals, etc., to collect on the pad surface. Higher areas of a tank and cabinet are warmed and dried by the operating transformer, but the lowest areas in contact with the pad remain cool. Also, the sill and tank surfaces in contact with the pad are most likely to have the paint scratched. To address this, manufacturers sometimes offer hybrid transformers, where the cabinet sill, hood, or the tank base may be selectively made from stainless steel.

2.2.2.8.3 *Composites*

There have been many attempts to conquer the corrosion tendencies of transformers by replacing metal structures with reinforced plastics. One of the more successful is a one-piece composite hood for single-phase pad-mounted transformers (Figure 2.2.3).

2.2.2.9 Modern Processing

2.2.2.9.1 *Adhesive Bonding*

Today's distribution transformers almost universally use a kraft insulating paper that has a diamond pattern of epoxy adhesive on each side. Each finished coil is heated prior to assembly. The heating drives out any moisture that might be absorbed in the insulation. Bringing the entire coil to the elevated temperature also causes the epoxy adhesive to bond and cure, making the coil into a solid mass, which is more capable of sustaining the high thermal and mechanical stresses that the transformer might encounter under short-circuit current conditions while in service. Sometimes the application of heat is combined with clamping of the coil sides to ensure intimate contact of the epoxy-coated paper with the conductors as the epoxy cures. Another way to improve adhesive bonding in the high-voltage winding is to flatten round wire as the coil is wound. This produces two flat sides to contact adhesive on the layer paper above and below the conductor. It also improves the space factor of the conductor cross section, permitting more actual conductor to fit within the same core window. Flattened conductor is less likely to "gutter" or fall into the spaces in the previous layer, damaging the layer insulation. Figure 2.2.4 shows a cross section of enameled round wire after flattening.

FIGURE 2.2.3   Single-phase transformer with composite hood. (By permission of ABB Inc., Jefferson City, MO.)

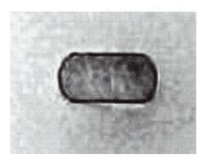

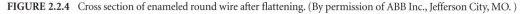

FIGURE 2.2.4   Cross section of enameled round wire after flattening. (By permission of ABB Inc., Jefferson City, MO.)

2.2.2.9.2   Vacuum Processing

With the coil still warm from the bonding process, transformers are held at a high vacuum while oil flows into the tank. The combination of heat and vacuum assures that all moisture and all air bubbles have been removed from the coil, providing electrical integrity and a long service life. Factory processing with heat and vacuum is impossible to duplicate in the field or in most service facilities. Transformers, if opened, should be exposed to the atmosphere for minimal amounts of time, and oil levels should never be taken down below the tops of the coils. All efforts must be taken to keep air bubbles out of the insulation structure.

2.2.3   General Transformer Design

2.2.3.1   Liquid-Filled vs. Dry Type

The vast majority of distribution transformers on utility systems today are liquid-filled. Liquid-filled transformers offer the advantages of smaller size, lower cost, and greater overload capabilities compared with dry types of the same rating.

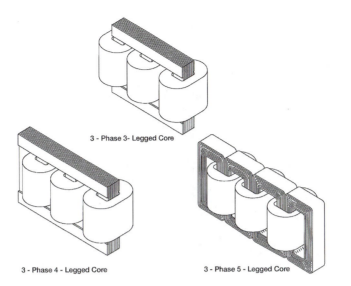

3 - Phase 3- Legged Core

3 - Phase 4 - Legged Core

3 - Phase 5 - Legged Core

FIGURE 2.2.5 Three- and four-legged stacked cores and five-legged wound core. (From IEEE C57.105-1978, IEEE Guide for Application of Transformer Connections in Three-Phase Distribution Systems, copyright 1978 by the Institute of Electrical and Electronics Engineers, Inc. The IEEE disclaims any responsibility or liability resulting from the placement and use in the described manner. Information is reprinted with the permission of the IEEE.)

2.2.3.2 Stacked vs. Wound Cores

Stacked-core construction favors the manufacturer that makes a small quantity of widely varying special designs in its facility. A manufacturer that builds large quantities of identical designs will benefit from the automated fabrication and processing of wound cores. Figure 2.2.5 shows three-phase stacked and wound cores.

2.2.3.3 Single Phase

The vast majority of distribution transformers used in North America are single phase, usually serving a single residence or as many as 14 to 16, depending on the characteristics of the residential load. Single-phase transformers can be connected into banks of two or three separate units. Each unit in a bank should have the same voltage ratings but need not supply the same kVA load.

2.2.3.3.1 Core-Form Construction

A single core loop linking two identical winding coils is referred to as core-form construction. This is illustrated in Figure 2.2.6.

2.2.3.3.2 Shell-Form Construction

A single winding structure linking two core loops is referred to as shell-form construction. This is illustrated in Figure 2.2.7.

2.2.3.3.3 Winding Configuration

Most distribution transformers for residential service are built as a shell form, where the secondary winding is split into two sections with the primary winding in between. This so-called LO-HI-LO configuration results in a lower impedance than if the secondary winding is contiguous. The LO-HI configuration is used where the higher impedance is desired and especially on higher-kVA ratings where higher impedances are mandated by standards to limit short-circuit current. Core-form transformers are always built LO-HI because the two coils must always carry the same currents. A 120/240 V service using a core-form in the LO-HI-LO configuration would need eight interconnected coil sections. This is considered too complicated to be commercially practical. LO-HI-LO and LO-HI configurations are illustrated in Figure 2.2.8.

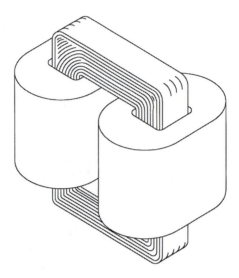

SINGLE - PHASE CORE - TYPE TRANSFORMER

FIGURE 2.2.6 Core-form construction. (From IEEE C57.105-1978, IEEE Guide for Application of Transformer Connections in Three-Phase Distribution Systems, copyright 1978 by the Institute of Electrical and Electronics Engineers, Inc. The IEEE disclaims any responsibility or liability resulting from the placement and use in the described manner. Information is reprinted with the permission of the IEEE.)

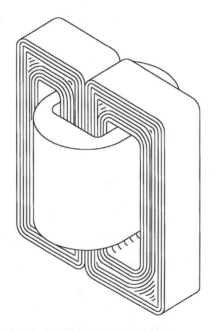

SINGLE - PHASE SHELL - TYPE TRANSFORMER

FIGURE 2.2.7 Shell-form construction. (From IEEE C57.105-1978, IEEE Guide for Application of Transformer Connections in Three-Phase Distribution Systems, copyright 1978 by the Institute of Electrical and Electronics Engineers, Inc. The IEEE disclaims any responsibility or liability resulting from the placement and use in the described manner. Information is reprinted with the permission of the IEEE.)

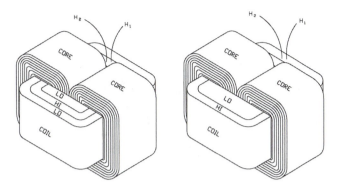

FIGURE 2.2.8 LO-HI-LO and LO-HI configurations. (By permission of ABB Inc., Jefferson City, MO.)

2.2.3.4 Three Phase

Most distribution transformers built and used outside North America are three phase, even for residential service. In North America, three-phase transformers serve commercial and industrial sites only. All three-phase distribution transformers are said to be of core-form construction, although the definitions outlined above do not hold. Three-phase transformers have one coaxial coil for each phase encircling a vertical leg of the core structure. Stacked cores have three or possibly four vertical legs, while wound cores have a total of four loops creating five legs or vertical paths: three down through the center of the three coils and one on the end of each outside coil. The use of three vs. four or five legs in the core structure has a bearing on which electrical connections and loads can be used by a particular transformer. The advantage of three-phase electrical systems in general is the economy gained by having the phases share common conductors and other components. This is especially true of three-phase transformers using common core structures. See Figure 2.2.5.

2.2.3.5 Duplex and Triplex Construction

Occasionally, utilities will require a single tank that contains two completely separate core-coil assemblies. Such a design is sometimes called a duplex and can have any size combination of single-phase core-coil assemblies inside. The effect is the same as constructing a two-unit bank with the advantage of having only one tank to place. Similarly, a utility may request a triplex transformer with three completely separate and distinct core structures (of the same kVA rating) mounted inside one tank.

2.2.3.6 Serving Mixed Single- and Three-Phase Loads

The utility engineer has a number of transformer configurations to choose from, and it is important to match the transformer to the load being served. A load that is mostly single phase with a small amount of three phase is best served by a bank of single-phase units or a duplex pair, one of which is larger to serve the single-phase load. A balanced three-phase load is best served by a three-phase unit, with each phase's coil identically loaded (ABB, 1995).

2.2.4 Transformer Connections

2.2.4.1 Single-Phase Primary Connections

The primary winding of a single-phase transformer can be connected between a phase conductor and ground or between two phase conductors of the primary system (IEEE, 2000).

2.2.4.1.1 Grounded Wye Connection

Those units that must be grounded on one side of the primary are usually only provided with one primary connection bushing. The primary circuit is completed by grounding the transformer tank to the grounded system neutral. Thus, it is imperative that proper grounding procedure be followed when the transformer is installed so that the tank never becomes "hot." Since one end of the primary winding is always grounded, the manufacturer can economize the design and grade the high-voltage insulation. Grading provides less insulation at the end of the winding closest to ground. A transformer with graded insulation usually cannot be converted to operate phase-to-phase. The primary-voltage designation on the nameplate of a graded insulation transformer will include the letters, "GRDY," as in "12470 GRDY/7200," indicating that it must be connected phase-to-ground on a grounded wye system.

2.2.4.1.2 Fully Insulated Connection

Single-phase transformers supplied with fully insulated (not graded) coils and two separate primary connection bushings may be connected phase-to-phase on a three-phase system or phase-to-ground on a grounded wye system as long as the proper voltage is applied to the coil of the transformer. The primary-voltage designation on the nameplate of a fully insulated transformer will look like 7200/12470Y, where 7200 is the coil voltage. If the primary voltage shows only the coil voltage, as in 2400, then the bushings can sustain only a limited voltage from the system ground, and the transformer must be connected phase-to-phase.

2.2.4.2 Single-Phase Secondary Connections

Distribution transformers will usually have two, three, or four secondary bushings, and the most common voltage ratings are 240 and 480, with and without a mid-tap connection. Figure 2.2.9 shows various single-phase secondary connections.

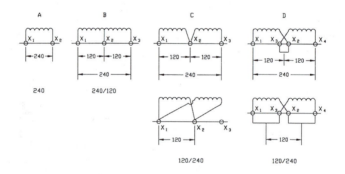

FIGURE 2.2.9 Single-phase secondary connections. (By permission of ABB Inc., Jefferson City, MO.)

2.2.4.2.1 Two Secondary Bushings

A transformer with two bushings can supply only a single voltage to the load.

2.2.4.2.2 Three Secondary Bushings

A transformer with three bushings supplies a single voltage with a tap at the midpoint of that voltage. This is the common three-wire residential service used in North America. For example, a 120/240 V secondary can supply load at either 120 or 240 V as long as neither 120-V coil section is overloaded. Transformers with handholes or removable covers can be internally reconnected from three to two bushings in order to serve full kVA from the parallel connection of coil sections. These are designated 120/240 or 240/480 V, with the smaller value first. Most pad-mounted distribution transformers are permanently and completely sealed and therefore cannot be reconnected from three to two bushings. The secondary voltage for permanently sealed transformers with three bushings is 240/120 V or 480/240 V.

2.2.4.2.3 Four Secondary Bushings

Secondaries with four bushings can be connected together external to the transformer to create a mid-tap connection with one bushing in common, or a two-bushing connection where the internal coil sections are paralleled. The four-bushing secondary will be designated as 120/240 or 240/480 V, indicating that a full kVA load can be served at the lower voltage. The distinction between 120/240 and 240/120 V must be carefully followed when pad-mounted transformers are being specified.

2.2.4.3 Three-Phase Connections

When discussing three-phase distribution transformer connections, it is well to remember that this can refer to a single three-phase transformer or single-phase transformers interconnected to create a three-phase bank. For either an integrated transformer or a bank, the primary or secondary can be wired in either delta or wye connection. The wye connections can be either grounded or ungrounded. However, not all combinations will operate satisfactorily, depending on the transformer construction, characteristics of the load, and the source system. Detailed information on three-phase connections can be found in the literature (ABB, 1995; IEEE, 1978a). Some connections that are of special concern are listed below.

2.2.4.3.1 *Ungrounded Wye–Grounded Wye*

A wye–wye connection where the primary neutral is left floating produces an unstable neutral where high third-harmonic voltages are likely to appear. In some Asian systems, the primary neutral is stabilized by using a three-legged core and by limiting current unbalance on the feeder at the substation.

2.2.4.3.2 *Grounded Wye–Delta*

This connection is called a grounding transformer. Unbalanced primary voltages will create high currents in the delta circuit. Unless the transformer is specifically designed to handle these circulating currents, the secondary windings can be overloaded and burn out. Use of the ungrounded wye–delta is suggested instead.

2.2.4.3.3 *Grounded Wye–Grounded Wye*

A grounded wye–wye connection will sustain unbalanced voltages, but it must use a four- or five-legged core to provide a return path for zero-sequence flux.

2.2.4.3.4 *Three-Phase Secondary Connections–Delta*

Three-phase transformers or banks with delta secondaries will have simple nameplate designations such as 240 or 480. If one winding has a mid-tap, say for lighting, then the nameplate will say 240/120 or 480/240, similar to a single-phase transformer with a center tap. Delta secondaries can be grounded at the mid-tap or any corner.

2.2.4.3.5 *Three-Phase Secondary Connections–Wye*

Popular voltages for wye secondaries are 208Y/120, 480Y/277, and 600Y/347

2.2.4.4 Duplex Connections

Two single-phase transformers can be connected into a bank having either an open-wye or open-delta primary along with an open-delta secondary. Such banks are used to serve loads that are predominantly single phase but with some three phase. The secondary leg serving the single-phase load can have a mid-tap, which may be grounded.

2.2.4.5 Other Connections

For details on other connections such as T-T and zigzag, consult the listed references (IEEE, 2002b; ABB, 1995; IEEE, 1978a).

2.2.4.6 Preferred Connections

In the earliest days of electric utility systems, it was found that induction motors drew currents that exhibited a substantial third harmonic component. In addition, transformers on the system that were operating close to the saturation point of their cores had third harmonics in the exciting current. One way to keep these harmonic currents from spreading over an entire system was to use delta-connected windings in transformers. Third-harmonic currents add up in-phase in a delta loop and flow around the loop, dissipating themselves as heat in the windings but minimizing the harmonic voltage distortion that might be seen elsewhere on the utility's system. With the advent of suburban underground systems in the 1960s, it was found that a transformer with a delta-connected primary was more prone to ferroresonance problems because of higher capacitance between buried primary cables and ground. An acceptable preventive was to go to grounded-wye-grounded-wye transformers on all but the heaviest industrial applications.

2.2.5 Operational Concerns

Even with the best engineering practices, abnormal situations can arise that may produce damage to equipment and compromise the continuity of the delivery of quality power from the utility.

2.2.5.1 Ferroresonance

Ferroresonance is an overvoltage phenomenon that occurs when charging current for a long underground cable or other capacitive reactance saturates the core of a transformer. Such a resonance can result in voltages as high as five times the rated system voltage, damaging lightning arresters and other equipment and possibly even the transformer itself. When ferroresonance is occurring, the transformer is likely to produce loud squeals and groans, and the noise has been likened to the sound of steel roofing being dragged across a concrete surface. A typical ferroresonance situation is shown in Figure 2.2.10 and consists of long underground cables feeding a transformer with a delta-connected primary. The transformer is unloaded or very lightly loaded and switching or fusing for the circuit operates one phase at a time. Ferroresonance can occur when energizing the transformer as the first switch is closed, or it can occur if one or more distant fuses open and the load is very light. Ferroresonance is more likely to occur on systems with higher primary voltage and has been observed even when there is no cable present. All of the contributing factors — delta or wye connection, cable length, voltage, load, single-phase switching — must be considered together. Attempts to set precise limits for prevention of the phenomenon have been frustrating.

2.2.5.2 Tank Heating

Another phenomenon that can occur to three-phase transformers because of the common core structure between phases is tank heating. Wye–wye-connected transformers that are built on four- or five-legged cores are likely to saturate the return legs when zero-sequence voltage exceeds about 33% of the normal line-to-neutral voltage. This can happen, for example, if two phases of an overhead line wrap together and are energized by a single electrical phase. When the return legs are saturated, magnetic flux is then

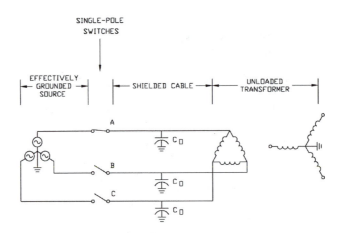

FIGURE 2.2.10 Typical ferroresonance situation. (From IEEE C57.105-1978, IEEE Guide for Application of Transformer Connections in Three-Phase Distribution Systems, copyright 1978 by the Institute of Electrical and Electronics Engineers, Inc. The IEEE disclaims any responsibility or liability resulting from the placement and use in the described manner. Information is reprinted with the permission of the IEEE.)

forced out of the core and finds a return path through the tank walls. Eddy currents produced by magnetic flux in the ferromagnetic tank steel will produce tremendous localized heating, occasionally burning the tank paint and boiling the oil inside. For most utilities, the probability of this happening is so low that it is not economically feasible to take steps to prevent it, other than keeping trees trimmed. A few, with a higher level of concern, purchase only triplex transformers, having three separate core-coil assemblies in one tank.

2.2.5.3 Polarity and Angular Displacement

The phase relationship of single-phase transformer voltages is described as "polarity." The term for voltage phasing on three-phase transformers is "angular displacement."

2.2.5.3.1 Single-Phase Polarity

The polarity of a transformer can either be additive or subtractive. These terms describe the voltage that may appear on adjacent terminals if the remaining terminals are jumpered together. The origin of the polarity concept is obscure, but apparently, early transformers having lower primary voltages and smaller kVA sizes were first built with additive polarity. When the range of kVAs and voltages was extended, a decision was made to switch to subtractive polarity so that voltages between adjacent bushings could never be higher than the primary voltage already present. Thus the transformers built to ANSI standards today are additive if the voltage is 8660 or below and the kVA is 200 or less; otherwise they are subtractive. This differentiation is strictly a U.S. phenomenon. Distribution transformers built to Canadian standards are all additive, and those built to Mexican standards are all subtractive. Although the technical definition of polarity involves the relative position of primary and secondary bushings, the position of primary bushings is always the same according to standards. Therefore, when facing the secondary bushings of an additive transformer, the X1 bushing is located to the right (of X3), while for a subtractive transformer, X1 is farthest to the left. To complicate this definition, a single-phase pad-mounted transformer built to ANSI standard Type 2 will always have the X2 mid-tap bushing on the lowest right-hand side of the low-voltage slant pattern. Polarity has nothing to do with the internal construction of the transformer windings but only with the routing of leads to the bushings. Polarity only becomes important when transformers are being paralleled or banked. Single-phase polarity is illustrated in Figure 2.2.11.

2.2.5.3.2 Three-Phase Angular Displacement

The phase relation of voltage between H1 and X1 bushings on a three-phase distribution transformer is referred to as angular displacement. ANSI standards require that wye–wye and delta–delta transformers have 0° displacement. Wye–delta and delta–wye transformers will have X1 lagging H1 by 30°. This difference in angular displacement means that care must be taken when three-phase transformers are paralleled to serve large loads. Sometimes the phase difference is used to advantage, such as when supplying power to 12-pulse rectifiers or other specialized loads. European standards permit a wide variety of displacements, the most common being Dy11. This IEC designation is interpreted as Delta primary–wye secondary, with X1 lagging H1 by $11 \times 30° = 330°$, or leading by 30°. The angular displacement of Dy11 differs from the ANSI angular displacement by 60°. Three-phase angular displacement is illustrated in Figure 2.2.12.

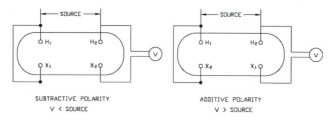

FIGURE 2.2.11 Single-phase polarity. (Adapted from IEEE C57.12.90-1999. The IEEE disclaims any responsibility or liability resulting from the placement and use in the described manner. Information is reprinted with the permission of the IEEE.)

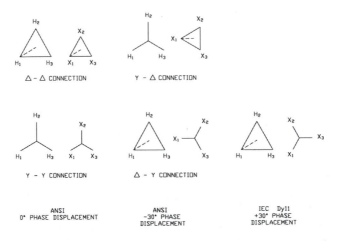

FIGURE 2.2.12 Three-phase angular displacement. (Adapted from IEEE C57.105-1978. The IEEE disclaims any responsibility or liability resulting from the placement and use in the described manner. Information is reprinted with the permission of the IEEE.)

2.2.6   Transformer Locations

2.2.6.1   Overhead

With electric wires being strung at the tops of poles to keep them out of the reach of the general public, it is obvious that transformers would be hung on the same poles, as close as possible to the high-voltage source conductors. Larger units are often placed on overhead platforms in alleyways, or alongside buildings, or on ground-level pads protected by fencing. Overhead construction is still the most economical choice in rural areas, but it has the disadvantage of susceptibility to ice and wind storms. The public no longer perceives overhead wiring as a sign of progress, instead considering it an eyesore that should be eliminated from view.

2.2.6.2   Underground

Larger cities with concentrated commercial loads and tall buildings have had underground primary cables and transformers installed in below-grade ventilated vaults since the early part of the 20th century. By connecting many transformers into a secondary network, service to highly concentrated loads can be maintained even though a single transformer may fail. In a network, temporary overloads can be shared among all the connected transformers. The use of underground distribution for light industrial and commercial and residential service became popular in the 1960s, with the emphasis on beautification that promoted fences around scrap yards and the elimination of overhead electric and telephone lines. The most common construction method for residential electric services is underground primary cables feeding a transformer placed on a pad at ground level. The problems of heat dissipation and corrosion are only slightly more severe than overheads, but they are substantially reduced compared with transformers confined in below-grade ventilated vaults. Since pad mounts are intended to be placed in locations that are frequented by the general public, the operating utility has to be concerned about security of the locked cabinet covering the primary and secondary connections to the transformer. The industry has established standards for security against unauthorized entry and vandalism of the cabinet and for locking provisions (ANSI/NEMA, 1999). Another concern is the minimization of sharp corners or edges that may be hazardous to children at play, and that also has been addressed by standards. The fact that pad-mounted transformers can operate with surface temperatures near the boiling point of water is a further concern that is voiced from time to time. One argument used to minimize the danger of burns is to point out that it is no more hazardous to touch a hot transformer than it is to touch the hood of an automobile on a sunny day. From a scientific standpoint, research has shown that people will pull away after touching a hot object in a much shorter time than it takes to sustain a burn injury. The point above which persons might be burned is about 150°C (Hayman, 1973). See Section 2.2.7 for a detailed description of underground transformers.

2.2.6.3 Directly Buried

Through the years, attempts have been made to place distribution transformers directly in the ground without a means of ventilation. A directly buried installation may be desirable because it is completely out of sight and cannot be damaged by windstorms, trucks and automobiles, or lawn mowers. There are three major challenges when directly buried installations are considered: the limited operational accessibility, a corrosive environment, and the challenge of dissipating heat from the transformer. The overall experience has been that heat from a buried transformer tends to dry out earth that surrounds it, causing the earth to shrink and create gaps in the heat-conduction paths to the ambient soil. If a site is found that is always moist, then heat conduction may be assured, but corrosion of the tank or of cable shields is still a major concern. Within the last several years, advances in encapsulation materials and techniques have fostered development of a solid-insulation distribution transformer that can be installed in a ventilated vault or directly buried using thermal backfill materials while maintaining loadability comparable with overhead or pad-mounted transformers. For further information, see Section 2.2.7.4, Emerging Issues.

2.2.6.4 Interior Installations

Building codes generally prohibit the installation of a distribution transformer containing mineral oil inside or immediately adjacent to an occupied building. The options available include use of a dry-type transformer or the replacement of mineral oil with a less-flammable coolant. See Section 2.2.2.6, Coolants.

2.2.7 Underground Distribution Transformers

Underground transformers are self-cooled, liquid-filled, sealed units designed for step-down operation from an underground primary-cable supply. They are available in both single- and three-phase designs. Underground transformers can be separated into three subgroups: those designed for installation in roomlike vaults, those designed for installation in surface-operable enclosures, and those designed for installation on a pad at ground level.

2.2.7.1 Vault Installations

The vault provides the required ventilation, access for operation, maintenance, and replacement, while at the same time providing protection against unauthorized entry. Vaults used for transformer installations are large enough to allow personnel to enter the enclosure, typically through a manhole and down a ladder. Vaults have been used for many decades, and it is not uncommon to find installations that date back to the days when only paper-and-lead-insulated primary cable was available. Transformers for vault installations are typically designed for radial application and have a separate fuse installation on the source side.

Vaults can incorporate many features:

- Removable top sections for transformer replacement
- Automatic sump pumps to keep water levels down
- Chimneys to increase natural air flow
- Forced-air circulation

Transformers designed for vault installation are sometimes installed in a room inside a building. This, of course, requires a specially designed room to limit exposure to fire and access by unauthorized personnel and to provide sufficient ventilation. Both mineral-oil-filled units and units with one of the less-flammable insulating oils are used in these installations. These installations are also made using dry-type or pad-mounted transformers.

Transformers for vault installation are manufactured as either subway transformers or as vault-type transformers, which, according to C57.12.40 (ANSI, 2000a), are defined as follows:

- Vault-type transformers are suitable for occasional submerged operation.
- Subway transformers are suitable for frequent or continuous submerged operation.

From the definitions, the vault type should only be used when a sump pump is installed, while the subway-type could be installed without a sump pump. The principal distinction between *vault* and *subway* is their corrosion resistance. For example, the 1994 version of the network standard, C57.12.40, required the auxiliary coolers to have a corrosion-resistance equivalence of not less than 5/16 in. of copper-bearing steel for subway transformers but only 3/32 in. for vault-type transformers. In utility application, vault and subway types may be installed in the same type of enclosure, and the use of a sump pump is predicated more on the need for quick access for operations than it is on whether the transformer is a vault or subway type.

2.2.7.1.1   Transformers for Vault Installation

2.2.7.1.1.1 Network Transformers — As defined in IEEE C57.12.80 (IEEE, 2002b), network transformers (see Figure 2.2.13) are designed for use in vaults to feed a variable-capacity system of interconnected secondaries. They are three-phase transformers that are designed to connect through a network protector to a secondary network system. Network transformers are typically applied to serve loads in the downtown areas of major cities. National standard C57.12.40 (ANSI, 2000a) details network transformers. The standard kVA ratings are 300, 500, 750, 1000, 1500, 2000, and 2500 kVA. The primary voltages range from 2,400 to 34,500 V. The secondary voltages are 216Y/125 or 480Y/277.

Network transformers are built as either vault type or subway type. They incorporate a primary switch with open, closed, and ground positions. Primary cable entrances are made by one of the following methods:

- Wiping sleeves or entrance fittings for connecting to lead cables — either one three-conductor or three single-conductor fittings or sleeves
- Bushing wells or integral bushings for connecting to plastic cables — three wells or three bushings

2.2.7.1.1.2 Network Protectors — Although not a transformer, the network protector is associated with the network transformer. The protector is an automatic switch that connects and disconnects the transformer from the secondary network being served. The protector connects the transformer when power flows from the primary circuit into the secondary network, and it disconnects upon reverse power flow from the secondary to the primary. The protector is described in C57.12.44 (IEEE, 2000c). The protector is typically mounted on the secondary throat of the network transformer, as shown in Figure 2.2.13.

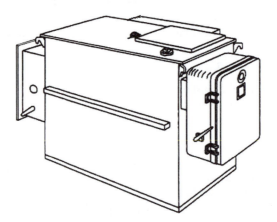

FIGURE 2.2.13 Network transformer with protector. (By permission of Pacific Gas & Electric Company, San Francisco, CA.)

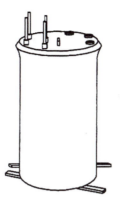

FIGURE 2.2.14   Single-phase subway. (By permission of Pacific Gas & Electric Company, San Francisco, CA.)

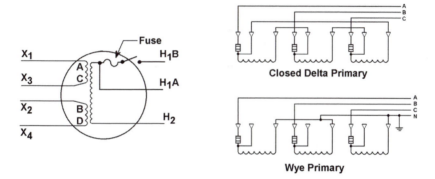

FIGURE 2.2.15   Three-bushing subway. (By permission of Pacific Gas & Electric Company, San Francisco, CA.)

2.2.7.1.1.3 Single-Phase Subway or Vault Types — These are round single-phase transformers designed to be installed in a vault and capable of being banked together to provide three-phase service (Figure 2.2.14). These can be manufactured as either subway-type or vault-type transformers. They are typically applied to serve small- to medium-sized commercial three-phase loads. The standard kVA ratings are 25, 37.5, 50, 75, 100, 167, and 250 kVA. Primary voltages range from 2,400 to 34,500 V, with the secondary voltage usually being 120/240. Four secondary bushings allow the secondary windings to be connected in parallel for wye connections or in series for delta connections. The secondary can be either insulated cables or spades. The units are designed to fit through a standard 36-in.-diameter manhole. They are not specifically covered by a national standard, however they are very similar to the units in IEEE C57.12.23 (IEEE, 2002c). Units with three primary bushings or wells, and with an internal primary fuse (Figure 2.2.15), allow for connection in closed-delta, wye, or open-wye banks. They can also be used for single-phase phase-to-ground connections.

Units with two primary bushings or wells, and with two internal primary fuses (Figure 2.2.16) allow for connection in an open-delta or an open-wye bank. This construction also allows for single-phase line-to-line connection.

2.2.7.1.1.4 Three-Phase Subway or Vault Types — These are rectangular-shaped three-phase transformers that can be manufactured as either subway-type or vault-type. Figure 2.2.17 depicts a three-phase vault. These are used to supply large three-phase commercial loads. Typically they have primary-bushing well terminations on one of the small sides and the secondary bushings with spades on the opposite end. These are also designed for radial installation and require external fusing. They can be manufactured in any of the standard three-phase kVA sizes and voltages. They are not detailed in a national standard.

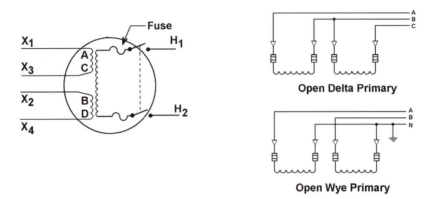

FIGURE 2.2.16 Two-bushing subway. (By permission of Pacific Gas & Electric Company, San Francisco, CA.)

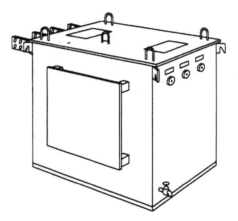

FIGURE 2.2.17 Three-phase vault. (By permission of Pacific Gas & Electric Company, San Francisco, CA.)

2.2.7.2 Surface-Operable Installations

The subsurface enclosure provides the required ventilation as well as access for operation, maintenance, and replacement, while at the same time providing protection against unauthorized entry. Surface-operable enclosures have grade-level covers that can be removed to gain access to the equipment. The enclosures typically are just large enough to accommodate the largest size of transformer and allow for proper cable bending. Transformers for installation in surface-operable enclosures are manufactured as submersible transformers, which are defined in C57.12.80 (IEEE, 2002b) as "so constructed as to be successfully operable when submerged in water under predetermined conditions of pressure and time." These transformers are designed for loop application and thus require internal protection. Submersible transformers are designed to be connected to an underground distribution system that utilizes 200-A-class equipment. The primary is most often #2 or 1/0 cables with 200-A elbows. While larger cables such as 4/0 can be used with the 200-A elbows, it is not recommended. The extra stiffness of 4/0 cable makes it very difficult to avoid putting strain on the elbow-bushing interface, which may lead to early failure. The operating points of the transformer are arranged on or near the cover. The installation is designed to be hot-stick operable by a person standing at ground level at the edge of the enclosure. There are three typical variations of submersible transformers.

2.2.7.2.1 Single-Phase Round Submersible

Single-phase round transformers (Figure 2.2.18) have been used since the early 1960s. These transformers are typically applied to serve residential single-phase loads. These units are covered by C57.12.23 (IEEE,

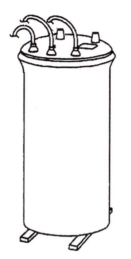

FIGURE 2.2.18 Single-phase round. (By permission of Pacific Gas & Electric Company, San Francisco, CA.)

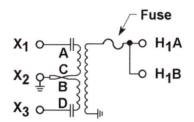

FIGURE 2.2.19 Two-primary bushing. (By permission of Pacific Gas & Electric Company, San Francisco, CA.)

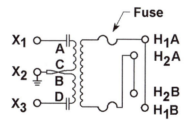

FIGURE 2.2.20 Four-primary bushing. (By permission of Pacific Gas & Electric Company, San Francisco, CA.)

1992). They are manufactured in the normal single-phase kVA ratings of 25, 37.5, 50, 75, 100, and 167 kVA. Primary voltages are available from 2,400 through 24,940 GrdY/14,400, and the secondary is 240/120 V. They are designed for loop-feed operation with a 200-A internal bus connecting the two bushings. Three low-voltage cable leads are provided through 100 kVA, while the 167-kVA size has six. They commonly come in two versions — a two-primary-bushing unit (Figure 2.2.19) and a four-primary-bushing unit (Figure 2.2.20) — although only the first is detailed in the standard. The two-bushing unit is for phase-to-ground-connected transformers, while the four-bushing unit is for phase-to-phase-connected transformers. As these are designed for application where the primary continues on after feeding through the transformer, the transformers require internal protection. The most common method is to use a secondary breaker and an internal nonreplaceable primary-expulsion fuse element. These units are designed for installation in a 36-in.-diameter round enclosure. Enclosures have been made of fiberglass

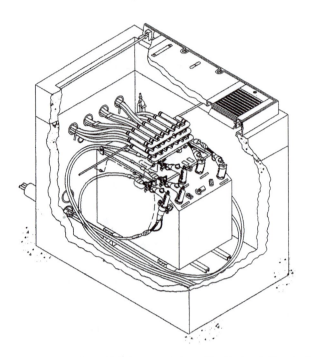

FIGURE 2.2.21 Four-bushing horizontal installed. (By permission of Pacific Gas & Electric Company, San Francisco, CA.)

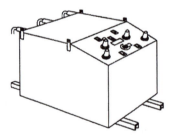

FIGURE 2.2.22 Four-bushing horizontal. (By permission of Pacific Gas & Electric Company, San Francisco, CA.)

or concrete. Installations have been made with and without a solid bottom. Those without a solid bottom simply rest on a gravel base.

2.2.7.2.2 *Single-Phase Horizontal Submersible*

Functionally, these are the same as the round single-phase. However, they are designed to be installed in a rectangular enclosure, as shown in Figure 2.2.21. Three low-voltage cable leads are provided through 100 kVA, while the 167-kVA size has six. They are manufactured in both four-primary-bushing designs (Figure 2.2.22) and in six-primary-bushing designs (Fig. 2.2.23). As well as the normal single-phase versions, there is also a duplex version. This is used to supply four-wire, three-phase, 120/240-V services from two core-coil assemblies connected open-delta on the secondary side. The primary can be either open-delta or open-wye. Horizontal transformers also have been in use since the early 1960s. These units are not specifically covered by a national standard. The enclosures used have included treated plywood, fiberglass, and concrete. The plywood and fiberglass enclosures are typically bottomless, with the transformer resting on a gravel base.

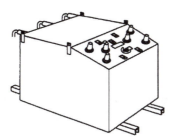

FIGURE 2.2.23 Six-bushing horizontal. (By permission of Pacific Gas & Electric Company, San Francisco, CA.)

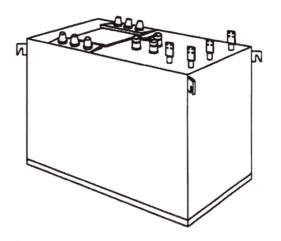

FIGURE 2.2.24 Three-phase submersible. (By permission of Pacific Gas & Electric Company, San Francisco, CA.)

2.2.7.2.3 *Three-Phase Submersible*

The three-phase surface-operable units are detailed in C57.12.24 (ANSI, 2000b). Typical application for these transformers is to serve three-phase commercial loads from loop-feed primary underground cables. Primary voltages are available from 2,400 through 34,500 V. The standard three-phase kVA ratings from 75 to 1000 kVA are available with secondary voltage of 208Y/120 V. With a 480Y/277-V secondary, the available sizes are 75 to 2500 kVA. Figure 2.2.24 depicts a three-phase submersible.

Protection options include:

- Dry-well current-limiting fuses with an interlocked switch to prevent the fuses from being removed while energized
- Submersible bayonet fuses with backup, under-oil, partial-range, current-limiting fuses, or with backup internal nonreplaceable primary-expulsion fuse elements.

These are commonly installed in concrete rectangular boxes with removable cover sections.

2.2.7.3 **Vault and Subsurface Common Elements**

2.2.7.3.1 *Tank Material*

The substrate and coating should meet the requirements detailed in C57.12.32 (ANSI, 2002a). The smaller units can be constructed out of 400-series or 300-series stainless steels or out of mild carbon steel. In general, 300-series stainless steel outperforms 400-series stainless steel, which significantly outperforms mild carbon steel. Most of the small units are manufactured out of 400-series stainless steel, since it is significantly less expensive than 300-series. Stainless steels from the 400-series with a good coating have been found to give satisfactory field performance. Due to lack of material availability, many of the larger units cannot be manufactured from 400-series stainless. With the choice then being limited to mild carbon steel or the very expensive 300-series stainless, most of the large units are constructed out of mild carbon steel.

2.2.7.3.2 Temperature Rating

Kilovoltampere ratings are based on not exceeding an average winding temperature rise of 55°C and a hottest-spot temperature rise of 70°C. However, they are constructed with the same 65°C rise insulation systems used in overhead and pad-mounted transformers. This allows for continuous operation at rated kVA provided that the enclosure ambient air temperature does not exceed 50°C and the average temperature does not exceed 40°C. Utilities commonly restrict loading on underground units to a lower limit than they do with pad-mounted or overhead units.

2.2.7.3.3 Siting

Subsurface units should not be installed if any of the following conditions apply:

- Soil is severely corrosive.
- Heavy soil erosion occurs.
- High water table causes repeated flooding of the enclosures.
- Heavy snowfall occurs.
- A severe mosquito problem exists.

2.2.7.3.4 Maintenance

Maintenance mainly consists of keeping the enclosure and the air vents free of foreign material. Dirt allowed to stay packed against the tank can lead to accelerated anaerobic corrosion, resulting in tank puncture and loss of mineral oil.

2.2.7.4 Emerging Issues

2.2.7.4.1 Water Pumping

Pumping of water from subsurface enclosures has been increasingly regulated. In some areas, water with any oily residue or turbidity must be collected for hazardous-waste disposal. Subsurface and vault enclosures are often subject to runoff water from streets. This water can include oily residue from vehicles. So even without a leak from the equipment, water collected in the enclosure may be judged a hazardous waste.

2.2.7.4.2 Solid Insulation

Transformers with solid insulation are commercially available for subsurface distribution applications (see Figure 2.2.25) with ratings up 167 kVA single-phase and 500 kVA three-phase. The total encapsulation of what is essentially a dry-type transformer allows it to be applied in a subsurface environment (direct buried or in a subsurface vault). The solid insulation distribution transformer addresses problems often associated with underground and direct buried transformers. See Sections 2.2.6.2 and 2.2.6.3. Such installations can be out of sight, below grade, and not subject to corrosion and contamination. Pad-mounted and pole-mounted versions are also available.

2.2.8 Pad-Mounted Distribution Transformers

Pad-mounted transformers are the most commonly used type of transformer for serving loads from underground distribution systems. They offer many advantages over subsurface, vault, or subway transformers.

- Installation: less expensive to purchase and easier to install
- Maintenance: easier to maintain
- Operability: easier to find, less time to open and operate
- Loading: utilities often assign higher loading limits to pad-mounted transformers as opposed to surface-operable or vault units.

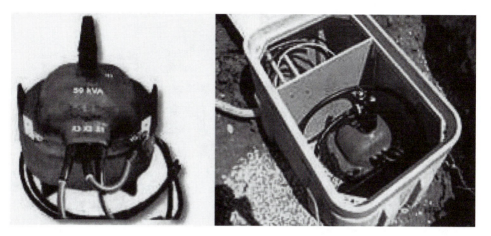

FIGURE 2.2.25 Solid-insulation distribution transformer. (By permission of ABB Inc., Quebec.)

Many users and suppliers break distribution transformers into just two major categories – overhead and underground, with pad-mounted transformers included in the underground category. The IEEE standards, however, divide distribution transformers into three categories – overhead, underground, and pad-mounted.

Pad-mounted transformers are manufactured as either:

- Single-phase or three-phase units: Single-phase units are designed to transform only one phase. Three-phase units transform all three phases. Most three-phase transformers use a single-, three-, four-, or five-legged core structure, although duplex or triplex construction is used on occasion.
- Loop or radial units: Loop-style units have the capability of terminating two primary conductors per phase. Radial-style units can only terminate one primary cable per phase. The primary must end at a radial-style unit, but from a loop style it can continue on to serve other units.
- Live-front or dead-front units: Live-front units have the primary cables terminated in a stress cone supported by a bushing. Thus the primary has exposed energized metal, or "live," parts. Dead-front units use primary cables that are terminated with high-voltage separable insulated connectors. Thus the primary has all "dead" parts — no exposed energized metal.

2.2.8.1 Single-Phase Pad-Mounted Transformers

Single-phase pad-mounted transformers are usually applied to serve residential subdivisions. Most single-phase transformers are manufactured as clamshell, dead-front, loop-type with an internal 200-A primary bus designed to allow the primary to loop through and continue on to feed the next transformer. These are detailed in the IEEE Standard C57.12.25 (ANSI, 1990). The standard assumes that the residential subdivision is served by a one-wire primary extension. It details two terminal arrangements for loop-feed systems: Type 1 (Figure 2.2.26) and Type 2 (Figure 2.2.27). Both have two primary bushings and three secondary bushings. The primary is always on the left facing the transformer bushings with the cabinet hood open, and the secondary is on the right. There is no barrier or division between the primary and secondary. In the Type 1 units, both primary and secondary cables rise directly up from the pad. In Type 2 units, the primary rises from the right and crosses the secondary cables that rise from the left. Type 2 units can be shorter than the Type 1 units, since the crossed cable configuration gives enough free cable length to operate the elbow without requiring the bushing to be placed as high. Although not detailed in the national standard, there are units built with four and with six primary bushings. The four-bushing unit is used for single-phase lines, with the transformers connected phase-to-phase. The six-primary-bushing units are used to supply single-phase loads from three-phase taps. Terminating all of the phases in the transformer allows all of the phases to be sectionalized at the same location. The internal single-phase transformer can be connected either phase-to-phase or phase-to-ground. The

FIGURE 2.2.26 Typical Type 1 loop-feed system. (By permission of ABB Inc., Raleigh, NC.)

FIGURE 2.2.27 Typical Type 2 loop-feed system. (By permission of ABB Inc., Raleigh, NC.)

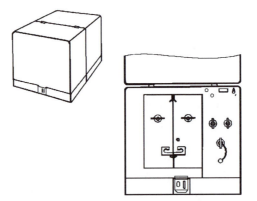

FIGURE 2.2.28 Single-phase live front. (By permission of Pacific Gas & Electric Company, San Francisco, CA.)

six-bushing units also allow the construction of duplex pad-mounted units that can be used to supply small three-phase loads along with the normal single-phase residential load. In those cases, the service voltage is four-wire, three-phase, 120/240 V.

Cabinets for single-phase transformers are typically built in the clamshell configuration with one large door that swings up, as shown in Figure 2.2.26 and Figure 2.2.27. Older units were manufactured with two doors, similar to the three-phase cabinets. New installations are almost universally dead front; however, live-front units (see Figure 2.2.28) are still purchased for replacements. These units are also built with clamshell cabinets but have an internal box-shaped insulating barrier constructed around the primary connections.

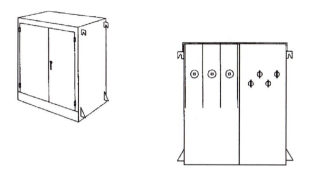

FIGURE 2.2.29 Radial-style live front. (By permission of Pacific Gas & Electric Company, San Francisco, CA.)

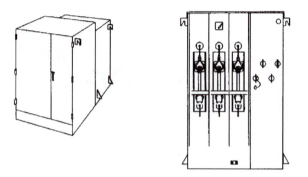

FIGURE 2.2.30 Loop-style live front. (By permission of Pacific Gas & Electric Company, San Francisco, CA.)

2.2.8.2 Three-Phase Pad-Mounted Transformers

Three-phase pad-mounted transformers are typically applied to serve commercial and industrial three-phase loads from underground distribution systems. Traditionally, there have been two national standards that detailed requirements for pad-mounted transformers — one for live front (ANSI C57.12.22) and one for dead front (IEEE C57.12.26). The two standards have now been combined into one for all pad mounts, designated IEEE C57.12.34.

2.2.8.3 Live Front

Live-front transformers are specified as radial units and thus do not come with any fuse protection. See Figure 2.2.29. The primary compartment is on the left, and the secondary compartment is on the right, with a rigid barrier separating them. The secondary door must be opened before the primary door can be opened. Stress-cone-terminated primary cables rise vertically and connect to the terminals on the end of the high-voltage bushings. Secondary cables rise vertically and are terminated on spades connected to the secondary bushings. Units with a secondary of 208Y/120 V are available up to 1000 kVA. Units with a secondary of 480Y/277 V are available up to 2500 kVA.

Although not detailed in a national standard, there are many similar types available. A loop-style live front (Figure 2.2.30) can be constructed by adding fuses mounted below the primary bushings. Two primary cables are then both connected to the bottom of the fuse. The loop is then made at the terminal of the high-voltage bushing, external to the transformer but within its primary compartment.

2.2.8.4 Dead Front

Both radial- and loop-feed dead-front pad-mounted transformers are detailed in the standard. Radial-style units have three primary bushings arranged horizontally, as seen in Figure 2.2.31. Loop-style units have six primary bushings arranged in a V pattern, as seen in Figure 2.2.32 and Figure 2.2.33. In both,

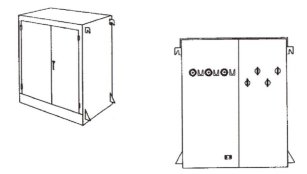

FIGURE 2.2.31 Radial-style dead front. (By permission of Pacific Gas & Electric Company, San Francisco, CA.)

FIGURE 2.2.32 Small loop-style dead front. (By permission of ABB Inc., Raleigh, NC.)

FIGURE 2.2.33 Large loop-style dead front. (By permission of ABB Inc., Raleigh, NC.)

the primary compartment is on the left, and the secondary compartment is on the right, with a rigid barrier between them. The secondary door must be opened before the primary door can be opened. The primary cables are terminated with separable insulated high-voltage connectors, commonly referred to as 200-A elbows, specified in IEEE Standard 386. These plug onto the primary bushings, which can be either bushing wells with an insert, or they can be integral bushings. Bushing wells with inserts are preferred, as they allow both the insert and elbow to be easily replaced. Units with a secondary of 208Y/120 V are available up to 1000 kVA. Units with a secondary of 480Y/277 V are available up to 2500 kVA.

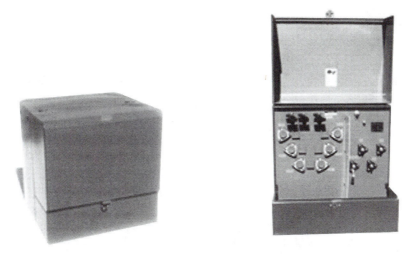

FIGURE 2.2.34 Mini three-phase in clamshell cabinet. (By permission of ABB Inc., Raleigh, NC.)

2.2.8.5 Additional Ratings

In addition to what is shown in the national standards, there are other variations available. The smallest size in the national standards is the 75 kVA unit. However, 45 kVA units are also manufactured in the normal secondary voltages. Units with higher secondary voltages, such as 2400 and 4160Y/2400, are manufactured in sizes up to 3750 kVA. There is a new style being produced that is a cross between single- and three-phase units. A small three-phase transformer is placed in a six-bushing loop-style clamshell cabinet, as seen in Figure 2.2.34. These are presently available from 45 to 150 kVA in both 208Y/120 and 480Y/277V secondaries.

2.2.8.6 Pad-Mount Common Elements

2.2.8.6.1 Protection

Most distribution transformers include some kind of primary overcurrent protection. For a detailed discussion, see Section 2.2.13, Transformer Protection.

2.2.8.6.2 Primary Conductor

Pad-mounted transformers are designed to be connected to an underground distribution system that utilizes 200-A-class equipment. The primary is most often #2 or 1/0 cables with 200-A elbows or stress cones. It is recommended that larger cables such as 4/0 not be used with the 200-A elbows. The extra stiffness of 4/0 cable makes it very difficult to avoid putting strain on the elbow-bushing interface, leading to premature elbow failures.

2.2.8.6.3 Pad

Pads are made out of various materials. The most common is concrete, which can be either poured in place or precast. Concrete is suitable for any size pad. Pads for single-phase transformers are also commonly made out of fiberglass or polymer-concrete.

2.2.8.6.4 Enclosure

There are two national standards that specify the requirements for enclosure integrity for pad-mounted equipment: C57.12.28 (ANSI/NEMA, 1999) for normal environments and C57.12.29 (ANSI, 1991) for coastal environments. The tank and cabinet of pad-mounted transformers are commonly manufactured out of mild carbon steel. When applied in corrosive areas, such as near the ocean, they are commonly made out of 300- or 400-series stainless steel. In general, 300-series stainless steel will outperform 400-series stainless steel, which significantly outperforms mild carbon steel in corrosive applications.

2.2.8.6.5 *Maintenance*

Maintenance mainly consists of keeping the enclosure rust free and in good repair so that it remains tamper resistant, i.e., capable of being closed and locked so that it resists unauthorized entry.

2.2.8.6.6 *Temperature Rating*

The normal temperature ratings are used. The kilovoltampere ratings are based on not exceeding an average winding temperature rise of 65°C and a hottest-spot temperature rise of 80°C over a daily average ambient of 30°C.

2.2.9   Transformer Losses

2.2.9.1   No-Load Loss and Exciting Current

When alternating voltage is applied to a transformer winding, an alternating magnetic flux is induced in the core. The alternating flux produces hysteresis and eddy currents within the electrical steel, causing heat to be generated in the core. Heating of the core due to applied voltage is called no-load loss. Other names are iron loss or core loss. The term "no-load" is descriptive because the core is heated regardless of the amount of load on the transformer. If the applied voltage is varied, the no-load loss is very roughly proportional to the square of the peak voltage, as long as the core is not taken into saturation. The current that flows when a winding is energized is called the "exciting current" or "magnetizing current," consisting of a real component and a reactive component. The real component delivers power for no-load losses in the core. The reactive current delivers no power but represents energy momentarily stored in the winding inductance. Typically, the exciting current of a distribution transformer is less than 0.5% of the rated current of the winding that is being energized.

2.2.9.2   Load Loss

A transformer supplying load has current flowing in both the primary and secondary windings that will produce heat in those windings. Load loss is divided into two parts, I^2R loss and stray losses.

2.2.9.2.1 *I^2R Loss*

Each transformer winding has an electrical resistance that produces heat when load current flows. Resistance of a winding is measured by passing dc current through the winding to eliminate inductive effects.

2.2.9.2.2 *Stray Losses*

When alternating current is used to measure the losses in a winding, the result is always greater than the I^2R measured with dc current. The difference between dc and ac losses in a winding is called "stray loss." One portion of stray loss is called "eddy loss" and is created by eddy currents circulating in the winding conductors. The other portion is generated outside of the windings, in frame members, tank walls, bushing flanges, etc. Although these are due to eddy currents also, they are often referred to as "other strays." The generation of stray losses is sometimes called "skin effect" because induced eddy currents tend to flow close to the surfaces of the conductors. Stray losses are proportionally greater in larger transformers because their higher currents require larger conductors. Stray losses tend to be proportional to current frequency, so they can increase dramatically when loads with high-harmonic currents are served. The effects can be reduced by subdividing large conductors and by using stainless steel or other nonferrous materials for frame parts and bushing plates.

2.2.9.3   Harmonics and DC Effects

Rectifier and discharge-lighting loads cause currents to flow in the distribution transformer that are not pure power-frequency sine waves. Using Fourier analysis, distorted load currents can be resolved into components that are integer multiples of the power frequency and thus are referred to as harmonics. Distorted load currents are expected to be high in the 3rd, 5th, 7th, and sometimes the 11th and 13th harmonics, depending on the character of the load.

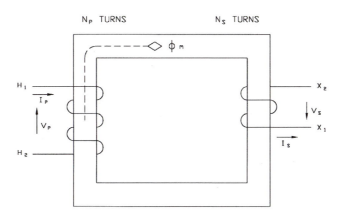

FIGURE 2.2.35 Two-winding transformer schematic. (By permission of ABB Inc., Jefferson City, MO.)

2.2.9.3.1 Odd-Ordered Harmonics

Load currents that contain the odd-numbered harmonics will increase both the eddy losses and other stray losses within a transformer. If the harmonics are substantial, then the transformer must be derated to prevent localized and general overheating. ANSI standards suggest that any transformer with load current containing more than 5% total harmonic distortion should be loaded according to the appropriate ANSI guide (IEEE, 1998).

2.2.9.3.2 Even-Ordered Harmonics

Analysis of most harmonic currents will show very low amounts of even harmonics (2nd, 4th, 6th, etc.) Components that are even multiples of the fundamental frequency generally cause the waveform to be nonsymmetrical about the zero-current axis. The current therefore has a zeroth harmonic or dc-offset component. The cause of a dc offset is usually found to be half-wave rectification due to a defective rectifier or other component. The effect of a significant dc current offset is to drive the transformer core into saturation on alternate half-cycles. When the core saturates, exciting current can be extremely high, which can then burn out the primary winding in a very short time. Transformers that are experiencing dc-offset problems are usually noticed because of objectionably loud noise coming from the core structure. Industry standards are not clear regarding the limits of dc offset on a transformer. A recommended value is a dc current no larger than the normal exciting current, which is usually 1% or less of a winding's rated current (Galloway, 1993).

2.2.10   Transformer Performance Model

A simple model will be developed to help explain performance characteristics of a distribution transformer, namely impedance, short-circuit current, regulation, and efficiency.

2.2.10.1   Schematic

A simple two-winding transformer is shown in the schematic diagram of Figure 2.2.35. A primary winding of N_P turns is on one side of a ferromagnetic core loop, and a similar coil having N_S turns is on the other. Both coils are wound in the same direction with the starts of the coils at H_1 and X_1, respectively. When an alternating voltage V_P is applied from H_2 to H_1, an alternating magnetizing flux φ_m flows around the closed core loop. A secondary voltage $V_S = V_P \times N_S/N_P$ is induced in the secondary winding and appears from X_2 to X_1 and very nearly in phase with V_P. With no load connected to X_1–X_2, I_P consists of only a small current called the magnetizing current. When load is applied, current I_S flows out of terminal X_1 and results in a current $I_P = I_S \times N_S/N_P$ flowing into H_1 in addition to magnetizing current. The ampere-turns of flux due to current $I_P \times N_P$ cancels the ampere-turns of flux due to current $I_S \times N_S$, so only the magnetizing flux exists in the core for all the time the transformer is operating normally.

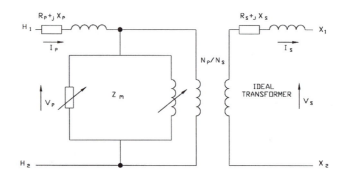

FIGURE 2.2.36 Complete transformer equivalent circuit. (By permission of ABB Inc., Jefferson City, MO.)

2.2.10.2 Complete Equivalent Circuit

Figure 2.2.36 shows a complete equivalent circuit of the transformer. An ideal transformer is inserted to represent the current- and voltage-transformation ratios. A parallel resistance and inductance representing the magnetizing impedance are placed across the primary of the ideal transformer. Resistance and inductance of the two windings are placed in the H_1 and X_1 legs, respectively.

2.2.10.3 Simplified Model

To create a simplified model, the magnetizing impedance has been removed, acknowledging that no-load loss is still generated and magnetizing current still flows, but it is so small that it can be ignored when compared with the rated currents. The R and X values in either winding can be translated to the other side by using percent values or by converting ohmic values with a factor equal to the turns ratio squared $(N_P/N_S)^2$. To convert losses or ohmic values of R and X to percent, use Equation 2.2.1 or Equation 2.2.2:

$$\%R = \frac{Load\ Loss}{10 \cdot kVA} = \frac{\Omega_{(R)} \cdot kVA}{kV^2} \tag{2.2.1}$$

$$\%X = \frac{AW}{10 \cdot kVA} = \frac{\Omega_{(L)} \cdot kVA}{kV^2} \tag{2.2.2}$$

where AW is apparent watts, or the scalar product of applied voltage and exciting current in units of amperes. Once the resistances and inductances are translated to the same side of the transformer, the ideal transformer can be eliminated and the percent values of R and X combined. The result is the simple model shown in Figure 2.2.37. A load, having power factor $\cos\theta$, may be present at the secondary.

2.2.10.4 Impedance

The values of %R and %X form the legs of what is known as the "impedance triangle." The hypotenuse of the triangle is called the transformer's impedance and can be calculated using Equation 2.2.3.

$$\%Z = \sqrt{\%R^2 + \%X^2} \tag{2.2.3}$$

A transformer's impedance is sometimes called "impedance volts" because it can be measured by shorting the secondary terminals and applying sufficient voltage to the primary so that rated current flows in each winding. The ratio of applied voltage to rated voltage, times 100, is equal to the percent impedance.

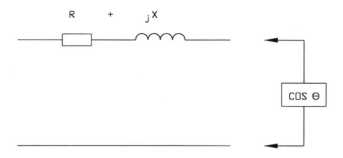

FIGURE 2.2.37 Simplified transformer model. (By permission of ABB Inc., Jefferson City, MO.)

2.2.10.5 Short-Circuit Current

If the load (right) side of the model of Figure 2.2.37 is shorted and rated voltage from an infinite source is applied to the left side, the current I_{SC} will be limited only by the transformer impedance:

$$I_{SC} = 100 \times I_R / \%Z \qquad (2.2.4)$$

For example, if the rated current, I_R, is 100 A and the impedance is 2.0%, the short-circuit current will be $100 \times 100/2 = 5000$ A.

2.2.10.6 Percent Regulation

When a transformer is energized with no load, the secondary voltage will be exactly the primary voltage divided by the turns ratio (N_P/N_S). When the transformer is loaded, the secondary voltage will be diminished by an amount determined by the transformer impedance and the power factor of the load. This change in voltage is called regulation and is actually defined as the rise in voltage when the load is removed. One result of the definition of regulation is that it is always a positive number. The primary voltage is assumed to be held constant at the rated value during this process. The exact calculation of percent regulation is given in Equation 2.2.5:

$$\%reg = \left(L^2 \cdot \left(\%R^2 + \%X^2 \right) + 200 \cdot L \cdot \left(\%X \cdot sin\theta + \%R \cdot cos\theta \right) + 10000 \right)^{0.5} - 100 \qquad (2.2.5)$$

where $\cos\theta$ is the power factor of the load and L is per unit load on the transformer. The most significant portion of this equation is the cross products, and since %X predominates over %R in the transformer impedance and $\cos\theta$ predominates over $\sin\theta$ for most loads, the percent regulation is usually less than the impedance (at $L = 1$). When the power factor of the load is unity, then $\sin\theta$ is zero and regulation is much less than the transformer impedance. A much simpler form of the regulation calculation is given in Equation 2.2.6. For typical values, the result is the same as the exact calculation out to the fourth significant digit or so.

$$\%reg \cong L \cdot \left(\%R * \cos\theta + \%X * \sin\theta + \frac{\left(\%X * \cos\theta - \%R * \sin\theta \right)^2}{200} \right) \qquad (2.2.6)$$

2.2.10.7 Percent Efficiency

As with any other energy conversion device, the efficiency of a transformer is the ratio of energy delivered to the load divided by the total energy drawn from the source. Percent efficiency is expressed as:

$$\%Efficiency = \frac{L \cdot kVA \cdot \cos\theta \cdot 10^5}{L \cdot kVA \cdot \cos\theta \cdot 10^3 + NL + L^2 \cdot LL} \qquad (2.2.7)$$

where cos θ is again the power factor of the load, therefore kVA · cos θ is real energy delivered to the load. NL is the no-load loss, and LL is the load loss of the transformer. Most distribution transformers serving residential or light industrial loads are not fully loaded all the time. It is assumed that such transformers are loaded to about 50% of nameplate rating on the average. Thus efficiency is often calculated at $L = 0.5$, where the load loss is about 25% of the value at full load. Since a typical transformer will have no-load loss of around 25% of load loss at 100% load, then at $L = 0.5$, the no-load loss will equal the load loss and the efficiency will be at a maximum.

2.2.11 Transformer Loading

2.2.11.1 Temperature Limits

According to ANSI standards, modern distribution transformers are to operate at a maximum 65°C average winding rise over a 30°C ambient air temperature at rated kVA. One exception to this is submersible or vault-type distribution transformers, where a 55°C rise over a 40°C ambient is specified. The bulk oil temperature near the top of the tank is called the "top oil temperature," which cannot be more than 65°C over ambient and will typically be about 55°C over ambient, 10°C less than the average winding rise.

2.2.11.2 Hottest-Spot Rise

The location in the transformer windings that has the highest temperature is called the "hottest spot." Standards require that the hottest-spot temperature not exceed 80°C rise over a 30°C ambient, or 110°C. These are steady-state temperatures at rated kVA. The hottest spot is of great interest because, presumably, this is where the greatest thermal degradation of the transformer's insulation system will take place. For calculation of thermal transients, the top-oil rise over ambient air and the hottest-spot rise over top oil are the parameters used.

2.2.11.3 Load Cycles

If all distribution loads were constant, then determining the proper loading of transformers would be a simple task. Loads on transformers, however, vary through the hours of a day, the days of a week, and through the seasons of the year. Insulation aging is a highly nonlinear function of temperature that accumulates over time. The best use of a transformer, then, is to balance brief periods of hottest-spot temperatures slightly above 110°C with extended periods at hottest spots well below 110°C. Methods for calculating the transformer loss-of-life for a given daily cycle are found in the ANSI Guide for Loading (IEEE, 1995). Parameters needed to make this calculation are the no-load and load losses, the top-oil rise, the hottest-spot rise, and the thermal time constant.

2.2.11.4 Thermal Time Constant

Liquid-filled distribution transformers can sustain substantial short-time overloads because the mass of oil, steel, and conductor takes time to come up to a steady-state operating temperature. Time constant values can vary from two to six hours, mainly due to the differences in oil volume vs. tank surface for different products.

2.2.11.5 Loading Distribution Transformers

Utilities often assign loading limits to distribution transformers that are different from the transformer's nameplate kVA. This is based on three factors: the actual ambient temperature, the shape of the load curve, and the available air for cooling. For example, one utility divides its service territory into three temperature situations for different ambient temperatures: summer interior, summer coastal, and winter. The transformer installations are divided into three applications for the available air cooling: overhead or pad-mounted, surface operable, and vault. The load shape is expressed by the peak-day load factor, which is defined as the season's peak kVA divided by the average kVA and then expressed as a percentage. Table 2.2.1 shows the assigned capabilities for a 100-kVA transformer. Thus this utility would assign the

TABLE 2.2.1 Assigned Capabilities for a 100-kVA Transformer

Transformer			Peak-Day Load Factor									
Location	Temperature District	kVA	10%	20%	30%	40%	50%	60%	70%	80%	90%	100%
Overhead or pad-mounted	Summer interior	100	205	196	187	177	168	159	149	140	131	122
	Summer coastal	100	216	206	196	186	176	166	156	146	136	126
	Winter	100	249	236	224	211	198	186	173	160	148	135
Surface operable	Summer interior	100	147	140	133	127	120	113	107	100	93	87
	Summer coastal	100	154	147	140	133	126	119	111	104	97	90
	Winter	100	178	169	160	151	142	133	124	115	105	96
Vault	Summer interior	100	173	164	156	147	139	130	122	113	105	96
	Summer coastal	100	182	173	164	155	146	137	127	118	109	100
	Winter	100	185	176	166	157	147	138	128	119	110	100

same 100-kVA transformer a peak capability of 87 to 249 kVA depending on its location, the season, and the load-shape.

2.2.12 Transformer Testing

2.2.12.1 Design Tests

Tests that manufacturers perform on prototypes or production samples are referred to as "design tests." These tests may include sound-level tests, temperature-rise tests, and short-circuit-current withstand tests. The purpose of a design test is to establish a design limit that can be applied by calculation to every transformer built. In particular, short-circuit tests are destructive and may result in some invisible damage to the sample, even if the test is passed successfully. The ANSI standard calls for a transformer to sustain six tests, four with symmetrical fault currents and two with asymmetrical currents. One of the symmetrical shots is to be of long duration, up to 2 s, depending on the impedance for lower ratings. The remaining five shots are to be 0.25 s in duration. The long-shot duration for distribution transformers 750 kVA and above is 1 s. The design passes the short-circuit test if the transformer sustains no internal or external damage (as determined by visual inspection) and minimal impedance changes. The tested transformer also has to pass production dielectric tests and experience no more than a 25% change in exciting current (Bean et al., 1959).

2.2.12.2 Production Tests

Production tests are given to and passed by each transformer made. Tests to determine ratio, polarity or phase-displacement, iron loss, load loss, and impedance are done to verify that the nameplate information is correct. Dielectric tests specified by industry standards are intended to prove that the transformer is capable of sustaining unusual but anticipated electrical stresses that may be encountered in service. Production dielectric tests may include applied-voltage, induced-voltage, and impulse tests.

2.2.12.2.1 Applied-Voltage Test

Standards require application of a voltage of (very roughly) twice the normal line-to-line voltage to each entire winding for one minute. This checks the ability of one phase to withstand voltage it may encounter when another phase is faulted to ground and transients are reflected and doubled.

2.2.12.2.2 Induced-Voltage Test

The original applied-voltage test is now supplemented with an induced-voltage test. Voltage at higher frequency (usually 400 Hz) is applied at twice the rated value of the winding. This induces the higher voltage in each winding simultaneously without saturating the core. If a winding is permanently grounded on one end, the applied-voltage test cannot be performed. In this case, many ANSI product standards specify that the induced primary test voltage be raised to 1000 plus 3.46 times the rated winding voltage (Bean et al., 1959).

2.2.12.2.3 Impulse Test

Distribution lines are routinely disturbed by voltage surges caused by lightning strokes and switching transients. A standard 1.2×50-μs impulse wave with a peak equal to the BIL (basic impulse insulation level) of the primary system (60 to 150 kV) is applied to verify that each transformer will withstand these surges when in service.

2.2.13 Transformer Protection

Distribution transformers require some fusing or other protective devices to prevent premature failure while in service. Circuit breakers at the substation or fusing at feeder taps or riser poles may afford some protection for individual transformers, but the most effective protection will be at, near, or within each transformer.

2.2.13.1 Goals of Protection

Transformer-protection devices that limit excessive currents or prevent excessive voltages are intended to achieve the following:

- Minimize damage to the transformer due to overloads
- Prevent transformer damage caused by secondary short circuits
- Prevent damage caused by faults within the transformer
- Minimize the possibility of damage to other property or injury to personnel
- Limit the extent or duration of service interruptions or disturbances on the remainder of the system

The selection of protection methods and equipment is an economic decision and may not always succeed in complete achievement of all of the goals listed above. For example, the presence of a primary fuse may not prevent longtime overloads that could cause transformer burnout.

2.2.13.2 Separate Protection

Distribution transformers may have fused cutouts on the same pole to protect an overhead transformer or on a nearby pole to protect a pad-mounted transformer. Sometimes a separate pad-mounted cabinet is used to house protection for larger pad-mounted and submersible transformers.

2.2.13.3 Internal Protection

When protection means are located within the transformer, the device can react to oil temperature as well as primary current. The most common internal protective devices are described below.

2.2.13.3.1 Protective Links

Distribution transformers that have no other protection are often supplied with a small high-voltage-expulsion fuse. The protective link is sized to melt at from six to ten times the rated current of the transformer. Thus it will not protect against longtime overloads and will permit short-time overloads that may occur during inrush or cold-load-pickup phenomena. For this reason, they are often referred to as "fault-sensing" links. Depending on the system voltage, protective links can safely interrupt faults of 1000 to 3000 A. Internal protective links are about the size of a small cigar.

2.2.13.3.2 Dual-Sensing or Eutectic Links

High-voltage fuses made from a low-melting-point tin alloy melt at 145°C and thus protect a transformer by detecting the combination of overload current and high oil temperature. A eutectic link, therefore, prevents longtime overloads but allows high inrush and cold-load-pickup currents. A similar device called a "dual element" fuse uses two sections of conductor that respond separately to current and oil temperature with slightly better coordination characteristics.

2.2.13.3.3 Current-Limiting Fuses
Current-limiting fuses can be used if the fault current available on the primary system exceeds the interrupting ratings of protective links. Current-limiting fuses can typically interrupt 40,000- to 50,000-A faults and do so in less than one half of a cycle. The interruption of a high-current internal fault in such a short time will prevent severe damage to the transformer and avoid damage to surrounding property or hazard to personnel that might otherwise occur. Full-range current-limiting fuses can be installed in small air switches or in dry-well canisters that extend within a transformer tank. Current-limiting fuses cannot prevent longtime overloads, but they can open on a secondary short circuit, so the fuse must be easily replaceable. Current-limiting fuses are considerably larger than expulsion fuses.

2.2.13.3.4 Bayonets
Pad-mounts and submersibles may use a primary link (expulsion fuse) that is mounted internally in the transformer oil but that can be withdrawn for inspection of the fuse element or to interrupt the primary feed. This device is called a bayonet and consists of a probe with a cartridge on the end that contains the replaceable fuse element. Fuses for bayonets may be either fault sensing or dual sensing.

2.2.13.3.5 Combination of Bayonet and Partial-Range Current-Limiting Fuses
The most common method of protection for pad-mounted distribution transformers is the coordinated combination of a bayonet fuse (usually dual sensing) and a partial-range current-limiting fuse (PRCL). The PRCL only responds to a high fault current, while the bayonet fuse is only capable of interrupting low fault currents. These fuses must be coordinated in such a way that any secondary fault will melt the bayonet fuse. Fault currents above the bolted secondary fault level are assumed to be due to internal faults. Thus the PRCL, which is mounted inside the tank, will operate only when the transformer has failed and must be removed from service.

2.2.13.4 Coordination of Protection

As applied to overcurrent protection for distribution transformers, the term *coordination* means two things:

1. A fuse must be appropriately sized for the transformer. A fuse that is too large will not prevent short-circuit currents that can damage the transformer coils. A fuse that is too small may open due to normal inrush currents when the transformer is energized or may open due to short-time overload currents that the transformer is capable of handling.
2. Transformer protection must fit appropriately with other protection means located upstream, downstream, or within the transformer. For example, a secondary oil circuit breaker should be coordinated with a primary fuse so that any short circuit on the transformer secondary will open the breaker before the primary fuse melts.

Where two fuses are used to protect a transformer, there are two methods of achieving coordination of the pair: "matched melt" and "time-current-curve crossover coordination" (TCCCC).

2.2.13.4.1 Matched Melt
An example of matched-melt coordination is where a cutout with an expulsion fuse and a backup current-limiting fuse are used to protect an overhead transformer. The two fuses are sized so that the expulsion fuse always melts before or at the same time as the current-limiting fuse. This permits the current-limiting fuse to help clear the fault if necessary, and the cutout provides a visible indication that the fault has occurred.

2.2.13.4.2 TCCC Coordination of Bayonet and Partial-Range Current-Limiting Fuses
TCCCC is much more common for pad-mounted and self-protected transformers, where the fuses are not visible. The TCCCC method is described as follows:

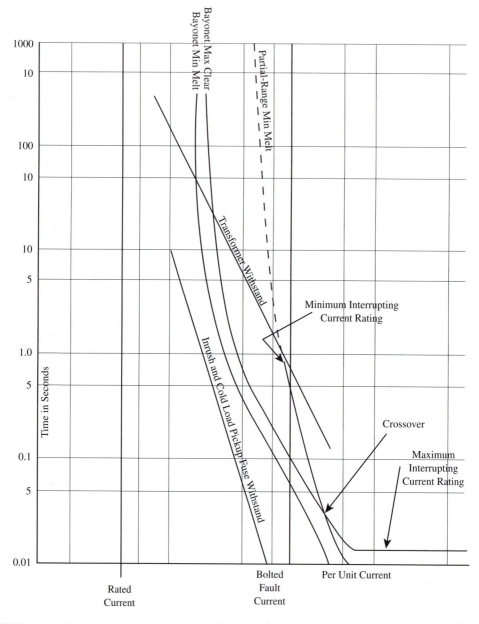

FIGURE 2.2.38 Time–current-curve crossover coordination. (By permission of ABB Inc., Jefferson City, MO.)

2.2.13.4.2.1 Fuse Curves — The main tool used for coordination is a graph of time vs. current for each fuse or breaker, as seen in Figure 2.2.38. The graph is displayed as a log–log plot and has two curves for any particular fuse. The first curve is called the minimum-melt curve, and this represents time-current points where the fuse element just starts to melt. The other curve is a plot of points at longer times (to the right of the minimum-melt curve). The latter curve is called the maximum-clear or sometimes the average-clear curve. The maximum-clear curve is where the fuse can be considered open and capable of sustaining full operating voltage across the fuse without danger of restrike. Even if a fuse has melted due to a fault, the fault current continues to flow until the maximum-clear time has passed. For expulsion fuses, there is a maximum interrupting rating that must not be exceeded unless a current-limiting or other backup fuse is present. For partial-range current-limiting fuses, there is a minimum interrupting current. Above that minimum current, clearing occurs in about 0.25 cycles, so the maximum-clear curve is not actually needed for most cases.

2.2.13.4.2.2 Transformer Characteristics — Each transformer has characteristics that are represented on the time-current curve to aid in the coordination process:

- Rated current = primary current at rated kVA
- Bolted fault current (I_{SC}) = short-circuit current in the primary with secondary shorted
- Inrush and cold-load-pickup curve:
  - Inrush values are taken as 25 times rated current at 0.01 s and 12 times rated current at 0.1 s.
  - Cold-load-pickup values are presumed to be six times rated current at 1 s and three times rated current at 10 s.

- Through-fault duration or short-circuit withstand established by IEEE C57.109. For most transformers, the curve is the plot of values for $I^2t = 1250$ or 50 times rated current at 0.5 s, 25 times rated current at 2 s, and 11.2 times rated current at 10 s. Values longer than 10 s are usually ignored.

2.2.13.4.2.3 Fuse Coordination Steps — Select an expulsion fuse such that:

- The minimum-melt curve falls entirely to the right of the inrush/cold-load-pickup curve. For most fuses, the minimum-melt curve will always be to the right of 300% of rated load, even for very long times.
- The maximum-clear curve will fall entirely to the left of the through-fault-duration curve at 10 s and below.

Select a partial-range current-limiting (PRCL) fuse such that its minimum-melt curve:

- Crosses the expulsion-fuse maximum-clear curve to the right of the bolted fault line, preferably with a minimum 25% safety margin
- Crosses the expulsion-fuse maximum-clear curve at a current higher than the PRCL minimum interrupting rating
- Crosses the expulsion-fuse maximum-clear curve at a current below the maximum interrupting rating of the expulsion fuse. It is not a critical issue if this criterion is not met, since the PRCL will quickly clear the fault anyway.

There are additional considerations, such as checking for a longtime recross of the two fuse characteristics or checking for a recross at a "knee" in the curves, as might occur with a dual-sensing fuse or a low-voltage circuit breaker with a high-current magnetic trip.

2.2.13.4.3 Low-Voltage Oil-Breaker Coordination

The coordination of an oil breaker with an expulsion fuse is slightly different than the previous example. The oil-breaker current duty is translated to the high-voltage side and is sized in a manner similar to the expulsion fuse in the previous example. The expulsion fuse is then selected to coordinate with the breaker so that the minimum melt falls entirely to the right of the breaker's maximum clear for all currents less than the bolted fault current. This ensures that the breaker will protect against all secondary faults and that the internal expulsion fuse will only open on an internal fault, where current is not limited by the transformer impedance.

2.2.13.5 Internal Secondary Circuit Breakers

Secondary breakers that are placed in the bulk oil of a transformer can protect against overloads that might otherwise cause thermal damage to the conductor-insulation system. Some breakers also have magnetically actuated trip mechanisms that rapidly interrupt the secondary load in case of a secondary fault. When properly applied, secondary breakers should limit the top-oil temperature of a transformer to about 110°C during a typical residential load cycle. Breakers on overhead transformers are often equipped with a red signal light. When this light is on, it signifies that the transformer has come close to tripping the breaker. The light will not go off until a lineman resets the breaker. The lineman can also

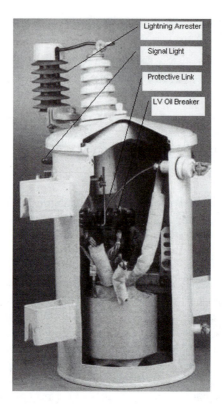

FIGURE 2.2.39 Cutaway showing CSP components. (By permission of ABB Inc., Jefferson City, MO.)

set the breaker on its emergency position, which allows the transformer to temporarily supply a higher overload until the utility replaces the unit with one having a higher kVA capacity. The secondary oil breaker is also handy to disconnect load from a transformer without touching the primary connections.

2.2.13.6 CSP®[2] Transformers

Overhead transformers that are built with the combination of secondary breaker, primary protective link, and external lightning arrester are referred to generically as CSPs (completely self-protected transformers). This protection package is expected to prevent failures caused by excessive loads and external voltage surges, and to protect the system from internal faults. The breaker is furnished with a signal light and an emergency control as described above. The protective link is often mounted inside the high-voltage bushing insulator, as seen in Figure 2.2.39.

2.2.13.7 Protection Philosophy

CSP transformers are still in use, especially in rural areas, but the trend is away from secondary breakers to prevent transformer burnouts. Continued growth of residential load is no longer a foregone conclusion. Furthermore, utilities are becoming more sophisticated in their initial transformer sizing and are using computerized billing data to detect a transformer that is being overloaded. Experience shows that modern distribution transformers can sustain more temporary overload than a breaker would allow. Most utilities would rather have service to their customers maintained than to trip a breaker unnecessarily.

2.2.13.8 Lightning Arresters

Overhead transformers can be supplied with primary lightning arresters mounted nearby on the pole structure, on the transformer itself, directly adjacent to the primary bushing, or within the tank. Pad-

[2] CSP® is a registered trademark of ABB Inc., Raleigh, NC.

mounted transformers can have arresters too, especially those at the end of a radial line, and they can be inside the tank, plugged into dead-front bushings, or at a nearby riser pole, where primary lines transition from overhead to underground.

2.2.14 Economic Application

2.2.14.1 Historical Perspective

Serious consideration of the economics of transformer ownership did not begin until the oil embargo of the early 1970s. With large increases in the cost of all fuels, utilities could no longer just pass along these increases to their customers without demonstrating fiscal responsibility by controlling losses on their distribution systems.

2.2.14.2 Evaluation Methodology

An understanding soon developed that the total cost of owning a transformer consisted of two major parts, the purchase price and the cost of supplying thermal losses of the transformer over an assumed life, which might be 20 to 30 years. To be consistent, the future costs of losses have to be brought back to the present so that the two costs are both on a present-worth basis. The calculation methodologies were published first by Edison Electric Institute and recently updated in the form of a proposed ANSI standard (IEEE, 2001).The essential part of the evaluation method is the derivation of *A* and *B* factors, which are the utility's present-worth costs for supplying no-load and load losses, respectively, in the transformer as measured in $/W.

2.2.14.3 Evaluation Formula

The proposed ANSI guide for loss evaluation expresses the present value of the total owning cost of purchasing and operating a transformer as follows (in its simplest form):

$$TOC = Transformer\ Cost + A \times No\ Load\ Loss + B \times Load\ Loss \tag{2.2.8}$$

where

A = loss-evaluation factor for no-load loss, $/W
B = loss-evaluation factor for load loss, $/W

The guide develops in detail the calculation of *A* and *B* factors from utility operating parameters as shown in Equation 2.2.9 and Equation 2.2.10, respectively:

$$A = \frac{SC + EC \times HPY}{FCR \times 1000} \tag{2.2.9}$$

$$B = \frac{\left[(SC \times RF) + (EC \times LSF \times HPY)\right] \times (PL)^2}{FCR \times 1000} \tag{2.2.10}$$

where   SC = GC + TD
SC = avoided cost of system capacity
GC = avoided cost of generation capacity
TD = avoided cost of transmission and distribution capacity
EC = avoided cost of energy
HPY = hours per year
FCR = levelized fixed-charge rate
RF = peak responsibility factor
LSF = transformer loss factor
PL = equivalent annual peak load

With the movement to deregulate electric utilities in the U.S., most utilities have now chosen to neglect elements of system cost that no longer may apply or to abandon entirely the consideration of the effects of transformer losses on the efficiency of their distribution system. Typical loss evaluation factors in the year 2003 are $A = \$2.50/W$ and $B = \$0.80/W$.

References

ABB, Distribution Transformer Guide, Distribution Transformer Division, ABB Power T&D Co., Raleigh, NC, 1995, pp. 40–70.

ANSI, Requirements for Pad-Mounted, Compartmental-Type, Self-Cooled, Single-Phase Distribution Transformers with Separable Insulated High Voltage: High Voltage (34,500 GrdY/19,920 V and Below); Low Voltage (240/120 V, 167 kVA and Smaller), C57.12.25-1990, Institute of Electrical and Electronics Engineers, Piscataway, NJ, 1990.

ANSI, Standard for Switchgear and Transformers: Pad-Mounted Equipment — Enclosure Integrity for Coastal Environments, C57.12.29-1991, Institute of Electrical and Electronics Engineers, Piscataway, NJ, 1991.

ANSI, Standard for Requirements for Secondary Network Transformers — Subway and Vault Types (Liquid Immersed) Requirements, IEEE C57.12.40-2000, Institute of Electrical and Electronics Engineers, Piscataway, NJ, 2000a.

ANSI, Underground-Type Three-Phase Distribution Transformers: 2500 kVA and Smaller; High-Voltage, 34,500 GrdY/19,920 Volts and Below; Low Voltage, 480 Volts and Below — Requirements, C57.12.24-2000, Institute of Electrical and Electronics Engineers, Piscataway, NJ, 2000b.

ANSI, Submersible Equipment — Enclosure Integrity, C57.12.32-2002, Institute of Electrical and Electronics Engineers, Piscataway, NJ, 2002a.

ANSI/NEMA, Pad-Mounted Equipment — Enclosure Integrity, ANSI/NEMA C57.12.28-1999, National Electrical Manufacturers Association, Rosslyn, VA, 1999.

Bean, R.L., Chackan, N., Jr., Moore, H.R., and Wentz, E.C., Transformers for the Electric Power Industry, Westinghouse Electric Corp. Power Transformer Division, McGraw-Hill, NY, 1959, pp. 338–340.

Claiborne, C.C., ABB Electric Systems Technology Institute, Raleigh, NC, personal communication, 1999.

Galloway, D.L., Harmonic and DC Currents in Distribution Transformers, presented at 46th Annual Power Distribution Conference, Austin, TX, 1993.

Hayman, J.L., E.I. duPont de Nemours & Co., letter to Betty Jane Palmer, Westinghouse Electric Corp., Jefferson City, MO, October 11, 1973.

IEEE, Guide for Application of Transformer Connections in Three-Phase Distribution System, IEEE C57.105-1978, section 2, Institute of Electrical and Electronics Engineers, Piscataway, NJ, 1978a.

IEEE, Standard Terminology for Power and Distribution Transformers, IEEE C57.12.80-2002, clause 2.3, Institute of Electrical and Electronics Engineers, Piscataway, NJ, 2002b.

IEEE, Standard for Transformers: Underground-Type, Self-Cooled, Single-Phase Distribution Transformers with Separable, Insulated, High-Voltage Connectors; High Voltage (24,940 GrdY/14,400 V and Below) and Low Voltage (240/120 V, 167 kVA and Smaller), C57.12.23-2002, Institute of Electrical and Electronics Engineers, Piscataway, NJ, 2002c.

IEEE, Standard Requirements for Secondary Network Protectors, C57.12.44-2000, Institute of Electrical and Electronics Engineers, Piscataway, NJ, 2000c.

IEEE, Guide for Loading Mineral-Oil-Immersed Transformers, IEEE C57.91-1995, Institute of Electrical and Electronics Engineers, Piscataway, NJ, 1995, p. iii.

IEEE, Recommended Practice for Establishing Transformer Capability when Supplying Nonsinusoidal Load Currents, C57.110-1998, Institute of Electrical and Electronics Engineers, Piscataway, NJ, 1998.

IEEE, Standard General Requirements for Liquid-Immersed Distribution, Power, and Regulating Transformers, C57.12.00-2000, Institute of Electrical and Electronics Engineers, Piscataway, NJ, 2000d.

IEEE, Guide for Distribution Transformer Loss Evaluation, C57.12.33, Draft 8-2001, Institute of Electrical and Electronics Engineers, Piscataway, NJ, 2001.

Myers, S.D., Kelly, J.J., and Parrish, R.H., *A Guide to Transformer Maintenance,* footnote 12, Transformer Maintenance Division, S.D. Myers, Akron, OH, 1981.

Oommen, T.V. and Claiborne, C.C., Natural and Synthetic High Temperature Fluids for Transformer Use, internal report, ABB Electric Systems Technology Institute, Raleigh, NC, 1996.

Palmer, B.J., History of Distribution Transformer Core/Coil Design, Distribution Transformer Engineering Report No. 83-17, Westinghouse Electric, Jefferson City, MO, 1983.

Powel, C.A., General considerations of transmission, in *Electrical Transmission and Distribution Reference Book*, ABB Power T&D Co., Raleigh, NC, 1997, p. 1.

2.3   Phase-Shifting Transformers

Gustav Preininger

2.3.1   Introduction

The necessity to control the power flow rose early in the history of the development of electrical power systems. When high-voltage grids were superimposed on local systems, parallel-connected systems or transmission lines of different voltage levels became standard. Nowadays large high-voltage power grids are connected to increase the reliability of the electrical power supply and to allow exchange of electrical power over large distances. Complications, attributed to several factors such as variation in power-generation output and/or power demand, can arise and have to be dealt with to avoid potentially catastrophic system disturbances. Additional tools in the form of phase-shifting transformers (PSTs) are available to control the power flow to stabilize the grids. These may be justified to maintain the required quality of the electrical power supply.

To transfer electrical power between two points of a system, a difference between source voltage (V_S) and load voltage (V_L) in quantity and/or in phase angle is necessary. See Figure 2.3.1. Using the notation of Figure 2.3.1, it follows that:

$$\mathbf{Z} = R + jX = Z * e^{j\gamma_Z} \tag{2.3.1}$$

$$Z = \sqrt{R^2 + X^2} \tag{2.3.2}$$

$$\gamma_Z = \arctan(\frac{X}{R}) \tag{2.3.3}$$

$$\mathbf{V}_S = V_S * (\cos\gamma_S + j\sin\gamma_S), \quad \mathbf{V}_L = V_L * (\cos\gamma_L - j\sin\gamma_L) \tag{2.3.4}$$

$$\Delta\mathbf{V} = \mathbf{V}_S - \mathbf{V}_L \tag{2.3.5}$$

$$\Delta\mathbf{V} = (V_S * \cos\gamma_S - V_L * \cos\gamma_L) + j(V_S * \sin\gamma_S - V_L * \sin\gamma_L) = \Delta V * e^{-j\gamma_\Delta} \tag{2.3.6}$$

$$\Delta V = \sqrt{V_S^2 - 2 * V_S * V_L * \cos(\gamma_S - \gamma_L) + V_L^2} \tag{2.3.7}$$

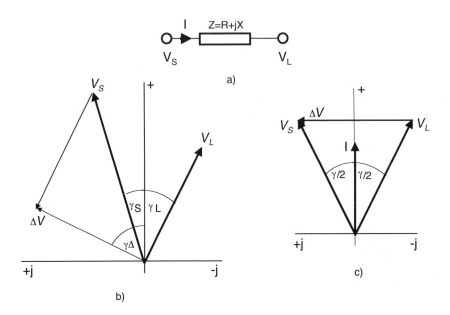

FIGURE 2.3.1 Power transfer.

$$\gamma_\Delta = \arctan\left(\frac{V_S * \sin\gamma_S - V_L * \sin\gamma_L}{V_S * \cos\gamma_S - V_L * \cos\gamma_L}\right) \qquad (2.3.8)$$

$$I = \frac{\Delta V}{Z} * e^{j(\gamma_\Delta - \gamma_Z)} \qquad (2.3.9)$$

For symmetrical conditions $V_S = V_L = V$, and $\gamma_S = \gamma/2$, and $\gamma_L = -\gamma/2$, R<<X, then

$$\Delta V = V * 2 * \sin(\gamma/2) \qquad (2.3.10)$$

$$\gamma_\Delta = \pi/2 \qquad (2.3.11)$$

$$I = \frac{V * 2 * \sin(\gamma/2)}{X} \qquad (2.3.12)$$

2.3.2 Basic Principle of Application

Because of the predominantly inductive character of the power system, an active power flow between source and load must be accomplished with a phase lag between the terminals. Phase-shifting transformers are a preferred tool to achieve this goal. Two principal configurations are of special interest: (1) the power flow between transmission systems operating in parallel where one system includes a PST and (2) where a single transmission line which includes a PST is connecting two otherwise independent power systems. The latter is in fact a special case of the first, but it has become more important nowadays for the interconnection of large systems. For the following considerations, it is assumed that the ohmic resistance R is small compared with the reactance X and thus has been neglected.

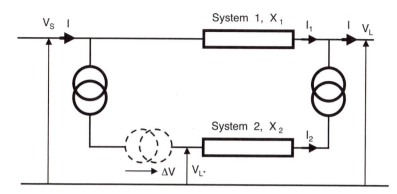

FIGURE 2.3.2 Parallel systems.

One practical basic situation is that a location where power is needed (load side) is connected to the source side through two systems that need not necessarily have the same rated voltage level. See Figure 2.3.2. Without any additional measure, the currents I_1 and I_2 would be distributed in proportion to the ratio of the impedances of the systems,

$$I_1 = I \times X_2/(X_1 + X_2)$$

$$I_2 = I \times X_1/(X_1 + X_2) \qquad (2.3.13)$$

and there is no doubt that system 2 would take only a small part of the load because of the additional impedances of the two transformers in that branch. If the power flow in system 2 should be increased, an additional voltage ΔV must be introduced to compensate the increased voltage drop in system 2. Presuming that active power should be supplied to the load side and considering the inductive character of the systems, this voltage must have a 90° phase lag to the line-to-ground voltages of the system (V_L). In principle, the source of ΔV could be installed in each of the two systems. Figure 2.3.3 shows the voltage diagrams of both options. Figure 2.3.3a corresponds to Figure 2.3.2 with the PST installed in system 2, the system with the higher impedance. The additional voltage reduces the voltage drop in system 2 to that of system 1. The voltage at the output or load side of the PST V_L* leads the voltage at the input or source side V_S. Per definition, this is called an advanced phase angle. If the PST were installed in system 1 (Figure 2.3.3b), the additional voltage would increase the voltage drop to that of system 2. In this case, the load-side voltage V_L* lags the source side voltage V_S, and this is defined as retard phase angle. As can also be seen from the diagrams, an advanced phase angle minimizes the total angle between source and load side.

The second important application is the use of a PST to control the power flow between two large independent grids (Figure 2.3.4). An advanced phase angle is necessary to achieve a flow of active power from system 1 to system 2.

2.3.3 Load Diagram of a PST

So far, the PST has been considered as a black box under a no-load condition, and only its phase-shifting effect has been discussed. Now phase-shifting is quite a normal condition for a transformer and can be found in every transformer that incorporates differently connected primary and secondary windings. But this has no effect, as the aspects of the connected systems follow the same change of the phase angle, and only the voltage drop across the transformer is of interest. PSTs operate between systems having the same frequency and phase sequence. The voltages can differ in magnitude and phase angle. To develop the load-diagram of a PST, the unit has to be split up into two parts: the one presenting an ideal transformer without losses, which accomplishes the phase shift, and the other that is a transformer with a 1:1 turn-to-turn ratio and an equivalent impedance (Figure 2.3.5a).

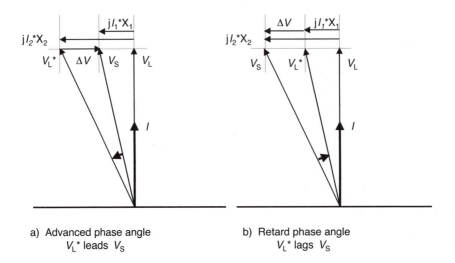

a) Advanced phase angle
V_L^* leads V_S

b) Retard phase angle
V_L^* lags V_S

FIGURE 2.3.3 No-load voltage diagram of parallel systems.

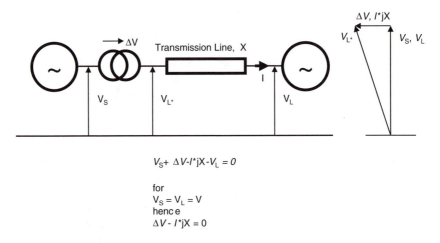

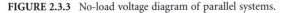

$V_S + \Delta V - I^*jX - V_L = 0$

for
$V_S = V_L = V$
hence
$\Delta V - I^*jX = 0$

FIGURE 2.3.4 Connection of two systems.

The diagram is developed beginning at the load side, where voltage V_L and current I_L are known. Adding the voltage drop, $I^*R_T + I^*jX_T$, to voltage V_L results in voltage V_L^*, which is an internal, not measurable, voltage of the PST. This voltage is turned either clockwise or counterclockwise, and as a result the source voltages $V_{S,retard}$ or $V_{S,advanced}$ are obtained, which are necessary to produce voltage V_L and current I_L at the load side. Angle α determines the phase-shift at the no-load condition, either as retard phase-shift angle $\alpha_{(r)}$ or as advanced phase-shift angle $\alpha_{(a)}$. Under no load, the voltages V_L and V_L^* are identical, but under load, V_L^* is shifted by the load angle β. As a result, the phase angles under load are not the same as under no load. The advanced phase-shift angle is reduced to $\alpha^*_{(a)}$, whereas the retard phase-shift angle is increased to $\alpha^*_{(r)}$. The advanced phase angle under load is given by

$$\alpha^*_{(a)} = \alpha_{(a)} - \beta \tag{2.3.14}$$

and the retard phase angle under load is given by

$$\alpha^*_{(r)} = \alpha_{(r)} + \beta \tag{2.3.15}$$

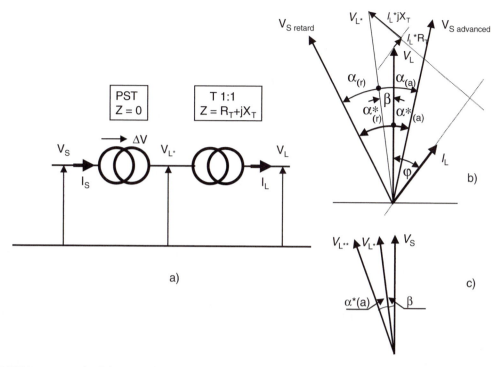

FIGURE 2.3.5 On-load diagram of a PST.

This is very important for the operation of a PST. Because the phase angle determines the voltage across the PST, an increase of the load phase angle in a retard position would mean that the PST is overexcited; therefore the retard-load phase-shift angle should be limited with the no-load angle.

The load angle β can be calculated from

$$\beta = \arctan\left(\frac{z_T * \cos\gamma_Z}{100 + z_T * \sin\gamma_Z}\right) \tag{2.3.16}$$

where all quantities are per unit (p.u.) and z_T = transformer impedance.

In reality, the PST does not influence the voltages neither at the source nor at the load side because it is presumed that the systems are stable and will not be influenced by the power flow. This means that V_S and V_L coincide and V_{L^*} would be shifted by $\alpha^*_{(a)}$ to $V_{L^{**}}$ (Figure 2.3.5c). As for developing the PST load diagram, a certain load has been assumed. The advanced phase-shift angle is a measure of the remaining available excess power.

2.3.4 Total Power Transfer

The voltages at the source side (V_S) and at the load side (V_L) are considered constant, i.e., not influenced by the transferred power, and operating synchronously but not necessarily of the same value and phase angle. To calculate the power flow it has been assumed that the voltages at source side (V_S) and load side (V_L) and the impedance (Z) are known.

$$V_S = V_S^*(\cos\gamma_S + j\sin\gamma_S) \tag{2.3.17}$$

$$V_L = V_L^*(\cos\gamma_L + j\sin\gamma_L) \tag{2.3.18}$$

$$Z = jX \qquad (2.3.19)$$

Then the current becomes

$$I_L = \frac{1}{Z}*(V_S - V_L) = \frac{1}{X}*\left((V_S*\sin\gamma_L - V_L*\sin\gamma_L) - j(V_S*\cos\gamma_S - V_L*\cos\gamma_L)\right) \qquad (2.3.20)$$

and the power at source (S_S) and load (S_L) side can be calculated by multiplying the respective voltage with the conjugate complex current:

$$S_S = \frac{V_S}{X}*V_L*\sin(\gamma_S - \gamma_L) + j\frac{V_S}{X}*(V_S - V_L*\cos(\gamma_S - \gamma_L)) \qquad (2.3.21)$$

$$S_L = \frac{V_S}{X}*V_L*\sin(\gamma_S - \gamma_L) + j\frac{V_S}{X}(V_S*\cos(\gamma_S - \gamma_L) - V_L) \qquad (2.3.22)$$

Because a mere inductive impedance has been assumed, only the reactive power changes.
Symmetrical conditions ($V_S = V_L = V$, $\gamma_S = 0$, and $\gamma_L = -\gamma$) are very common:

$$S_S = \frac{V^2}{X}*(\sin\gamma + j(1-\cos\gamma)) = \frac{2*V^2*\sin(\gamma/2)}{X}*(\cos(\gamma/2) + j\sin(\gamma/2)) \qquad (2.3.23)$$

$$S_L = \frac{V^2}{X}*(\sin\gamma - j(1-\cos\gamma)) = \frac{2*V^2*\sin(\gamma/2)}{X}*(\cos(\gamma/2) - j\sin(\gamma/2)) \qquad (2.3.24)$$

This solution can be considered as a basic load ($S_{L0} = P_0 + jQ_0$) that exists only when the magnitude and/or phase angle of the source and load voltages are different. If a PST is installed in this circuit with an advanced phase-shift angle α, the transferred load can be calculated by substituting $\alpha + \gamma$ as angle and adding the PST impedance X_T to X. By introducing the basic load in the result, the power flow can be calculated as a function of α.

$$P_L(\alpha) = P_{L0}*\cos\alpha - Q_{L0}*\sin\alpha + \frac{V^2}{X}*\sin\alpha \qquad (2.3.25)$$

$$Q_L(\alpha) = P_{L0}*\sin\alpha + Q_{L0}*\cos\alpha + \frac{V^2}{X}*(1-\cos\alpha) \qquad (2.3.26)$$

Figure 2.3.6 shows the variation of the additional power flow for symmetrical conditions, no previous load, constant impedance X, and maximum additional load 1 p.u. ($P_0 = Q_0 = 0$, $V^2/X = 1$). As can be seen from Equation 2.3.25 and Equation 2.3.26 and Figure 2.3.6, a mere ohmic power flow is not possible in the symmetrical case.

In Figure 2.3.7a and Figure 2.3.7b, the variation of the total power flow with the phase-shift angle is plotted for 1 p.u. additional load and constant impedance, depending on different previous loads. The most effective active power transfer can be obtained in the case of a capacitive load.

Another problem is the determination of the necessary voltage difference (value and angle) when V_S, the load $S_0 = P_0 + jQ_0$, and the impedance X are known:

$$V_L = \sqrt{\frac{V_S^2 - 2*Q_0*X}{2} + \sqrt{(\frac{V_S^2 - 2*Q_0*X)}{2})^2 - (P_0^2 + Q_0^2)*X^2)}} \qquad (2.3.27)$$

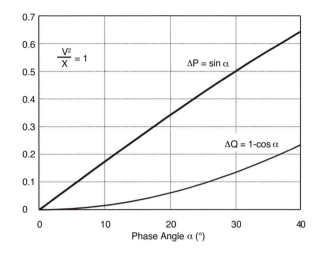

FIGURE 2.3.6 Additional power flow.

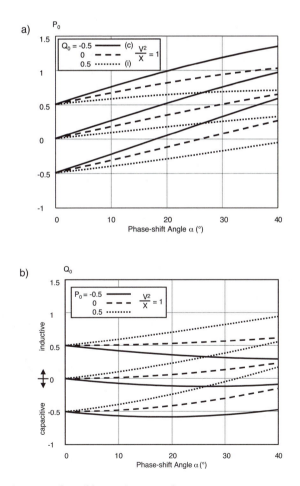

FIGURE 2.3.7 (a) Active power flow. (b) Reactive power flow.

$$V_{La} = V_S * V_L^2 * \frac{V_L^2 + Q_0 * X}{(P_0 * X)^2 + (V_L^2 + Q_0 * X)^2} \tag{2.3.28}$$

$$V_{Lr} = -V_S * V_L^2 * \frac{P_0 * X}{(P_0 * X)^2 + (V_L^2 + Q_0 * X)^2} \tag{2.3.29}$$

$$\gamma_L = \arctan \frac{-P_0 * X}{(P_0 * X)^2 + (V_L^2 + Q_0 * X)^2} \tag{2.3.30}$$

2.3.5 Types of Phase-Shifting Transformers

2.3.5.1 General Aspects

The general principle to obtain a phase shift is based on the connection of a segment of one phase with another phase. To obtain a 90° additional voltage ΔV, the use of delta-connected winding offers the simplest solution. Figure 2.3.8 shows a possible arrangement and is used to introduce a few basic definitions. The secondary winding of phase $V_2 - V_3$ is split up into two halves and is connected in series with phase V_1. By designing this winding as a regulating winding and using on-load tap changers (OLTC), ΔV and the phase-shift angle can be changed under load. The phasor diagram has been plotted for no-load conditions, i.e., without considering the voltage drop in the unit. It also should be noted that the currents in the two halves of the series winding are not in phase. This is different from normal power transformers and has consequences with respect to the internal stray field.

From the connection diagram (Figure 2.3.8a), the following equations can be derived:

$$V_{S1} = V_{10} + (\Delta V_1/2) \tag{2.3.31}$$

$$V_{L1} = V_{10} - (\Delta V_1/2) \tag{2.3.32}$$

$$\Delta V_1 = V_{S1} - V_{L1} \tag{2.3.33}$$

From the phasor diagram (Figure 2.3.8b) follows ($V_{S1} = V_{L1} = V$):

$$V_0 = V * \cos(\alpha/2) \tag{2.3.34}$$

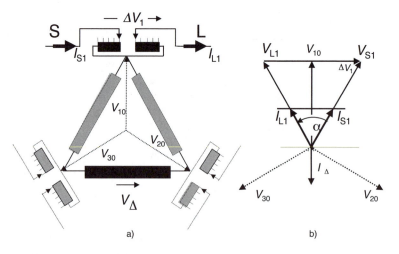

FIGURE 2.3.8 Single core symmetrical PST.

$$\Delta V = V * 2 * \sin(\alpha / 2) \tag{2.3.35}$$

$$V_\Delta = V * \cos(\alpha / 2) * \sqrt{3} \tag{2.3.36}$$

and with $I_S = I_L = I$, the part of the current that is transferred to the exciting winding becomes

$$I_\Delta = \frac{\Delta V}{V_\Delta} * I * \cos(\alpha / 2) = I * \frac{2}{\sqrt{3}} * \sin(\alpha / 2) \tag{2.3.37}$$

The throughput power can be calculated from

$$P_{SYS} = 3 * V * I \tag{2.3.38}$$

and the rated design power, which determines the size of the PST, becomes

$$P_T = 3 * \Delta V * I = P_{SYS} * 2 * \sin(\alpha / 2) \tag{2.3.39}$$

A third kind of power (P_Δ) is the power that is transferred into the secondary circuit. This power is different from P_T because a part of the primary current is compensated between the two parts of the series winding itself. In two-core designs (see Equation 2.3.33), this power determines also the necessary breaking capability of the OLTC.

$$P_\Delta = V_\Delta * I_\Delta = \frac{1}{3} * P_{SYS} * \sin \alpha \tag{2.3.40}$$

In addition to the transferred power, the phase-shift angle is also important. A phase-shift angle of 20° means that the PST has to be designed for 34.8% of the throughput power, and an angle of 40° would require 68.4%. In this respect, it has to be considered that the effective phase-shift angle under load is smaller than the no-load phase-shift angle. In the optimum case when the load power factor is close to 1, a PST impedance of 15% would reduce the load phase-shift angle by 8.5° (Equations 2.3.14 and 2.3.16).

In practice, various solutions are possible to design a PST. The major factors influencing the choice are:

- Throughput power and phase-shift angle requirement
- Rated voltage
- Short-circuit capability of the connected systems
- Shipping limitations
- Load tap-changer performance specification

In addition, preferences of a manufacturer as to the type of transformer (core or shell) or type of windings and other design characteristics may also play a role. Depending on the rating, single- or two-core designs are used. Two-core designs may require either a one-tank or a two-tank design.

2.3.5.2 Single-Core Design

Symmetrical conditions are obtained with the design outlined in Figure 2.3.8a. Figure 2.3.9a and Figure 2.3.9b show the general connection diagrams with more details of the regulating circuit.

The advantage of the single-core design is simplicity and economy. But there are also a number of disadvantages. The OLTCs are connected to the system and directly exposed to all overvoltages and through faults. The voltage per OLTC step and the current are determined by the specification and do not always permit an optimal economical choice of the OLTC. The short-circuit impedance of the PST varies between a maximum and zero. Therefore, it can not be planned that the PST will contribute to the limitation of fault currents in the system.

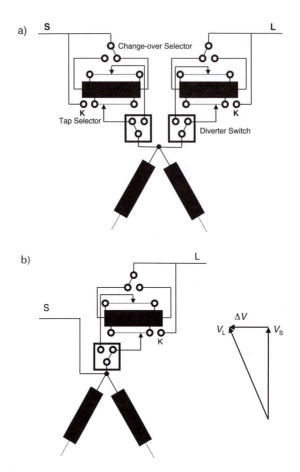

FIGURE 2.3.9 (a) Single-core symmetrical PST (b) Single-core unsymmetrical PST.

The advantage of the symmetrical design (Figure 2.3.9a) is that the phase-shift angle is the only parameter that influences the power flow. The design needs two single-phase OLTCs (for low ratings, one two-phase OLTC may be used instead) per phase or two three-phase OLTCs.

Figure 2.3.9b shows an unsymmetrical solution. Only one-half of the regulating windings is used. The number of necessary OLTCs is reduced, but the ratio between source voltage and load voltage changes with the phase-shift angle and additionally influences the power flow.

A solution that often is used for transformers interconnecting two systems is shown in Figure 2.3.10. The tap winding of a regulating transformer can be connected to a different phase, causing a voltage shift between the regulated winding and the other windings of the unit. The regulated winding normally is connected to the source side, but indirect regulation of the load-side is also possible. The change from the normal regulating transformer state to the phase-shifting state is possible in the middle position of the OLTC without the need to switch off the unit.

Another solution of a symmetrical PST, the delta-hexagonal phase-shifting transformer, is shown in Figure 2.3.11.

2.3.5.3 Two-Core Design

The most commonly used circuit for a two-core design is shown in Figure 2.3.12. This configuration consists of a series unit and a main unit. For smaller ratings and lower voltages, two-core PSTs can be built into a single tank, while larger ratings and higher-voltage PSTs require a two-tank design.

The advantage of a two-core design is the flexibility in selecting the step voltage and the current of the regulating winding. These can be optimized in line with the voltage and current ratings of the OLTC.

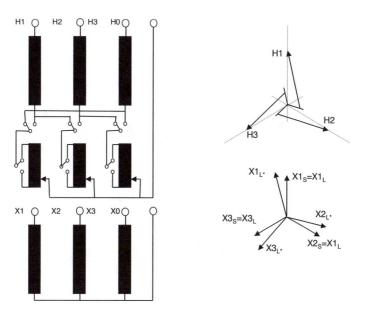

FIGURE 2.3.10 Regulating transformer with PST effect.

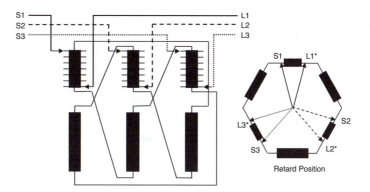

FIGURE 2.3.11 Delta-hexagonal PST.

Since OLTCs have limited current ratings and step voltages per phase as well as limited switching capacity, they are the main limiting features for the maximum possible rating of PSTs. More than one OLTC per phase may have to be utilized for very large ratings. Up to a certain rating, three-pole OLTCs can be used; for higher ratings, three single-pole OLTCs are necessary. The OLTC insulation level is independent of the system voltage and can be kept low.

The short-circuit impedance is the sum of the impedances of the main and series transformers. Because the impedance of the series unit is constant and independent from the phase angle, the unit can be designed to be self-protecting, and the variation of the impedance with the phase-shift angle can be kept small when the impedance of the main unit is kept low.

2.3.5.4 Quadrature Booster Transformers

Quadrature booster transformers are a combination of a regulating power- or auto-transformer with a phase-shifting transformer. The PST, which can be a single- or two-core design, is supplied from the regulated side of the power transformer (Figure 2.3.13). By this method, the output voltage can be adjusted in a four-quadrant (magnitude and phase) relationship.

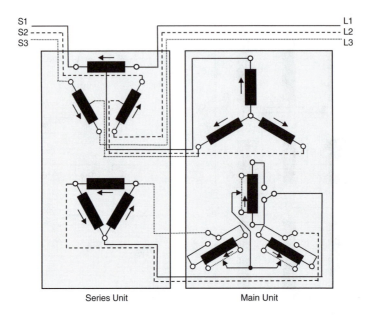

FIGURE 2.3.12 Two-core PST.

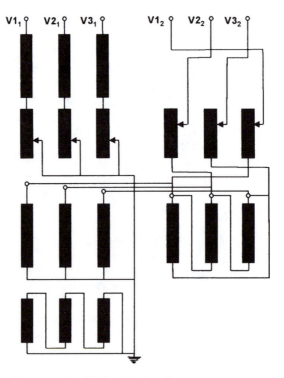

FIGURE 2.3.13 Quadrature booster — simplified connection diagram.

2.3.6 Details of Transformer Design

In general, the design characteristics of PSTs do not differ from ordinary power transformers. In the symmetrical case, however, the phase-angle difference of the currents flowing through the two parts of the series winding has to be recognized. The additional magnetic field excited by the self-compensating

FIGURE 2.3.14 Quadrature booster set (300 MVA, 50 Hz, 400±12*1.25%/115±12*1.45° kV).

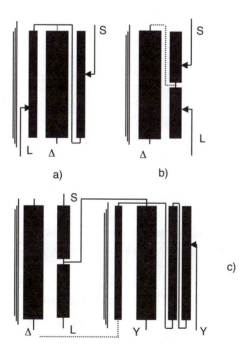

FIGURE 2.3.15 Winding arrangements.

components in the series winding influences mechanical forces, additional losses, and the short-circuit impedance. Figure 2.3.15 shows schematically variations of the physical winding arrangements in PSTs.

In Figure 2.3.15a, a double concentric design of a single-phase PST is shown. This arrangement does not offer any problems with respect to the phase lag between currents and is a standard winding arrangement in shell-type transformers. In core types, the arrangement of the connecting leads from the innermost regulating winding need some attention, but this does not present a real obstacle to the use

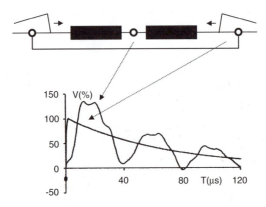

FIGURE 2.3.16 Lightning impulse response (bypassed PST).

of this design. On the other hand, the axial arrangement of the regulating winding in core-type transformers, as shown in Figure 2.3.15b, offers the advantage of direct access and saves space. But because
the pattern of the stray field is more complicated, thorough field calculations have to be performed using
the appropriate computer programs.

It is possible that a client may operate the PST in a bypassed condition. In this case, the source and
load terminals are directly connected phasewise. In this state, a lightning impulse would penetrate the
series winding from both ends at the same time. Two traveling waves would meet in the middle of the
winding and would theoretically be reflected to double the amplitude (Figure 2.3.16). Therefore the series
winding must either be designed with a high internal capacitance or be protected by external or internal
surge arresters. A high internal capacitance can be obtained using any of the measures that are used to
improve the lightning-impulse voltage distribution along the winding, e.g., interleaving or shielding.
Again, the axial arrangement is more advantageous and easier to make self-protecting.

Figure 2.3.15c shows the arrangement of a two-core design with two coarse and one fine tap winding,
which is a variation of the circuit drawn in Figure 2.3.12 (see also Section 2.3.7, Details of On-Load Tap-
Changer Application). If a two-tank solution has to be used, the connection between the main and series
unit requires an additional set of six high-voltage bushings, which means that a total of nine high-voltage
bushings would have to be arranged on the series unit. Because a short circuit between the main and
series units could destroy the regulating winding, a direct and encapsulated connection between the two
tanks is preferred. This requires a high degree of accuracy in the mechanical dimensions and the need
for experienced field engineers to assemble both units on site. If required, special oil-tight insulation
systems allow the separation of the two tanks without the need to drain the oil in one or both units. In
the latter case, an extra oil-expansion system is needed for the connecting tubes. Figure 2.3.17 shows a
double-tank PST design at a testing site.

2.3.7 Details of On-Load Tap-Changer Application

OLTCs are subject to numerous limits. The most essential limit is, of course, the current-interrupting or
current-breaking capability. In addition to this limit, the voltage per step and the continuous current are
also limited. The product of these two limits is generally higher than the capability limit, so the maximum
voltage per step and the maximum current cannot be utilized at the same time. Table 2.3.1, Table 2.3.2,
and Table 2.3.3 show examples for design power, voltage per step, and system current as functions of
phase-shift angle, throughput power, system voltage, and number of voltage steps.

These limits (e.g., 4000 V/step, 2000 A) determine the type of regulation (the need for only a fine tap
winding or a combination of fine and coarse tap windings) and the number of parallel branches. But it
must be noted that the mutual induction resulting from the use of a coarse/fine-tap winding configuration
also has to be taken into account and may influence the decision.

FIGURE 2.3.17 Two-tank PST at a test site (coolers not assembled) (650 MVA, 60 Hz, 525/525±20*1.2° kV).

TABLE 2.3.1 Design Power of PSTs as a Function of Throughput Power and Phase-Shift Angle, MVA

	Design Power, MVA	Phase-Shift Angle α°				
		10	20	30	40	50
	100	17.4	34.7	51.8	68.4	84.5
Throughput	250	43.6	86.8	129.4	171.0	211.3
Power, MVA	500	87.2	173.6	258.8	342.5	422.6
	750	130.7	260.5	388.2	513.0	633.9
	1000	174.3	347.3	517.6	684.0	845.2

TABLE 2.3.2 Step Voltage as a Function of System Voltage and Phase-Shift Angle for a 16-Step OLTC, V

	Step Voltage, V	Phase-Shift Angle α°				
	Steps:16	10	20	30	40	50
	69.0	434	865	1,289	1,703	2,104
	115.0	723	1,441	2,148	2,839	3,507
System	138.0	868	1,729	2,578	3,406	4,209
Voltage, kV	161.0	1,013	2,018	3,007	3,974	4,910
	230.0	1,447	2,882	4,296	5,677	7,015
	345.0	2,170	4,324	6,444	8,516	10,522
	500.0	3,145	6,266	9,339	12,342	15,250
	765.0	4,812	9,587	14,289	18,883	23,332

Another problem that also is not specific to PSTs, but may be more significant, is the possible altering of the potential of the regulating winding when the changeover selector is operated. In this moment, the tap selector is positioned at tap "K" (see Figure 2.3.18), and the regulating winding is no longer fixed to the potential of the excitation winding. The new potential is determined by the ratio of the capacitances

TABLE 2.3.3 System Current as a Function of System Voltage and Throughput Power, A

	System Current, A	Throughput Power, MVA				
		100	250	500	750	1000
	69.0	837	2092	4184	6276	8367
	115.0	502	1255	2510	3765	5020
System	138.0	418	1046	2092	3138	4184
Voltage,	161.0	359	897	1793	3974	3586
kV	230.0	251	628	1255	2690	2510
	345.0	167	418	837	1883	1673
	500.0	115	289	577	1255	1155
	765.0	75	189	377	866	755

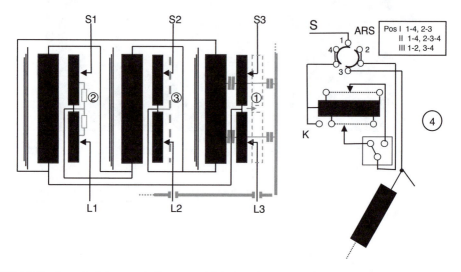

FIGURE 2.3.18 Control of the tap-winding potential during changeover selector operation.

between the regulating winding and the excitation winding and between the regulating winding and ground (location 1 in Figure 2.3.18). The resulting differential voltage stresses the switching distance of the opening changeover selector and may cause discharges during the operation. This can be prevented either by connecting the regulating winding to the excitation winding by a resistor (location 2) or by a static shield (location 3) or by using a double-reversing (advanced-retard) changeover selector switch (ARS switch, location 4).

When the voltage per step cannot be kept within the OLTC limit, either a second OLTC must be used (Figure 2.3.19a), or the number of steps can be doubled by using an additional coarse tap winding. Two or more of those windings are also possible (Figure 2.3.19b). An appropriate multiple changeover selector must be used.

Figure 2.3.20 shows a two-tank PST design with three coarse and one fine tap winding at a field site. This design was forced by the limitation of the mutual induction between the fine and coarse tap windings.

2.3.8   Other Aspects

2.3.8.1   Connection

PSTs can be operated in parallel or series connection or a combination of these two. It is assumed that all n units are of equivalent design.

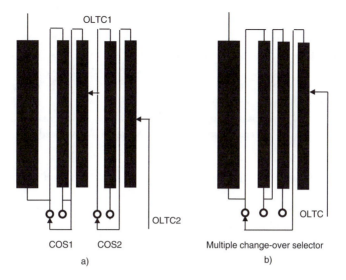

FIGURE 2.3.19 Use of more than one OLTC and/or multiple changeover selector.

FIGURE 2.3.20 Two-tank PST with three coarse- and one fine-tap winding at field site (650 MVA, 60 Hz, 525/525±12*1.2°kV).

2.3.8.1.1 *Parallel Connection*

- The throughput power of a single unit is equal to 1/n of the total required throughput power.
- The phase-shift angle of all units is equal to the required phase-shift angle.
- The impedance of a single unit in ohms must be n times the required total impedance in ohms.
- The impedance of a single unit in percent, referred to 1/n of the total throughput power, is equal to the total impedance in percent.
- When one unit is lost, the total throughput power decreases to $(n - 1)/n$, and the impedance in ohms increases to $n/(n - 1)$ of the original value.

2.3.8.1.2 Series Connection

- The throughput power of a single unit is equal to the total throughput power.
- The phase-shift angle of a single unit is equal to 1/n of the total phase-shift angle.
- The impedance of a single unit in ohms is 1/n of the total impedance.
- The impedance of a single unit in percent, referred to the throughput power, is 1/n of the total impedance in percent.
- When one unit is lost or bypassed, the throughput power remains constant, and the total phase-shift angle and the total impedance in ohms and in percent are reduced by (n − 1)/n.

2.3.8.2 Tests

In case of a two-tank design, the PST should be completely assembled with the two units connected together, as in service.

Auxiliary bushings may be necessary to make inner windings accessible for measurements of resistances, losses, and temperatures or to allow dielectric tests.

2.3.8.2.1 Special Dielectric Tests

When bypassing of the unit is a specified operating condition, a special lightning impulse and a special switching impulse has to be performed with both terminals of the series winding connected.

2.3.8.2.2 No-Load Phase-Shift Angle

From Figure 2.3.5, the phase shift angle can be calculated as

$$\alpha = \arccos \frac{V_S^2 + V_L^2 - V_{S-L}^2}{2 * V_S * V_L} \tag{2.3.41}$$

with V_S, V_L, and V_{S-L} being the absolute values of the voltages of corresponding source and load terminals to ground and between them, respectively.

To differentiate between advanced and retard angle, the following criteria can be used:
advanced phase-shift angle:

$$V_{S1} - V_{S2} > V_{S1} - V_{L2}$$

retard phase-shift angle:

$$V_{S1} - V_{S2} < V_{S1} - V_{L2}$$

The tolerance depends on the accuracy of the voltages. ANSI/IEEE C57.135 recommends that this be 1%. This considers the worst case, when V_S and V_L are at the upper tolerance limit (1.005) and V_{S-L} is at the lower limit (0.995).

References

ANSI/IEEE, Guide for the Application, Specification and Testing of Phase-Shifting Transformers, ANSI/IEEE C57.135-2001, Institute of Electrical and Electronics Engineers, Piscataway, NJ, 2001.

Brown, F.B., Lundquist, T.G. et al., The First 525 kV Phase Shifting Transformer — Conception to Service, presented at 64th Annual Conference of Doble Clients, 1997.

Krämer, A., *On-Load Tap-Changers for Power Transformers*, Maschinenfabrik Reinhausen, Regensburg, Germany, 2000.

Seitlinger, W., Phase Shifting Transformers — Discussion of Specific Characteristics, CIGRE Paper 12-306, CIGRE, Paris, 1998.

2.4 Rectifier Transformers

Sheldon P. Kennedy

Power electronic circuits can convert alternating current (ac) to direct current (dc). These are called rectifier circuits. Power electronic circuits can also convert direct current to alternating current. These are called inverter circuits. Both of these circuits are considered to be converters. A transformer that has one of its windings connected to one of these circuits, as a dedicated transformer, is a converter transformer, or rectifier transformer. IEC standards refer to these transformers as converter transformers, while IEEE standards refer to these transformers as rectifier transformers. Because it is IEEE practice to refer to these transformers as rectifier transformers, that same term is used throughout this discussion.

Transformers connected to circuits with a variety of loads, but which may contain some electronic circuits that produce harmonics, are not considered to be rectifier transformers. However, they may have harmonic heating effects similar to rectifier transformers. Those transformers are covered under IEEE Recommended Practice for Establishing Transformer Capability when Supplying Non-Sinusoidal Load Currents, ANSI/IEEE C57.110.

Electronic circuits provide many types of control today, and their use is proliferating. These circuits are generally more efficient than previous types of control, and they are applied in many types of everyday use. Rectifier circuits are used to provide high-current dc for electrochemical processes like chlorine production as well as copper and aluminum production. They are also used in variable-speed-drive motor controls, transit traction applications, mining applications, electric furnace applications, higher-voltage laboratory-type experiments, high-voltage direct-current power transmission (HVDC), static precipitators, and others. While HVDC transmission and static precipitators are not directly covered in this chapter, much of the basic information still applies.

2.4.1 Background and Historical Perspective

Rectifier transformers can be liquid-immersed, dry-type, or cast-coil technology. Dry-type transformers were primarily used in distribution-voltage classes. Impregnation systems have improved with the development of vacuum pressure impregnation (VPI) technology. These types of transformers have been developed to 34-kV and 46-kV classes, although basic-impulse-insulation levels (BIL) are often less than in liquid-immersed transformers. Cast-coil technology has developed as a more rugged, nonliquid-filled technology. Both of these types of transformers — dry type and cast coil — are limited by voltage and kVA size. They have advantages over liquid-filled transformers for fire protection, since they have no liquids to ignite. However, liquid-immersed transformers can be built to all voltage levels and current levels. High-fire-point fluids can be used for fire-protection considerations. Auxiliary cooling can be utilized to cool larger levels of power loss developed in higher-current applications.

The early rectifiers were pool-cathode mercury rectifiers. These had high levels of short-circuit failures on transformers and suffered from arc-backs. When one phase faulted, all phases would dump through the faulted phase. So on a six-phase transformer with 10% impedance, instead of 10 times rated current during a fault, it could develop up to 60 times rated current. Usually the fault would not be this high, but it could still be in the area of 40 times rated current. This is an extremely high fault current for a transformer to withstand. Transformers had to be built very ruggedly and were extremely heavy compared with most transformers, which greatly increased the cost of these systems. They also had the disadvantage of the environmental problems associated with mercury. These transformers were built to comply with ANSI/IEEE C57.18-1964, Pool Cathode Mercury-Arc Rectifier Transformer. The latest revision of this standard was 1964, but it was reaffirmed later.

Semiconductor rectifiers advanced higher in voltage and current capability, and finally semiconductor technology developed to the point that pool-cathode mercury-arc rectifiers were replaced. Semiconductor rectifiers also brought the ability of control with thyristors in addition to diodes, without the use of magnetic devices such as the saturable core reactors and amplistats that had been used for this purpose.

They did have the advantage of little harmonic production, but they are less efficient than semiconductor devices. Nevertheless, they are still used in many applications today with diodes.

The new problem presented by semiconductor technology was harmonic current. The operation of the semiconductor rectifier produces harmonic voltages and currents. The harmonic currents are at higher frequencies than the fundamental frequency of the transformer. These higher-frequency currents cause high levels of eddy-current losses and other stray losses in other parts of the transformer. This can create potentially high temperatures, which degrade the insulation of the transformer and can cause early failure of the transformer.

Problems of failures were reported to the IEEE Power Engineering Society Transformer Committee. These were typically hottest-spot failures in the windings. In 1981 a new standard began development. Rather than just updating the old mercury-arc rectifier standard, a new standard was created — ANSI/IEEE C57.18.10-1998, Practices and Requirements for Semiconductor Power Rectifier Transformers. In a similar time frame, IEC developed its own standard, IEC 61378-1-1997, Convertor Transformers — Part 1: Transformers for Industrial Applications. The IEEE standard took an inordinate amount of time to develop due to the development of new products, new terminology, the definition of harmonics, the estimated harmonic effects on losses and heating, individual company practices, conflicting standards, harmonization with the IEC's efforts to develop a standard, and development of appropriate test methods, to name a few of the obstacles.

Important developments occurred during the preparation of these standards. Some involve the specification of transformers, some involve performance characteristics and calculations, and some involve the testing of transformers.

2.4.2 New Terminology and Definitions

At least two new terms were defined in ANSI/IEEE C57.18.10. Both are important in the specification of rectifier transformers.

2.4.2.1 Fundamental kVA

The traditional IEEE method for rating a rectifier transformer has always been the root-mean-square (rms) kVA drawn from the primary line. This is still the method used to develop all of the tables and figures given in ANSI/IEEE C57.18.10, Clause 10. However, the IEC converter transformer standards define the kVA by the fundamental kVA drawn from the primary line. The rms-rated kVA method is based on the rms equivalent of a rectangular current waveshape based on the dc rated load commutated with zero commutating angle. The fundamental kVA method is based on the rms equivalent of the fundamental component of the line current. There are pros and cons to both methods. IEC allows only the fundamental-kVA-method rating with an "in some countries" clause to accommodate North American practice. The logic behind this rating is that the transformer manufacturer will only be able to accurately test losses at the fundamental frequency. The manufacturer can not accurately test losses with the complex family of harmonics present on the system. Therefore, according to IEC, it is only proper to rate the transformer at the fundamental frequency. Transformer rating and test data will then correspond accurately.

The traditional IEEE rms-kVA rating method will not be exactly accurate at test. However, it does represent more accurately what a user sees as meter readings on the primary side of the transformer. Users feel strongly that this is a better method, and this is what their loading is based on.

ANSI/IEEE C57.18.10 allows for both kVA methods. It is important for a user to understand the difference between these two methods so that the user can specify which rating is wanted.

2.4.2.2 Harmonic Loss Factor

The term *harmonic-loss factor*, F_{HL}, was developed by IEEE and IEC as a method to define the summation of harmonic terms that can be used as a multiplier on winding eddy-current losses and other stray losses. These items are separated into two factors, winding eddy-current harmonic-loss factor, $F_{HL\text{-}WE}$, and the other-stray-loss harmonic-loss factor, $F_{HL\text{-}OSL}$. These are used as multipliers of their respective losses as measured at test at the fundamental frequency. Both factors can be normalized to either the fundamental current or the rms current.

These terms are similar to the values used by the Underwriters Laboratories (UL) *K*-factor multiplier, except that stray losses are amplified by a lesser factor than the winding eddy-current losses. The term $F_{HL\text{-}WE}$ comes mathematically closest to being like the term *K*-factor. It must be noted that the term *K*-factor was never an IEEE term but only a UL definition. The new IEEE Recommended Practice for Establishing Transformer Capability when Supplying Non-Sinusoidal Load Currents, ANSI/IEEE C57.100, gives a very good explanation of these terms and comparisons to the UL definition of *K*-factor. The Transformers Committee of the IEEE Power Engineering Society has accepted the term *harmonic-loss factor* as more mathematically and physically correct than the term *K*-factor. *K*-factor is used in UL standards, which are safety standards. IEEE standards are engineering standards.

In their most simple form, these terms for harmonic-loss factor can be defined as follows:

$$F_{HL\text{-}WE} = \sum_{1}^{n} I_h(pu)^2 h^2 \tag{2.4.1}$$

and

$$F_{HL\text{-}OSL} = \sum_{1}^{n} I_h(pu)^2 h^{0.8} \tag{2.4.2}$$

where
   $F_{HL\text{-}WE}$ = winding eddy-current harmonic-loss factor
   $F_{HL\text{-}OSL}$ = other-stray-loss harmonic-loss factor
      I_h = harmonic component of current of the order indicated by the subscript h
      h = harmonic order

As is evident, the primary difference is that the other stray losses are only increased by a harmonic exponent factor of 0.8. Bus-bar, eddy-current losses are also increased by a harmonic exponent factor of 0.8. Winding eddy-current losses are increased by a harmonic exponent factor of 2. The factor of 0.8 or less has been verified by studies by manufacturers in the IEC development and has been accepted in ANSI/IEEE C57.18.10. Other stray losses occur in core clamping structures, tank walls, or enclosure walls. On the other hand, current-carrying conductors are more susceptible to heating effects due to the skin effect of the materials. Either the harmonic spectrum or the harmonic-loss factor must be supplied by the specifying engineer to the transformer manufacturer.

2.4.3   Rectifier Circuits

Rectifier circuits often utilize multiple-circuit windings in transformers. This is done to minimize harmonics on the system or to subdivide the rectifier circuit to reduce current or voltage to the rectifier. Since different windings experience different harmonics on multiple-circuit transformers, the kVA ratings of the windings do not add arithmetically. Rather, they are rated on the basis of the rms current carried by the winding. This is an area where the rms-kVA rating of the windings is important. The fundamental-kVA rating would add arithmetically, since harmonics would not be a factor.

Rectifier circuits can be either single way or double way. Single-way circuits fire only on one side of the waveform and therefore deliver dc. Double-way circuits fire on both side of the waveform.

A variety of common rectifier circuits are shown in Figure 2.4.1 through Figure 2.4.11. Table 9 in ANSI/IEEE C57.18.10 shows the properties of the more common circuits, including the currents and voltages of the windings. The dc winding — the winding connected to the rectifier — is usually the secondary winding, unless it is an inverter transformer. The ac winding — the winding connected to the system — is usually the primary winding, unless it is an inverter transformer.

A six-pulse single-way transformer would have to have two secondary windings, like a Circuit 45 transformer (Figure 2.4.5), which has a delta primary winding with double-wye secondary windings. A six-pulse double-way transformer can be a simple two-winding transformer, like a Circuit 23 transformer (Figure 2.4.1), which is a simple delta-wye transformer. This is due to the number of pulses each circuit has over the normal period of a sine wave.

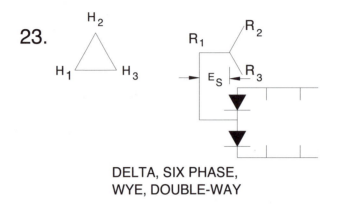

DELTA, SIX PHASE,
WYE, DOUBLE-WAY

FIGURE 2.4.1 ANSI Circuit 23. (From ANSI/IEEE C57.18.10-1998. © IEEE 1998. With permission.)

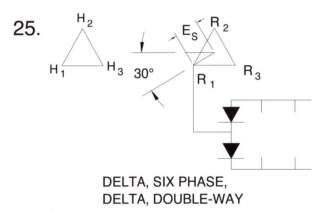

DELTA, SIX PHASE,
DELTA, DOUBLE-WAY

FIGURE 2.4.2 ANSI Circuit 25. (From ANSI/IEEE C57.18.10-1998. © IEEE 1998. With permission.)

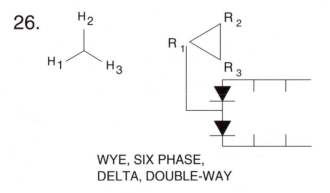

WYE, SIX PHASE,
DELTA, DOUBLE-WAY

FIGURE 2.4.3 ANSI Circuit 26. (From ANSI/IEEE C57.18.10-1998. © IEEE 1998. With permission.)

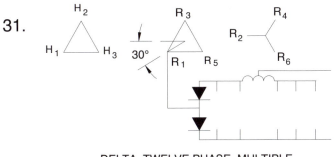

DELTA, TWELVE PHASE, MULTIPLE
DELTA-WYE, DOUBLE-WAY

FIGURE 2.4.4 ANSI Circuit 31. (From ANSI/IEEE C57.18.10-1998. © IEEE 1998. With permission.)

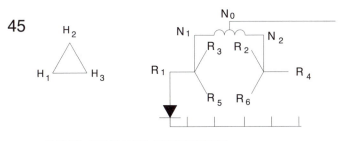

DELTA, SIX PHASE, DOUBLE WYE

FIGURE 2.4.5 ANSI Circuit 45. (From ANSI/IEEE C57.18.10-1998. © IEEE 1998. With permission.)

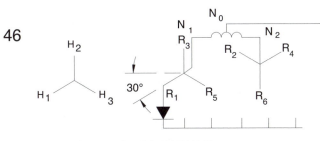

WYE, SIX PHASE, DOUBLE WYE

FIGURE 2.4.6 ANSI Circuit 45. (From ANSI/IEEE C57.18.10-1998. © IEEE 1998. With permission.)

2.4.4 Commutating Impedance

Commutating impedance is defined as one-half the total impedance in the commutating circuit expressed in ohms referred to the total secondary winding. It is often expressed as percent impedance on a secondary kVA base. For wye, star, and multiple-wye circuits, this is the same as derived in ohms on a phase-to-neutral voltage basis. With diametric and zigzag circuits, it must be expressed as one-half the total due to both halves being mutually coupled on the same core leg or phase. This is not to be confused with the short-circuit impedance, i.e., the impedance with all secondary windings shorted. Care must be taken when expressing these values to be careful of the kVA base used in each. The commutating impedance is the impedance with one secondary winding shorted, and it is usually expressed on its own kVA base, although it can also be expressed on the primary kVA base if desired.

Care must be taken when specifying these values to the transformer manufacturer. The impedance value, whether it is commutating impedance or short-circuit impedance, and kVA base are extremely important.

Use ANSI/IEEE C57.18.10 as a reference for commutating impedance. The tables of circuits in this reference are also useful.

2.4.5 Secondary Coupling

Three-winding transformers with one primary winding and two secondary windings, such as Circuit 31 (Figure 2.4.4), Circuit 45 (Figure 2.4.5), and Circuit 46 (Figure 2.4.6), can be constructed as tightly coupled secondary windings or as loosely coupled or uncoupled secondary windings. The leakage reactance is common to the tightly coupled secondary windings but is independent for the uncoupled secondary windings.

If two separate transformers were constructed on separate cores, the couplings would be completely independent of one another, i.e., uncoupled. When the windings of one of these transformers are combined on a common core, there can be varying degrees of coupling, depending on the transformer construction.

Some transformers must be made with tightly coupled windings. This is commonly accomplished by winding the secondary windings in an interleaved or bifilar fashion. This is almost always done with low-voltage, high-current windings. Higher-voltage windings would have high-voltage stresses from this type

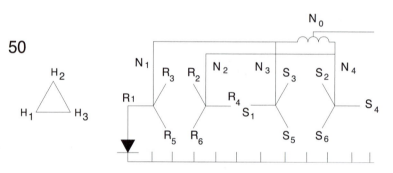

DELTA, SIX PHASE, PARALLEL DOUBLE WYE
WITH SINGLE INTERPHASE TRANSFORMER.

FIGURE 2.4.7 ANSI Circuit 50. (From ANSI/IEEE C57.18.10-1998. © IEEE 1998. With permission.)

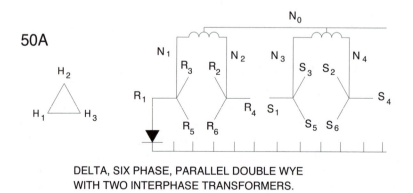

DELTA, SIX PHASE, PARALLEL DOUBLE WYE
WITH TWO INTERPHASE TRANSFORMERS.

FIGURE 2.4.8 ANSI Circuit 50A. (From ANSI/IEEE C57.18.10-1998. © IEEE 1998. With permission.)

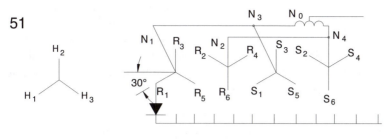

WYE, SIX PHASE, PARALLEL DOUBLE WYE
WITH SINGLE INTERPHASE TRANSFORMER.

FIGURE 2.4.9 ANSI Circuit 51. (From ANSI/IEEE C57.18.10-1998. © IEEE 1998. With permission.)

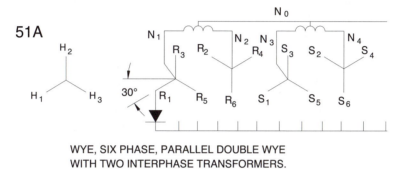

WYE, SIX PHASE, PARALLEL DOUBLE WYE
WITH TWO INTERPHASE TRANSFORMERS.

FIGURE 2.4.10 ANSI Circuit 51A. (From ANSI/IEEE C57.18.10-1998. © IEEE 1998. With permission.)

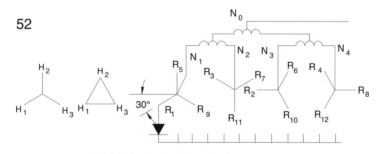

WYE DELTA, TWELVE PHASE, QUADRUPLE WYE

FIGURE 2.4.11 ANSI Circuit 52. (From ANSI/IEEE C57.18.10-1998. © IEEE 1998. With permission.)

of construction. However, the higher-current windings also benefit from the harmonic cancellation resulting from the close coupling of the secondary windings. This type of construction is usually required with Circuit 45 and 46 transformers and other single-way double-wye transformers. Such transformers would suffer from three-pulse harmonics without the tight coupling of the secondary windings.

Secondary coupling is used in all types of traction-duty transformers. This offers the advantage of low commutating impedance with high short-circuit impedance. Harmonic cancellation also is accomplished in the secondary windings.

The degree of coupling affects the magnitude of the fault current that can be produced at short circuit. If secondary circuits are paralleled, unbalanced commutation impedances can produce high fault currents. The degree of coupling can also affect the voltage regulation at high overloads, as can be seen in traction service or high-pulse duty.

The degree of coupling also influences the voltage regulation when overloads are high, say in the range of 200% or higher. When developing these values, the impedance can be found mostly in the primary windings or in the secondary windings, depending on the transformer construction. Equations 2.4.3, 2.4.4, and 2.4.5 will bear this out when used on different types of three-winding rectifier transformers. With some constructions, all of the impedance is in the secondary winding, and the primary winding can actually calculate as negative impedance. This can be beneficial in high overload conditions for 12-pulse converters.

Three-winding transformers are usually not completely coupled or uncoupled in practicality. The degree of coupling varies between one and zero, where zero means the secondary windings are completely uncoupled. When the coupling is a one, the secondary windings are completely coupled. Anything between is partial coupling, and the precise value gives an appraisal of the degree of coupling. If a particular coupling is required, it must be specified to the transformer manufacturer.

Three-winding transformers with one delta secondary winding and one wye secondary winding (an ANSI Circuit 31) are the transformers where the coupling can be accomplished with a coupling from nearly zero to nearly unity. The common types of construction for three-winding rectifier transformer windings are shown in Figure 2.4.12a through Figure 2.4.12e.

To determine the degree of coupling, a series of impedance tests are required as explained in ANSI/IEEE C57.18.10, Clause 8.6.3. An impedance test is done where each secondary is shorted in turn. For commutating impedance, this is done at the secondary kVA rating or base. However, for the purposes of this calculation, it should be calculated at the primary kVA base in order to keep all terms in Equations 2.4.3, 2.4.4, and 2.4.5 on the same base. Finally, the short-circuit impedance test is performed, where both secondary windings are shorted and primary current of sufficient magnitude is applied to the transformer to obtain the primary kVA rating.

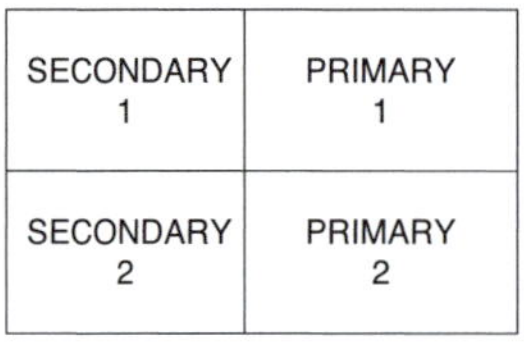

FIGURE 2.4.12A Two-tier or stacked-coil arrangement.

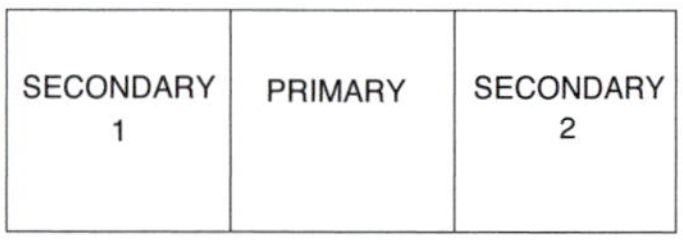

FIGURE 2.4.12B   Concentric-coil arrangement.

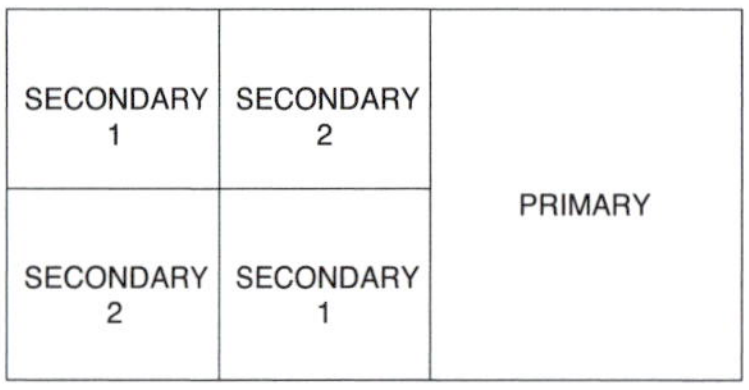

FIGURE 2.4.12C Split stacked-secondary arrangement

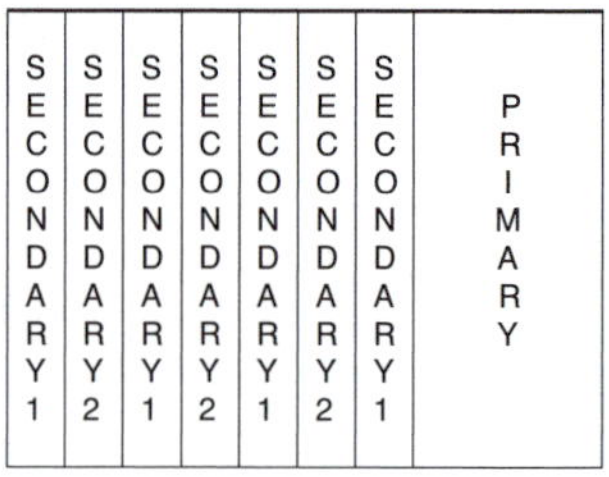

FIGURE 2.4.12D Interleaved or bifilar secondary arrangement.

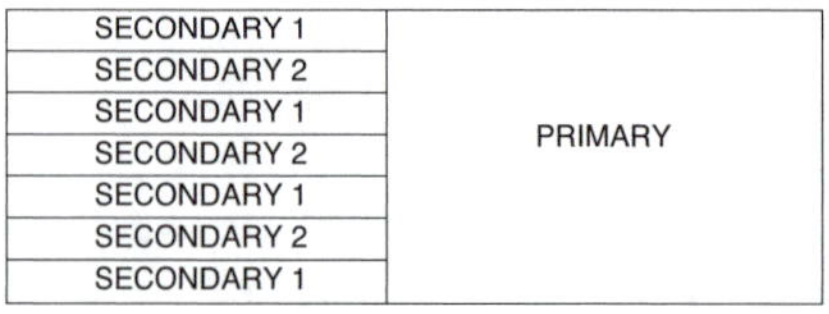

FIGURE 2.4.12E Interleaved helical secondary arrangement.

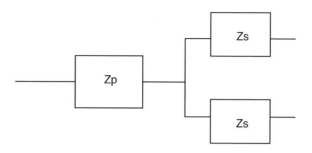

FIGURE 2.4.13 Impedance diagram of a three-winding transformer.

For the purposes of this discussion and Figure 2.4.13, we will assume that the two secondary imped-ances are equal. In reality, the two values are usually slightly different. This would make the calculations somewhat more complicated. The degree of coupling may be somewhat different based on this simpli-fication.

The variables depicted in Figure 2.4.13 are as follows:

Z_p = primary impedance
Z_s = secondary impedance
K = coupling factor

From the results of the impedance tests, we have the short-circuit impedance and the commutating impedance, both on a primary kVA base. We can then express the two terms as follows:
The commutating impedance is

$$Z_c = Z_p + Z_s \qquad (2.4.3)$$

The short-circuit impedance is

$$Z_{sc} = Z_p + Z_s/2 \qquad (2.4.4)$$

From the results of the tests, solve the two simultaneous equations (Equation 2.4.3 and Equation 2.4.4) for the value of the primary impedance. This value and the commutating impedance (Equation 2.4.3) are the values needed to calculate the degree of coupling.
The coupling factor is

$$K = Z_p/Z_c \qquad (2.4.5)$$

If the coupling factor is zero, the secondary windings are completely uncoupled, and the primary impedance is zero. When the coupling factor is unity, the secondary windings are completely coupled and the secondary impedance is zero. Values in between indicate partial coupling.

The derived values can be further broken down into reactance and resistance, which gives a closer definition of the coupling factor if required.

Each of the transformer types shown in Figure 2.4.12 gives a different coupling factor, and only the interleaved types provide close coupling. The coil design should be matched to the coupling desired by the user. The designs in Figure 2.4.12a and Figure 2.4.12b are loosely coupled. The design in Figure 2.4.12c is rather tightly coupled, while the designs in Figure 2.4.12d and Figure 2.4.12e are very tightly coupled. The user must specify to the transformer manufacturer the type of coupling desired.

2.4.6   Generation of Harmonics

Semiconductor power converters for both ac and dc conversions have long been known to produce harmonic currents that affect other electrical equipment. IEEE 519, *IEEE Recommended Practices and Requirements for Harmonic Control in Electric Power Systems*, is an excellent source for discussion of the generation and amplitudes of the harmonic currents discussed.

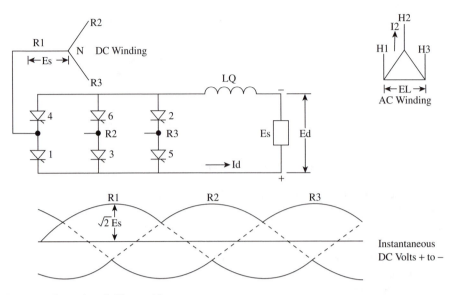

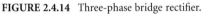

FIGURE 2.4.14 Three-phase bridge rectifier.

Figure 2.4.14 shows a three-phase bridge rectifier. A periodic switching converts electrical power from one form to another. A dc voltage with superimposed high-frequency ripple is produced. The ripple voltage consists of the supply voltage with a frequency of multiples of six times the fundamental frequency. This can be done using either diode rectifiers or thyristors.

Energy balance and Fourier analysis of the square waves confirms that each 6n harmonic in the dc voltage requires harmonic currents of frequencies of 6n + 1 and 6n − 1 in the ac line. The magnitude of the harmonic current is essentially inversely proportional to the harmonic number, or a value of 1/h.

This is true of all converters. Once the number of pulses is determined from a converter, the harmonics generated will begin to be generated on either side of the pulse number. So a six-pulse converter begins to generate harmonics on the 5th and 7th harmonic, then again on the 11th and 13th harmonic, and so on. A 12-pulse converter will begin to generate harmonics on the 11th and 13th harmonic, and again on the 23rd and 25th harmonic, and so on. In order to reduce harmonics, topologies are developed to increase the number of pulses. This can be done using more-sophisticated converters or by using multiple converters with phase shifts. From this, we can express the above in a general formula for all pulse converters as:

$$h = kq \pm 1 \qquad (2.4.6)$$

where

k = any integer
q = converter pulse number

Table 2.4.1 shows a variety of theoretical harmonics that are produced based on the pulse number of the rectifier system. It should be emphasized that these are theoretical maximum magnitudes. In reality, harmonics are generally reduced due to the impedances in the transformer and system, which block the flow of harmonics. In fact, harmonics may need to be expressed at different values, depending on the load. On the other hand, due to system interaction and circuit topology, sometimes the harmonic current values are greater than the ideal values for certain harmonic values. The interaction of devices such as filters and power-factor correction capacitors can have a great affect on the harmonic values. These interactions are not always predictable. This is also one of the problems faced by filter designers. This has increasingly led toward multisecondary transformers that cancel the harmonics at their primary terminals, although the transformer generally still has to deal with the harmonics.

TABLE 2.4.1 Theoretical Harmonic Currents Present in Input Current to a Typical Static Power Rectifier, Per Unit of the Fundamental Current

Harmonic Order	Rectifier Pulse Number						
	6	12	18	24	30	36	48
5	0.2000	—	—	—	—	—	—
7	0.1429	—	—	—	—	—	—
11	0.0909	0.0909	—	—	—	—	—
13	0.0769	0.0769	—	—	—	—	—
17	0.0588	—	0.0588	—	—	—	—
19	0.0526	—	0.0526	—	—	—	—
23	0.0435	0.0435	—	0.0435	—	—	—
25	0.0400	0.0400	—	0.0400	—	—	—
29	0.0345	—	—	—	0.0345	—	—
31	0.0323	—	—	—	0.0323	—	—
35	0.0286	0.0286	0.0286	—	—	0.0286	—
37	0.0270	0.0270	0.0270	—	—	0.0270	—
41	0.0244	—	—	—	—	—	—
43	0.0233	—	—	—	—	—	—
47	0.0213	0.0213	—	0.0213	—	—	0.0213
49	0.0204	0.0204	—	0.0204	—	—	0.0204

A quick study of Table 2.4.1 shows that the harmonic currents are greatly reduced as the pulse number increases. Of course, the cost of the rectifiers and transformers increases as the complexity of the design increases. One must determine what level of power quality one can afford or what level is necessary to meet the limits given by IEEE 519.

2.4.7 Harmonic Spectrum

One requirement of ANSI/IEEE C57.18.10 is that the specifying engineer supply the harmonic load spectrum to the transformer manufacturer. There are too many details of circuit operation, system parameters, and other equipment, such as power-factor correction capacitors, for the transformer manufacturer to be able to safely assume a harmonic spectrum with complete confidence. The requirement to specify the harmonic spectrum is of utmost importance. Indeed, the problem of harmonic heating was the primary reason for the creation of ANSI/IEEE C57.18.10, which shows the theoretical harmonic spectrum similar to that presented in Table 2.4.1, also in theoretical values. The 25th harmonic was used as the cutoff point, since the theoretical values were given in the table. Table 1 of ANSI/IEEE C57.18.10 provides very conservative values. Consider what would happen if we used the theoretical values of harmonic current and allowed the spectrum to go on and on. The harmonic-loss factor from such an exercise would be infinity for any pulse system. A reasonable cutoff point and accurate harmonic spectrum are necessary to properly design transformers for harmonic loads. While it may be prudent to be somewhat conservative, the specifying engineer must recognize the cost of being overly conservative. If the transformer spectrum is underestimated, the transformer may overheat. If the harmonic spectrum is estimated at too high a value, the transformer will be overdesigned, and the user will invest more capital in the transformer than is warranted.

It is also not possible to simply specify the total harmonic distortion (THD) of the current in order to specify the harmonics. Many different harmonic-current spectra can give the same THD, but the harmonic-loss factor may be different for each of them.

Total percent harmonic distortion, as it relates to current, can be expressed as:

$$I_{THD} = [\sum_{h=2}^{H} I_h^2/I_1^2] \times 100\% \qquad (2.4.7)$$

where

I_h = current for the hth harmonic, expressed in per-unit terms

I_1 = fundamental frequency current, expressed in per-unit terms

Limits of allowable total harmonic distortion are given in IEEE 519, Recommended Practices and Requirements for Harmonic Control in Electric Power Systems.

Likewise, total harmonic distortion as it relates to voltage can be expressed as:

$$V_{THD} = [\sum_{h=2}^{H} V_h^2/V_1^2] \times 100\% \tag{2.4.8}$$

where

V_h = voltage for the hth harmonic, expressed in per-unit terms

V_1 = fundamental frequency current, expressed in per-unit, terms

Again, IEEE 519 also addresses total harmonic distortion and its limits.

Typically, voltage harmonics do not affect a rectifier transformer. Voltage and current harmonics usually do not create a core-heating problem. However, if there is dc current in the secondary waveform, the core can go into saturation. This results in high vibration, core heating, and circulating currents, since the core can no longer hold the flux. Nevertheless, the normal effect of harmonics is a noisier core as it reacts to different load frequencies.

Assume a harmonic spectrum as shown in Table 2.4.2.

TABLE 2.4.2 Theoretical Harmonic Spectrum

Harmonic	Harmonic, Per Unit Current
1	1.0000
5	0.1750
7	0.1000
11	0.0450
13	0.0290
17	0.0150
19	0.0100
23	0.0090
25	0.0080

If we look at the theoretical spectrum shown in Table 2.4.2 and compare it with an example spectrum in Table 2.4.3, we can see that the effects of the harmonic currents are quite different. The harmonic-loss factor, F_{HL}, is calculated for both the theoretical spectrum and the example spectrum in Table 2.4.3.

The results in Table 2.4.3 dramatically show the reality of many harmonic spectra. The winding eddy- and stray-loss multiplier from the example harmonic spectrum is much less than the theoretical value would indicate. This was one of the failings of rating transformers using the UL K-factor and then assigning an arbitrary value based on service. While this approach may be conservative and acceptable in a safety standard, it is not an engineering solution to the problem. The values of F_{HL} above demonstrate the need to have a reasonable harmonic spectrum for applications. Many site-specific installations measure and determine their harmonic spectra. For ease of specification, many specifying engineers use a standard spectrum that may not be applicable in all installations. This practice runs the risk of underspecifying or overspecifying the transformer. Underspecifying the harmonic spectrum results in overheated transformers and possible failures. Overspecifying the harmonic spectrum results in overbuilt and more costly capital equipment.

TABLE 2.4.3 Comparison of Harmonic-Loss Factor for the Theoretical Spectrum and an Example Spectrum

Harmonic h	I_h Theoretical	I_h Example	h^2	I_h^2 Theoretical	I_h^2 Example	$I_h^2 h^2$ Theoretical	$I_h^2 h^2$ Example
1	1.0000	1.0000	1	1.0000	1.0000	1.0000	1.0000
5	0.2000	0.1750	25	0.0400	0.3063	1.0000	0.7656
7	0.1429	0.1000	49	0.0204	0.0100	1.0000	0.4900
11	0.0909	0.0450	121	0.0083	0.0020	1.0000	0.2450
13	0.0769	0.0290	169	0.0059	0.0008	1.0000	0.1421
17	0.0588	0.0150	289	0.0035	0.0002	1.0000	0.0650
19	0.0526	0.0100	361	0.0028	0.0001	1.0000	0.0361
23	0.0435	0.0090	529	0.0019	0.0001	1.0000	0.0428
25	0.0400	0.0080	625	0.0016	0.0001	1.0000	0.0400
					$F_{HL} = \Sigma$	9.0000	2.8266

2.4.8 Effects of Harmonic Currents on Transformers

To better understand how harmonic currents affect transformers one must first understand the basic construction. For power transformers up to about 50 MVA, the typical construction is core form. The low-voltage winding is generally placed next to the core leg, with the high-voltage winding wound concentrically over the low-voltage winding. For some high-current transformers, these windings may be reversed, with the low-voltage winding wound on the outside over the high-voltage coil. The core and coils are held together with core clamps, and the core and coil is generally enclosed by a tank or enclosure. See Figure 2.4.15 for this construction and a view of leakage field around the transformer.

Losses in the transformer can be broken down into core loss, no-load loss, and load loss. Load losses can be further broken down into I²R loss and stray loss. Stray loss can be further broken down into eddy-current losses and other stray losses. Electromagnetic fields from the ac currents produce voltages across conductors, causing eddy currents to flow in them. This increases the conductor loss and operating temperature. Other stray losses are due to losses in structures other than the windings, such as core clamps and tank or enclosure walls.

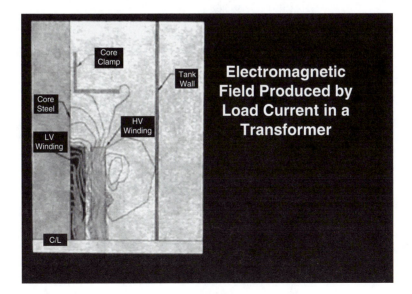

FIGURE 2.4.15 Transformer construction and electromagnetic leakage field.

The region of maximum eddy-current losses is the upper region of the winding, near the high–low barrier. The same usually exists at the bottom of the transformer winding as well, but it is typically the upper region that has the most damaging effects, as it is in a higher ambient temperature of liquid or air.

Core-loss components can be broken down into core eddy loss, hysteresis loss, and winding-excitation loss. These losses are a function of the grade of core steel, the lamination thickness, the type of core and joint, the operating frequency, the destruction factor during manufacture, and the core induction. Harmonic currents can create harmonic voltage distortions and somewhat increase the core loss, the exciting current, and sound levels while leading to potential core-saturation problems. However, this is not considered to be the main cause of problems in rectifier transformers. ANSI/IEEE C57.18.10 does not calculate any effect on the core loss by the harmonic currents.

Other stray losses are generally proportional to the current squared times the harmonic frequency order to the 0.8 power, as shown earlier in Equation 2.4.2. Metallic parts will increase in temperature, and load loss will increase. These losses are generally not detrimental to the life of the transformer as long as the insulating system is not damaged. The metallic parts typically affected are the core clamps, winding clamping structures, and tank or enclosure walls. The use of nonmagnetic materials, magnetic shields, conductive shields, increased magnetic clearances, and interleaving of high-current buswork are useful methods in reducing the stray losses that are amplified by the harmonic currents.

Eddy-current losses in the windings are affected mostly by harmonic currents. The eddy-current loss is proportional to the square of the load current and the square of the harmonic frequency, as shown earlier in Equation 2.4.1. These losses are increased in the hottest-spot area of the winding and can lead to early insulation failure. The transformer designer must make efforts to reduce the winding eddy-current losses due to the harmonic amplification of these losses. Careful winding and impedance balances, dimensioning of the conductors, and transposition of the conductors are useful methods in this effort.

I^2R losses increase as the rms current of the transformer increases. A transformer with a higher harmonic spectrum will draw more current from the system.

To see the results on the losses of the transformer, consider the following example for a 5000-kVA rectifier transformer with the harmonic spectrum shown in Table 2.4.4.

From the design or test losses, we can determine how many watts will be generated by each loss component and what needs to be cooled. For this example we will assume that the transformer has the following loss data:

Core loss = 9,000 W
I^2R loss = 30,000 W
Winding eddy-current loss = 6,000 W
Bus loss = 1,000 W
Other stray loss = 4,000 W
Total sinusoidal loss = 50,000 W

TABLE 2.4.4 Example of a Harmonic Spectrum for the Example Problem

Harmonic Order	Per Unit Load Current
1	1.000
5	0.180
7	0.120
11	0.066
13	0.055
17	0.029
19	0.019
23	0.009
25	0.005

TABLE 2.4.5 Calculated Harmonic Loss Factors

Harmonic Order (h)	Fundamental, p.u. A	Fundamental, p.u. A^2	h^2	h$^{0.8}$	F_{HL_WE}	$F_{HL\text{-}OSL}$ and Bus Loss
1	1.000	1.000000	1	1.0000	1.0000	1.0000
5	0.180	0.032400	25	3.6238	0.8100	0.1174
7	0.120	0.014400	49	4.7433	0.7056	0.0683
11	0.066	0.004356	121	6.8095	0.5271	0.0297
13	0.055	0.003025	169	7.7831	0.5112	0.0235
17	0.029	0.000841	289	9.6463	0.2430	0.0081
19	0.019	0.000361	361	10.5439	0.1303	0.0038
23	0.009	0.000081	529	12.2852	0.0428	0.0010
25	0.005	0.000025	625	13.1326	0.156	0.0003
Σ	—	1.055489	—	—	3.9856	1.2521

We can determine the harmonic loss factors in Table 2.4.5.

$$\text{rms p.u. current} = \sqrt{1.055489} = 1.02737 \tag{2.4.9}$$

From the results of the calculations in Table 2.4.5, we can now calculate the service losses under harmonic load conditions as follows:

1. Core loss remains unchanged from the fundamental kVA, per ANSI/IEEE C57.18.10.
2. The rms kVA would be 1.02737 times the fundamental kVA.
3. The fundamental I^{2}R losses would be increased by 1.055.
4. Multiply the fundamental winding eddy-current losses by 3.9856.
5. Multiply the fundamental other stray losses and bus-bar eddy-current losses by 1.2521.

The results of these calculations are shown in Table 2.4.6. The losses to be considered for thermal dimensioning would be the service losses, not the fundamental losses of Table 2.4.4. When the temperature-rise test is performed, it must be done with the calculated service losses per ANSI/IEEE C57.18.10. These calculated service losses should be a conservative estimate of the actual losses. Actual losses will probably be slightly less.

These calculated service losses would require significantly more radiators for a liquid-filled unit or the winding ducts for a dry-type or cast-coil transformer than the fundamental losses would indicate. More detailed examples of these types of loss calculations are given in the Annex of ANSI/IEEE C57.18.10.

2.4.9 Thermal Tests

The losses calculated by using the harmonic-loss factors are used when the thermal tests are performed. The thermal tests are performed per ANSI/IEEE C57.18.10. The methods are similar to ANSI/IEEE C57.12.90 and ANSI/IEEE C57.12.91, except instead of the fundamental or rms current, the current must be sufficient to produce the losses with harmonics considered.

TABLE 2.4.6 Calculated Service Losses

	Fundamental Losses	F_{HL} Multiplier	Service Losses
Core loss	9,000 W	1.0000	9,000 W
I^{2}R loss	30,000 W	1.0555	31,655 W
Winding eddy-current loss	6,000 W	3.9856	22,914 W
Bus loss	1,000 W	1.2521	1,252 W
Other stray loss	4,000 W	1.2521	5,008 W
Total losses	50,000 W	—	69,829 W

In the case of liquid-filled transformers, sufficient current is applied to the transformer to obtain the calculated service losses with harmonics. Once the top-oil temperature is obtained, the core loss is backed out. The rated-current portion of the test is performed with current sufficient to produce the load losses with harmonics included.

In the case of dry-type transformers, the core-loss portion of the test is the same as a standard transformer. When the load portion of the test is preformed, the rated-current portion of the test is performed with current sufficient to produce the load losses with the harmonics included.

Once the thermal tests are completed, there are possibly some adjustments necessary to the calculated temperature rise. This is especially true in the case of multiple-secondary-winding transformers. If the harmonic-loss factor for the secondary winding is different from the primary winding, the secondary-winding rise may need to be adjusted upward based on the losses it should have had. The primary-winding rise is then adjusted downward.

Temperature-rise tests of single-way transformers, such as ANSI Circuit 45 (Figure 2.4.5), are at best a compromise, unless the rectifier is available. Standard short-circuit tests used to thermally test the transformer will not even energize the interphase portions of the transformer. Either the primary windings or the secondary windings will be either greatly overloaded or underloaded. Care must be taken not to damage the transformer during the thermal tests. For instance, attempts to circulate currents in both secondary windings may cause the primary winding to be overloaded.

This is a significant change from the mercury-arc rectifier transformer standard, ANSI/IEEE C57.18. That standard did not require the use of calculated harmonic losses in the thermal test. Transformers were tested with the normal fundamental sinusoidal losses and were cooled to a fixed number of degrees below the normal temperature rise, based on the service class and type of transformer. These were shown in of Table 8 (limits of temperature rise) of that standard.

2.4.10   Harmonic Cancellation

Some harmonics can be cancelled, depending on how the windings are constructed or on the transformer circuit selected. Cancellation considerations are vital to the proper design and cooling of the transformer windings.

If the secondary windings are interleaved, the harmonic currents still exist for the converter pulse of the secondary, but the effects of the harmonics are reduced to the next pulse converter level. This is true of secondary windings constructed like Figure 2.4.12d and Figure 2.4.12e of Circuit 31 (Figure 2.4.4), Circuit 45 (Figure 2.4.5), or Circuit 46 (Figure 2.4.6) secondary windings and the like. Circuit 45 and 46 secondary windings will still carry three-pulse harmonics, but the harmonic-loss factor affecting the secondary-winding eddy-current losses is reduced to that of a six-pulse transformer. For a Circuit 31 transformer with six-pulse secondary windings, it will still carry six-pulse secondary harmonic currents, but the secondary-winding eddy-current losses will only be affected by 12-pulse harmonics.

If primary windings are made of paralleled sections, they have the same harmonic currents and harmonic-loss effects as their opposing secondary windings. This is the case for Figure 2.4.12a, a popular Circuit 31 transformer. All windings of this figure have six-pulse harmonics in the winding currents and the effect of the harmonics on the secondary-winding eddy-current losses. At the primary terminals of the transformer, the unit will have 12-pulse harmonics. For the user, this is the important feature. The winding construction is usually the choice of the transformer manufacturer. However, some specifying engineers will request particular constructions in order to achieve their desired system characteristics.

All of the harmonic cancellations discussed so far have used simple delta and wye windings in order to achieve them. Additional phase shifts are used to achieve higher pulse orders. This is accomplished by using extended delta windings, zigzag wye windings, and polygon windings. These are shown in Figure 2.4.16 through Figure 2.4.18. The windings are extended the required amount to produce the degree of shift desired. These are usually done at $7^{1}/_{2}$, 10, 15, and 20°, although more increments are possible in order to achieve the desired converter pulse.

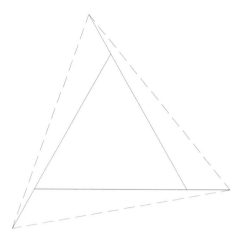

FIGURE 2.4.16 Extended delta-winding shift.

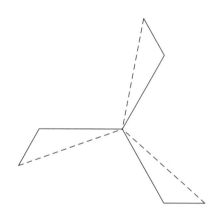

FIGURE 2.4.17 Zigzag wye-winding shift.

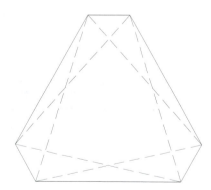

FIGURE 2.4.18 Polygon winding shift.

The type of winding combinations required for the converter pulse desired can be quite comprehensive. A table of typical winding shifts is shown in Table 2.4.7, but other combinations are acceptable. The table shows the required number of secondary windings needed with a single-primary-winding transformer. This is usually used with higher-voltage, low-current secondary windings. It can also be used as the required number of primary-winding shifts and transformers, when the transformers have the same single-secondary-winding phase relationship. This is usually used with high-current secondary windings.

TABLE 2.4.7 Required Number of Six-Pulse Windings and Connections

Pulse Number	Number of Six-Pulse Windings and Typical Connections							
	1	2	3	4	5	6	7	8
6	delta or wye	—	—	—	—	—	—	—
12	delta	wye	—	—	—	—	—	—
18	delta	wye −10°	wye +10°	—	—	—	—	—
18	delta +10°	delta −10°	wye	—	—	—	—	—
18	delta +20°	delta	delta −20°	—	—	—	—	—
24	delta +15°	delta	delta −15°	wye	—	—	—	—
24	delta	wye +15°	wye	wye −15°	—	—	—	—
24	delta +7½°	delta −7½°	wye +7½°	wye −7½°	—	—	—	—
30	delta +12°	delta	delta −12°	wye +6°	wye −6°	—	—	—
36	delta +10°	delta	delta −10°	wye +10°	wye −10°	wye	—	—
48	delta +15°	delta +7½°	delta	delta −7½°	delta −15°	wye −7½°	wye	wye +7½°

Using different phase shifts on the single winding of the transformer, whether the primary or secondary winding, can increase the number of phase shifts. For instance, two 12-pulse transformers can make a 24-pulse system by using a delta primary on one transformer and a wye primary on the other. In cases where it may be desirable to have an interchangeable spare, it is sometimes beneficial to use two 15° phase-shifted primary windings. The spare transformer can then be made with a reconnectable winding for ±15° shift.

It is important to note that harmonic cancellation is generally not perfect. This is due to several factors, such as unbalanced loading, inaccurate phase shifts, differences in commutating impedances, and tap changes. That may be acceptable at some times but not at others. It is common to assume a 5% residual of lower harmonics to accommodate these realities.

When the phase shift is incorporated in the primary winding, the degree of shift will vary somewhat as taps are changed on the transformer unless a tap changer is used in the main part of the shifted winding and the extended part of the shifted winding. Even then, there may be a slight shift. These problems may vary by about a degree of shift over the tap range of most transformers (Figures 2.4.19, 2.4.20, 2.4.21, 2.4.22, and 2.4.23).

2.4.11 DC Current Content

If dc current is present in either the supply side or the load side of the transformer windings, it must be specified to the transformer manufacturer at the time of quotation. Some rectifier circuits, such as cyclo-converters, have the possibility of dc current in the load current. A small amount of dc current can saturate the core of a transformer. The effects of this may be core and core-joint overheating, core-clamp heating from fields and circulating currents, winding hot spots, and even tank heating. Noise and vibration are also often present.

FIGURE 2.4.19 A 12-pulse Circuit 31 5450-kVA, 4160-V delta primary to 2080-V delta and wye secondaries, cast-coil transformer in case. (Photo courtesy of Niagara Transformer Corp.)

FIGURE 2.4.20 A 24-pulse dry-type transformer. 7000-kVA drive duty transformer, to be located in a tunnel, made of two 3500-kVA 12-pulse transformers. Each core and coil has 1100-V delta and wye secondary windings. Each core and coil has a 13,200-V extended-delta primary winding. One core and coil has its primary winding shifted −15°, while the other core and coil has its primary winding shifted +15°. One reconnectable primary spare core and coil can be used to replace either core and coil assembly. (Photo courtesy of Niagara Transformer Corp.)

2.4.12 Transformers Energized from a Converter/Inverter

Transformers energized from a converter/inverter are often subject to considerably distorted voltages. If voltage harmonics are known to be above the limits of IEEE 519, they must be specified. Variable-frequency applications are generally considered to be at constant volts per hertz. If the volt-per-hertz ratio is variable, the degree of variation must be specified. The flux density of the core is the governing factor, not the maximum value of the sinusoidal voltage.

FIGURE 2.4.21 A 36-pulse system made up of 12 transformers with six types of six-pulse transformers for high-current electrochemical rectifier duty. Each transformer is 11,816-kVA OFWF with 426-V delta secondary windings. Each transformer has a 34,500-V primary with the shifts shown in Table 2.4.7 for a 36-pulse system. (Photo courtesy of Niagara Transformer Corp.)

FIGURE 2.4.22 A 5000-kVA 18-pulse transformer ready for shipment. Motor drive duty, 13,800-V delta primary to secondaries of 1400-V delta, 1400-V delta +20°, and 1400-V delta −20°. (Photo courtesy of Niagara Transformer Corp.)

2.4.13 Electrostatic Ground Shield

It is usually desirable to have an electrostatic ground shield between the primary and secondary windings. The electrostatic ground shield provides capacitive decoupling of the primary and secondary windings. Generally, the winding connected to the rectifier circuit is ungrounded. Without the presence of the electrostatic ground shield, transients on the primary side transfer to the secondary side of the transformer. These may be approximately 50% of the magnitude of the primary transient if there are no grounds in the system. This is high enough to fail secondary windings and core insulation or to cause rectifier-circuit failures. The other normally considered advantage to the system is the minimization of high-frequency disturbances to the primary system due to the rectifier.

FIGURE 2.4.23 A 37,500-kVA OFAF regulating autotransformer with LTC and saturable core reactors and dc power supplies. The high voltage is dual voltage for 13,800-V wye and 23,000-V wye. The LTC and saturable-core reactors permit the secondary voltage to range from 14,850-V wye down to 7,150-V wye. The LTC makes coarse taps, while the saturable-core reactors provide infinite variability between taps. This transformer powers downstream to an electrochemical service diode rectifier. The transformer weighs over 170,000 lb, as it includes the main transformer, a series transformer, a preventive autotransformer, and saturable core reactors. The dc power supplies provide control to the saturable-core reactors. (Photo courtesy of Niagara Transformer Corp.)

2.4.14 Load Conditions

Load conditions generally are categorized by the service to which the transformer will be subjected. ANSI/IEEE C57.18.10 gives the limits of rectifier transformer winding temperatures for defined load cycles. These limits are the hottest-spot temperature limits of the applicable insulation systems. These are the same limits that one may have with a standard power transformer that is not subject to loads rich in harmonics. However, the harmonic losses are to be included in the calculation and test values used to determine thermal capability for rectifier transformers. The standard service rating classes are as follows:

1. Electrochemical service
2. Industrial service
3. Light traction or mining service
4. Heavy traction service
5. Extra-heavy traction service
6. User-defined service

User-defined service is a catch-all for load patterns not defined in items 1 through 5.

2.4.15 Interphase Transformers

Interphase transformers help to combine multiple rectifier outputs. They may be external or internal to the rectifier transformer. The interphase transformer supports ac voltage differences between the rectifier outputs. They cannot balance steady-state differences in dc voltage, since they only provide support to ac voltage differences. The interphase-transformer windings carry both ac and dc currents. The windings are in opposition so as to allow dc current to flow, but this causes opposing ampere-turns on the core. The core usually has to be gapped for the expected dc current unbalances and to be able to support the expected magnetizing current from the ripple voltage. Excellent sources on this are Shaefer (1965) and Paice (1996, 2001) listed in the references below.

References

ANSI/IEEE, Pool Cathode Mercury-Arc Rectifier Transformer, ANSI/IEEE C57.18-1964, Institute of Electrical and Electronics Engineers, Piscataway, NJ, 1964.

Blume, L.F. et al., *Transformer Engineering*, 2nd ed., John Wiley & Sons, New York, 1951.

Crepaz, S., Eddy current losses in rectifier transformers, *IEEE Trans. Power Appar. Syst.*, PAS-89, p. 1651, 1970.

Dwight, H.B., *Electrical Coils and Conductors*, McGraw-Hill, New York, 1945.

Forrest, J. Alan C., Harmonic load losses in HVDC converter transformers, *IEEE Trans. Power Delivery*, 6, 6(1) pp. 153–157, 1991.

General Electric, *Power Converter Handbook*, Canadian General Electric Co., Peterborough, Ontario, Canada, 1976.

IEC, Convertor Transformers — Part 1: Transformers for Industrial Applications, IEC 61378-1-1997, International Electrotechnical Commission, Geneva, 1997.

IEEE, Standard Practices and Requirements for Semiconductor Power Rectifiers, ANSI/IEEE C34.2-1968, Institute of Electrical and Electronics Engineers, Piscataway, NJ, 1968.

IEEE, Guide for Harmonic Control and Reactive Compensation of Static Power Converters, IEEE Std. 519-1992, Institute of Electrical and Electronics Engineers, Piscataway, NJ, 1992.

IEEE, Standard Test Code for Dry-Type Distribution and Power Transformers, ANSI/IEEE C57.12.91-1996, Institute of Electrical and Electronics Engineers, Piscataway, NJ, 1996.

IEEE, Standard Practices and Requirements for Semiconductor Power Rectifier Transformers, IEEE C57.18.10-1998, Institute of Electrical and Electronics Engineers, Piscataway, NJ, 1998.

IEEE, Recommended Practice for Establishing Transformer Capability When Supplying Non-Sinusoidal Load Currents, IEEE C57.110-1999, Institute of Electrical and Electronics Engineers, Piscataway, NJ, 1999.

IEEE, Standard Test Code for Liquid-Immersed Distribution, Power, and Regulating Transformers, ANSI/IEEE C57.12.90-1999, Institute of Electrical and Electronics Engineers, Piscataway, NJ, 1999.

Kennedy, S.P. and Ivey, C.L., Application, design and rating of transformers containing harmonic currents, presented at 1990 Annual Pulp and Paper Technical Conference, Portland OR, 1990.

Kennedy, S.P., Design and application of semiconductor rectifier transformers, *IEEE Trans. Ind. Applic.*, July/August 2002, Vol. 38(4), pp. 927–933, 2002.

Kline, A.D., Transformers in SCR converter circuits, Conference Record of the Industry Applications Society, IEEE, 1981 Annual Meeting, pp. 456-458, New York.

MIT electrical engineering staff, *Magnetic Circuits and Transformers*, John Wiley & Sons, New York, 1949.

Paice, D.A., *Power Electronic Converter Harmonics: Multiphase Methods for Clean Power*, IEEE Press, New York, 1996.

Paice, D.A., *Power Electronic Converter Design and Application, Multipulse Methods and Engineering Issues*, Paice and Associates, Palm Harbor, FL, 2001.

Ram, B.S. et al., Effects of harmonics on converter transformer load losses, *IEEE Trans. Power Delivery*, 3, Vol. 3(3), p. 1059, 1988.

Schaefer, J., *Rectifier Circuits: Theory and Design*, John Wiley & Sons, New York, 1965.

UL, Dry-Type General Purpose and Power Transformers, UL 1561, Underwriters' Laboratories, Northbrook, IL, 1999.

2.5   Dry-Type Transformers

Paulette A. Payne

A dry-type transformer is one in which the insulating medium surrounding the winding assembly is a gas or dry compound. Basically, any transformer can be constructed as "dry" as long as the ratings, most especially the voltage and kVA, can be economically accommodated without the use of insulating oil or other liquid media. This section covers single- and three-phase, ventilated, nonventilated, and sealed dry-type transformers with voltage in excess of 600 V in the highest-voltage winding.

Many perceptions of dry-type transformers are associated with the class of design by virtue of the range of ratings or end-use applications commonly associated with that form of construction Of course, the fundamental principles are no different from those encountered in liquid-immersed designs, as discussed in other chapters. Considerations involving harmonics are especially notable in this regard. Consequently, this chapter is brief, expounding only on those topics that are particularly relevant for a transformer because it is "dry."

Dry-type transformers compared with oil-immersed are lighter and nonflammable. Increased experience with thermal behavior of materials, continued development of materials and transformer design have improved transformer thermal capability. Upper limits of voltage and kVA have increased. Winding insulation materials have advanced from protection against moisture to protection under more adverse conditions (e.g., abrasive dust and corrosive environments).

2.5.1 Transformer Taps

Transformers may be furnished with voltage taps in the high-voltage winding. Typically two taps above and two taps below rated voltage are provided, yielding a 10% total tap voltage range (ANSI/IEEE, 1981 [R1989]; ANSI/IEEE C57.12.52-1981 [R1998]).

2.5.2 Cooling Classes for Dry-Type Transformers

American and European cooling-class designations are indicated in Table 2.5.1. Cooling classes for dry-type transformers are as follows (IEEE, 100, 1996; ANSI/IEEE, C57.94-1982 (R-1987)):

Ventilated — Ambient air may circulate, cooling the transformer core and windings
Nonventilated — No intentional circulation of external air through the transformer
Sealed — Self-cooled transformer with hermetically sealed tank
Self-cooled — Cooled by natural circulation of air
Force-air cooled — Cooled by forced circulation of air
Self-cooled/forced-air cooled — A rating with cooling by natural circulation of air and a rating with cooling by forced circulation of air.

2.5.3 Winding Insulation System

General practice is to seal or coat dry-type transformer windings with resin or varnish to provide protection against adverse environmental conditions that can cause degradation of transformer windings. Insulating media for primary and secondary windings are categorized as follows:

Cast coil — The winding is reinforced or placed in a mold and cast in a resin under vacuum pressure. Lower sound levels are realized as the winding is encased in solid insulation. Filling the winding with resin under vacuum pressure eliminates voids that can cause corona. With a solid insulation system, the winding has superior mechanical and short-circuit strength and is impervious to moisture and contaminants.

TABLE 2.5.1 Cooling Class Designation

Cooling Class	IEEE Designation (ANSI/IEEE 57.12.01-1989 [R1998])	IEC Designation (IEC 60726-1982 [Amend. 1-1986])
Ventilated self-cooled	AA	AN
Ventilated forced-air cooled	AFA	AF
Ventilated self-cooled/forced-air cooled	AA/FA	ANAF
Nonventilated self-cooled	ANV	ANAN
Sealed self-cooled	GA	GNAN

Vacuum-pressure encapsulated — The winding is embedded in a resin under vacuum pressure. Encapsulating the winding with resin under vacuum pressure eliminates voids that can cause corona. The winding has excellent mechanical and short-circuit strength and provides protection against moisture and contaminants.

Vacuum-pressure impregnated — The winding is permeated in a varnish under vacuum pressure. An impregnated winding provides protection against moisture and contaminants.

Coated — The winding is dipped in a varnish or resin. A coated winding provides some protection against moisture and contaminants for application in moderate environments.

Below are two photographs of dry-type transformer assemblies.

FIGURE 2.5.1   A cast-resin encapsulated low-voltage (600 V) coil for a 1500-kVA transformer. (Photo courtesy of ABB.)

FIGURE 2.5.2 Three-phase 1500-kVA dry-type transformer, 15-kV-class primary, 600-V secondary. Three cooling fans can be seen on the base, below the coils. (Photo courtesy of ABB.)

2.5.4 Application

Nonventilated and sealed dry-type transformers are suitable for indoor and outdoor applications (ANSI/IEEE, 57.94-1982 [R-1987]). As the winding is not in contact with the external air, it is suitable for applications, e.g., exposure to fumes, vapors, dust, steam, salt spray, moisture, dripping water, rain, and snow.

   Ventilated dry-type transformers are recommended only for dry environments unless designed with additional environmental protection. External air carrying contaminants or excessive moisture could degrade winding insulation. Dust and dirt accumulation can reduce air circulation through the windings (ANSI/IEEE, 57.94-1982 [R-1987]). Table 2.5.2 indicates transformer applications based upon the process employed to protect the winding insulation system from environmental conditions.

2.5.5 Enclosures

All energized parts should be enclosed to prevent contact. Ventilated openings should be covered with baffles, grills, or barriers to prevent entry of water, rain, snow, etc. The enclosure should be tamper resistant. A means for effective grounding should be provided (ANSI/IEEE, C2-2002). The enclosure should provide protection suitable for the application, e.g., a weather- and corrosion-resistant enclosure for outdoor installations.

   If not designed to be moisture resistant, ventilated and nonventilated dry-type transformers operating in a high-moisture or high-humidity environments when deenergized should be kept dry to prevent moisture ingress. Strip heaters can be installed to switch on manually or automatically when the transformer is deenergized for maintaining temperature after shutdown to a few degrees above ambient temperature.

TABLE 2.5.2 Transformer Applications

Winding Insulation System	Cast Coil	Encapsulated	Impregnated or Coated	Sealed Gas
Harsh environments[a]	Yes	Yes	Yes	Yes
Severe climates[b]	Yes	Yes		Yes
Load cycling	Yes	Yes	Yes	Yes
Short circuit	Yes	Yes	Yes	Yes
Nonflammability	Yes	Yes	Yes	Yes
Outdoor	Yes	Yes	Yes[c]	Yes
Indoor	Yes	Yes	Yes	Yes

[a] Fumes, vapors, excessive or abrasive dust, steam, salt spray, moisture, or dripping water.
[b] Extreme heat or cold, moisture.
[c] If designed for installation in dry environments.

TABLE 2.5.3 Usual Operating Conditions for Transformers (ANSI/IEEE, C57.12.01-1989 [R1998])

Temperature of cooling air	$\leq 40°C$
24 hr average temperature of cooling air	$\leq 30°C$
Minimum ambient temperature	$\geq -30°C$
Load current[a]	Harmonic factor ≤ 0.05 per unit
Altitude[b]	≤ 3300 ft (1000 m)
Voltage[c] (without exceeding limiting temperature rise)	• Rated output kVA at 105% rated secondary voltage, power factor ≥ 0.80 • 110% rated secondary voltage at no load

[a] Any unusual load duty should be specified to the manufacturer.
[b] At higher altitudes, the reduced air density decreases dielectric strength; it also increases temperature rise, reducing capability to dissipate heat losses (ANSI/IEEE, C57.12.01-1989 [R1998]).
[c] Operating voltage in excess of rating may cause core saturation and excessive stray losses, which could result in overheating and excessive noise levels (ANSI/IEEE, C57.94-1982 [R1987], C57.12.01-1989 [R1998]).

2.5.6 Operating Conditions

The specifier should inform the manufacturer of any unusual conditions to which the transformer will be subjected. Dry-type transformers are designed for application under the usual operating conditions indicated in Table 2.5.3.

Gas may condense in a gas-sealed transformer left deenergized for a significant period of time at low ambient temperature. Supplemental heating may be required to vaporize the gas before energizing the transformer (ANSI/IEEE, C57.94-1982 [R1987]).

2.5.7 Limits of Temperature Rise

Winding temperature-rise limits are chosen so that the transformer will experience normal life expectancy for the given winding insulation system under usual operating conditions. Table 2.5.4 indicates the limits of temperature rise for the thermal insulation systems most commonly applied. Operation at rated load and loading above nameplate will result in normal life expectancy. A lower average winding temperature rise, 80°C rise for 180°C temperature class and 80°C or 115°C rise for 220°C temperature class, may be designed providing increased life expectancy and additional capacity for loading above nameplate rating.

TABLE 2.5.4 Limits of Temperature Rise for Commonly Applied Thermal Insulation Systems

Insulation System Temperature Class (°C)	Winding Hottest-Spot Temperature Rise (°C) for Normal Life Expectancy		Average Winding Temperature Rise (°C)[a]
	Continuous Operation at Rated Load[a]	Loading above Nameplate Rating[b]	
150	110	140	80
180	140	170	115
220	180	210	150

[a] ANSI/IEEE Standard C57.12.01-1989 (R1998).
[b] ANSI/IEEE Standard C57.96-1999.

2.5.8 Accessories

The winding-temperature indicator can be furnished with contacts to provide indication and/or alarm of winding temperature approaching or in excess of maximum operating limits.

For sealed dry-type transformers, a gas-pressure switch can be furnished with contacts to provide indication and/or alarm of gas-pressure deviation from recommended range of operating pressure.

2.5.9 Surge Protection

For transformers with exposure to lightning or other voltage surges, protective surge arresters should be coordinated with transformer basic lightning impulse insulation level, BIL. The lead length connecting from transformer bushing to arrester—and from arrester ground to neutral—should be minimum length to eliminate inductive voltage drop in the ground lead and ground current (ANSI-IEEE, C62.2-1987 [R1994]). Table 2.5.5 provides transformer BIL levels corresponding to nominal system voltage. Lower BIL levels can be applied where surge arresters provide appropriate protection. At 25 kV and above, higher BIL levels may be required due to exposure to overvoltage or for a higher protective margin (ANSI/IEEE, C57.12.01-1989 [R1998]).

TABLE 2.5.5 Transformer BIL Levels

Nominal System Voltage	BIL (kV)
1,200 and lower	10
2,500	20
5,000	30
8,700	45
15,000	60
25,000	110
34,500	150

References

ANSI/IEEE Standard C57.12.50-1981 (R1998), Requirements for Ventilated Dry-Type Distribution Transformers, 1 to 500 kVA, Single-Phase, and 15 to 500 kVA, Three-Phase, with High-Voltage 601 to 34500 Volts, Low-Voltage 120 to 600 Volts.

ANSI/IEEE Standard C57.12.52-1981 (R1998), Requirements for Sealed Dry-Type Power Transformers, 501 kVA and Larger, Three-Phase, with High-Voltage 601 to 34500 Volts, Low-Voltage 208Y/120 to 4160 Volts.

ANSI/IEEE Standard C57.94-1982 (R-1987), IEEE Recommended Practice for Installation, Application, Operation, and Maintenance of Dry-Type General Purpose Distribution and Power Transformers.

ANSI/IEEE Standard C57.12.01-1989 (R1998), Standard General Requirements for Dry-Type Distribution and Power Transformers Including Those with Solid Cast and/or Resin-Encapsulated Windings.

ANSI/IEEE Standard C57.96-1999, Guide for Loading Dry-Type Distribution and Power Transformers.

ANSI/IEEE C57.110-1998, Recommended Practice for Establishing Transformer Compatibility When Supplying Non-Sinusoidal Currents.

ANSI/IEEE C62.2-1987 (R1994), Guide for the Application of Gapped Silicon-Carbide Surge Arresters for Alternating Current Systems.

ANSI/IEEE C2-2002, National Electrical Safety Code.

IEC Standard 60726-1982 (Amendment 1-1986), Dry-Type Power Transformers, 1st ed.

IEEE 519-1992, Recommended Practices and Requirements for Harmonic Control in Electrical Power Systems.

IEEE Standard 100-1996, Standard Dictionary of Electrical and Electronics Terms, 6th ed.

NEMA Standard TP1-2002, Guide for Determining Energy Efficiency for Distribution Transformers.

2.6 Instrument Transformers

Randy Mullikin

This section covers the fundamental basics and theory of operation of instrument transformers. Common types of instrument transformers and construction highlights will be discussed. Application features and characteristics of instrument transformers will be covered without providing details of three-phase circuit fundamentals, fault analysis, or the operation and selection of protective devices and measuring instruments. Though incomplete, this section covers the common practices used in industry over the last 30 years.

2.6.1 Overview

Instrument transformers are primarily used to provide isolation between the main primary circuit and the secondary control and measuring devices. This isolation is achieved by magnetically coupling the two circuits. In addition to isolation, levels in magnitude are reduced to safer levels.

Instrument transformers are divided into two categories: voltage transformers (VT) and current transformers (CT). The primary winding of the VT is connected in parallel with the monitored circuit, while the primary winding of the CT is connected in series (see Figure 2.6.1). The secondary windings proportionally transform the primary levels to typical values of 120 V and 5 A. Monitoring devices such as wattmeters, power-factor meters, voltmeters, ammeters, and relays are often connected to the secondary circuits.

2.6.2 Transformer Basics

An ideal transformer (see Figure 2.6.2) magnetically induces from the primary circuit a level exactly proportional to the turns ratio into the secondary circuit and exactly opposite in phase, regardless of

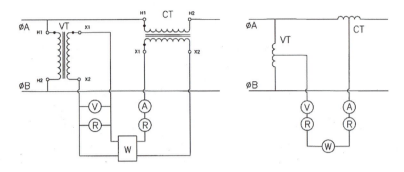

FIGURE 2.6.1 Typical wiring and single-line diagram.

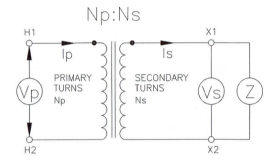

FIGURE 2.6.2 Ideal transformer.

the changes occurring in the primary circuit. A review of the general relationships of the ideal case yields the transformation ratio

$$V_P/V_S = N_P/N_S \qquad (2.6.1)$$

and the law of conservation of energy yields

$$V_P I_P = V_S I_S \qquad (2.6.2)$$

$$I_P N_P = I_S N_S \qquad (2.6.3)$$

where

    V_P = primary-terminal voltage
    V_S = secondary-terminal voltage
    I_P = primary current
    I_S = secondary current
    N_P = primary turns
    N_S = secondary turns

2.6.2.1 Core Design

In practice, the use of steel core material is a major factor in forcing the transformer to deviate from the ideal. The available core steels offer differing properties around which to design. The most common type of steel used is electrical-grade silicon-iron. This material offers low losses at high flux densities, but it has low initial permeability. Exotic materials, such as nickel-iron, offer high initial permeability and low

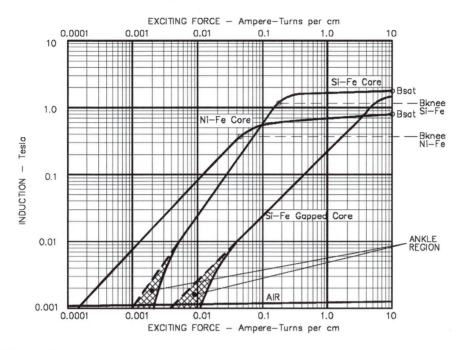

FIGURE 2.6.3 Typical characteristic curve.

losses, but they have much lower saturation levels. These exotics are often used when extremely high accuracy is desired, but they are cost prohibitive in standard products. A typical excitation characteristic for instrument transformers is shown in Figure 2.6.3. There are three areas of interest: the ankle region, the knee point, and saturation. The ankle region is at the lowest permeability and flux levels. Due to the uncertainties in this region, performance will deviate from core to core. Steel manufacturers never guarantee performance in this region. As a result, the manufacturer must have tight control over its annealing process. The exotic core steels have a well-defined characteristic as low as 0.001 T. The knee represents the maximum permeability and is the beginning of the saturation zone. The area between the ankle and the knee is the linear portion, where performance is predictable and repeatable. Saturation is the point at which no additional flux enters the core.

There are occasions when the designer would like the benefits of both silicon-iron and nickel-iron. A core using two or more different types of steel is called a composite core. The ratio of each depends on the properties desired and the overall cost. The most commonly used of the nickel-iron family are the 49% and 80% varieties. Table 2.6.1 shows some typical characteristics as well as some trade names. Figure 2.6.4 demonstrates how two cores of equal turns, area, and magnetic path are added to become one. The voltages will add, and the net saturation flux is the sum of one half of the individual saturation density of each core, i.e., $^{1}/_{2}$ 1.960 + $^{1}/_{2}$ 0.760 = 1.360 T.

Steel type and lamination thickness are additional factors in reducing losses. Other materials, such as noncrystalline or amorphous metals, offer lower losses but are difficult to fabricate. However, new developments in processing are improving steel quality. New compositions that claim to perform like nickel — at a fraction of its cost — are now surfacing. But regardless of the improvement offered by these materials, there are some inherent properties that must be overcome:

1. Some portion of the primary energy is required to establish the magnetic flux that is required to induce the secondary winding.
2. The magnetization of the core is nonlinear in nature.

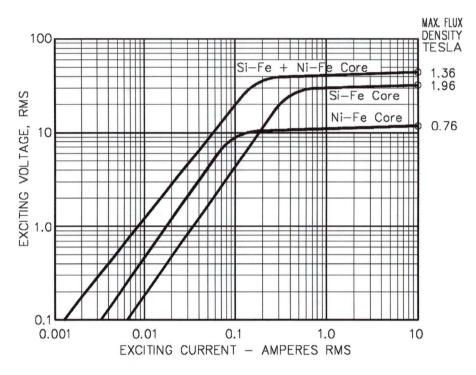

FIGURE 2.6.4 Typical composite core.

TABLE 2.6.1 Core Materials

Material (0.009–0.014 in.)	Saturation Density, T	Initial Permeability, μ @ 0.004 T	Maximum Permeability, μ @ T	Trade Names
80% Ni/Fe/Mo	0.760	35,000	150 k @ 0.4 T	SuperPerm 80[a] Hymu 80 Mu-Metal[b] 4–79 Permalloy[c]
49% Ni/Fe/Mo	1.200	5,000	60 k @ 0.4 T	Super Perm 49[a] High Perm 49[c] 48-Ni 4750[b]
3% Si/Fe – M3	1.900	8,000	51 k @ 1.15 T	Microsil[a] Silectron[d] Oriented T-S[e] Hipersil[a]
3% Si/Fe – M4	1.900	7,200	48 k @ 1.15 T	
3% Si/Fe – M6	1.900	6,700	40 k @ 1.15 T	

[a] Magnetic Metals Co.
[b] Allegheny Ludlum Steel.
[c] Carpenter Steel.
[d] Arnold Engineering.
[e] Armco (now AK Steel).

From statement 1, the primary energy required to magnetize the core is a product of the flux in the core and the magnetic reluctance of the core. This energy is called the magnetomotive force, mmf, and is defined as follows:

$$\text{mmf} = \phi \Re = k_1 \left[\frac{Z_S I_S}{N_S f} \right] k_2 \left[\frac{\text{mmp}}{A_C \mu_C} \right] = k_1 k_2 \left[\frac{Z_S I_S \text{ mmp}}{N_S f A_C \mu_C} \right] \tag{2.6.4}$$

where

ϕ = flux in the core
$\Re$ = magnetic reluctance
k_1 = constant of proportionality
k_2 = constant of proportionality
Z_S = secondary impedance
mmp = core mean magnetic path
I_S = secondary current
A_C = core cross-sectional area
N_S = number of secondary turns
μ_C = permeability of core material
f = frequency, Hz

The magnetic reluctance, in terms of Ohm's law, is analogous to resistance and is a function of the core type used. An annular or toroidal core, one that is a continuous tape-wound core, has the least amount of reluctance. A core with a straight cut through all of its laminations, thereby creating a gap, exhibits high reluctance. Minimizing gaps in core constructions reduces reluctance. Figure 2.6.5 shows some of the more common core and winding arrangements. Generally — after the steel material is cut, stamped, or wound — it undergoes a stress-relief anneal to restore the magnetic properties that may have been altered during fabrication. After the annealing process, the core is constructed and insulated.

From statement 2, the core permeability, μ_C, changes with flux density, ϕ/A_C. Neglecting leakage flux, we can now see the error-producing elements. From Equation 2.6.4, an increase in any of the elements in the denominator will decrease errors, while an increase in Z_S and mmf will increase errors.

There are also other contributing factors that, based on the construction of the instrument transformer, can introduce errors. The resistance of the windings, typically of copper wire and/or foil, introduces voltage drops (see Figure 2.6.6). Moreover, the physical geometry and arrangement of the windings — with respect to each other and the core — can introduce inductance and, sometimes, capacitance, which has an effect on magnetic leakage, reducing the flux linkage from the primary circuit and affecting performance. A winding utilizing all of its magnetic path will have the lowest reactance. Figure 2.6.5 shows some typical winding arrangements and leakage paths.

Figure 2.6.6 illustrates an equivalent transformer circuit, where

V_P = primary-terminal voltage
V_S = secondary-terminal voltage
E_P = primary-induced voltage
E_S = secondary-induced voltage
I_P = primary current
I_S = secondary current
N_P = primary turns
N_S = secondary turns
R_P = primary-winding resistance
R_S = secondary-winding resistance
X_P = primary-winding reactance
X_S = secondary-winding reactance

R_{ex} = wattful magnetizing component
Z = secondary burden (load
X_{ex} = wattless magnetizing componen
I_{ex} = magnetizing current

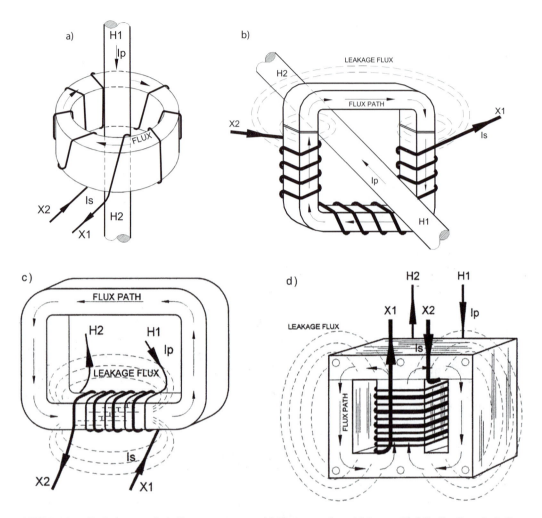

FIGURE 2.6.5 Typical core and winding arrangements. (a) Tape-wound toroidal core with fully distributed winding. Note the absence of leakage flux, which exists but is considered to be negligible. Sensitivity to primary-conductor position is also negligible. The primary can also consist of several turns. (b) Cut core with winding partially distributed. Leakage flux, in this case, depends on the location of the primary conductor. As the conductor moves closer to the top of the core, the leakage flux increases. (c) Distributed-gap core with winding distributed on one leg only. This type has high leakage flux but good coupling, since the primary and secondary windings occupy the same winding space. (d) Laminated "EI" core, shell-type. This type has high leakage flux, since a major portion of magnetic domains are against the direction of the flux path. The orientation is not all in the same direction, as it is with a tape-wound core.

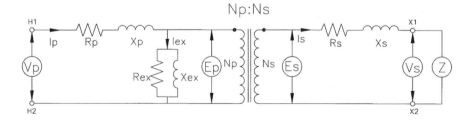

FIGURE 2.6.6 Equivalent transformer circuit.

Those factors not directly related to construction would be (1) elements introduced from the primary circuit, such as harmonics, that account for hysteresis and eddy-current losses in the core material and (2) fault conditions that can cause magnetic saturation. Most all can be compensated for by careful selection of core and winding type. It is also possible to offset error by adjusting turns on one of the windings, preferably the one with the higher number of turns, which will provide better control. It is also possible to compensate using external means, such as resistance capacitance and inductance (RCL) networks, but this limits the transformer to operation to a specific load and can have adverse effects over the operating range.

2.6.2.2 Burdens

The burden of the instrument transformer is considered to be everything connected externally to its terminals, such as monitoring devices, relays, and pilot wiring. The impedance values of each component, which can be obtained from manufacturer data sheets, should be added algebraically to determine the total load. The units of measurement must be the same and in the rectangular form R + jX. Table 2.6.2 shows typical ranges of burdens for various devices used.

For the purpose of establishing a uniform basis of test, a series of standard burdens has been defined for calibrating VTs and CTs. The burdens are inductive and designated in terms of VA. All are based on 120 V and 5 A at 60 Hz. They can be found in Table 2.6.3.

2.6.2.3 Relative Polarity

The instantaneous relative polarity of instrument transformers may be critical for proper operation in metering and protection schemes. The basic convention is that as current flows into the H1 terminal it

TABLE 2.6.2 Typical Burden Values for Common Devices

Device	Voltage Transformers		Current Transformers	
	Burden, VA	Power Factor	Burden, VA	Power Factor
Voltmeter	0.1–20	0.7–1.0	—	—
Ammeter	—	—	0.1–15	0.4–1.0
Wattmeter	1–20	0.3–1.0	0.5–25	0.2–1.0
P.F. meter	3–25	0.8–1.0	2–6	0.5–0.95
Frequency meter	1–50	0.7–1.0	—	—
kW·h meter	2–50	0.5–1.0	0.25–3	0.4–0.95
Relays	0.1–50	0.3–1.0	0.1–150	0.3–1.0
Regulator	50–100	0.5–0.9	10–180	0.5–0.95

TABLE 2.6.3 Standard Metering and Relaying Class Burdens

Typical Use	Voltage Transformers			Current Transformers		
	Burden Code	Burden, VA	Power Factor	Burden Code	Burden, VA	Power Factor
Metering	W	12.5	0.1	B0.1	2.5	0.9
Metering	X	25	0.7	B0.2	5	0.9
Metering	M	35	0.2	B0.5	12.5	0.9
Metering	—	—	—	B0.9	22.5	0.9
Metering	—	—	—	B1.8	45	0.9
Relaying	Y	75	0.85	B1.0	25	0.5
Relaying	Z	200	0.85	B2.0	50	0.5
Relaying	ZZ	400	0.85	B4.0	100	0.5
Relaying	—	—	—	B8.0	200	0.5

TABLE 2.6.4 Instrument Transformer Standards

Country	CT Standard	VT Standard
U.S.	IEEE C57.13	IEEE C57.13
Canada	CAN-C13-M83	CAN-C13-M83
IEC.	60044-1 (formerly 185)	60044-2 (formerly 186)
U.K.	BS 3938	BS 3941
Australia	AS 1675	AS 1243
Japan	JIS C 1731	JIS C 1731

flows out of the X1 terminal, making this polarity subtractive. These terminals are identified on the transformer by name and/or a white dot.

2.6.2.4 Industry Standards

In the U.S., the utility industry relies heavily on IEEE C57.13, Requirements for Instrument Transformers. This standard establishes the basis for the test and manufacture of all instrument transformers used in this country. It defines the parameters for insulation class and accuracy class. The burdens listed in Table 2.6.3 are defined in IEEE C57.13. Often, standards for other electrical apparatus that may use instrument transformers have adopted their own criteria based on IEEE C57.13. These standards, along with utility practices and the National Electric Code, are used in conjunction with each other to ensure maximum safety and system reliability. The industrial market may also coordinate with Underwriters Laboratories. As the marketplace becomes global, there is a drive for standard harmonization with the International Electrotechnical Commission (IEC), but we are not quite there yet. It is important to know the international standards in use, and these are listed in Table 2.6.4. Most major countries originally developed their own standards. Today, many are beginning to adopt IEC standards to supersede their own.

2.6.2.5 Accuracy Classes

Instrument transformers are rated by performance in conjunction with a secondary burden. As the burden increases, the accuracy class may, in fact, decrease. For revenue-metering use, the coordinates of ratio error and phase error must lie within a prescribed parallelogram, as seen in Figure 2.6.7 and Figure 2.6.8 for VTs and CTs, respectively. This parallelogram is based on a 0.6 system power factor (PF). The ratio error (RE) is converted into a ratio correction factor (RCF), which is simply

$$RCF = 1 - (RE/100) \tag{2.6.5}$$

The total-error component is the transformer correction factor (TCF), which is the combined ratio and phase-angle error. The limits of phase-angle error are determined from the following relationship:

$$TCF = RCF \pm \left[\frac{PA \ \tan\theta}{3438} \right] \tag{2.6.6}$$

where

      TCF = transformer correction factor
      RCF = ratio correction factor
        PA = phase-angle error, min
         θ = supply-system PF angle
       + = for VTs only (see Figure 2.6.7)
       − = for CTs only (see Figure 2.6.8)
   3438 = minutes of angle in 1 rad

Therefore, using 0.6 system power factor ($\theta = 53°$) and substituting in Equation 2.6.6, the relationship for VTs is

$$TCF = RCF + (PA/2600) \tag{2.6.7}$$

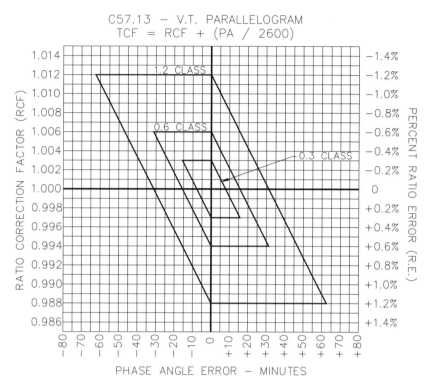

FIGURE 2.6.7 Accuracy coordinates for VTs.

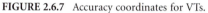

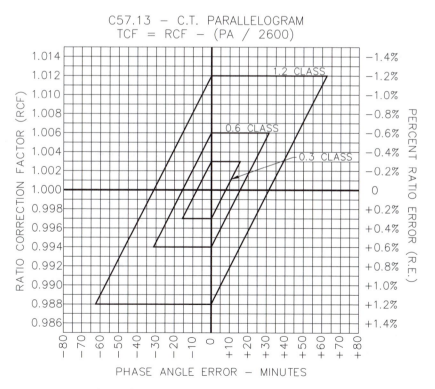

FIGURE 2.6.8 Accuracy coordinates for CTs.

and for CTs is

$$\text{TCF} = \text{RCF} - (\text{PA}/2600) \qquad (2.6.8)$$

The TCF is mainly applied when the instrument transformer is being used to measure energy usage. From Table 2.6.5, the limits of TCF are also the same as RCF. A negative RE will yield an RCF > 1, while a positive RE will yield an RCF < 1. The "adopted" class in Table 2.6.5 is extrapolated from these relationships and is recognized in industry.

The accuracy-class limits of the CT apply to the errors at 100% of rated current up through the rating factor of the CT. At 10% of rated current, the error limits permitted are twice that of the 100% class. There is no defined requirement for the current range between 10% and 100%, nor is there any requirement below 10%. There are certain instances in which the user is concerned about the errors at 5% and will rely on the manufacturer's guidance. Because of the nonlinearity in the core and the ankle region, the errors at low flux densities are exponential. As the current and flux density increase, the errors become linear up until the core is driven into saturation, at which point the errors increase at a tremendous rate (see Figure 2.6.9).

Trends today are driving accuracy classes to 0.15%. Although not yet recognized by IEEE C57.13, manufacturers and utilities are establishing acceptable guidelines that may soon become part of the standard. With much cogeneration, the need to meter at extremely low currents with the same CT used

TABLE 2.6.5 Accuracy Classes

	Accuracy Class	RCF Range	Phase Range, min	TCF Range
New	0.15	1.0015–0.9985	± 7.8	1.0015–0.9985
IEEE C57.13	0.3	1.003–0.997	± 15.6	1.003–0.997
IEEE C57.13	0.6	1.006–0.994	± 31.2	1.006–0.994
IEEE C57.13	1.2	1.012–0.988	± 62.4	1.012–0.988
IEEE C57.13	2.4	1.024–0.976	± 124.8	1.024–0.976
Adopted	4.8	1.048–0.952	± 249.6	1.048–0.952

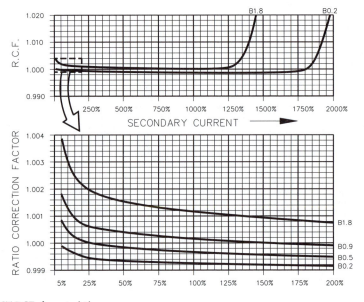

FIGURE 2.6.9 CT RCF characteristic curve.

for regular loads has forced extended-range performance to be constant from rating factor down to 1% of rated current. This is quite a deviation from the traditional class.

In the case of the VT, the accuracy-class range is between 90% and 110% of rated voltage for each designated burden. Unlike the CT, the accuracy class is maintained throughout the entire range. The manufacturer will provide test data at 100% rated voltage, but it can furnish test data at other levels if required by the end user. The response is somewhat linear over a long range below 90%. Since the normal operating flux densities are much higher than in the CT, saturation will occur much sooner at voltages above 110%, depending on the overvoltage rating.

Protection, or relay class, is based on the instrument transformer's performance at some defined fault level. In VTs it may also be associated with an under- and overvoltage condition. In this case, the VT may have errors as high as 5% at levels as low as 5% of rated voltage and at the VT overvoltage rating. In CTs, the accuracy is based on a terminal voltage developed at 20 times nominal rated current. The limits of RCF are 0.90 to 1.10, or 10% RE from nominal through 20 times nominal. This applies to rated burden or any burden less than rated burden.

2.6.2.6 Insulation Systems

The insulation system is one of the most important features of the instrument transformer, establishing its construction, the insulation medium, and the unit's overall physical size. The insulation system is determined by three major criteria: dielectric requirements, thermal requirements, and environmental requirements.

Dielectric requirements are based on the source voltage to which the instrument transformer will be connected. This source will define voltage-withstand levels and basic impulse-insulation levels (BIL). In some cases, the instrument transformer may have to satisfy higher levels, depending on the equipment with which they are used. Equipment such as power switchgear and isolated-phase bus, for instance, use instrument transformers within their assembly, but they have test requirements that differ from the instrument-transformer standard. It is not uncommon to require a higher BIL class for use in a highly polluted environment. See Table 2.6.6A and Table 2.6.6B.

TABLE 2.6.6A Low- and Medium-Voltage Dielectric Requirements

Class, kV	Instrument Transformers (IEEE C57.13)		Other Equipment Standards [a]	
	BIL, kV	Withstand Voltage, kV	BIL, kV	Withstand Voltage, kV
0.6	10	4	—	2.2
1.2	30	10	—	—
2.4	45	15	—	—
5.0	60	19	60	19
8.7	75	26	75/95	26/36
15.0	95/110	34	95/110	36/50
25.0	125/150	40/50	125/150	60
34.5	200	70	150	80
46	250	95	—	—
69	350	140	350	160

[a] IEEE C37.06, C37.20.1, C37.20.2, C37.20.3, C37.23.

TABLE 2.6.6B High-Voltage Dielectric Requirements

Class, kV	Instrument Transformers (IEEE C57.13)		Other Equipment Standards (IEEE C37.06)	
	BIL, kV	Withstand Voltage, kV	BIL, kV	Withstand Voltage, kV
115	450/550	185/230	550	215/260
138	650	275	650	310
161	750	325	750	365
230	900/1050	395/460	900	425
345	1300	575	1300	555
500	1675/1800	750/800	1800	860
765	2050	920	2050	960

TABLE 2.6.7 Materials/Construction for Low- and Medium-Voltage Classes

Class, kV	Indoor Applications Materials/Construction	Outdoor Applications Materials/Construction
0.6	Tape, varnished, plastic, cast, or potted	Cast or potted
1.2 - 5.0	Plastic, cast	Cast
8.7	Cast	Cast
15.0	Cast	Cast or tank/oil/porcelain
25.0	Cast	Cast or tank/oil/porcelain
34.5	Cast	Cast or tank/oil/porcelain
46	Not commonly offered	Cast or tank/oil/porcelain
69	Not commonly offered	Cast or tank/oil/porcelain

Note: the term *cast* can imply any polymeric material, e.g., butyl rubber, epoxy, urethane, etc. *Potted* implies that the unit is embedded in a metallic housing with a casting material.

Environmental requirements will help define the insulation medium. In indoor applications, the instrument transformer is protected from external weather elements. In outdoor installations, the transformer must endure all weather conditions from extremely low temperatures to severe UV radiation and be impervious to moisture penetration. The outer protection can range from fabric or polyester tape, varnish treatment, or thermoplastic housings to molding compounds, porcelain, or metal enclosures. Table 2.6.7 identifies, by voltage rating, the commonly used materials and construction types.

All installations above 69 kV are typically for outdoor service and are of the tank/oil/SF$_6$/porcelain construction type.

2.6.2.7 Thermal Ratings

An important part of the insulation system is the temperature class. For instrument transformers, only three classes are generally defined in the standard, and these are listed in Table 2.6.8A. This rating is coordinated with the maximum continuous current flow allowable in the instrument transformer that will limit the winding heat rise accordingly. Of course, other classes can be used to fit the application, especially if the instrument transformer is part of an apparatus that has a higher temperature class, e.g., when used under hot transformer oil or within switchgear, bus compartments, and underground network devices, where ambient temperatures can be 65 to 105°C. In these cases, a modest temperature rise can change the insulation-system rating. These apply to the instrument transformer under the most extreme continuous conditions for which it is rated. The insulation system used must be coordinated within its designated temperature class (Table 2.6.8B). It is not uncommon for users to specify a higher insulation system even though the unit will never operate at that level. This may offer a more robust unit at a higher price than normally required, but can also provide peace of mind.

TABLE 2.6.8A Temperature Class (IEEE C57.13)

Temperature Class	30°C Ambient		55°C Ambient
	Temperature Rise	Hot-Spot Temperature Rise	Temperature Rise
105°C	55°C	65°C	30°C
120°C	65°C	80°C	40°C
150°C	80°C	110°C	55°C

TABLE 2.6.8B Temperature Class (General)

Temperature Class	Hot-Spot Temperature Rise @ 30°C Ambient (40°C Maximum)
Class 90 (O)	50°C
Class 105 (A)	65°C
Class 130 (B)	90°C
Class 155 (F)	115°C
Class 180 (H)	140°C
Class 220 (C)	180°C

2.6.2.8 Primary Winding

The primary winding is subjected to the same dynamic and thermal stresses as the rest of the primary system when large short-circuit currents and voltage transients are present. It must be sized to safely carry the maximum continuous current without exceeding the insulation system's temperature class.

2.6.3 Voltage Transformer

The voltage transformer (VT) is connected in parallel with the circuit to be monitored. It operates under the same principles as power transformers, the significant differences being power capability, size, operating flux levels, and compensation. VTs are not typically used to supply raw power; however, they do have limited power ratings. They can often be used to supply temporary 120-V service for light-duty maintenance purposes where supply voltage normally would not otherwise be available. In switchgear compartments, they may be used to drive motors that open and close circuit breakers. In voltage regulators, they may power a tap-changing drive motor. The power ranges are from 500 VA and less for low-voltage VT, 1–3 kVA for medium-voltage VT, and 3–5 kVA for high-voltage VT. Since they have such low power ratings, their physical size is much smaller. The performance characteristics of the VT are based on standard burdens and power factors, which are not always the same as the actual connected burden. It is possible to predict, graphically, the anticipated performance when given at least two reference points. Manufacturers typically provide this data with each VT produced. From that, one can construct what is often referred to as the VT circle diagram, or fan curve, shown in Figure 2.6.10. Knowing the ratio-error and phase-error coordinates, and the values of standard burdens, the graph can be produced to scale in terms of VA and power factor. Other power-factor lines can be inserted to pinpoint actual circuit conditions. Performance can also be calculated using the same phasor concept by the following relationships, provided that the value of the unknown burden is less than the known burden. Two coordinates must be known: at zero and at one other standard burden value.

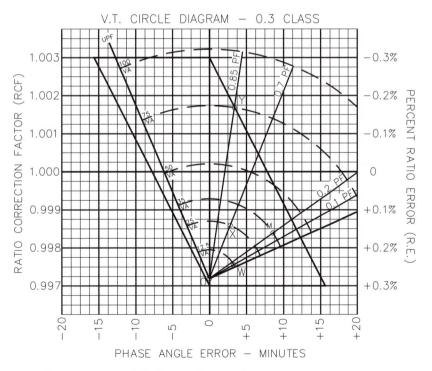

FIGURE 2.6.10 Voltage transformer circle diagram (fan curves).

$$\text{RCF}_x = \left[\frac{B_x}{B_t}\right] \left[(\text{RCF}_t - \text{RCF}_0)\cos(\theta_t - \theta_x) + (\gamma_t - \gamma_0)\sin(\theta_t - \theta_x)\right] \qquad (2.6.9)$$

$$\gamma_x = \left[\frac{B_x}{B_t}\right] \left[(\gamma_t - \gamma_0)\cos(\theta_t - \theta_x) - (\text{RCF}_t - \text{RCF}_0)\sin(\theta_t - \theta_x)\right] \qquad (2.6.10)$$

where

RCF_x = RCF of new burden
RCF_t = RCF of known burden
RCF_0 = RCF at zero burden
γ_x = phase error of new burden, radians (to obtain γ_x in minutes, multiply value from Equation 2.6.7 by 3438)
γ_t = phase error of known burden, radians
γ_0 = phase error at zero burden, radians
B_x = new burden
B_t = known burden
θ_x = new burden PF angle, radians
θ_t = known burden PF angle, radians

2.6.3.1 Overvoltage Ratings

The operating flux density is much lower than in a power transformer. This is to help minimize the losses and to prevent the VT from possible overheating during overvoltage conditions. VTs are normally designed to withstand 110% rated voltage continuously unless otherwise designated. IEEE C57.13 divides VTs into groups based on voltage and application. Group 1 includes those intended for line-to-line or line-to-ground connection and are rated 125%. Group 3 is for units with line-to-ground connection only and with two secondary windings. They are designed to withstand 173% of rated voltage for 1 min, except for those rated 230 kV and above, which must withstand 140% for the same duration. Group 4 is for line-to-ground connections with 125% in emergency conditions. Group 5 is for line-to-ground connections with 140% rating for 1 min. Other standards have more stringent requirements, such as the Canadian standard, which defines its Group 3 VTs for line-to-ground connection on ungrounded systems to withstand 190% for 30 sec to 8 h, depending on ground-fault protection. This also falls in line with the IEC standard.

2.6.3.2 VT Compensation

The high-voltage windings are always compensated to provide the widest range of performance within an accuracy class. Since there is compensation, the actual turns ratio will vary from the rated-voltage ratio. For example, say a 7200:120-V, 60:1 ratio is required to meet 0.3 class. The designer may desire to adjust the primary turns by 0.3% by removing them from the nominal turns, thus reducing the actual turns ratio to, say 59.82:1. This will position the no-load (zero burden) point to the bottommost part of the parallelogram, as shown in Figure 2.6.10. Adjustment of turns has little to no effect on the phase-angle error.

2.6.3.3 Short-Circuit Operation

Under no normal circumstance is the VT secondary to be short-circuited. The VT must be able to withstand mechanical and thermal stresses for 1 sec with full voltage applied to the primary terminals without suffering damage. In most situations, this condition would cause some protective device to operate and remove the applied voltage, hopefully in less than 1 sec. If prolonged, the temperature rise would far exceed the insulation limits, and the axial and radial forces on the windings would cause severe damage to the VT.

2.6.3.4   VT Connections

VTs are provided in two arrangements: dual or two-bushing type and single-bushing type. Two-bushing types are designed for line-to-line connection, but in most cases can be connected line-to-ground with reduced output voltage. Single-bushing types are strictly for line-to-ground connection. The VT should never be connected to a system that is higher than its rated terminal voltage. As for the connection between phases, polarity must always be observed. Low- and medium-voltage VTs may be configured in delta or wye. As the system voltages exceed 69 kV, only single-bushing types are available. Precautions must be taken when connecting VT primaries in wye on an ungrounded system. (This is discussed further in Section 2.6.3.5, Ferroresonance.) Primary fusing is always recommended. Indoor switchgear types are often available with fuse holders mounted directly on the VT body.

2.6.3.5   Ferroresonance

VTs with wye-connected primaries on three-wire systems that are ungrounded can resonate with the distributed line-to-ground capacitance (see Figure 2.6.11). Under balanced conditions, line-to-ground voltages are normal. Momentary ground faults or switching surges can upset the balance and raise the line-to-ground voltage above normal. This condition can initiate a resonant oscillation between the primary windings and the system capacitance to ground, since they are effectively in parallel with each other. Higher current flows in the primary windings due to fluctuating saturation, which can cause overheating. The current levels may not be high enough to blow the primary fuses, since they are generally sized for short-circuit protection and not thermal protection of the VT. Not every disturbance will cause ferroresonance. This phenomenon depends on several factors:

- Initial state of magnetic flux in the cores
- Saturation characteristics (magnetizing impedance) of the VT
- Air-core inductance of the primary winding
- System circuit capacitance

One technique often used to protect the VT is to increase its loading resistance by (1) connecting a resistive load to each of the secondaries individually or (2) connecting the secondaries in a delta-configuration and inserting a load resistance in one corner of the delta. This resistance can be empirically approximated by Equation 2.6.11,

$$R_{delta} = (100 \times L_A)/N^2 \tag{2.6.11}$$

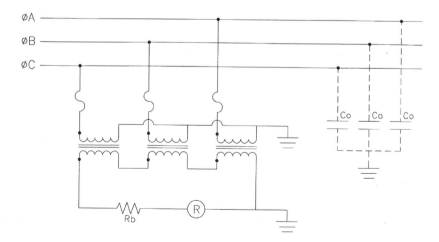

FIGURE 2.6.11   VTs wye-connected on ungrounded system.

where

R_{delta} = loading resistance, ohms

L_A = VT primary inductance during saturation, mH

N = VT turns ratio, N_P/N_S

This is not a fix-all solution, as ferroresonance may still occur, but this may reduce the chances of it happening. The loading will have an effect on VT errors and may cause it to exceed 0.3%, but that is not critical for this scheme, since it is seldom used for metering.

2.6.3.6 VT Construction

The electromagnetic wound-type VT is similar in construction to that of the power transformer. The magnetic circuit is a core-type or shell-type arrangement, with the windings concentrically wound on one leg of the core. A barrier is placed between the primary and secondary winding(s) to provide adequate insulation for its voltage class. In low-voltage applications it is usually a two-winding arrangement, but in medium- and high-voltage transformers, a third (tertiary) winding is often added, isolated from the other windings. This provides more flexibility for using the same VT in metering and protective purposes simultaneously. As mentioned previously, the VT is available in single- or dual-bushing arrangements (Figure 2.6.12 a,b,c). A single bushing has one lead accessible for connection to the high-voltage conductor, while the other side of the winding is grounded. The grounded terminal (H2) may be accessible somewhere on the VT body near the base plate. There is usually a grounding strap connected from it to the base, and it can be removed to conduct field power-factor tests. In service, the strap must always be connected to ground. Some medium-voltage transformers are solidly grounded and have no H2 terminal access. The dual-bushing arrangement has two live terminal connections, and both are fully rated for the voltage to which it is to be connected.

2.6.3.7 Capacitive-Coupled Voltage Transformer

The capacitive-coupled voltage transformer (CCVT) is primarily a capacitance voltage divider and electromagnetic VT combined. Developed in the early 1920s, it was used to couple telephone carrier current with the high-voltage transmission lines. The next decade brought a capacitive tap on many high-voltage bushings, extending its use for indication and relaying. To provide sufficient energy, the divider output had to be relatively high, typically 11 kV. This necessitated the need for an electromagnetic VT to step the voltage down to 120 V. A tuning reactor was used to increase energy transfer (see Figure 2.6.13). As transmission voltage levels increased, so did the use of CCVTs. Its traditional low cost versus the conventional VT, and the fact that it was nearly impervious to ferroresonance due to its low flux density, made it an ideal choice. It proved to be quite stable for protective purposes, but it was not adequate for revenue metering. In fact, the accuracy has been known to drift over time and temperature ranges. This would often warrant the need for routine maintenance. CCVTs are commonly used in 345- to 500-kV systems. Improvements have been made to better stabilize the output, but their popularity has declined.

Another consideration with CCVTs is their transient response. When a fault reduces the line voltage, the secondary output does not respond instantaneously due to the energy-storing elements. The higher the capacitance, the lower is the magnitude of the transient response. Another element is the ferroresonance-suppression circuit, usually on the secondary side of the VT. There are two types, active and passive. Active circuits, which also contain energy-storing components, add to the transient. Passive circuits have little effect on transients. The concern of the transient response is with distance relaying and high-speed line protection. This transient may cause out-of-zone tripping, which is not tolerable.

2.6.3.8 Optical Voltage Transducer

A new technology, optical voltage transducers, is being used in high-voltage applications. It works on the principle known as the Kerr effect, by which polarized light passes through the electric field produced by the line voltage. This polarized light, measured optically, is converted to an analog electrical signal proportional to the voltage in the primary conductor. This device provides complete isolation, since there is no electrical connection to the primary conductor. With regard to its construction, since there is no

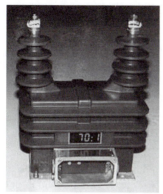

FIGURE 2.6.12 (a) 15-kV dual-bushing outdoor VT, (b) 69-kV single-bushing VT, (c) 242-kV single-bushing VT. (Photos courtesy of Kuhlman Electric Corp.)

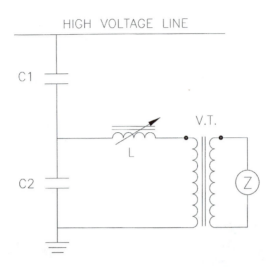

FIGURE 2.6.13 CCVT simplified circuit.

magnetic core and windings, its physical size and weight is significantly smaller than the conventional wound-type high-voltage VT. And with the absence of a core, there are no saturation limits or overvoltage concerns. The full line-to-ground voltage is applied across the sensor. It is still required to satisfy the system BIL rating. It also must have a constant and reliable light source and a means of detecting the absence of this light source. The connection to and from the device to the control panel is via fiber-optic cables. These devices are available for use in the field. High initial cost and the uncertainty of its performance will limit its use.

2.6.4 Current Transformer

The current transformer (CT) is often treated as a "black box." It is a transformer that is governed by the laws of electromagnetic induction:

$$\varepsilon = k\, \beta\, A_C\, N\, f \tag{2.6.12}$$

where

ε = induced voltage
β = flux density
A_C = core cross-sectional area
N = turns
f = frequency
k = constant of proportionality

As previously stated, the CT is connected in series with the circuit to be monitored, and it is this difference that leads to its ambiguous description. The primary winding is to offer a constant-current source of supply through a low-impedance loop. Because of this low impedance, current passes through it with very little regulation. The CT operates on the ampere-turn principle (Faraday's law): primary ampere-turns = secondary ampere-turns, or

$$I_P\, N_P = I_S\, N_S \tag{2.6.13}$$

Since there is energy loss during transformation, this loss can be expressed in ampere-turns: primary ampere-turns − magnetizing ampere-turns = secondary ampere-turns, or

$$I_P\, N_P - I_{ex}\, N_P = I_S\, N_S \tag{2.6.14}$$

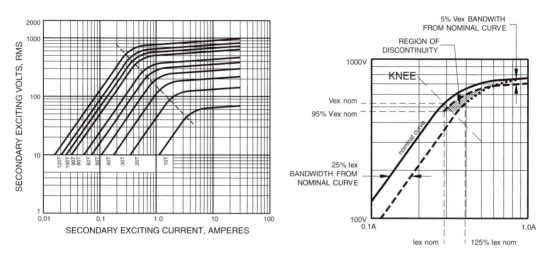

FIGURE 2.6.14 Left) Saturation curve for a multiratio CT; right) saturation curve discontinuity of tolerances.

The CT is not voltage dependant, but it is voltage limited. As current passes through an impedance, a voltage is developed (Ohm's law, $V = I \times Z$). As this occurs, energy is depleted from the primary supply, thus acting like a shunt. This depletion of energy results in the CT errors. As the secondary impedance increases, the voltage proportionally increases. Thus the limit of the CT is magnetic saturation, a condition when the core flux can no longer support the increased voltage demand. At this point, nearly all of the available energy is going into the core, leaving none to support the secondary circuit.

2.6.4.1 Saturation Curve

The saturation curve, often called the secondary-excitation curve, is a plot of secondary-exciting voltage versus secondary-exciting current drawn on log-log paper. The units are in rms with the understanding that the applied voltage is sinusoidal. This characteristic defines the core properties after the stress-relief annealing process. It can be demonstrated by test that cores processed in the same manner will always follow this characteristic within the specified tolerances. Figure 2.6.14 shows a typical characteristic of a 600:5 multiratio CT. The knee point is indicated by the dashed line. Since the voltage is proportional to the turns, the volts-per-turn at the knee is constant. The tolerances are 95% of saturation voltage for any exciting current above the knee point and 125% of exciting current for any voltage below the knee point. These tolerances, however, can create a discontinuity about the knee of the curve, which is illustrated in Figure 2.6.14. Since the tolerance is referenced at the knee point, it is possible to have a characteristic that is shifted to the right of the nominal, within tolerance below the knee point. But careful inspection shows that a portion of the characteristic will exceed the tolerance above the knee point. For this reason, manufacturers' typical curves may be somewhat conservative to avoid this situation in regards to field testing. Some manufacturers will provide actual test data that may provide the relay engineer with more useful information. Knowing the secondary-winding resistance and the excitation characteristic, the user can calculate the expected RCF under various conditions. Using this type of curve is only valid for nonmetering applications. The required voltage needed from the CT must be calculated using the total circuit impedance and the anticipated secondary-current level. The corresponding exciting current is read from the curve and used to approximate the anticipated errors.

$$V_{ex} = I_{Sf} \, Z_t = I_{Sf} \, \sqrt{(R_s + R_B)^2 + X_B} \tag{2.6.15}$$

$$RCF = (I_{Sf} + I_{ex})/I_{Sf} \tag{2.6.16}$$

$$\%RE = I_{ex}/I_{Sf} \times 100 \tag{2.6.17}$$

where

V_{ex} = secondary-excitation voltage required at fault level
I_{sf} = secondary fault current (primary fault current/turns ratio)
I_{ex} = secondary-exciting current at V_{ex}, obtained from curve
Z_T = total circuit impedance, in ohms
R_s = secondary winding resistance, ohms
R_B = secondary burden resistance, ohms
X_B = secondary burden reactance, ohms
RCF = ratio correction factor
RE = ratio error

In the world of protection, the best situation is to avoid saturation entirely. This can be achieved by sizing the CT knee-point voltage to be greater than V_{ex}, but this may not be the most practical approach. This could force the CT physical size to substantially increase as well as cause dielectric issues. There must be some reasonable trade-offs to reach a desirable condition. Equation 2.6.15 provides the voltage necessary to avoid ac saturation. If there is an offset that will introduce a dc component, then the system X/R ratio must be factored in:

$$V_{ex} = I_{sf} Z_T [1 + (X/R)] \tag{2.6.18}$$

And if the secondary burden is inductive, Equation 2.6.18 is rewritten as

$$V_{ex} = I_{sf} Z_T \{1 + [X/R (R_S + R_B)/Z_T]\} \tag{2.6.19}$$

The saturation factor, K_S, is the ratio of the knee-point voltage to the required secondary voltage V_{ex}. It is an index of how close to saturation a CT will be in a given application. K_S is used to calculate the time a CT will saturate under certain conditions:

$$T_S = -\frac{X}{\omega R} \ln \left[1 - \frac{K_S - \frac{1}{X}}{R} \right] \tag{2.6.20}$$

where

T_S = time to saturate
$\omega = 2\pi f$, where f = system frequency
K_S = saturation factor (V_k/V_{ex})
R = primary system resistance at point of fault
X = primary system reactance at point of fault
ln = natural log function

2.6.4.2   CT Rating Factor

The continuous-current rating factor is given at a reference ambient temperature, usually 30°C. The standard convention is that the average temperature rise will not exceed 55°C for general-purpose use, but it can be any rise shown in Table 2.6.8. From this rating factor, a given CT can be derated for use in higher ambient temperatures from the following relationship:

$$\frac{RF_{NEW}^2}{RF_{STD}^2} = \frac{85°C - AMB_{NEW}}{85°C - 30°C} \tag{2.6.21}$$

which can be simplified and rewritten as

$$RF_{NEW} = RF_{STD} \sqrt{\frac{85°C - AMB_{NEW}}{55°C}} \tag{2.6.22}$$

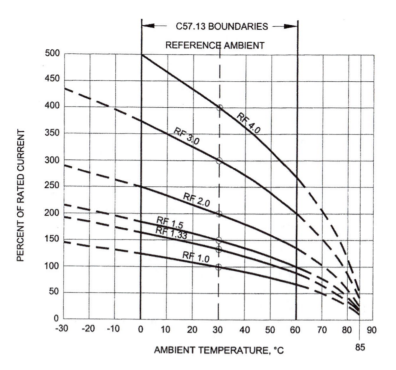

FIGURE 2.6.15 CT derating chart.

where

    RF_{NEW} = desired rating factor at some other ambient temperature

    RF_{STD} = reference rating factor at 30°C

    Amb_{NEW} = desired ambient temperature <85°C

Equation 2.6.22 is valid only for 55°C rise ratings and maximum ambient less than 85°C.

The rating factor ensures that the CT will not exceed its insulation-class rating. This expression follows the loading curves of Figure 1 in IEEE C57.13 (see Figure 2.6.15). It assumes that the average winding-temperature rise is proportional to the current squared and that core watt losses are insignificant under continuous operating conditions.

Conversely, this expression will work for ambient temperatures lower than 30°C, but in this situation the rating factor will be higher. The concern is not with exceeding the insulation system but, rather, with the chance of increasing the errors of the CT accuracy limits. A number of factors can affect this, such as core material and compensation methods used. It is best to consult the manufacturer for overrating performances.

Another consideration is the stated rating factor versus the actual temperature rise. In most cases, the stated rating factor may be quite less than the permitted 55°C rise allowance. For example, the actual rating factor may be 2.67, but the stated rating factor is 2.0. If the actual rise data were known, then Equation 22 could be rewritten as

$$RF_{NEW} = RF_{ACTUAL} \sqrt{\frac{85°C - AMB_{NEW}}{\Delta T_{RISE}}} \tag{2.6.23}$$

where

    RF_{ACTUAL} = rating factor at actual temperature rise

    ΔT_{RISE} = actual measured temperature rise

This may be more useful when operating a CT at higher ambient, where there is a need for the maximum rating factor. For example, if the CT has an actual rating of 2.67 based on temperature-rise data but is only rated 2.0, and if it is desired to use the CT at 50°C ambient with the stated rating factor 2.0, then this unit should work within its insulation rating and within its stated accuracy class.

If a higher-temperature-class insulation system is provided, then the rise must be in compliance with that class per Table 2.6.8. In some cases, the temperature class is selected for the environment rather than the actual temperature rise of the CT.

2.6.4.3 Open-Circuit Conditions

The CT functions best with the minimum burden possible, which would be its own internal impedance. This can only be accomplished by applying a short circuit across the secondary terminals. Since the core mmf acts like a shunt, with no load connected to its secondary, the mmf becomes the primary current, thus driving the CT into hard saturation. With no load on the secondary to control the voltage, the winding develops an extremely high peak voltage. This voltage can be in the thousands, or even tens of thousands, of volts. This situation puts the winding under incredible stress, ultimately leading to failure. This could result in damage to other equipment or present a hazard to personnel. It is for this reason that the secondary circuit should never be open. It must always have a load connected. If it is installed to the primary but not in use, then the terminals should be shorted until it is to be used. Most manufacturers ship CTs with a shorting strap or wire across the secondary terminals. The CT winding must be able to withstand 3500 V_{peak} for 1 min under open-circuit conditions. If the voltage can exceed this level, then it is recommended that overvoltage protection be used.

2.6.4.4 Overvoltage Protection

Under load, the CT voltage is limited. The level of this voltage depends on the turns and core cross-sectional area. The user must evaluate the limits of the burdens connected to ensure equipment safety. Sometimes protective devices are used on the secondary side to maintain safe levels of voltage. These devices are also incorporated to protect the CT during an open-circuit condition. In metering applications, it is possible for such a device to introduce a direct current (dc) across the winding that could saturate the core or leave it in some state of residual flux. In high-voltage equipment, arrestors may be used to protect the primary winding from high voltage-spikes produced by switching transients or lightning.

2.6.4.5 Residual Magnetism

Residual magnetism, residual flux, or remanence refers to the amount of stored, or trapped, flux in the core. This can be introduced during heavy saturation or with the presence of some dc component. Figure 2.6.16 shows a typical B-H curve for silicon-iron driven into hard saturation. The point at which the curve crosses zero force, identified by $+B_{res}$, represents the residual flux. If at some point the CT is disconnected from the source, this flux will remain in the magnetic core until another source becomes present. If a fault current drove the CT into saturation, when the supply current resumed normal levels, the core would contain some residual component. Residual flux does not gradually decay but remains constant once steady-state equilibrium has been reached. Under normal conditions, the minor B-H loop must be high enough to remove the residual component. If it is not, then it will remain present until another fault occurrence takes place. The effective result could be a reduction of the saturation flux. However, if a transient of opposite polarity occurs, saturation is reduced with the assistance of the residual. Conversely, the magnitude of residual is also based on the polarity of the transient and the phase relationship of the flux and current. Whatever the outcome, the result could cause a delayed response to the connected relay.

It has been observed that in a tape-wound toroidal core, as much as 85% of saturation flux could be left in the core as residual component. The best way to remove residual flux is to demagnetize the core. This is not always practical. The user could select a CT with a relay class that is twice that required. This may not eliminate residual flux, but it will certainly reduce the magnitude. The use of hot-rolled steel may inherently reduce the residual component to 40 to 50% of saturation flux. Another way of reducing residual magnetism is to use an air-gapped core. Normally, the introduction of a gap that is, say, 0.01% of core circumference could limit the residual flux to about 10% of the saturation flux. Referring to Figure 2.6.16, a typical B-H loop for an air-gapped core is shown. The drawback is significantly higher exciting current and lower saturation levels, as can be seen in Figure 2.6.3. To overcome the high exciting current, the core would be made larger. That — coupled with the gap — would increase its overall cost. This type of core construction is often referred to as a linearized core.

2.6.4.6 CT Connections

As previously mentioned, some devices are sensitive to the direction of current flow. It is often critical in three-phase schemes to maintain proper phase shifting. Residually connected CTs in three-phase ground-fault scheme (Figure 2.6.17A) sum to zero when the phases are balanced. Reversed polarity of a CT could cause a ground-fault relay to trip under a normal balanced condition. Another scheme to detect zero-sequence faults uses one CT to simultaneously monitor all leads and neutral (Figure 2.6.17C). In differential protection schemes (Figure 2.6.17B), current-source phase and magnitude are compared. Reverse polarity of a CT could effectively double the phase current flowing into the relay, thus causing a nuisance tripping of a relay. When two CTs are driving a three-phase ammeter through a switch, a reversed CT could show 1.73 times the monitored current flowing in the unmonitored circuit.

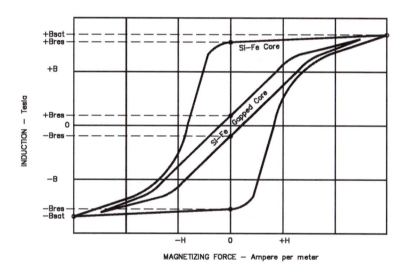

FIGURE 2.6.16 Typical B-H curve for Si-Fe steel.

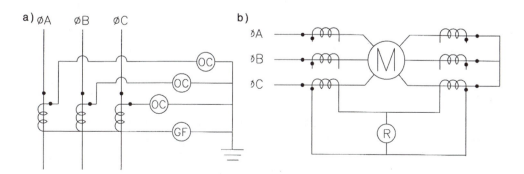

FIGURE 2.6.17 (a) Overcurrent and ground-fault protection scheme. (b) Differential protection scheme.

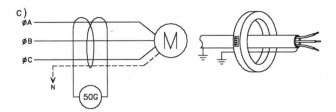

FIGURE 2.6.17 (c) Zero-sequence scheme with all three-phase leads going through the window of the CT. This connection, as well as the residual connection in Figure 2.6.17a, will cancel out the positive- and negative-sequence currents leaving only the zero-sequence current to the 50G device. Sometimes the ground or neutral lead will be included. The diagram on the right shows sheathed cable. It is important that one ground point go back through the window to avoid the possibility of a shorted electrical turn via the ground path.

TABLE 2.6.9 Total Burden on CT in Fault Conditions

CT Secondary Connections	Three-Phase or Ph-to-Ph Fault	Phase-to-GND Fault
Wye-connected at CT	$R_S + R_L + Z_R$	$R_S + 2R_L + Z_R$
Wye-connected at switchboard	$R_S + 2R_L + Z_R$	$R_S + 2R_L + Z_R$
Delta-connected at CT	$R_S + 2R_L + 3Z_R$	$R_S + 2R_L + 2Z_R$
Delta-connected at switchboard	$R_S + 3R_L + 3Z_R$	$R_S + 2R_L + 2Z_R$

Note: R_S = CT secondary-winding resistance + CT lead resistance, ohms; R_L = one-way circuit lead resistance, ohms; Z_R = relay impedance in secondary current path, ohms

The wye-connected secondary circuit is the most commonly used. The CT will reproduce positive-, negative-, and zero-sequence elements as they occur in the primary circuit. In the delta-connected secondary, the zero-sequence components are filtered and left to circulate in the delta. This is a common scheme for differential protection of delta-wye transformer. A general rule of thumb is to connect the CT secondaries in wye when they are on the delta winding of a transformer and, conversely, connect the secondaries in delta when they are on the wye winding of the transformer. See Table 2.6.9.

2.6.4.7 Construction

There are four major types of CT: window-type CT, which includes bushing-type (BCT); bar-type CT; split-core-type CT; and wound-type CT, the latter having both a primary and secondary winding. There is ongoing development and limited use of optical-type current transformers (OCT), which rely on the principles of light deflection.

2.6.4.7.1 Window-Type CT

The window-type CT is the simplest form of instrument transformer. It is considered to be an incomplete transformer assembly, since it consists only of a secondary winding wound on its core. The most common type is that wound on a toroidal core. The secondary winding is fully distributed around the periphery of the core. In special cases when taps are employed, they are distributed such that any connection made would utilize the entire core periphery. Windings in this manner ensure optimum flux linkage and distribution. Coupling is almost impervious to primary-conductor position, provided that the return path is sufficiently distanced from the outer periphery of the secondary winding (see Figure 2.6.5a). The effects of stray flux are negligible, thus making this type of winding a low-reactance design. The primary winding in most cases is a single conductor centrally located in the window. A common application is to position the CT over a high-voltage bushing, hence the name BCT. Nearly all window-type CTs manufactured today are rated 600-V class. In practice, they are intended to be used over insulated conductors when the conductor voltage exceeds 600 V. It is common practice to utilize a 600-V-class window type in conjunction with air space between the window and the conductor on higher-voltage systems. Such use may be seen in isolated-phase bus compartments. There are some window types that can be rated for higher voltages as stand-alone units, for use with an integral high-voltage stress shield,

FIGURE 2.6.18 Bar-type CT. (Photo courtesy of Kuhlman Electric Corp.)

FIGURE 2.6.19 Split-core type CT with secondary winding on three legs of core. (Photo courtesy of Kuhlman Electric Corp.)

or for use with an insulating sleeve or tube made of porcelain or some polymeric material. Window-type CTs generally have a round window opening but are also available with rectangular openings. This is sometimes provided to fit a specific bus arrangement found in the rear of switchgear panels or on draw-out-type circuit breakers. This type of CT is used for general-purpose monitoring, revenue metering and billing, and protective relaying.

2.6.4.7.2 Bar-Type CT

The bar-type CT is, for all practical purposes, a window-type CT with a primary bar inserted straight through the window (Figure 2.6.18). This bar assembly can be permanently attached or held in place with brackets. Either way, the primary conductor is a single turn through the window, fully insulated from the secondary winding. The bar must be sized to handle the continuous current to be passed through it, and it must be mechanically secured to handle high-level short-circuit currents without incurring damage. Uses are the same as the window-type.

2.6.4.7.3 Split-Core Type

The split-core type CT is a special case of window-type CT. Its winding and core construction is such that it can hinge open, or totally separate into two parts (Figure 2.6.19). This arrangement is ideal to use in cases where the primary conductor cannot be opened or broken. However, because of this cut, the winding is not fully distributed (see Figure 2.6.5b). Often only 50% of the effective magnetic path is used. Use of this type should be with discretion, since this construction results in higher-than-normal errors. There is also some uncertainty in repeatability of performance from installation to installation. The reassembly of the core halves is critical. It is intended for general-purpose monitoring and temporary installations.

2.6.4.7.4   *Wound-Type CT*

The wound-type CT has a primary winding that is fully insulated from the secondary winding. Both windings are permanently assembled on the core (Figure 2.6.5c). The insulation medium used — whether polymeric, oil, or even air — in conjunction with its rated voltage class dictates the core and coil construction. There are several core types used, from a low-reactance toroid to high-reactance cut-cores and laminations. The distribution of the cut(s) and gap(s) helps control the magnetizing losses. The manner in which the windings are arranged on the core affects the reactance, since it is a geometric function. Generally, the windings do not utilize the magnetic path efficiently. The proper combination of core type and coil arrangement can greatly reduce the total reactance, thus reducing errors (Figures 2.6.21 and 2.6.22).

The auxiliary CT is a wound-type (see Figure 2.6.5d) used in secondary circuits for totalizing, summation, phase shifting, isolation, or to change ratio. They are typically 600-V class (Figure 2.6.23), since they are used in the low-voltage circuit. When applying auxiliary CTs, the user must be aware of its reflected impedance on the main-line CT.

2.6.4.8   Optical Current Transducers

A relatively new technology, optical current transducers are being used in high-voltage applications. It works on the principle known as the Faraday effect, by which a polarized light beam, passing around a current-carrying conductor, deflects in the presence of magnetic fields. The angular deflection is proportional to the length of path through the dielectric medium, the magnetic field strength, and the cosine of the angle between the direction of the light beam and the direction of the magnetic field. This angle of deflection, measured optically, is converted to an electrical signal that is proportional to the current flowing in the primary conductor. This device provides complete isolation, since there is no electrical connection to the primary conductor. With regard to its construction, since there is no magnetic core and windings, its physical size and weight is significantly smaller than the conventional wound-type high-voltage CT. And with the absence of a core, there are no saturation limits and no mechanism for failure. It is still required to satisfy the system BIL rating. It also must have a constant and reliable light source and a means to detect the absence of this light source. The connection to and from the device to the control panel is via fiber-optic cables. These devices have been used in the field over the last ten years. High initial cost and the uncertainty of its performance has limited its use. However, it may prove to be a viable option in many high-voltage applications.

2.6.4.9   Proximity Effects

Current flowing in a conductor will induce magnetic flux through the air. This flux is inversely proportional to the distance squared, $B \Leftrightarrow 1/d^2$. As current increases, the flux increases. Considering the case in a three-phase bus compartment, each bus is equally spaced from the other. If CTs are mounted on each phase, then it is possible that the flux from adjacent conductor fields link to the adjacent CTs. Often, the distance is sufficient that stray flux linkage is almost negligible. But at higher current levels it can cause problems, especially in differential protection schemes. This stray flux can cause localized saturation in segments of the core, and this saturation can cause heating in the winding and increase the CT errors. This same effect can be seen when return paths are also in close proximity to a CT. In the case of draw-out circuit breakers rated 2000 A and above, the phase distances and return paths are close together, causing problems with CTs mounted on the stabs. Sometimes magnetic shunts or special winding arrangements are incorporated to offset the effects. Another concern is with CTs mounted over large generator bushings. The distances are typically adequate for normal operation, but overcurrent situations may lead to misoperation of protective devices. Consequently, CTs used in this application rated above 10,000 A are shielded. The shield can be external, such as cast-aluminum or cast-copper housings. These housings are of large cross section. In the presence of high stray-flux

FIGURE 2.6.20   (a) Small 600-V-class window-type CT mounted over a bushing on a 15-kV recloser. (b) 15-kV-class window-type CTs with porcelain sleeve mounted on substation structure. (c) 600-V-class window-type CTs mounted over a 15-kV bus inside a metal enclosure. (d) Large 600-V-class window-type CT (slipover) mounted over a high-voltage bushing. (e) 8.7-kV-class window-type CT with rectangular opening. (Photos courtesy of Kuhlman Electric Corp.)

FIGURE 2.6.21 15-kV wound-type CT cast in epoxy resin. (Photo courtesy of Kuhlman Electric Corp.)

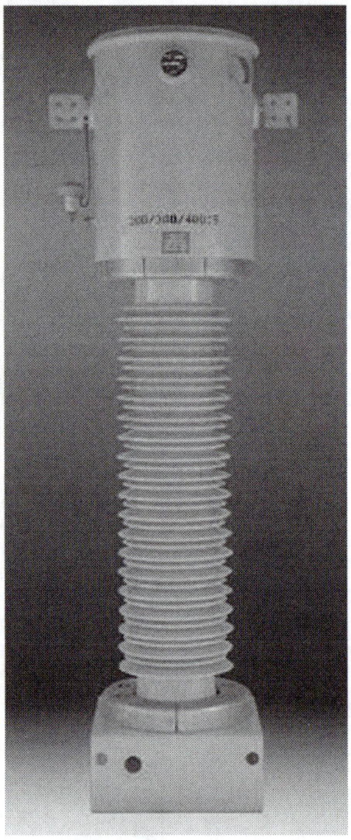

FIGURE 2.6.22 High-voltage wound-type CT in combination steel tank, oil, and porcelain construction. (Photo courtesy of Kuhlman Electric Corp.)

FIGURE 2.6.23 00-V, indoor-class auxiliary CT frame shell-type laminated "EI" core. (Photo courtesy of Kuhlman Electric Corp.)

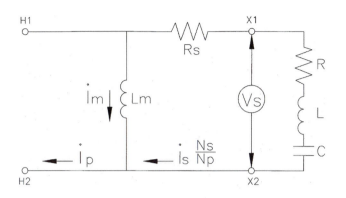

FIGURE 2.6.24 Linear-coupler equivalent circuit.

fields, large eddy currents flow in the housings, producing high temperatures. The shield can also be internal as an integral part of the secondary winding. This technique eliminates the eddy-current problems. Both techniques are effective in protecting the core from stray flux, but neither will make it immune from it.

2.6.4.10 Linear Coupler

The linear coupler (LC), or Rogowski coil, is a current transformer that utilizes a nonmagnetic core, e.g., wood, plastic, or paper, which acts only as a form for the winding. It is typically a window-type construction. With an air core, the magnetizing components of error have been removed, thus offering linear response and no possibility of saturation. They do not produce secondary currents that would be provided by an ideal CT. If the magnetizing impedance approached infinity, the secondary current would approach the ideal, $i_S = -i_P (N_P/N_S)$, as seen in Figure 2.6.24. The low permeability of the core prevents a high-magnetizing inductance, thus providing considerable divergence of performance from that of a conventional CT. Consequently, protective equipment must be designed to present essentially infinite impedance to the LC and operate as mutual inductors. Closely matching the LC impedance will provide maximum power transfer to the device. LC outputs are typically defined by V_S/I_P, e.g., 5 V per 1000 A. Since the load is high impedance, the LC can be safely open-circuited, unlike the conventional iron-core CT.

Because coupling is important and there is no iron to direct the flux, the window size is made as small as possible to accommodate the primary conductor. Positioning is critical, and the return conductor and adjacent phases must be far enough away from the outer diameter of the LC so that stray flux is not introduced into the winding.

2.6.4.11 Direct-Current Transformer

The basic direct-current transformer (DCT) utilizes two coils, referred to as elements, that require external ac (alternating current) excitation (Figure 2.6.25). The elements are window-type CTs that fit over the dc bus. The elements are connected in opposition such that the instantaneous ac polarity of one is always in opposition to the other. The ac flux in one element opposes the dc flux in the primary bus and desaturates the core, while in the other element the ac flux aids the dc flux and further saturates the core. This cycle is repeated during the other (opposite polarity) half of the ac cycle. The need to rectify is due to the square-wave output. The direction of current flow in the primary is not important.

Proximity of the return conductor may have an effect on accuracy due to local saturation. For best results, all external influences should be kept to a minimum.

The output waveform contains a commutation notch at each half-cycle of the applied exciting voltage. These notches contribute to the errors and can interfere with the operation of fast-acting devices. Ideally,

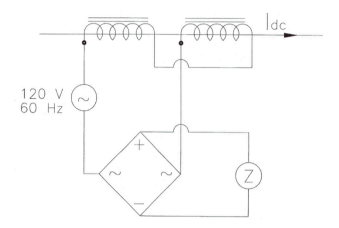

FIGURE 2.6.25 Standard two-element connection.

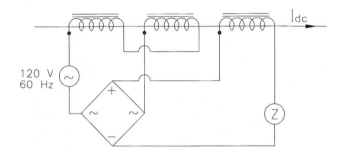

FIGURE 2.6.26 Three-element connection.

the core material should have a square B-H characteristic, which will minimize the notches. There are several connection schemes that can eliminate the notches but will increase the overall cost.

A simple approach could be the three-element connection (Figure 2.6.26), where the third element acts as a smoothing choke to the notches. This also increases overall frequency response by providing ac coupling between the primary and the output circuits.

Finally, there is the Hall-effect device. This solid-state chip is inserted into the gap of an iron core whose area is much larger than the device itself. The core has no secondary winding. The Hall effect requires a low-voltage dc source to power and provides an output proportional to the dc primary current and the flux linked to the core and gap.

2.6.4.12 Slipover CT Installations

CTs are often installed on existing systems as power requirements increase. One of the most common retrofit installations is the application of a window-type or external slipover CT, which is mounted over the high-voltage bushing of a power transformer or circuit breaker (Figure 2.6.27). To maintain the integrity of the insulation system, adequate strike clearances must be observed. It is also important to protect the CT from a high-voltage flashover to ground. This can be done by placing the CT below the ground plane or bushing flange. When using an external slipover CT, a ground shield can be placed on top of the unit and connected to ground. When grounding the shield, it is important that the lead is routed such that it does not make a shorted electrical turn around the CT.

A slipover, when used internally, is called a bushing CT or BCT. Its insulation is not suitable for direct outdoor use and requires protection from the weather. In high-voltage switchgear, BCTs are fitted over the high-voltage bushing in a similar manner as described above, but they are enclosed inside a metallic

FIGURE 2.6.27 00-V-class slipover CT installed on high-voltage bushing with ground shield. Note the ground lead going back through the CT window to avoid shorted electrical turn. (Photo courtesy of Kuhlman Electric Corp.)

FIGURE 2.6.28 High-voltage circuit breaker with BCTs mounted underneath a metallic cover. (Photo courtesy of Kuhlman Electric Corp.)

cover that protects them from the weather while providing a ground shield (Figure 2.6.28). The BCT can also be inside the apparatus, mounted off of the high-voltage bushing, either above or submerged in the oil. Either way, special materials must be used for exposure to oil.

2.6.4.13 Combination Metering Units

The last major assembly is the combination metering unit, which consists of a single VT and a single CT element within one common housing. These are typically available in 15-kV class and up. In most cases this is more economical than using single-housed elements and can save space. With two primary terminals, it mimics the conventional wound-type CT, thus simplifying the installation. The H1 (line) terminal is a common junction point for both the CT and VT elements. The H2 (load) terminal completes the current loop. The VT element is connected from the H1 terminal to ground. All secondary terminals are isolated and located inside the secondary-terminal box.

Optical metering units are available for use in high-voltage substations. They utilize the same principles as previously discussed, with the current and voltage transducers housed in a common structure.

2.6.4.14 Primary Metering Units

Primary metering units are single CT and VT elements assembled on a common bracket for pole-mounted installations or inside a pad-mount compartment for ground installations. They are typically molded

FIGURE 2.6.29   5-kV three-phase/four-wire metering rack with 3 VT and 3 CT. (Photo courtesy of Kuhlman Electric Corp.)

FIGURE 2.6.30   34.5-kV three-phase oil-filled metering unit. (Photo courtesy of Kuhlman Electric Corp.)

products, 5 kV through 34.5 kV (Figure 2.6.29), but they are also available in oil-filled units (Figure 2.6.30). It is a prewired assembly with all of the elements' secondary leads connected to a common junction box for easy access. The elements can be in any combination needed to provide accurate energy measurement. For three-phase/four-wire installations use either $2^1/_2$- or 3-element arrangements, and for three-phase/three-wire installations use 2-element arrangements. There are even single-phase assemblies available for special applications.

When measuring energy usage for the purposes of revenue billing, and knowing the RCF and phase-angle readings of each element, the watts or watt-hours can be corrected by multiplying the reading by the product of $[\text{TCF}_{CT}][\text{TCF}_{VT}]$.

2.6.4.15 New Horizons

With deregulation of the utility industry, the buying and selling of power, the leasing of power lines, etc., the need for monitoring power at the transmission and distribution level will increase. There will be a need to add more metering points within existing systems beginning at the generator. The utilities will want to add this feature at the most economical cost. The use of window-type slipover CTs for revenue metering will drive the industry toward improved performance. The need for higher accuracies at all levels will be desired and can be easily obtained with the use of low-burden solid-state devices. Products will become more environmentally safe and smaller in size. To help the transformer industry fulfill these needs, steel producers will need to make improvements by lowering losses and increasing initial permeability and developing new composites.

References

Burke, H.E., *Handbook of Magnetic Phenomena*, Van Nostrand Reinhold, New York, 1986.

Cosse, R.E., Jr. et al., eds., The Practice of Ground Differential Relaying, *IAS Trans.*, 30, 1472–1479, 1994.

Funk, D.G. and Beaty, H.W., *Standard Handbook for Electrical Engineers*, 12th ed., McGraw-Hill, New York, 1987.

Hague, B., *Instrument Transformers, Their Theory, Characteristics, and Testing*, Sir Isaac Pitman & Sons, London, 1936.

Hou, D. and Roberts, J., Capacitive Voltage Transformer Transient Overreach Concerns and Solutions for Distance Relaying, presented at 50th Annual Georgia Tech. Protective Relaying Conference, Atlanta, 1996.

Karlicek, R.F. and Taylor, E.R., Jr., Ferroresonance of Grounded Potential Transformers on Ungrounded Power Systems, *AIEE Trans.*, 607–618, 1959.

IEEE, Guide for Protective Relay Applications to Power System Buses, C37.97-1979, Institute of Electrical and Electronics Engineers, Piscataway, NJ, 1979.

IEEE, Guide for Protection of Shunt Capacitor Banks, C37.99-1980, Institute of Electrical and Electronics Engineers, Piscataway, NJ, 1980.

IEEE, Guide for Generator Ground Protection, C37.101-1985, Institute of Electrical and Electronics Engineers, Piscataway, NJ, 1985.

IEEE, Guide for AC Generator Protection, C37.102-1985, Institute of Electrical and Electronics Engineers, Piscataway, NJ, 1985.

IEEE, Guide for Protective Relay Applications to Power Transformers, C37.91-1985, Institute of Electrical and Electronics Engineers, Piscataway, NJ, 1985.

IEEE, Guide for AC Motor Protection, C37.96-1988, Institute of Electrical and Electronics Engineers, Piscataway, NJ, 1988.

IEEE, Requirements for Instrument Transformers, C57.13-1993, Institute of Electrical and Electronics Engineers, Piscataway, NJ, 1993.

IEEE, System Relaying Committee Working Group, Relay Performance Considerations with Low Ratio Current Transformers and High Fault Currents, *IAS Trans.*, 31, 392–404, 1995.

IEEE, Guide for the Application of Current Transformers Used for Protective Relaying Purposes, C37.110-1996, Institute of Electrical and Electronics Engineers, Piscataway, NJ, 1996.

Jenkins, B.D., *Introduction to Instrument Transformers*, George Newnes, London, 1967.

Moreton, S.D., A Simple Method for Determination of Bushing Current Transformer Characteristics, *AIEE Trans.*, 62, 581–585, 1943.

Pagon, J., Fundamentals of Instrument Transformers, paper presented at Electromagnetic Industries Conference, Gainesville, FL, 1975.

Settles, J.L. et al., eds., The Analytical and Graphical Determination of Complete Potential Transformer Characteristics, *AIEE Trans.*, 1213–1219, 1961.

Swindler, D.L. et al., eds., Modified Differential Ground Fault Protection for Systems Having Multiple Sources and Grounds, *IAS Trans.*, 30, 1490–1505, 1994.

Wentz, E.C., A Simple Method for Determination of Ratio Error and Phase Angle in Current Transformers, *AIEE Trans.*, 60, 949–954, 1941.

West, D.J., Current Transformer Application Guidelines, *IAS Annual*, 110–126, 1977.

Yarbrough, R.B., *Electrical Engineering Reference Manual*, 5th ed., Professional Publications, Belmont, CA, 1990.

2.7   Step-Voltage Regulators

Craig A. Colopy

2.7.1   Introduction

Requirements for electrical power become more challenging every day in terms of both quality and quantity. "Quality" means that consumers need stable voltage without distortions and interruptions. "Quantity" means that the user can draw as much load as needed without reducing the quality of the supply. These requirements come from industry as well as from domestic consumers, and each requirement influences the other. Maintaining voltage magnitude within a specified range is an important

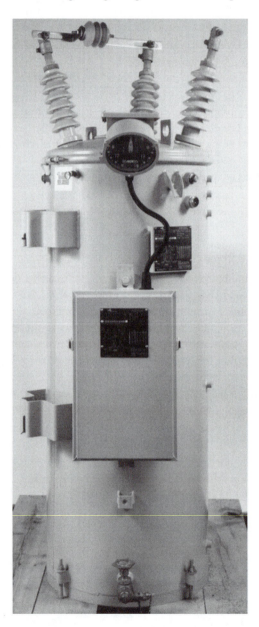

FIGURE 2.7.1   Single-phase step-voltage regulator.

component of power quality. Power transformers and feeders impose their own impedance, and the amount of voltage drop depends on the loads and, consequently, the currents that flow through them. Voltage magnitudes decrease along the feeder, which means that consumers at the end of the feeder will have the lowest voltage.

Distribution systems must be designed in such a way that voltage magnitudes always remain within a specified range as required by standards. This is accomplished through the use of voltage-control equipment and effective system design. Regulating power transformers (load-tap-changing transformers, or LTCs), three-phase step-voltage regulators, single-phase step-voltage regulators, and Auto-Boosters® are typical transformer-type equipment used to improve the voltage "profile" of a power system. Most of this section is dedicated to the single-phase step-voltage regulator, shown in Figure 2.7.1, that is used in substations and on distribution system feeders and laterals. Figure 2.7.2 shows the locations on a power system where voltage regulators are commonly applied.

Single-phase step-voltage regulators maintain a constant voltage by on-load voltage regulation wherever the voltage magnitude is beyond specified upper and lower limits. A common practice among utilities is to stay within preferred voltage levels and ranges of variation as set forth by ANSI 84.1, Voltage Rating for Electric Power Systems and Equipment, as shown in Table 2.7.1.

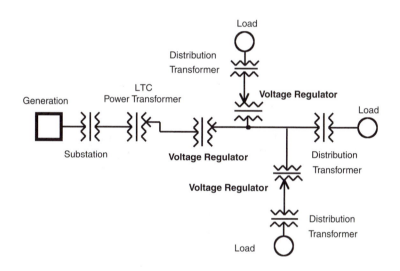

FIGURE 2.7.2 Power system.

TABLE 2.7.1 Voltage Rating for Electric Power Systems and Equipment

	Range A, V	Range B, V
Maximum allowable voltage	126 (125[a])	127
Voltage-drop allowance for primary distribution line	9	13
Minimum primary service voltage	117	114
Voltage-drop allowance for distribution transformer	3	4
Minimum secondary service voltage	114	110
Voltage-drop allowance for plant wiring	6 (4[b])	6 (4[b])
Minimum utilization voltage	108 (110[b])	104 (106[b])

[a] For utilization voltage of 120 through 600 V.
[b] For building wiring circuits supplying lighting equipment.
Source: ANSI 84.1, Voltage Rating for Electric Power Systems and Equipment. With permission.

Electric supply systems are to be designed and operated so that most service voltages fall within the range A limits. User systems are to be designed and operated so that, when the service voltages are within range A, the utilization voltages are within range A. Utilization equipment is to be designed and rated to give fully satisfactory performance within the range A limits for utilization voltages.

Range B is provided to allow limited excursion of voltage outside the range A limits that necessarily result from practical design and operating conditions. The supplying utility is expected to take action within a reasonable time (e.g., 2 or 3 min) to restore utilization voltages to range A limits.

The combination of step-voltage regulators and fixed-ratio power transformers is often used in lieu of load-tap-changing power transformers in the substation. To obtain constant voltage at some distance (load center) from the step-voltage regulator bank, a line-drop compensation feature in the control can be utilized.

Step-voltage regulators — single- and three-phase — are designed, manufactured, and tested in accordance with IEEE Std. C57.15, IEEE Standard Requirements, Technology, and Test Code for Step-Voltage Regulators. A step-voltage regulator is defined as "an induction device having one or more windings in shunt with and excited from the primary circuit, and having one or more windings in series between the primary circuit and the regulated circuit, all suitably adapted and arranged for the control of the voltage, or of the phase angle, or of both, of the regulated circuit in steps by means of taps without interrupting the load."

The most common step-voltage regulators manufactured today are single-phase, using reactive switching resulting in 32 5/8% voltage steps (16 boosting and 16 bucking the applied voltage), providing an overall ±10% regulation. They are oil-immersed and typically use ANSI Type II insulating oil in accordance with ANSI/ASTM D-3487.

Although not required by the IEEE standard, which now recognizes 65°C rise, most manufacturers today design and manufacture step-voltage regulators rated 55°C average winding rise over a 30°C average ambient and use a sealed-tank-type construction. The gases generated from arcing in oil, as a result of normal operation of the load-tap changer, are vented through a pressure-relief device located on the tank above the oil level. Thermally upgraded paper insulation designed for an average winding rise of 65°C, along with the use of a sealed-tank-type system, allows for a 12% increase in load over the nameplate 55°C rise kVA rating.

Many regulators with a continuous-current rating of 668 A and below can be loaded in excess of their rated ampere load if the range of voltage regulation is limited at a value less than the normal ±10% value. Table 2.7.2 shows the percent increase in ampere load permitted on each single-phase step-voltage regulator when the percent regulation range is limited to discrete values less than ±10%.

Some regulators have limitations in this increased current capacity due to tap-changer ampacity limitations. In those cases, the maximum tap-changer capacity is shown on the regulator nameplate. Limiting the percent regulation range is accomplished by setting limit switches in the position indicator of the regulator to prevent the tap changer from traveling beyond a set position in either raise or lower directions. It should be recognized, however, that although the regulators can be loaded by these additional amounts without affecting the regulator's normal coil-insulation longevity, when the percent regulation is decreased, the life of the tap-changer contacts will be adversely affected.

TABLE 2.7.2 Increase in Ampere Load Permitted on Single-Phase Step-Voltage Regulators for Regulation Range <±10%

Range of Voltage Regulation, %	Continuous Current Rating, %
10.0	100
8.75	110
7.5	120
6.25	135
5.0	160

2.7.2 Power Systems Applications

The following common types of circuits can be regulated using single-phase step-voltage regulators:

- A single-phase circuit
- A three-phase, four-wire, multigrounded wye circuit with three regulators
- A three-phase, three-wire circuit with two regulators
- A three-phase, three-wire circuit with three regulators

Figure 2.7.3 shows a voltage regulator that can be used to maintain voltage on a single-phase circuit or lateral off of a main feeder where the terminals are designated by the standard as S(ource), L(oad), and SL, as seen in Figure 2.7.4.

Regulators are designed to withstand severe fault currents and frequent switching or lightning surges that are encountered in substations or out on a main feeder. Three-phase power can be regulated by a bank of three single-phase step-voltage regulators connected in wye for a four-wire circuit or two or three single-phase regulators connected in an open- or closed-delta configuration for a three-wire circuit.

Wye-connected regulators (Figure 2.7.5 and Figure 2.7.6) work independently from each other. Regulators will regulate the voltage between the individual phases and neutral. It is not necessary for loads on each phase to be balanced. Unbalanced current will flow in the neutral wire, keeping the neutral reference from floating. In the case of a three-wire wye, the neutral can shift, so the regulator has no stable reference point from which to excite the controls of the regulator. This can cause overstressing of insulation and erratic regulator operation. Therefore, three regulators cannot be connected in ungrounded wye on a three-phase, three-wire circuit unless the neutral is stabilized.

One method of stabilization is to link the common SL connections back to the grounded secondary neutral of a substation transformer that is located in the same vicinity. If a substation transformer is not available, the alternative is to install a small grounding bank consisting of three transformers, each from one-third to two-thirds the kVA rating of the individual regulators. The rating within the range depends

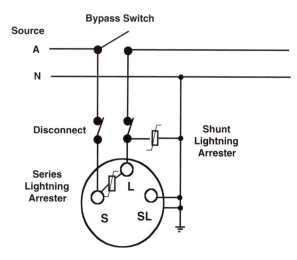

FIGURE 2.7.3 Voltage regulator connection in a single-phase circuit.

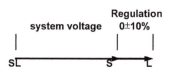

FIGURE 2.7.4 Phasor diagram of a voltage regulator regulating a single-phase circuit.

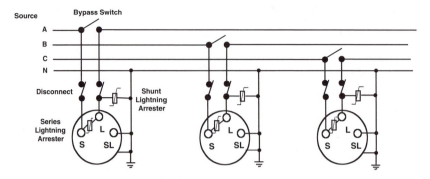

FIGURE 2.7.5 Connection of three regulators regulating a three-phase, four-wire, multigrounded wye circuit.

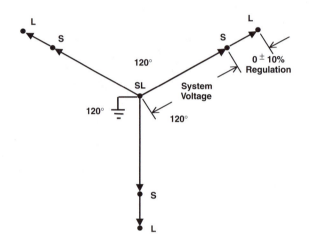

FIGURE 2.7.6 Phasor diagram of three regulators regulating a three-phase, four-wire, multigrounded wye circuit.

on the expected unbalance in load. Either method allows a path for the unbalanced current to flow. The primary windings are connected in a wye configuration tying their neutral back to the neutral of the three regulators, while the secondary windings are connected in a delta configuration.

Two single-phase regulators connected in an open-delta bank, as shown in Figure 2.7.7, allow for two of the phases to be regulated independent of each other, with the third phase tending to read the average of the other two. A 30° phase displacement between the regulator current and voltage, as shown in Figure 2.7.8, is a result of the open-delta connection. Depending on the phase rotation, one regulator has its current lagging the voltage, while the other has its current leading the voltage.

When a three-phase, three-wire circuit incorporates three single-phase voltage regulators in a closed-delta-connected bank as shown in Figure 2.7.9, the overall range of regulation of each phase is dependent on the range of regulation of each regulator. This type of connection will give approximately 50% more regulation (15% vs. 10%) than is obtained with two regulators in an open-delta configuration. A 30° phase displacement is also realized between the regulator current and voltage as a result of the delta connection. The phasor diagram in Figure 2.7.10 shows this. Depending on the phase rotation, the current will lag or lead the voltage, but the lag/lead relationship will be consistent for all three regulators.

Contributing to the effect is that the phase angle increases as the individual range of regulation of each regulator increases. A 4 to 6% shift in the phase angle will be realized with the regulators set at the same extreme tap position. A voltage improvement of 10% in the phase obtained with the regulator operation leads to a 5% voltage improvement in the adjacent phase. If all three regulators operate to the same extreme position, the overall effect increases the range of regulation to ±15%, as shown in the phasor diagram of Figure 2.7.10.

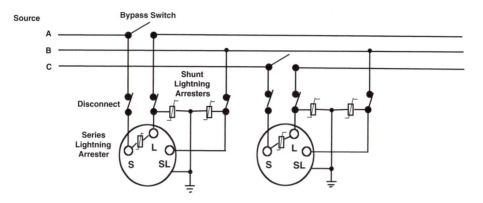

FIGURE 2.7.7 Connection of two voltage regulators regulating a three-phase, three-wire wye or delta circuit.

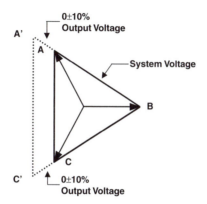

FIGURE 2.7.8 Phasor diagram of two voltage regulators regulating a three-phase circuit.

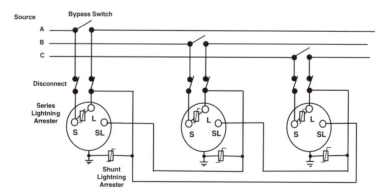

FIGURE 2.7.9 Connection of three voltage regulators in a three-phase, three-wire delta circuit.

Figure 2.7.11 reflects the power connection of a closed-delta bank of regulators. Current within the windings reaches a maximum of approximately minus or plus 5% of the line current as the regulators approach the full raise or lower positions, respectively. Because the operation of any regulator changes the voltage across the other two, it may be necessary for the other two to make additional tap changes to restore voltage balance. The 30° phase displacement between the regulator current and voltage at unity power-factor load for open- and closed-delta connections also affects the resulting arc energy and corresponding life of the tap-changer load-breaking contacts.

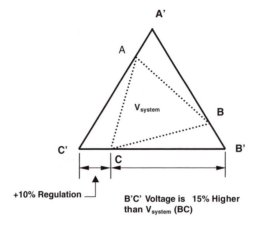

FIGURE 2.7.10 Phasor diagram of closed-delta-connected voltage regulators.

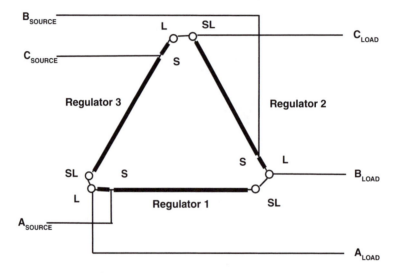

FIGURE 2.7.11 Power connection of closed-delta-connected voltage regulators.

2.7.3 Ratings

The rated-load current of a step-voltage regulator is determined by the following equation:

$$I_{Rated} = \frac{3\phi load}{V_{L-L} \times \sqrt{3}} \tag{2.7.1}$$

If the regulators are used in a single-phase circuit, four-wire grounded-wye circuit, or connected in a wye configuration in a three-wire system, the rated voltage of the regulator would be $V_{L-L}/\sqrt{3}$. If the regulators were connected in an open- or closed-delta configuration for a three-wire system, the rated voltage of the regulator would be V_{L-L}. As a result, the kVA rating of the regulator would be determined by the following equation:

$$kVA = (V_{Rated} \times I_{Rated} \times \text{per-unit range of regulation})/1000 \tag{2.7.2}$$

Thus the kVA rating of a step-voltage regulator is determined considering its range of voltage regulation. Single-phase voltage regulators are available in the common ratings shown in Table 2.7.3, where all entries are understood to be for ±10% range of voltage regulation.

TABLE 2.7.3 Commonly Available Ratings for Single-Phase Voltage Regulators

Rated Volts	BIL, kV	Rated, kVA	Load Current, A
2,500	60	50	200
		75	300
		100	400
		125	500
		167	668
		250	1000
		333	1332
		416	1665
5,000	75	50	100
		75	150
		100	200
		125	250
		167	334
		250	500
		333	668
		416	833
7,620	95	38.1	50
		57.2	75
		76.2	100
		114.3	150
		167	219
		250	328
		333	438
		416	548
		500	656
		667	875
		833	1093
13,800	95	69	50
		138	100
		207	150
		276	200
		414	300
		552	400
14,400	150	72	50
		144	100
		288	200
		333	200
		432	300
		576	400
		667	463
		833	578
19,920	150	100	50
		200	100
		333	200
		400	200
		667	300
		833	400

2.7.4   Theory

A step-voltage regulator is a tapped autotransformer. To understand how a regulator operates, one can start by comparing it with a two-winding transformer.

Figure 2.7.12 is a basic diagram of a transformer with a 10:1 turns ratio. If the primary winding or V_{source} has 1000 V applied, the secondary winding or V_{load} will have an output of 100 V (10%). These two independent windings can be connected so that their voltages aid or oppose one another. A voltmeter connected across the output terminals measures either the sum of the two voltages or the difference between them. The transformer becomes an autotransformer with the ability to raise (Figure 2.7.13) or lower (Figure 2.7.14) the primary or system voltage by 10%. This construction is similar to a "Type A" single-phase step-voltage regulator, as described later in this section.

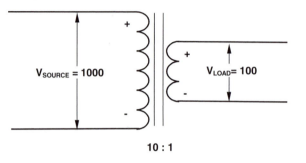

FIGURE 2.7.12   Transformer with 10:1 turns ratio.

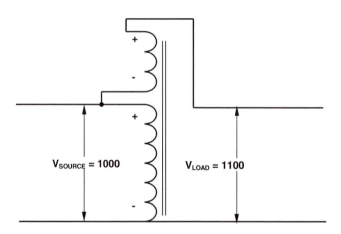

FIGURE 2.7.13   Step-up autotransformer.

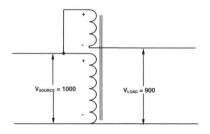

FIGURE 2.7.14   Step-down autotransformer.

FIGURE 2.7.15 Tap-changer position indicator.

In a voltage regulator, the equivalent of the high-voltage winding in a two-winding transformer would be referred to as the shunt winding. The low-voltage winding would be referred to as the series winding. The series winding is connected to the shunt winding in order to boost or buck the applied or primary voltage approximately 10%. The polarity of its connection to the shunt winding is accomplished by the use of a reversing switch on the internal motor-driven tap changer. Eight approximately $1\,^1/_4\%$ taps are added to the series winding to provide small voltage-adjustment increments. To go even further to provide fine voltage adjustment, such as 16 approximate 5/8% tap steps, a center tapped bridging reactor (preventive autotransformer) — used in conjunction with two movable contacts on the motorized tap changer — is utilized. In all, there are 33 positions that include neutral, 16 lower positions (1L, 2L, 3L, 4L, etc.), and 16 raise positions (1R, 2R, 3R, 4R, etc.). Figure 2.7.15 illustrates a dial that indicates the arrangement of the tap positions.

It is common practice to have the tap changer located in the same compartment as the core and coil. Dielectric practices consider the oil and insulation degradation due to the arc by-products. Figure 2.7.16 shows a typical load-break tap changer used in a single-phase step-voltage regulator.

The process of moving from one voltage-regulator tap to the adjacent voltage-regulator tap consists of closing the circuit at one tap before opening the circuit at the other tap. The movable tap-changer contacts move through stationary taps alternating in eight bridging and eight nonbridging (symmetrical) positions. Figure 2.7.17 shows the two movable tap-changer contacts on a symmetrical position, with the center tap of the reactor at the same potential. This is the case at tap position N (neutral) and all evenly numbered tap positions.

An asymmetrical position, as shown in Figure 2.7.18, is realized when one tap connection is open before transferring the load to the adjacent tap. At this juncture, all of the load current flows through one-half of the reactor, magnetizing the reactor, and the reactance voltage is introduced into the circuit for about 25 to 30 msec during the tap change.

Figure 2.7.19 shows the movable contacts in a bridging position; voltage change is one-half the $1\,^1/_4\%$ tap voltage of the series winding because of its center tap and movable contacts located on adjacent stationary contacts. This is the case at all oddly numbered tap positions.

Voltage phasor relations shown in Figure 2.7.20 represent a tap change. In this figure, S–SL represents source or unregulated voltage, and sections of the series winding are represented by S–TAP 1 and TAP 1–TAP 2. In operating the tap changer from TAP 1 to TAP 2, the load or regulated voltage has the following successive values:

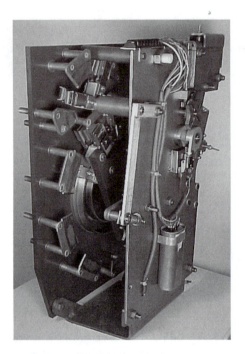

FIGURE 2.7.16 Single-phase step-voltage regulator tap changer.

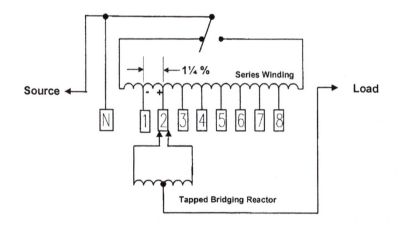

FIGURE 2.7.17 Two movable contacts on the same stationary contact (symmetrical position).

SL–TAP 2: both contacts are closed on TAP 2; symmetrical position as shown in Figure 2.7.17.

SL–a: one contact is closed on TAP 2; asymmetrical position as shown in Figure 2.7.18. TAP 2–a is the voltage drop across the reactor half.

SL–b: movable contacts are located on adjacent stationary contacts TAP 1 and TAP 2; bridging position as shown in Figure 2.7.19.

SL–c: one contact is closed on TAP 1, asymmetrical position. TAP 1–c is the voltage drop across the reactor half.

SL–TAP 1: both contacts are closed on TAP 1, symmetrical position.

TAP 1–c and TAP 2–a are reactance voltages introduced into the circuit by the reactor in the asymmetrical positions. TAP 1–e and TAP 2–d represent the total reactor voltage. TAP 2–e represents the voltage ruptured when bridging position is opened at TAP 2, while TAP 1–d represents the voltage ruptured if bridging position is opened at TAP 1.

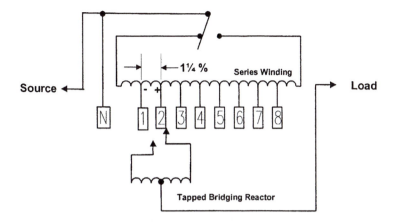

FIGURE 2.7.18 One movable contact on stationary contact (asymmetrical position).

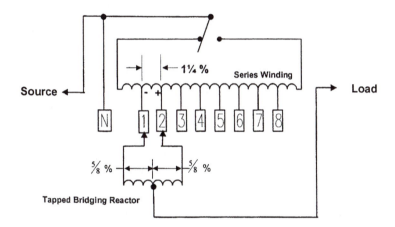

FIGURE 2.7.19 Two movable contacts on adjacent stationary contacts (bridging position).

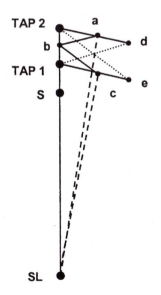

FIGURE 2.7.20 Voltage phasor diagrams involved in a tap change.

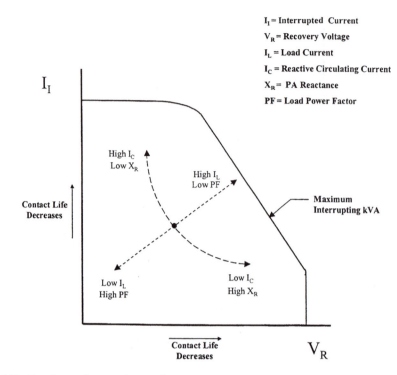

FIGURE 2.7.21 Tap-changer interruption envelope.

Figure 2.7.21 shows a typical tap-changer interruption envelope. Load current (I_L), tap voltage, reactor circulating current (I_C), and power factor (PF) are key variables affecting a tap changer's interrupting ability and its contact life. A circulating current (I_C), caused by the two contacts being at different positions (reactor energized with $1^1/_4$% tap voltage), is limited by the reactive impedance of this circuit. Two opposing requirements must be kept in mind when designing the amount of reactance for the value of the circulating current. First, the circulating current must not be excessive; second, the variation of reactance during the switching cycle should not be so large as to introduce undesirable fluctuations in the line voltage. The reactor has an iron core with gaps in the magnetic circuit to set this magnetizing circulating current between 25 and 60% of full-load current, thus providing an equitable compromise between no-load and load conditions. The value of this circulating current also has a decided effect on switching ability and contact life. The ideal reactor, from an arcing standpoint, would be one that has a closed magnetic circuit at no-load with an air gap that would increase in direct proportion to increase in load.

The voltage at the center tap is 5/8%, one-half of the $1^1/_4$% tap voltage of the series winding taps. Some regulators, depending upon the rating, use an additional winding (called an equalizer winding) in the bridging reactor circuit. The equalizer winding is a 5/8% voltage winding on the same magnetic circuit (core) as the shunt and series winding. The equalizer winding is connected into the reactor circuit opposite in polarity to the tap voltage. This is done so that the reactor is excited at 5/8% of line voltage on both the symmetrical and bridging positions. Figure 2.7.22 shows an equalizer winding incorporated into the main coil of a regulator.

Voltage regulators are designed and manufactured in two basic constructions, defined by IEEE standards as Type A and Type B. Type A step-voltage regulators have the primary circuit (source voltage) connected directly to the shunt winding of the regulator. The series winding is connected to the load side of the regulator and, by adjusting taps, changes the output voltage. With Type A construction, the core excitation varies with the source voltage because the shunt winding is connected across the primary circuit. A separate voltage transformer is used to provide voltage for the tap-changer and control. The

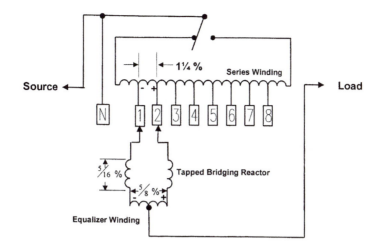

FIGURE 2.7.22 Equalizer winding incorporated into the main coil of a regulator.

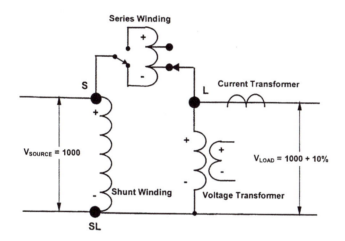

FIGURE 2.7.23 Voltage regulation on the load side (Type A).

maximum range of regulation of the "raise" side equals the maximum range of regulation of the "lower" side, with 10% being the minimum amount of regulation. See the schematic diagram in Figure 2.7.23.

Type B step-voltage regulators are constructed so that the primary circuit (source voltage) is applied by way of taps to the series winding of the regulator, which is connected to the source side of the regulator. With Type B construction, the core excitation is constant, since the shunt winding is connected across the regulated circuit. A control winding located on the same core as the series and shunt windings is used to provide voltage for the tap-changer and control. The maximum range of regulation of the "raise" side is higher than the maximum range of regulation of the "lower" side, with 10% being the minimum amount of regulation on the "raise" side. See the schematic diagram in Figure 2.7.24.

Usually the choice of Type A or Type B is that of the supplier. However, the user can specify that an identical regulator design be provided if the application can anticipate the need to parallel with another unit in the same substation. Paralleling of regulators that are not of identical design can cause excessive circulating current between them. This is true even for short-term operation during switching, when the units are placed on the same numerical tap position and the control function is disabled. Caution: Any paralleling of step-voltage regulators (as described in this chapter) that may operate, even momentarily, on differing tap positions requires the inclusion of supplemental system reactance to avoid excessive circulating current during operation.

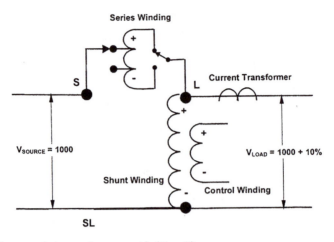

FIGURE 2.7.24 Voltage regulation on the source side (Type B).

FIGURE 2.7.25 Single-phase Auto-Booster® four-step voltage regulator.

2.7.5 Auto-Booster®

Another type of voltage regulator available in the industry, in addition to the 32-step single phase, is the Auto-Booster® single-phase four-step voltage regulator. This voltage-regulating device, shown in Figure 2.7.25, is mainly used for boosting voltage on laterals off the distribution system feeders. The voltage ratings range from 2,400 to 19,920, while the ampere ratings are either 50 A or 100 A. The design is a 65° winding rise Type B construction providing +10% or +6% regulation in four steps of $2^1/_2$% or $1^1/_2$%, respectively. The tap changer incorporates resistors to provide the impedance during the switching cycle.

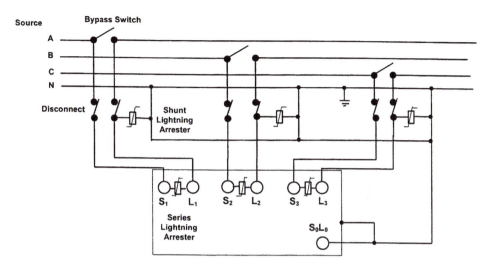

FIGURE 2.7.26 Connection of three-phase regulator regulating a three-phase, four-wire, multigrounded wye circuit.

2.7.6 Three-Phase Regulators

If feeders have similar load characteristics and load-center voltage requirements — and if the incoming supply voltage is balanced — then the distribution side of the substation bus can be regulated by a three-phase regulator. The choice of using a three-phase or three single-phase regulators depends on several factors, including cost and the need for unbalance correction. With a three-phase unit there is no capability to correct for voltage imbalances caused by unequal loading. If the load is predominantly three-phase, or consistently balanced, a three-phase unit may be the better choice. However, most rural distribution systems contain only a small percentage of balanced three-phase loads. Therefore three-phase regulators are less common in the industry than the single-phase step-voltage regulators. Chief among the benefits of single-phase regulation is the ability of the single-phase voltage regulators to correct system unbalance. Each phase is regulated to a given set point, irrespective of what is going on with the other two phases.

Basic theory behind the design of three-phase regulators is similar to single-phase regulators. Essentially, three single-phase regulators are located in the same tank with their tap changers ganged together and being operated and monitored by one control. The three-phase regulator has only one of its phases monitored to provide a current and voltage supply to the control. Thus all three phases are regulated based on the monitoring of one phase. This is fine for a system that has balanced loads on all three phases. Figure 2.7.26 shows a three-phase regulator connected to a four-wire system. The phasor diagram is similar to the one shown in Figure 2.7.5.

The three-phase voltage regulators, up to 13.8 kV, inclusive, with a continuous-current rating of 668 A and below, can be loaded in excess of their rated ampere load if the range of voltage regulation is limited at a value less than the normal ±10% value. Table 2.7.4 shows the percent increase in ampere load permitted on each three-phase step-voltage regulator when the percent regulation range is limited to values of ±10%, ±8_%, ±7_%, ±6_%, and ±5%.

Three-phase voltage regulators are available in the common ratings shown in Table 2.7.5.

2.7.7 Regulator Control

The purpose of the regulator control is to provide an output action as a result of changing input conditions, in accordance with preset values that are selected by the regulator user. The output action is the energization of the motorized tap changer to change taps to maintain the correct regulator output voltage. The changing input is the output-load voltage and current from an internal voltage transformer

TABLE 2.7.4 Increase in Ampere Load Permitted on Each Three-Phase Step-Voltage Regulator at Five Levels of Voltage Regulation

Range of Voltage Regulation, %	Continuous Current Rating, %
10.0	100
8.75	108
7.5	115
6.25	120
5.0	130

TABLE 2.7.5 Common Ratings for Three-Phase Voltage Regulators

Rated Volts	BIL, kV	Rated kVA	Load Current, A
7,620/13,200 wye	95	500	219
		750	328
		1000	437
		1500	656
		2000	874
		2500	1093
		3000	1312
19,920/34,500 wye	150	500	84
		750	126
		1000	167
		1500	251
		2000	335
		2500	418
		3000	502

or control winding and current transformer, as shown previously in Figure 2.7.23 and Figure 2.7.24, respectively. The preset values are the values the regulator user has selected as control parameters for the regulated voltage. Basic regulator control settings are:

- Set voltage
- Bandwidth
- Time delay
- Line-drop resistive and reactive compensation

2.7.7.1 Set Voltage

The control set voltage is dependent upon the regulator rating and the system voltage on which it is installed. The regulator nameplate shows the voltage transformer or shunt winding/control ratio that corresponds to the system voltage. The regulator load voltage is the product of this ratio and the control set voltage. If the winding ratio of the internal voltage transformer of the step-voltage regulator is the same as that of a typical distribution transformer on the system, the voltage level is simply the desired voltage, given on a 120-V base. However, the regulator voltage-transformer ratio is not always the same, and it may be necessary to calculate an equivalent value that corresponds to 120 V on the distribution transformer secondary. Equation 2.7.3 can be used to find this equivalent voltage value.

$$V = \frac{\left(\text{Distribution Transformer Ratio}\right)}{\left(\text{Regulator PT Ratio}\right)} \times 120 \qquad (2.7.3)$$

For example, if the distribution transformers are connected 7620/120, this gives a ratio of 63.5:1. If the regulator voltage-transformer ratio is 60:1, the equivalent voltage-level setting, from Equation 2.7.3, is found to be:

$$V = \frac{(63.5)}{(60)} \times 120 = 127 \text{ V} \tag{2.7.4}$$

Normally the operator sets the voltage level at a minimum value that is required at the location, which is usually above the optimum level, e.g., 122 V.

2.7.7.2 Bandwidth

The bandwidth is the total voltage range around the set-voltage value, which the control will consider as a satisfied condition. For example, a 2-V bandwidth on a 120-V setting means that the control will not activate a tap change until the voltage is above 121 V or below 119 V. The bandwidth is generally kept quite narrow in order to keep the voltage as close as possible to the desired level. Increasing the bandwidth may reduce the number of tap-change operations, but at the expense of voltage quality. The regulators in a substation or on a main feeder tend to have their bandwidths set at a smaller setting than units located on laterals that feed isolated loads. A minimum bandwidth of two times the volts per tap (5/8% or 0.75 V) of the regulator is recommended; this correlates to 1.5 V on most regulators.

2.7.7.3 Time Delay

The time delay is the period of time in seconds that the control waits from the time the voltage goes out of band to when power is applied to the tap-changer motor to make a tap change. Many voltage fluctuations are temporary in nature, such as those resulting from motor starting, or even from fault conditions. It is not desirable to have the step-voltage regulator change taps for these momentary voltage swings. By specifying a time-delay value of 15 or 30 sec, for example, the regulator will ignore the vast majority of these temporary voltage swings and only operate when the voltage change is more long term, such as from adding or subtracting load. A general recommendation is a minimum time delay of 15 sec. This length of time covers the vast majority of temporary voltage swings due to equipment starting, cold-load pickup, etc.

The time-delay setting also has another important benefit when attempting to coordinate two or more regulators in series along the line. One common situation would be to have a regulator bank at the substation providing whole-feeder regulation, with a line regulator out on the feeder to regulate a specific load. The substation regulators should be the first to respond to voltage changes on the system, with the line regulator adjusting as needed for its individual area. By setting the time delay of the line regulator higher that of the substation bank, the substation bank will respond first and regulate the voltage as best it can. If the substation regulation is sufficient to return the feeder voltage to within the bandwidth of the line regulator, that regulator will not need to operate. The suggested minimum time-delay difference between banks of regulators is 15 sec. This coordination between cascading banks of regulators, as shown in Figure 2.7.27, eliminates unnecessary hunting, thus improving efficiency.

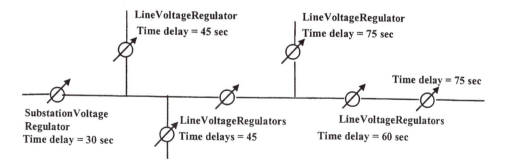

FIGURE 2.7.27 Cascading single-phase voltage regulators.

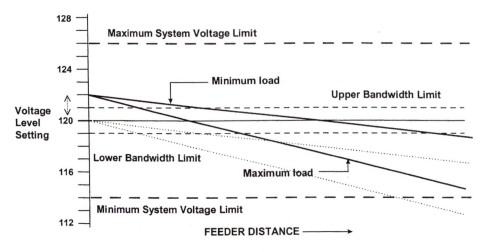

FIGURE 2.7.28 Voltage profile using voltage-level setting without LDC.

2.7.7.4 Line-Drop Resistive and Reactive Compensation

Quite often regulators are installed some distance from a theoretical load center or the location at which the voltage is regulated. This means the load will not be served at the desired voltage level due to the losses (voltage drop) on the line between the regulator and the load. Furthermore, as the load increases, line losses also increase, causing the lowest-voltage condition to occur during the time of heaviest loading. This is the least desirable time for this to occur.

To provide the regulator with the capability to regulate at a "projected" load center, a line-drop-compensation feature is incorporated in the control. Of all the devices in the step-voltage regulator control system, none is more important — and at the same time less understood — than line-drop compensation. This circuitry consists of a secondary supply from the internal current transformer, proportional to the load current, and resistive and reactive components through which this current flows. As the load current increases, the resulting secondary current flowing through these elements produces voltage drops, which simulate the voltage drops on the primary line. This causes the "sensed" voltage to be correspondingly altered; therefore, the control responds by operating upon this pseudo load-center voltage. To select the proper resistive and reactive values, the user must take into account several factors about the line being regulated.

When line-drop compensation is not used, the step-voltage regulator reads the voltage at its own terminals and compares it with a reference voltage level and bandwidth setting. If the sensed voltage is outside the set bandwidth, the regulator automatically responds to bring the load-side voltage in line with the programmed settings. This is the normal operation of a step-voltage regulator that is not using line-drop compensation. Adjusting the voltage level at the regulator to compensate for a low voltage away from the regulator near a load center is only a temporary solution for a specific load. Figure 2.7.28 shows an example of a power-system voltage profile using the voltage-level setting in lieu of line-drop compensation to improve voltage at a load center.

Therefore, the step-voltage regulator, without line-drop compensation, can hold a reasonably constant voltage at the regulator location only. This is not likely to be the best scenario for most applications that need a reliable voltage for the entire length of the feeder. The ideal situation is to provide the value of the voltage-level setting at the primary side of the distribution transformer for all consumers. Since this is not realistically attainable, the objective would be to provide each consumer a voltage as close as possible to the voltage-level value for all loads. To do this, the point on the line at which the voltage is regulated should not be at the regulator but at an area at the center of the majority of the consumers, the load center. It is not always possible to locate a regulator at the exact location where the regulation is most needed. Also, the system changes as loads are added and removed over time, and the desired point of

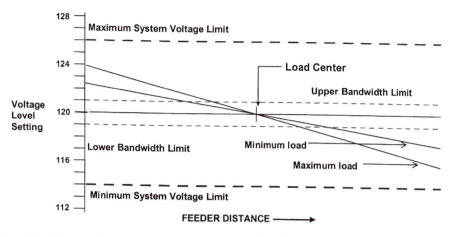

FIGURE 2.7.29 Voltage profile using voltage-level setting with LDC.

regulation may change. However, it may not be feasible to relocate the regulator bank. By having line-drop-compensation devices, the regulation point can be changed without having to physically move a bank of regulators.

In any mode, the regulator monitors a specific voltage and changes taps as needed based on that voltage level and the existing settings. With line-drop compensation, this monitored voltage can be modified in such a way as to simulate the voltage at some point further out on the distribution system. Knowing the peak-load current expected on the line and the size and length of the line to the load center, the voltage drop at the load center due to resistive and reactive components can be calculated. Inside the regulator control, the line-drop-compensation device will model that portion of the system between the load center and the step-voltage regulator. Figure 2.7.29 shows an example of a power-system voltage profile resulting from a regulator set up with the line-drop-compensation feature.

When line-drop compensation is used, the correct polarity of the resistance and reactance components is necessary for proper regulation. On four-wire wye-connected systems, the polarity selector is always set for +X and +R values. On delta-connected systems, however, the line current is 30° displaced from the line-to-line voltage (assuming 100% power factor). On open-delta-connected regulator banks, one regulator is 30° leading, the other is 30° lagging. On closed-delta regulator banks, all regulators in a given bank will be either leading or lagging. As a result of this displacement, the polarity of the appropriate resistive or reactive element must sometimes be reversed. The setting of the selector switch would be set on the +X+R, –X+R, or +X–R setting. A number of publications are available from manufacturers to assist in determining the variables needed and the resulting calculations.

2.7.8 Unique Applications

Most step-voltage regulators are installed in circuits with a well-defined power flow from source to load. However, some circuits have interconnections or loops in which the direction of power flow through the regulator may change. For optimum utility system performance, step-voltage regulators, installed on such circuits, have the capability of detecting this reverse power flow and then sensing and controlling the load-side voltage of the regulator, regardless of the direction of power flow.

In other systems, increasing levels of embedded (dispersed) generation pose new challenges to utilities in their use of step-voltage regulators. Traditionally, distribution networks have been used purely to transport energy from the transmission system down to lower voltage levels. A generator delivering electricity directly to the distribution network can reverse the normal direction of power flow in a regulator. Options in the electronic control of the step-voltage regulator are available for handling different types of scenarios that give rise to reverse power-flow conditions.

References

IEEE, Standard Requirements, Terminology, and Test Code for Step-Voltage Regulators, C57.15-1999, Institute of Electrical and Electronics Engineers, Piscataway, NJ, 1999.

ANSI/IEEE, IEEE Guide for Loading Liquid-Immersed Step-Voltage and Induction-Voltage Regulators, C57.95-1987, Institute of Electrical and Electronics Engineers, Piscataway, NJ, 1987.

Colopy, C.A., Grimes, S., and Foster, J.D., Proper Operation of Step Voltage Regulators in the Presence of Embedded Generation, presented at CIRED conference, Nice, France, June 1999.

Cooper Power Systems, How Step Voltage Regulators Operate, Bulletin 77006 2/93, Waukesha, WI, 1993.

Cooper Power Systems, Voltage regulator instruction and operating manuals and product catalogs, Waukesha, WI.

Cooper Power Systems, Distribution Voltage Regulation and Voltage Regulator Service Workshops, Waukesha, WI.

Day, T.R. and Down, D.A., The Effects of Open-Delta Line Regulation on Sensitive Earth Fault Protection of 3-Wire Medium Voltage Distribution Systems, Cooper Power Systems, Waukesha, WI, June 1995.

Foster, J.D., Bishop, M.T., and Down, D.A., The Application of Single-Phase Voltage Regulators on Three-Phase Distribution Systems, Paper 94 C2, presented at IEEE Conference, Colorado Springs, CO, April 1994.

General Electric Co., Voltage regulator instruction and operating manuals and product catalogs, Shreveport, LA.

Kojovic, L.A., Impact of DG on Voltage Regulation, presented at IEEE/PES summer meeting, Chicago, IL, 2002.

Kojovic, L.A., McCall, J.C., and Colopy, C.A., Voltage Regulator and Capacitor Application Considerations in Distribution Systems for Voltage Improvements, Systems Engineering Reference Bulletin SE9701, Franksville, WI, January 1997.

Siemens Energy and Automation, Voltage regulator instruction and operating manuals and product catalogs, Jackson, MS.

2.8   Constant-Voltage Transformers

Arindam Maitra, Anish Gaikwad, Ralph Ferraro, Douglas Dorr, and Arshad Mansoor

2.8.1   Background

Constant-voltage transformers (CVT) have been used for many years, primarily as a noise-isolation device. Recently, they have found value when applied as a voltage-sag protection device for industrial and commercial facilities. The purpose of this section in the handbook is to give power-system engineers and facility engineers who are unfamiliar with the CVT technology (also known as ferroresonant transformers) the insights and information necessary to determine the types of electric-service-supply events that CVTs can mitigate. Items covered in this chapter include operation, characteristics, applications, specifications, and sizing guidelines of CVTs. The goal here is not to duplicate information currently available but, rather, to collect information into a single location and then supplement it to provide:

- Adequate information and procedures to applications personnel in effectively selecting CVTs for voltage-sag ride-through protection
- Application notes to demonstrate how CVTs can improve process voltage-sag ride-through

2.8.1.1   History of Constant-Voltage Transformers

The industrial use of constant-voltage transformers (also called CVTs and ferroresonant transformers) goes back to the early 1940s. Living in the U.S. during the 1930s, Joseph G. Sola, a German-born engineer, discovered the CVT technology [1,2] based on a single transformer rather than an arrangement of

FIGURE 2.8.1 Typical constant-voltage transformer.

transformers, separate filters, and capacitors. This innovation provides several important advantages: its inherent robustness (CVT consists of just three or four windings and a high-reliability capacitor), its imperviousness to continuous short circuits (whether it is turned on into a short circuit or from full load), and its capability to maintain output-voltage stabilization on a cycle-to-cycle basis for significantly large overvoltages and undervoltages.

Sola was both a farsighted inventor and successful businessman. Internationally recognized as the pioneer of transformer magnetics technology, his inventions and subsequent refinements of other electronic equipment were considered revolutionary by the electrical industry. Sola's first transformers for furnace-ignition systems and neon lighting, based on the unique application of ferroresonant principles, led in 1938 to his invention of the CVT. This timely discovery was eagerly accepted by prime military contractors during World War II and established Sola as a world leader in voltage regulation. Sola was awarded a total of 55 U.S. patents during his lifetime, including five patents each for CVTs, Solatrons, and electronic power supplies, and 19 patents for high-intensity-discharge lighting ballast.

Throughout the last six decades, a series of applications has been found for products based on Sola's CVT technology. It is one of the most cost-effective ac power conditioners available.

2.8.1.2 What Is a Constant-Voltage Transformer?

A well-known solution for electrical "noise" in industrial plants has been the constant-voltage transformer, or CVT (see Figure 2.8.1).

The typical components of a CVT are shown in Figure 2.8.2. The magnetic shunt on the central core has the following effects on the core's reluctance. It reduces the reluctance of the core. This can be thought of as introducing more resistance in parallel to an existing resistance. The magnetic shunt in the CVT design allows the portion of the core below the magnetic shunt to become saturated while the upper portion of the core remains unsaturated. This condition occurs because of the presence of the air-gap between the magnetic shunt and the core limbs. Air has a much higher reluctance than the iron core. Therefore, most of the flux passes through the lower portion of the core, as shown by the thick lines in Figure 2.8.2. In terms of an electrical analogy, this configuration can be thought of as two resistances of unequal values in parallel. The smaller resistance carries the larger current, and the larger resistance carries the smaller current.

The CVT is designed such that:

- The lower portion of the central limb is saturated under normal operating conditions, and the secondary and the resonating windings operate in the nonlinear portion of the flux-current curve.
- Because of saturation in the central limb, the voltage in the secondary winding is not linearly related to the voltage in the primary winding.

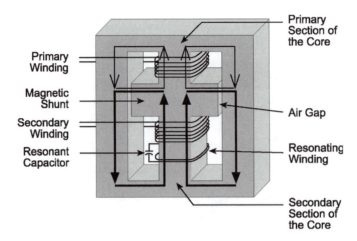

FIGURE 2.8.2 Components of a typical constant-voltage transformer.

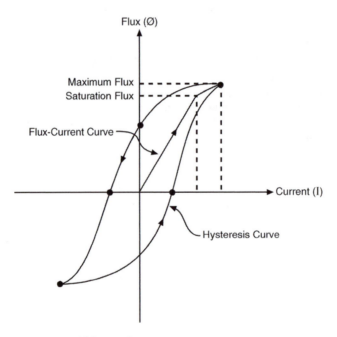

FIGURE 2.8.3 Flux-current curve with hysteresis curve.

There is ferroresonance between the resonating winding on the saturated core and the capacitor. This arrangement acts as a tank circuit, drawing power from the primary. This results in sustained, regulated oscillations at the secondary with the applied line frequency.

2.8.1.3   The Working of a Constant-Voltage Transformer

The working of a CVT can be explained with two physical phenomena:

- Saturation of the ferromagnetic core
- Ferroresonance

2.8.1.3.1 Saturation of Ferromagnetic Core

A flux-current curve is shown in Figure 2.8.3 [4]. In the linear portion of the curve, as the current increases, the magnetic-flux density increases. However, a point is reached where further increases in current yield smaller and smaller increases in flux density. This point is called the saturation point (see Figure 2.8.3) and is characterized by a dramatic reduction in the slope of the curve. In fact, the curve is no longer linear. Because the slope of the flux-current curve is proportional to the inductance of the coil, the reduction in the slope causes reduction in the inductance. The new inductance may be 100 times less than the inductance in the linear portion of the curve.

As shown on the hysteresis curve in Figure 2.8.3, the magnitude of current that causes the iron to go into saturation is not the same as the magnitude at which the iron comes out of saturation. The boundary between linear operation and saturated operation is not fixed but, instead, depends on the previous values of that current. The hysteresis phenomenon is a fundamental property of ferromagnetic materials. Hysteresis occurs because of residual flux density stored in the iron that must be overcome when the current changes direction. It should be noted that, for a particular value of flux, there are two values of current for which the core would be in saturation.

2.8.1.3.2 Ferroresonance

Most engineers are well versed with linear resonance (referred to simply as resonance). In a circuit, capacitive and inductive reactances (Z_C and Z_L, respectively) are calculated as shown in Equation 2.8.1 and Equation 2.8.2. Resonance occurs when the inductive and capacitive reactances of a circuit exactly balance. The resonating frequency is calculated as shown in Equation 2.8.3.

$$Z_C = \frac{1}{j\omega C} \tag{2.8.1}$$

$$Z_L = j\omega L \tag{2.8.2}$$

$$f = \frac{1}{2\pi\sqrt{LC}} \tag{2.8.3}$$

The inductance value used in Equation 2.8.2 and Equation 2.8.3 refers to the unsaturated or linear value of the inductor. However, the case when the core is saturated needs special consideration. If the inductor core is saturated, the relationship between flux (ϕ) and current (I) is no longer linear, and the inductance value is no longer a single value (L). The inductance in the nonlinear portion of the flux-current curve is not fixed and cannot be represented by a single value. Once the ferromagnetic inductance "enters" into saturation, it remains saturated until the current magnitude reduces. If the inductance when saturated causes a resonance (i.e., results in an inductive reactance that matches the capacitive reactance in the circuit), this phenomenon is called "ferroresonance." At the ferroresonance point, the current magnitude can increase dramatically, further driving the iron into saturation. If the current is constantly fed (as happens in a CVT), a stable resonant point is obtained [5]. If, at ferroresonance, the inductor value is L_S, the resonant frequency (f_S) is given by Equation 2.8.4.

$$f_S = \frac{1}{2\omega\sqrt{L_S C}} \tag{2.8.4}$$

It should be noted that $L_S \ll L$. The ferroresonant frequency can be either system frequency or a subharmonic of it. (Note: See the Addendum, Section 2.8.7, for tutorial description of a ferroresonant circuit.) If the current (I) drops below the saturation point, the inductance L_S "comes" out of saturation and the value changes back to L. Again, magnetic hysteresis causes the precise points where the inductor goes into and out of saturation to be different.

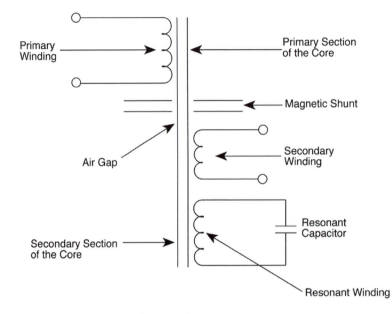

FIGURE 2.8.4 Schematic of a constant-voltage transformer.

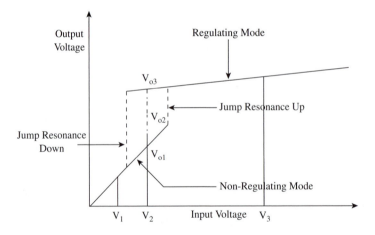

FIGURE 2.8.5 Output voltage vs. input voltage for constant-voltage transformer with jump resonance .

The salient features of ferroresonance are:

- Resonance occurs when the inductance is in saturation.
- As the value of inductance in saturation is not known precisely, a wide range of capacitances can potentially lead to ferroresonance at a given frequency.
- More than one stable, steady-state response is possible for a given configuration and parameter values [6].

A schematic of a constant-voltage transformer is shown in Figure 2.8.4.

In the ferroresonant circuit of a constant-voltage transformer, more than one steady-state response is possible for a given configuration and parameter values. This phenomenon (referred as "jump resonance") is described next.

The y-axis in Figure 2.8.5 is the CVT's output voltage at the secondary terminals, while the x-axis is the CVT's input primary voltage. There are three possible modes of behavior, depending on the level of

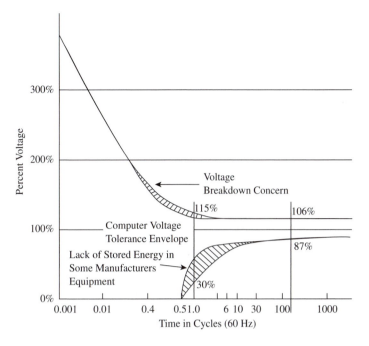

FIGURE 2.8.6 Voltage-tolerance envelope specified by the Computer and Business Equipment Manufacturers Association. Note: This is reproduced from IEEE Std. 446 (Orange Book).

the input voltage. At input voltage V_1, the CVT operates in nonsaturation mode. The secondary responds linearly to the primary supply. Note that this mode is used in a conventional transformer but is undesirable in a CVT.

At input voltage V_2, there are three outputs denoted by V_{o1}, V_{o2}, and V_{o3}. Outputs V_{o1} and V_{o3} are both stable states. Output V_{o1} corresponds to the normal state, whereas V_{o3} corresponds to ferroresonant state. Output V_{o2} is an unstable state that cannot be obtained in practice. Whether CVT output is V_{o1} or V_{o3} depends on the initial value of residual flux and voltage at the capacitor terminals. The phenomenon where the output voltage either suddenly changes to the regulating mode of operation at some value of the ascending input voltage, or suddenly drops out of the regulating mode of operation with descending input voltage, is called jump resonance [7]. The jump resonance is a factor during dynamic supply-voltage conditions. This is discussed in detail when the applications of a CVT are considered (Section 2.8.2). At input voltage V_3, the CVT operates in ferroresonant mode. The CVT must operate in this mode for proper operation.

2.8.1.4 Voltage Regulation on the Customer Side

The purpose of a voltage regulator is to maintain constant output voltage to the load in the face of variations in the input line voltage. In the past, however, voltage regulation was usually not a primary requirement for sensitive loads within end-user facilities. For instance, the Computer and Business Equipment Manufacturers Association (CBEMA) curve (see Figure 2.8.6) indicates that computer equipment should be able to handle steady-state voltage variations in the range of 87 to 106% of nominal. (Note that the CBEMA organization has been assumed by Information Technology Industry Council [ITIC] group.) However, with increased user-equipment sensitivity and the industry's dependency on sophisticated process-control devices in manufacturing, some types of equipment may have more stringent voltage-regulation requirements than the regulation tolerances specified. Also, the particular electric-service supply point may not be compatible with the connected electric load. In these cases, it is usually prudent to implement voltage regulation at the end-user's equipment level.

Many of the present voltage-regulation technologies can also provide mitigation of other power-quality problems (e.g., voltage-sag ride-through improvement or isolation for transient overvoltages). There are seven basic devices in use today. These include:

- Motor-actuated voltage regulator — Generally inexpensive, these devices can handle high kVA loads, but they are slow to respond to changes in the electric-service supply and can only correct for gradual load changes. See Section 2.7, Step-Voltage Regulators.
- Saturable reactor regulator — Relatively inexpensive and with a wide load-kVA range, these devices have a sluggish (five to ten cycles) response time, high output impedance, and are sensitive to a lagging-load power factor.
- Electronic tap-changing transformers — These devices use triacs or silicon-controlled rectifiers to change taps quickly on an autotransformer. They respond in 0.5 cycle and are insensitive to load power factor and voltage unbalances.
- Automatic voltage regulator — These devices function as an uninterruptible power supply with no energy storage. They have a fast response time (1 to 2 ms), but the need for a fully rated 60-Hz transformer can make their cost unacceptably high.
- Hybrid electronic voltage regulator — These devices use a series transformer and a power converter to accomplish the voltage-regulation function.
- Soft-switching automatic voltage regulator — These devices combine the high performance of active line filters with the lower cost of the more-conventional solutions. The electromagnetic interference generated by these units is low in spite of the high-frequency switching employed.
- Constant-voltage transformer (ferroresonant transformers) — Appropriate application of a CVT can handle most low-frequency disturbances, except for deep sags or outages. Detailed descriptions of these devices are provided in the subsequent sections.

2.8.1.5   What Constant-Voltage Transformers Can and Cannot Do

CVTs are attractive power conditioners because they are relatively maintenance-free; they have no batteries to replace or moving parts to maintain. They are particularly applicable to providing voltage-sag protection to industrial process-control devices such as programmable logic controllers (PLC), motor-starter coils, and the electronic control circuits of adjustable-speed drives.

Ongoing research [3] has demonstrated power-quality attributes of CVTs that include filtering voltage distortion and notched waveforms. Figure 2.8.7 depicts typical distorted and notched input voltages versus the filtered CVT output. Also, a CVT can practically eliminate oscillating transients caused by capacitor switching and can significantly dampen impulsive transients caused by lightning (see Figure 2.8.7).

To ensure full protection of sensitive electronic loads, CVTs may need to be coupled with other devices designed to mitigate dynamic disturbances. In addition, CVTs have been used for years for voltage isolation as well. Many plants install CVTs for voltage regulation. CVTs also offer protection for voltage swells. If properly sized, a CVT can regulate its output voltage during input voltage sags to 60% of nominal voltage for virtually any duration (see Figure 2.8.8).

However, many commercial and industrial facilities are not aware of most of the CVT's attractive features. At the same time, the CVT technology also has some negative characteristics, which in some applications may possibly outweigh its benefits. Some of these include:

- CVTs are not effective during momentary voltage interruptions or extremely deep voltage sags (generally below 50% of nominal).
- Because CVTs have relatively high output impedance, they produce large output drops during high current demands. As a result, conventionally sized CVTs cannot handle significant changes in current and are more attractive for constant, low-power loads.
- Because CVTs are physically large devices, it is not always practical to install this type of device in either a small-office or home environment.
- CVTs produce heat and noticeable operating hum and are sensitive to line-frequency variations.
- CVTs can have relatively poor efficiency and high losses for light loading conditions.

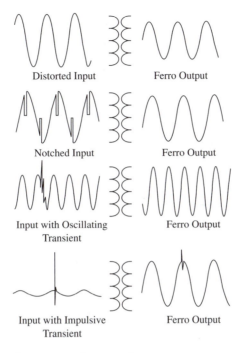

FIGURE 2.8.7 Filtering ability of a constant-voltage transformer.

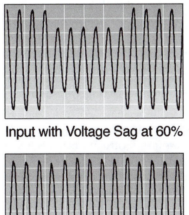

FIGURE 2.8.8 Voltage regulations with a constant-voltage transformer during a voltage sag to 60%.

2.8.2 Applications

Constant-voltage transformers have proven to be a reliable means of enhancing voltage-sag tolerance of industrial process-control elements such as relays, contactors, solenoids, dc power supplies, programmable logic controllers (PLCs), and motor starters. As mentioned earlier, while the CVT tends to work well for voltage sags, they are not a good solution for momentary interruptions. Additionally,

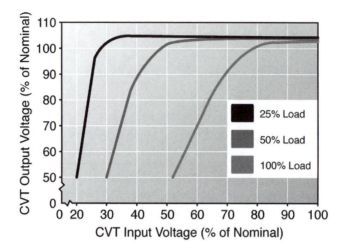

FIGURE 2.8.9 Typical performance of a constant-voltage transformer (CVT) as a function of load; as the CVT load increases, the ability of the CVT to regulate its output voltage decreases

there are sizing and fault-protection issues associated with proper application of the CVT, particularly when the output loads demand high-inrush starting currents. Specific issues that will be addressed in the subsequent sections include:

- Relationship between CVT input and output voltage under steady-state and dynamic supply conditions
- Relationship between CVT sizing and load size to enhance sag tolerance of a given load
- Impact of load inrush current on CVT sizing; output performance during various loading conditions
- Effect of different vendor designs in enhancing voltage-sag ride-through; three-phase designs versus single-phase designs

2.8.2.1 Application Considerations — Sizing Guidelines

Because the type of loads connected to a single CVT can range widely, the startup and steady-state operational characteristics of each load must be well understood before deciding on the appropriate power rating of a CVT. A load draws inrush current when it is first turned on or when it cycles on and off during normal process operation. If a CVT is sized [8] without considering the inrush currents of all connected loads, the CVT may be inadequately sized for the inrush current. Thus, during the startup or cycling of a connected load, the CVT output voltage may sag, causing other sensitive loads connected to the same CVT output voltage to shut down.

The ability of a CVT to regulate its output voltage is generally based upon two characteristics of the connected loads, both of which are related to current and both of which must be determined to properly size a CVT. The first characteristic is the amount of steady-state current drawn by all connected loads during their normal operation. As shown in Figure 2.8.9, the lower the ratio between the actual current drawn by the connected loads and the rated current of the CVT, the better the CVT can regulate its output voltage during dynamic load-switching events. As an illustration, a 1-kVA CVT loaded to 1 kVA will not mitigate voltage sags nearly as well as the same CVT loaded to 500 VA, and performance is even better if the same 1-kVA CVT is only loaded to 250 VA. Moreover, according to results of CVT testing at the Electric Power Research Institute's Power Electronics Application Center (known as EPRI PEAC), a CVT rated at less than 500 VA may not be able to handle even moderate inrush current. Therefore, a minimum CVT rating of 500 VA is recommended.

The second characteristic of a CVT load is the load's inrush current. Values for inrush current and steady-state current of the connected loads will enable a CVT to be properly sized.

TABLE 2.8.1 Typical Sizing Worksheet for a Constant-Voltage Transformer

CVT Circuit or Load	Measured Steady-State Current, A rms	Measured Peak Inrush Current, A (1 msec)
Programmable logic controller	0.16	14.8
Programmable logic controller	0.36	10.8
5-V, 12-V power supply	1.57	29.1
24-V power supply	1.29	14.4
NEMA size 3 motor starter	0.43	9.9
NEMA size 0 motor starter	0.13	3.1
Ice-cube relay	0.05	0.2
Master control relay	0.09	1.8
Computational Section		
Sum of steady-state rms currents	4.08	
Circuit voltage	× 120	
Steady-state load VA	= 490	× 2.5 = 1225
Highest peak inrush current	29.1	
Circuit voltage	× 120	
Inrush load VA	= 3492	× 0.5 = 1746

Note: Use the larger of the steady-state load VA (1225 in this example) and the inrush load VA (1746 in this example) to determine the CVT size.

A procedure to find the proper CVT size is described below:

1. Measure or estimate the total steady-state current drawn by the load and multiply this value by the circuit voltage to get steady-state VA. For optimum regulation during input-voltage sags, the VA rating of the CVT should be at least 2.5 times the steady-state VA calculated.
2. Measure the highest peak inrush current and multiply this value by the circuit voltage to get the worst-case inrush VA for all loads. For good sag regulation of the CVT output voltage during load starting or cycling, the VA rating of the CVT should be at least half of the maximum inrush VA. Recommended size of the CVT is based upon the larger of the two VA-rating calculations.
3. Add together all the steady-state currents and then multiply the resulting value by the circuit voltage to get the combined steady-state VA of all CVT loads. Then, select the highest peak-inrush-current measurement and multiply this value by the circuit voltage to get the worst-case inrush VA for all loads. For optimum regulation during input-voltage sags, the VA rating of the CVT should be at least 2.5 times the steady-state VA. For example, if the steady-state VA calculation is 490 VA, then the recommended size of the CVT would be 1225 VA or more. For good sag regulation of the CVT output voltage during load starting or cycling, the VA rating of the CVT should be at least half of the maximum inrush VA calculated. For example, if the maximum inrush VA is 3.49 kVA, then the optimum size of the CVT would be 1.75 kVA or more. A typical sizing worksheet for CVTs with measured data and calculations is shown in Table 2.8.1.

2.8.2.2 Application Considerations — Output Performance Under Varying Supply Conditions

A series of structured tests have been performed using a 1000-VA, 120-V, single-phase CVT. The following results of these tests provide insight on the operational characteristics of CVTs. These tests were designed to characterize the regulation performance of a constant-voltage transformer during amplitude and frequency variations in the ac input voltage and to determine its ability to filter voltage distortion and notching. The tests were performed at the EPRI PEAC power-quality test facility [9]. The CVT was energized for more than 30 min before each test to stabilize its temperature.

2.8.2.2.1 Performance: Regulation

The CVT was tested for its ability to regulate variations in input voltage amplitude at both half- and full-load levels for the three load types given in Table 2.8.2. The mixed nonlinear load was a combination of purely resistive loads and a 187-VA nonlinear load with a 0.85 power factor, resulting in a composite

TABLE 2.8.2 Characteristics of Load Type Used in the Test

Measured Characteristic	Load Type					
	Resistive		Resistive/Inductive		Mixed Nonlinear	
	½ Load	Full Load	½ Load	Full Load	½ Load	Full Load
Apparent power, VA	435	953	475	948	603	993
True power factor	1.00	1.00	0.79	0.79	0.99	0.99
Displacement power factor	1.00	1.00	0.80	0.79	1.00	1.00

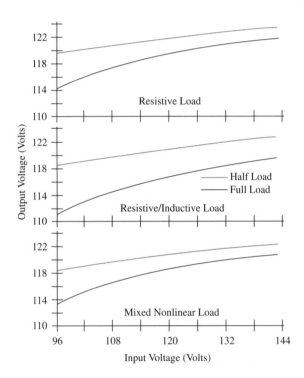

FIGURE 2.8.10 Output-voltage regulation with varying input ac voltage.

true power factor of 0.99. The amplitude of the ac input voltage was varied from 96 Vac (80% V nominal) to 144 Vac (120% V nominal) in 6-V increments.

The input voltage and resulting output voltage of the CVT for each load type are shown in Figure 2.8.10. The ferroresonant transformer effectively reduces or eliminates the effects of several kinds of voltage variations in the electric-service supply. When the amplitude of the input voltage was varied ±20%, the fully loaded transformer had an output voltage well within ANSI-C84.1 limits (from +6% to −13%).

2.8.2.2.2 *Performance: Frequency*
The CVT was half loaded with a purely resistive load. While the input voltage was fixed at 120 V, the frequency of the ac input voltage was varied from 50 Hz to 70 Hz in 1-Hz increments. The output voltage amplitude changed proportionally with the change in input frequency, ranging from 80% V nominal voltage at 50 Hz to 120% V nominal voltage at 70 Hz (see Figure 2.8.11).

2.8.2.2.3 *Performance: Harmonics*
To test the ability of the transformer to filter out harmonics, a distorted input voltage (15.29% V total harmonic distortion [THD]) with 15.1% third-harmonic content was applied. The resulting output harmonic distortion was 2.9%, with mostly fifth-harmonic content (2.3%). The distorted input voltage and the filtered output voltage is shown in Figure 2.8.12.

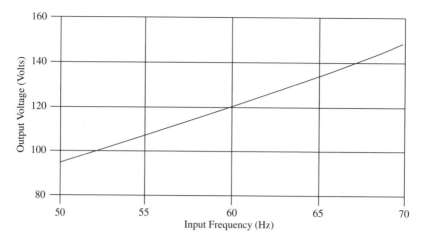

FIGURE 2.8.11 Output-voltage amplitude resulting from variations in frequency of the ac input voltage.

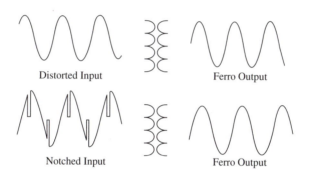

FIGURE 2.8.12 Output voltages resulting from distorted and notched input voltages at full load.

2.8.2.2.4 Performance: Notching

A notched voltage was applied to the input of the fully loaded transformer. As shown in Figure 2.8.12, the transformer successfully filtered the notched input voltage.

2.8.2.3 Application Considerations — Output Performance under Dynamic Supply Conditions

The objective of this application was to characterize the CVT performance during power system disturbances such as momentary interruptions, sags, phase shifts, capacitor switching, and lightning strikes [10]. The CVT was connected to a full, purely resistive load.

2.8.2.3.1 Performance: Voltage Interruption

The CVT was subjected to voltage interruptions lasting from 0.5 to 5 cycles and adjusted in 0.5-cycle increments. Switching from the normal supply voltage source to an open circuit created each interruption. The input and output voltages of the transformer during a three-cycle interruption are shown in Figure 2.8.13.

During the three-cycle interruption, the CVT dropped out of the regulating mode (refer to jump resonance shown in Figure 2.8.5), and the output voltage decreased as the resonant capacitor in the CVT discharged. When the input voltage returned to normal, an overshoot occurred on the output voltage as the resonant capacitor recharged. Note that the CVT does not act as an uninterruptible power supply, which is designed to eliminate the effect of an interruption on the electric service supply,

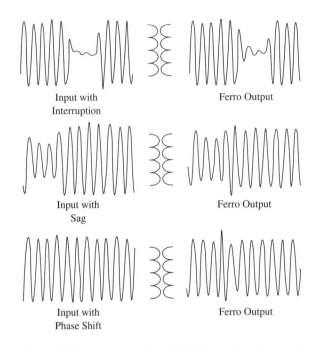

FIGURE 2.8.13 Output voltages (labeled ferro-output) resulting from voltage interruption, sag, and phase shift on the input.

but it does slightly decrease the effect of an interruption by reducing it to a deep voltage sag. In conclusion, a CVT can moderately mitigate, but not eliminate, the effects of momentary voltage interruptions.

2.8.2.3.2 Performance: Voltage Sag

To simulate a fault in the load's electric-service supply, the input voltage to the CVT was decreased to 90%, 70%, and 58% of nominal voltage for one, two, three, four, and five cycles. The variations in the output voltages were measured and recorded for each of the applied voltage sags. Figure 2.8.13 shows the CVT's input and output voltage waveforms down to 58% nominal voltage for sags lasting three cycles. In general, the input voltage sags produced output voltage sag having approximately the same duration but smaller sag depth. For example, during the nominal voltage sag to 58%, the CVT stayed in the regulation mode and reduced the effect of the sag to approximately 20% of nominal voltage lasting three cycles. Again, the recharging of the resonating capacitor caused an overshoot when the input voltage returned to normal.

2.8.2.3.3 Performance: Voltage Phase Shift

To simulate the effects of a large load being switched off near the end of a long electric-service supply feeder, the input voltage of the CVT was shifted forward 10° while the input and output voltages were monitored and recorded. The phase shift occurred at the positive peak of the input voltage (see Figure 2.8.13). The resulting phase shift in the input voltage caused the output voltage to briefly swell. The CVT is sensitive to voltage phase shifts in the electric-service supply.

2.8.2.3.4 Performance: Oscillating Transient

A transient caused by capacitor switching was simulated with a 500-Hz ring wave with a peak magnitude of 1 per unit and duration of 10 ms. The ring wave was applied to the positive peak of the input voltage. Figure 2.8.14 shows the input voltage with the ring-wave transient and the output voltage with a small impulsive transient.

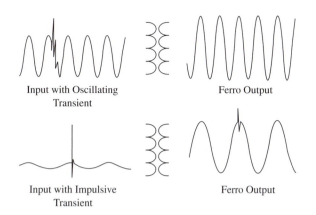

FIGURE 2.8.14 Output voltages resulting from oscillating and impulsive transients in the input voltages.

2.8.2.3.5 *Performance: Impulsive Transient*

To simulate a lightning strike, a 2-kV, 1.2/50-μs combination wave (as described in ANSI C62.41-1991) was applied to the positive peak of the input voltage. As shown in Figure 2.8.14, the CVT significantly damped and filtered the surge.

2.8.2.4 Application Considerations — CVT Electrical Characteristics during Linear and Nonlinear Loading

The objective of this application was to characterize the CVT as a load while the CVT supplied a simple linear load and while it supplied a complex nonlinear load [11]. In the following tests, the CVT was connected first to a simple linear load and then to a complex nonlinear load. An electric-service supply source with an average total harmonic distortion in the voltage of 3% supplied power to the CVT during all tests.

2.8.2.4.1 *Performance: Line Current Distortion*

A resistive linear load consisting of incandescent lamps was connected to the output of the CVT. The load was increased in ten equal increments from 0 to 8.3 A (output current rating of the CVT). Next, a bridge rectifier (such as the type that might be used in electric-vehicle battery chargers) was connected to the CVT. The rectifier and its resistive load (incandescent lamps) were the complex nonlinear load of the CVT. By adding lamps, this complex load was increased in ten equal increments from approximately 0.4 A (rectifier with no lamps connected) to 8.3 A. Figure 2.8.15 and Figure 2.8.16 show the line-current distortion during these tests compared with the line-current distortion for the same loads connected directly to the electric-service supply. At no load, the power consumption of the CVT was approximately 120 W (core losses only). With the full linear load, total losses increased to approximately 134 W (core losses plus load losses); with the full nonlinear load, total losses dropped to approximately 110 W.

Notice in Figure 2.8.15 and Figure 2.8.16 that, while the *y*-axis current-distortion magnitudes are significantly different, the absolute current-distortion values of the CVT's input current with either linear or nonlinear load is nearly identical. Current distortion at the CVT's input terminals was practically independent of the type of load connected to the output (approximately 40% at no loading to approximately 5% at full loading). When a linear, low-distortion load was connected to the CVT output, the CVT contributed to the current distortion at its input terminals from the electric-service power source, particularly during low loading. When a nonlinear, high-distortion load was connected, the CVT substantially reduced load-current distortion. When fully loaded, the CVT had relatively small power consumption and an efficiency of 85% to 90%. As opposed to most voltage regulators, the losses of the CVT decreased as the nonlinear load increased. The CVT also significantly affected the power factor of the load.

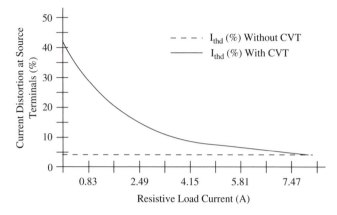

FIGURE 2.8.15 Line-current distortions for a linear load.

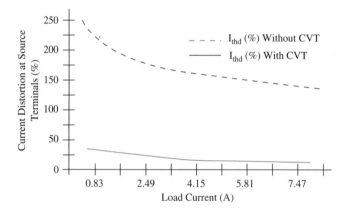

FIGURE 2.8.16 Line-current distortions for a nonlinear load.

2.8.2.4.2 *Performance: Power Factor*

For both linear and nonlinear loads, the size of the load affected the input power factor of the CVT. While the CVT was loaded at less than 40% of its output power rating (approximately 3.3 A), the power factor ranged from 0.65 to 0.95. While the CVT was loaded at greater than 40%, the power factor was greater than 0.95 for the linear load and greater than 0.90 for the nonlinear load. For the linear load, the power factor crossed from lagging to leading at approximately 60% load (approximately 5 A). Figure 2.8.17 and Figure 2.8.18 show the power factors for the linear and nonlinear load (without and with the CVT), respectively. The CVT significantly affected the power factor of the load. At low loading, the nonlinear load without the CVT had a power factor as low as 0.44. With the CVT, the total power factor of the nonlinear load ranged from 0.61 to near unity. However, when loaded at less than 50%, the CVT significantly reduced the power factor for the linear, resistive load, which normally has a unity power factor. Note that in most CVT applications, the aggregate facility loading is significantly small, so it would not be prudent to attempt any power-factor correction at individual CVT operating loads. Power-factor-correction initiatives should be accomplished at the electric service meter of the facility.

2.8.2.5 Application Considerations — Using Three-Phase Input

One of the drawbacks of using a CVT is its inability to protect equipment from voltage interruptions. A traditional CVT can protect equipment down to approximately 40% of nominal voltage. A company in the midwestern U.S. has introduced a prototype CVT that protects equipment from deep voltage sags and brief power interruptions. As shown in Figure 2.8.19, the ride-through transformer (RTT) is designed

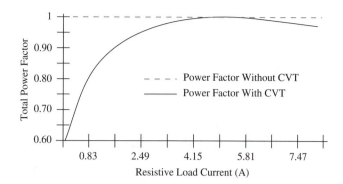

FIGURE 2.8.17 Power factor for a linear load.

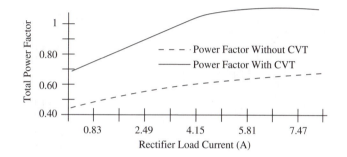

FIGURE 2.8.18 Power factor for a nonlinear load.

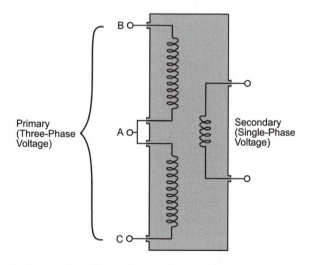

FIGURE 2.8.19 Schematic of a ride-through transformer.

to protect single-phase process controls. Unlike traditional CVTs, the RTT uses all three phases of supply voltage as its input. This enables the RTT to access energy in unsagged phases of the supply voltage during one- or two-phase voltage sags and interruptions.

EPRI PEAC tested the prototype, 1-kV, 480-V RTT [12] to determine its ability to protect process controls during single-phase, two-phase, and three-phase voltage sags and interruptions. The particular prototype acquired for testing was connected to a load bank that consisted of a mixture of 12 industrial

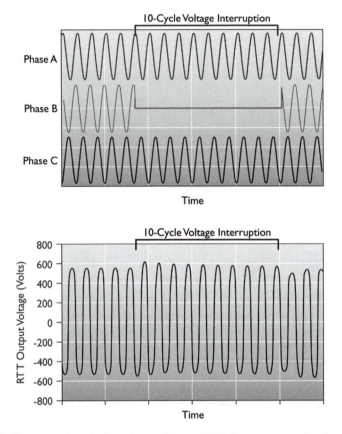

FIGURE 2.8.20A Performance of a ride-through transformer (RTT) during a ten-cycle voltage interruption and voltage sag. Voltage regulation of an RTT during a single-phase voltage interruption (top: input; bottom: output).

control components: ice-cube relays, motor starters, contactors, a programmable logic controller, a linear dc power supply, and a switch-mode power supply.

The test results revealed that the prototype RTT protected the connected process controls from most of the applied voltage sags and interruptions. Besides, it was observed that RTT performance greatly depended on the phase configuration (that is, single-, two-, or three-phase) of the voltage sags or interruption and, to a much lesser extent, on the loading of the RTT output. It was observed that the RTT performed like a typical CVT during three-phase voltage sags. Figure 2.8.20a and Figure 2.8.20b show the response of an RTT to phase-to-neutral and phase-to-phase sags.

To get the most out of a CVT with a three-phase input, the most trouble-free voltage phases of the electric-service supply will have to be determined. For example, if most voltage sags occur on phase A or B, then the center tap on the transformer primary should be connected to phase C. Although this prototype transformer promises to retail at a price substantially higher than the price of a traditional, single-phase CVT, the price differential can be greatly reduced by a reduction in size. Because the performance of a traditional CVT greatly depends upon loading, CVTs are often oversized for the connected load. A smaller but more loaded RTT should be able to perform as well as the derated, traditional CVT.

2.8.3 Procurement Considerations

This section describes what to look for when purchasing CVTs; typical prices of various CVT ratings; and typical sizes, weights, efficiencies, and service conditions of CVTs.

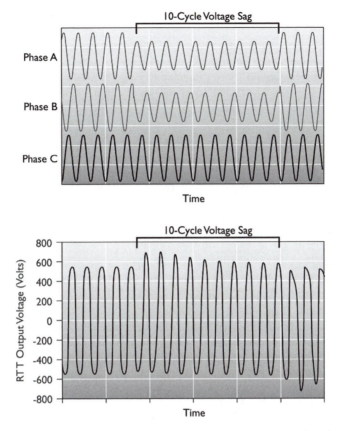

FIGURE 2.8.20B Performance of a ride-through transformer (RTT) during a ten-cycle voltage interruption and voltage sag. Voltage regulation of an RTT during a two-phase ten-cycle voltage sag to 60% of nominal (top: input; bottom: output).

2.8.3.1 What To Look for When Purchasing a Constant-Voltage Transformer

Most CVT manufacturers guarantee their products will meet their published specifications [2]. In general, most CVTs are specified to recognized industrial standards, and under well-defined conditions (agreed to by the CVT manufacturer) they may be capable of operating outside their published limits. Although CVTs are highly reliable and are effective at stabilizing variations in the electric-service supply, it would be prudent to review the following application issues with each CVT manufacturer before purchasing a CVT:

2.8.3.1.1 *Effect of High Input and High Output Voltages upon Operation of Input Protection Devices*

Establish the proper input fuse and/or circuit-breaker rating required for the application. Identify with the CVT manufacturer what the maximum primary rms current is under secondary short-circuit conditions for determining the rating of the fuse or circuit breaker. Be aware that sizing fuses and circuit breakers for conventional transformers is not applicable when sizing fuses and/or circuit breakers for CVTs. This is because CVTs are high-impedance units (typically ten times higher than a control transformer) compared with the equivalent-rated conventional transformer. Therefore, it is essential that the guidelines in the CVT manufacturer's specification should be consulted before finalizing the fuse sizing and/or circuit-breaker rating requirements.

With the correct input fuse or circuit breaker, the application should work fine until the protection opens at approximately 150% of the nominal input voltage. In some cases, after the circuit protection opens, the CVT's output voltage will rise with increasing input at approximately 20% of the increase (that is, the output voltage may go up 1% for each 5% the input voltage rises). This situation may "overvoltage" the load you want to protect. That's why this issue should be discussed with each CVT manufacturer.

2.8.3.1.2 Operating at Low Input Voltage
If the CVT is operated at significantly lower input voltage continuously, the output voltage will sag as the input voltage sags. Underloading the CVT unit will greatly improve the output voltage performance of the CVT. Discuss with each CVT manufacturer to what extend the CVT would have to be "underloaded" to prevent this situation from occurring.

2.8.3.1.3 Nonsinusoidal Input Voltage
If the correct pure fundamental-frequency voltage is applied to the CVT, it will operate. Also, most CVTs can tolerate an input-voltage THD up to 25% or even a square-wave input for short-term durations. Discuss with each CVT manufacturer how their CVT will perform with various levels of voltage THD versus application duration. It would also be good to request the CVT's range of input-frequency variance versus the CVT's output-voltage variance characteristics.

2.8.3.1.4 Overloading
Depending on the actual level of the CVT's input voltage, the CVT may deliver up to 50% more power than specified. Beyond this level, the CVT will protect itself by reducing the output voltage progressively until it reaches nearly zero. The CVT can be operated into a short circuit indefinitely. Whether the CVT can handle the inrush current of any single load or group of loads will depend on the sequencing of the loads and the aggregate inrush-current level. To establish the proper CVT rating for the intended application, discuss both the steady-state and dynamic inrush-current profiles with each CVT manufacturer.

2.8.3.1.5 Non-Unity-Power-Factor Loads
Linear inductive loads depress the CVT's output voltage and can usually be corrected by adding load capacitors. Linear capacitive loads have the opposite effect on CVTs. If the loads are nonlinear, ask each CVT manufacturer how its CVT will be affected.

2.8.3.1.6 Switching Lo ads
Ordinary switched-mode power supplies (SMPS) are particularly suited to be used with most CVTs. But when CVTs have "self-adjusting" arrangements, care must be taken because problems could arise with some dimmer or phased-controlled load circuits. In these kinds of applications, each CVT manufacturer should specifically document its CVT product's performance regarding the CVT's compatibility with dimmer and phased-controlled loads.

2.8.3.1.7 Low-Ambient-Temperature Performance
Most CVTs can tolerate ambient temperatures down to 0°C. Lower than 0°C, the capacitor becomes the limiting factor. If the ambient temperature of the intended CVT application is less than 0°C, discuss the application with each CVT manufacturer on what CVT modifications are possible for operating the CVT at less than 0°C.

2.8.3.1.8 High-Ambient-Temperature Performance
Most CVTs can tolerate ambient temperatures up to 50°C. Short-term operation at temperatures up to 70°C may be possible. Again, the capacitor becomes the limiting factor. If the intended CVT application is at temperatures higher than 50°C, discuss the application with each CVT manufacturer on what CVT modifications are possible for operating the CVT at greater than 50°C. Note that most CVT manufacturers report that for every 5°C above 50°C, expect the life expectancy of the CVT to be halved from the calculated 300,000-h mean time before failure (MTBF).

TABLE 2.8.3 Service Conditions

Service Conditions	Range
Ambient temperature	0–50°C
Relative humidity	20–90%
Altitude	0–1500 m
Type of cooling	Convection cooling is assumed

TABLE 2.8.4 Storage Conditions

Storage Conditions	Range
Storage temperature	–40 to +60°C
Relative humidity	5–90%
Altitude	0–2000 m

TABLE 2.8.5 Shipment Conditions

Shipment Conditions	Range
Temperature	–55°C to +60°C
Relative humidity	5–95%
Altitude	0–12,000 m

2.8.3.1.9 High-Humidity Performance

Most CVTs can operate at 90% relative humidity without problems. If the CVT has been stored at 100% relative humidity, it probably will require drying out before starting up. Discuss with each CVT manufacturer its proposed procedures for drying out CVTs and the consequences of operating the CVT continuously at relative humidity greater than 90%.

2.8.3.1.10 Failed-Capacitor Performance

If the CVT unit has several capacitors and one fails, the CVT may still provide reduced power. Shorted capacitors will stop the CVT's operation, but open-circuit capacitor failures can be tolerated. Problems will occur at CVT turn-on if the unit is operated at high input voltage and light loads when a capacitor has failed. If the CVT makes a "humping" or "motor-boating" noise, the CVT should be turned off and then on again. The effects of failed capacitors on each manufacturer's CVT may be different, so request each manufacturer to explain the CVT's operating scenario under a failed-capacitor condition.

2.8.4 Typical Service, Storage, and Shipment Conditions

Table 2.8.3 lists typical service conditions for ambient temperature, relative humidity, altitude, and type of cooling. If forced-air cooling is available or required, the direction of airflow, velocity, temperature, and volume flow per minute at the constant-voltage transformer are included in the specification. For liquid-cooled units, the type of coolant, rate of flow, and the inlet coolant temperature range are specified.

Table 2.8.4 lists storage temperature, relative humidity, and altitude ranges for a typical CVT.

Table 2.8.5 lists shipment conditions as it relates to temperature, relative humidity, and altitude ranges for a typical CVT.

2.8.5 Nameplate Data and Nomenclature

The following are important rating factors associated with a CVT:

- Input frequency or frequency range
- Input voltage range
- Input watts or volt-amperes
- Input current
- Output voltage
- Output current
- Output watts or volt-amperes
- Maximum and/or minimum ambient temperature
- Schematic diagram or connection information
- Maximum working voltage
- Resonant-capacitor information

2.8.6   New Technology Advancements

Because CVTs are primarily constructed with three major components — magnetic core, magnet wire, and a capacitor — a consensus exists among a number of leading CVT manufacturers that, while nothing revolutionary is expected in the near-future, their transformers will continue to be enhanced in many ways. Whatever CVT innovations do occur will essentially be in improving CVT assembly procedures and techniques. In general, advances in CVT design are focused on reducing CVT sizes and audible noise levels and increasing efficiencies at all load conditions.

It is worth mentioning that the CVT user market seems to be moving toward controlled CVTs because of their very precise output-voltage regulation, their ability to easily adjust the output voltage to exactly the desired reference required, and their extraordinary immunity to becoming unstable in certain loading applications. In a number of field situations, controlled[3] CVTS can be customer-adjusted for the specific application.

2.8.7   Addendum

The following is a tutorial description of the difference between a ferroresonant circuit and a linear circuit. The main differences between a ferroresonant circuit and a linear resonant circuit are, for a given ω:

- Resonance occurs when the inductance is in saturation.
- As the value of inductance in saturation is not known precisely, a wide range of capacitances can potentially lead to ferroresonance at a given frequency.
- The frequency of the voltage and current waves may be different from that of the sinusoidal voltage source.
- Initial conditions (initial charge on capacitors, remaining flux in the core of the transformers, switching instant) determine which steady-state response will result.

A study of the free oscillations of the circuit in Figure 2.8.21a illustrates this specific behavior. Losses are assumed negligible, and the simplified magnetization curve $\phi(i)$ of the iron-core coil is that represented in Figure 2.8.21b. Despite these simplifying assumptions, the corresponding waveforms (see Figure 2.8.21c) are typical of a periodic ferroresonance.

---

[3]The principles of operation for the controlled CVT are in that the transformer's output winding is on the same leg of the magnetic core as the resonant winding, and the resonant capacitor acts to maintain this core section at a high level of saturation, resulting in a fairly constant voltage. To provide a precise constant voltage, it is necessary to control this level of core saturation. This is frequently accomplished by shunting the resonant circuit with a solid-state switching device in series with an inductor.

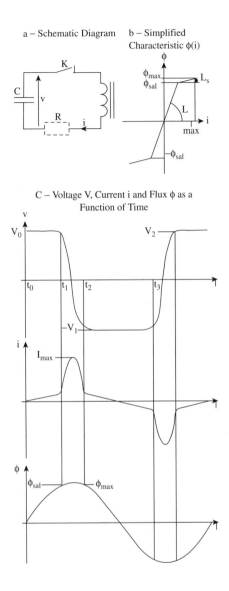

FIGURE 2.8.21

Originally, voltage at the capacitance terminals is assumed equal to V_0. At the instant t_0 switch K closes, a current i is created, and oscillates at the pulsation

$$\varpi_1 = \frac{1}{\sqrt{LC}} \tag{2.8.5}$$

The flux in the coil and voltage V at the capacitor terminals are then expressed as:

$$\phi = (V_0/\varpi_1)\,\sin(\varpi_1 t) \tag{2.8.6}$$

$$V = V_0\,\sin(\varpi_1 t) \tag{2.8.7}$$

If $(V_0/\omega_1) > \phi_{sat}$ at the end of time t_1, the flux ϕ reaches the saturation flux ϕ_{sat}, voltage V is equal to V_1, and the inductance of the saturated coil becomes L_S. As L_S is very small compared with L, the capacitor suddenly "discharges" across the coil in the form of an oscillation of pulsation

$$\varpi_2 = \frac{1}{\sqrt{L_S C}} \qquad (2.8.8)$$

The current and flux peak when the electromagnetic energy stored in the coil is equivalent to the electrostatic energy $1/2 CV^2$ restored by the capacitor.

At instant t_2, the flux returns to ϕ_{sat}, the inductance reassumes the value L, and since the losses have been ignored, voltage V, which has been reversed, is equal to $-V_1$.

At instant t_3, the flux reaches $-\phi_{sat}$ and voltage V is equal to $-V_2$. As ω_1 is in practice very small, we can consider $V_2 \approx V_1 \approx V_0$. Consequently, period T of the oscillation is included between $2\pi\sqrt{LC}$ in the *nonsaturated* case and $2\pi\sqrt{L_S C} + 2(t_3 - t_2)$ in the *saturated* case, where $(t_3 - t_2) \approx \dfrac{2\phi_{sat}}{V_0}$. The corresponding frequency f (f = 1/T) is thus such that:

$$\frac{1}{2\pi\sqrt{LC}} < f < \frac{1}{2\pi\sqrt{L_S C}}$$

This initial frequency depends on ϕ_{sat}, i.e., on the nonlinearity and the initial condition V_0. In practice, due to the losses Ri^2 in the resistance R, the amplitude of voltage V decreases ($V_2 < V_1 < V_0$). Because the flux varies as follows,

$$\Delta\phi = 2\phi_{stat} = \int_{t_2}^{t_3} v dt$$

a decrease of V results in a reduction in frequency. If the energy losses are supplied by the voltage source in the system, the frequency of the oscillations, as it decreases, can lock at the frequency of the source (if the initial frequency is greater than the power frequency) or even submultiple frequency of the source frequency (if the initial frequency is smaller than the power frequency).

Note that there can be four resonance types, namely fundamental mode, subharmonic mode, quasi-periodic mode, or chaotic mode (see Figure 2.8.22).

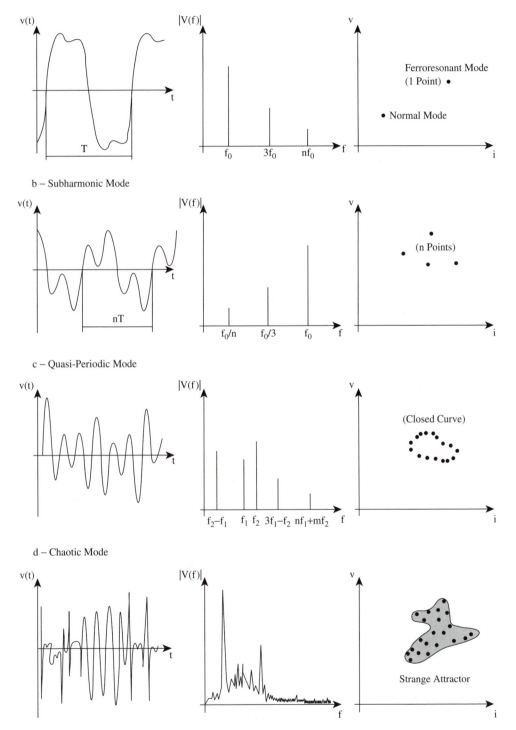

FIGURE 2.8.22 Waveforms typical of a periodic ferroresonance.

References

1. Sola/Hevi-Duty Corp., About Sola/Hevi-Duty, www.sola-hevi-duty.com/about/solahist.html, October 18, 2002.
2. Advance Galatrek, CVT Background Data, http://www.aelgroup.co.uk/hb/hb003.htm, October 18, 2002.
3. EPRI, System Compatibility Projects to Characterize Electronic Equipment Performance under Varying Electric Service Supply Conditions, EPRI PEAC, Knoxville, TN, May 1993.
4. Godfrey, S., Ferroresonance, http://www.physics.carleton.ca/courses/75.364/mp-1html/node7.html, October 18, 2002.
5. Cadicorp, Ferro-Resonance, Technical Bulletin 004a, www.cadicorp.com, October 18, 2002.
6. Groupe Schneider, Ferroresonance, No. 190, www.schneiderelectric.com, October 19, 2002.
7. IEEE, Standard for Ferroresonant Voltage Regulators, IEEE Std. 449-1998, Institute of Electrical and Electronics Engineers, Piscataway, NJ, 1998.
8. EPRI, Sizing Constant-Voltage Transformers to Maximize Voltage Regulation for Process Control Devices, PQTN Application No. 10, EPRI PEAC, Knoxville, TN, October 1997.
9. EPRI, Ferro-Resonant Transformer Output Performance under Varying Supply Conditions, PQTN Brief No. 13, EPRI PEAC, Knoxville, TN, May 1993.
10. EPRI, Ferro-Resonant Transformer Output Performance under Dynamic Supply Conditions, PQTN Brief No. 14, EPRI PEAC, Knoxville, TN, January 1994.
11. EPRI, Ferro-Resonant Transformer Input Electrical Characteristics during Linear and Nonlinear Loading, PQTN Brief No. 16, EPRI PEAC, Knoxville, TN, February 1994.
12. EPRI, Testing a Prototype Ferro-Resonant Transformer, EPRI PEAC, Knoxville, TN, unpublished.

2.9 Reactors

Richard F. Dudley, Michael Sharp, Antonio Castanheira, and Behdad Biglar

Reactors, like capacitors, are basic to and an integral part of both distribution and transmission power systems. Depending on their function, reactors are connected either in shunt or in series with the network. Reactors are connected either singularly (current-limiting reactors, shunt reactors) or in conjunction with other basic components such as power capacitors (shunt-capacitor-switching reactors, capacitor-discharge reactors, filter reactors).

Reactors are utilized to provide inductive reactance in power circuits for a wide variety of purposes, including fault-current limiting, inrush-current limiting (for capacitors and motors), harmonic filtering, VAR compensation, reduction of ripple currents, blocking of power-line carrier signals, neutral grounding, damping of switching transients, flicker reduction for arc-furnace applications, circuit detuning, load balancing, and power conditioning.

Reactors can be installed at any industrial, distribution, or transmission voltage level and can be rated for any current duty from a few amperes to tens of thousands of amperes and fault-current levels of up to hundreds of thousands of amperes.

2.9.1 Background and Historical Perspective

Reactors can be either dry type or oil immersed. Dry-type reactors can be of air-core or iron-core construction. In the past, dry-type air-core reactors were only available in open-style construction (Figure 2.9.1), their windings held in place by a mechanical clamping system and the basic insulation provided by the air space between turns. Modern dry-type air-core reactors (Figure 2.9.2) feature fully encapsulated windings with the turns insulation provided by film, fiber, or enamel dielectric. Oil-immersed reactors can be of gapped iron-core (Figure 2.9.3) or magnetically shielded construction. The application range

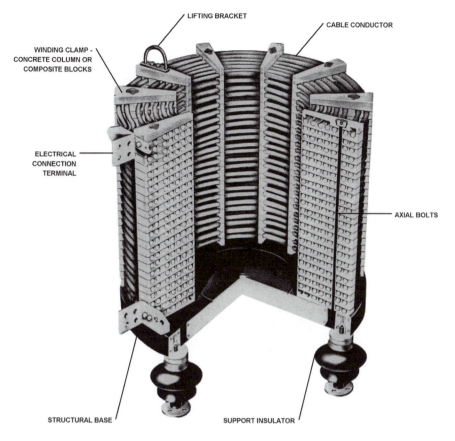

LIFTING BRACKET

CABLE CONDUCTOR

WINDING CLAMP -
CONCRETE COLUMN OR
COMPOSITE BLOCKS

ELECTRICAL
CONNECTION
TERMINAL

AXIAL BOLTS

STRUCTURAL BASE

SUPPORT INSULATOR

FIGURE 2.9.1 Open-style reactor.

for the different reactor technologies has undergone a major realignment from historical usage. In the past, dry-type air-core reactors (open-style winding technology) were limited to applications at distribution-voltage class. Modern dry-type air-core reactors (fully encapsulated with solid-dielectric-insulated windings) are employed over the full range of distribution and transmission voltages, including high voltage (HV) and extra high voltage (EHV) ac transmission voltage classes (high-voltage series reactors) and high-voltage direct-current (HVDC) systems (ac and dc filter reactors, smoothing reactors). Oil-immersed reactors are primarily used for EHV-shunt-reactor and for some HVDC-smoothing-reactor applications. Dry-type iron-core reactors (Figure 2.9.4) are usually used at low voltage and indoors for applications such as harmonic filtering and power conditioning (di /dt, smoothing, etc.). Applicable IEEE standards, such as IEEE C57.21-1990 (R 1995), IEEE C57.16-1996 (R 2001), and IEEE 1277-2000, reflect these practices. [6,8,9]

These standards provide considerable information not only concerning critical reactor ratings, operational characteristics, tolerances, and test code, but also guidance for installation and important application-specific considerations.

2.9.2 Applications of Reactors

2.9.2.1 General Overview

Reactors have always been an integral part of power systems. The type of technology employed for the various applications has changed over the years based on design evolution and breakthroughs in construction and materials. Dry-type air-core reactors have traditionally been used for current-limiting applications due to their inherent linearity of inductance vs. current. For this application, fully

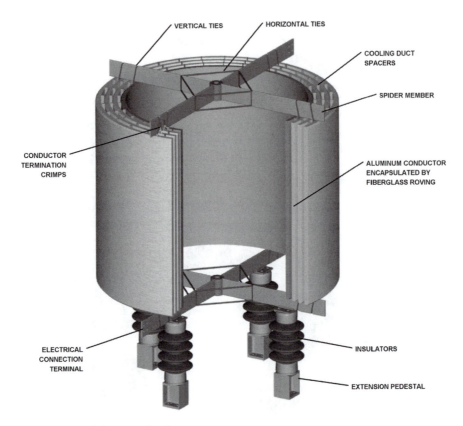

FIGURE 2.9.2 Modern fully encapsulated reactor.

encapsulated construction became the design of choice because its improved mechanical characteristics enabled the reactors to withstand higher fault currents. Conversely, high-voltage (HV) series reactors were initially oil-immersed shielded-core designs. However, beginning in the early 1970s, the requirements of these applications were also met by fully encapsulated dry-type air-core designs. Due to such developments, the latest revision of IEEE C57.16-1996 (R 2001), the series-reactor standard, is a now a dry-type air-core-reactor standard only.

The construction technology employed for modern shunt reactors, on the other hand, is more dependent on the applied voltage. Transmission-class shunt reactors are of oil-immersed construction, whereas tertiary connected, or lower-voltage direct-connect shunt reactors, utilize either dry-type air-core or oil-immersed construction. Hence ANSI/IEEE C57.21-1990 (R 1995) covers both oil-immersed and dry-type air-core shunt reactors.

A review of modern reactor applications is presented in the following subsections.

2.9.2.2 Current-Limiting Reactors

Current-limiting reactors are now used to control short-circuit levels in electrical power systems covering the range from large industrial power complexes to utility distribution networks to HV and EHV transmission systems.

Current-limiting reactors (CLR) are primarily installed to reduce the short-circuit current to levels consistent with the electromechanical withstand level of circuit components (especially transformers and circuit breakers) and to reduce the short-circuit voltage drop on bus sections to levels that are consistent with insulation-coordination practice. High fault currents on distribution or transmission systems, if not limited, can cause catastrophic failure of distribution equipment and can present a serious threat to the

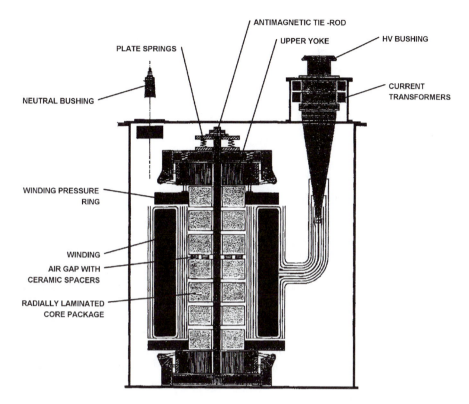

FIGURE 2.9.3   Gapped iron-core oil-immersed reactor.

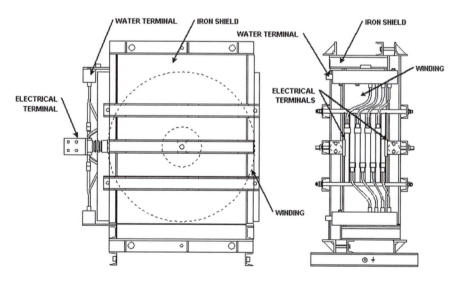

FIGURE 2.9.4   Dry-type iron-core reactor (water cooled).

safety of operating crews. In summary, current-limiting reactors are installed to reduce the magnitude of short-circuit currents in order to achieve one or more of the following benefits:

• Reduction of electromechanical loading and thermal stresses on transformer windings, thus extending the service life of transformers and associated equipment

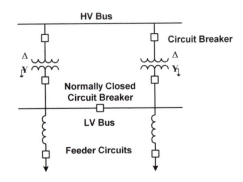

FIGURE 2.9.5 Typical phase-reactor connection.

- Improved stability of primary bus voltage during a fault event on a feeder
- Reduction of current-interrupting duty of feeder circuit breakers
- Reduction of line-to-line fault current to levels below those of line-to-ground faults or vice versa
- Protection of distribution transformer and all other downstream power equipment and devices from the propagation of initial fast-front voltage transients due to faults and/or circuit-breaker operations
- Reduced need for downstream protection devices such as reclosers, sectionalizers, and current-limiting fuses
- Ability to obtain complete control over the steady-state losses by meeting any specified Q-factor for any desired frequency, a feature that is particularly important for networks where high harmonic currents are to be damped without increasing the fundamental-frequency loss
- Increase in system reliability

Current-limiting reactors can be installed at different points in the power network, and as such, they are normally referred to by a name that reflects their location/application. The most common nomenclatures are:

- Phase reactors, installed in series with incoming or outgoing lines or feeders
- Bus-tie reactors, used to tie together two otherwise independent buses
- Neutral-grounding reactors, installed between the neutral of a transformer and ground
- Duplex reactors, installed between a single source and two buses

2.9.2.2.1 Phase Reactors
Figure 2.9.5 shows the location of phase reactors in the feeder circuits of a distribution system. Current-limiting reactors can also be applied at transmission-voltage levels, as shown in Figure 2.9.6, depicting 345-kV, 1100-A, 94.7-mH phase reactors installed in a U.S. utility substation.

The main advantage of the illustrated configuration is that it reduces line-to-line and phase-to-ground short-circuit current to any desired level at a strategic location in the distribution or transmission system.

Phase reactors are one of the most versatile embodiments of series-connected reactors in that they can be installed (1) in one feeder of a distribution system to protect one circuit or (2) at the output of a generator to limit fault contribution to the entire power grid or (3) anywhere in between. Any of the benefits listed in the Section 2.9.2.2, Current-Limiting Reactors, can be achieved with this type of current-limiting reactor.

The impedance of the phase reactor required to limit the 3Φ short-circuit current to a given value can be calculated using either Equation 2.9.1 or Equation 2.9.2.

$$X_{CLR} = V_{LL} \left[(1/I_{SCA}) - (1/I_{SCB}) \right] / \sqrt{3} \tag{2.9.1}$$

$$X_{CLR} = V^2_{LL} \left[(1/MVA_{SCA}) - (1/MVA_{SCB}) \right] \tag{2.9.2}$$

FIGURE 2.9.6 345-kV phase reactors.

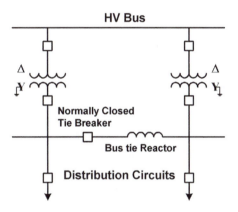

FIGURE 2.9.7 Typical bus-tie-reactor connection.

where

X_{CLR} = reactance of the current limiting reactor, Ω

V_{LL} = rated system voltage, kV

I_{SCA} or MVA_{SCA} = required value of the short-circuit current or power after the installation of the phase reactor, kA or MVA, respectively

I_{SCB} or MVA_{SCB} = available value of the short-circuit current or power before the installation of the phase reactor, kA or MVA, respectively

2.9.2.2.2 Bus-Tie Reactors

Bus-tie reactors are used when two or more feeders and/or power sources are connected to a single bus and it is desirable to sectionalize the bus without losing operational flexibility. Figure 2.9.7 illustrates the arrangement. As in the case of phase reactors, bus-tie reactors may be applied at any voltage level. Figure 2.9.8, shows an example of a 230-kV, 2000-A, 26.54-mH bus-tie reactor installed in a U.S. utility substation. The advantages of this configuration are similar to those associated with the use of phase reactors. An added benefit is that if the load is essentially balanced on both sides of the reactor under normal operating conditions, the reactor has negligible effect on voltage regulation or system losses.

The required reactor impedance is calculated using either Equation 2.9.1 or Equation 2.9.2 for 3Φ faults or Equation 2.9.3 for single line-to-ground faults.

FIGURE 2.9.8 230-kV bus-tie reactors.

A method to evaluate the merits of using either phase reactors or bus-tie reactors is presented in Section 2.9.3.2, Phase Reactors vs. Bus Tie Reactors.

2.9.2.2.3 *Neutral-Grounding Reactors*

Neutral-grounding reactors (NGR) are used to control single line-to-ground faults only. They do not limit line-to-line fault-current levels. They are particularly useful at transmission-voltage levels, when autotransformers with a delta tertiary are employed. Figure 2.9.9 shows a typical neutral-grounding reactor.

These transmission-station transformers can be a strong source of zero-sequence currents, and as a result, the ground-fault current may substantially exceed the 3Φ-fault current. These devices, normally installed between the transformer or generator neutral and ground, are effective in controlling single line-to-ground faults, since the system short-circuit impedance generally is largely reactive. NGRs reduce short-circuit stresses on station transformers for the most prevalent type of fault in an electrical system.

If the objective is to reduce the single line-to-ground (1Φ) fault, then Equations 2.9.14 and 2.9.14a must be used and, after algebraic manipulations, the required neutral reactor impedance, in Ohms, is calculated as:

$$X_{NGR} = \sqrt{3} \times V_{LL} \left[(1/I_{SCA}) - (1/I_{SCB}) \right]/3 \qquad (2.9.3)$$

where,

X_{NGR} = reactance of the neutral grounding reactor, Ω

V_{LL} = system line-to-line voltage kV

I_{SCA} = required single line-to-ground short circuit current after the installation of the neutral reactor, kA.

I_{SCB} = available single line-to-ground short curcuit current before installation of neutral reactor, kA

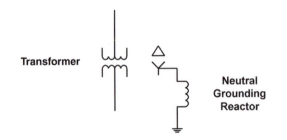

FIGURE 2.9.9 Typical neutral-grounding-reactor connection.

where the parameters are defined as before, with the exception that the short-circuit currents in question are the single line-to-ground fault currents expressed in units of kA.

A factor to be taken into consideration when applying NGRs is that the resulting X_0/X_1 may exceed a critical value ($X_0 > 10X_1$) and, as a result, give rise to transient overvoltages on the unfaulted phases. (For more information, see IEEE Std. 142-1991, Green Book.)[14]

Because only one NGR is required per three-phase transformer and because their continuous current is the system-unbalance current, the cost of installing NGRs is lower than that for phase CLRs. Operating losses are also lower than for phase CLRs, and steady-state voltage regulation need not be considered with their application.

The impedance rating of a neutral-grounding reactor can be calculated using Equation 2.9.1, provided that the short-circuit currents before and after the NGR installation are single line-to-ground faults.

Although NGRs do not have any direct effect on line-to-line faults, they are of significant benefit, since most faults start from line-to-ground, some progressing quickly to a line-to-line fault if fault-side energy is high and the fault current is not interrupted in time. Therefore, the NGR can contribute indirectly to a reduction in the number of occurrences of line-to-line faults by reducing the energy available at the location of the line-to-ground fault.

2.9.2.2.3.1 Generator Neutral-Grounding Reactors — The positive-, negative-, and zero-sequence reactances of a generator are not equal, and as a result, when its neutral is solidly grounded, its line-to-ground short-circuit current is usually higher than the three-phase short-circuit current. However generators are usually required to withstand only the three-phase short-circuit current, [1] and a grounding reactor or a resistor should be employed to lower the single line-to-ground fault current to an acceptable limit. Other reasons for the installation of a neutral-grounding device are listed below:

- A loaded generator can develop a third-harmonic voltage, and when the neutral is solidly grounded, the third-harmonic current can approach the generator rated current. Providing impedance in the grounding path can limit the third-harmonic current.
- When the neutral of a generator is solidly grounded, an internal ground fault can produce large fault currents that can damage the laminated core, leading to a lengthy and costly repair procedure.

The ratings of generator neutral-grounding reactors can be calculated as follows:

1. The reactance value required for limiting a single line-to-ground fault to the same value as a three-phase fault current can be calculated by the following formula:

$$X_{NGR} = \frac{X''_d - X_{m0}}{3} \qquad (2.9.4)$$

where

X_{NGR} = reactance of the neutral-grounding reactor, Ω
X_d'' = direct axis subtransient reactance of the machine, Ω
X_{m0} = zero-sequence reactance of the machine, Ω

Equation 2.9.4 assumes that the negative-sequence reactance is equal to X_d''. In the absence of complete information, the value of reactance calculated using Equation 2.9.4 is satisfactory. Equations presented in chapter 19 of the Westinghouse Electrical Transmission and Distribution Reference Book [2] can also be used when the negative-sequence reactance of the generator is not equal to X_d''.

2. The short-circuit-current rating of the grounding reactor is equal to the three-phase generator short-circuit current when the machine is an isolated generator or operating in a unit system (Figure 2.9.10). When more than one generator is connected to a shared bus and not all of them are grounded by a reactor with a reactance calculated from Equation 2.9.4, the short-circuit-current rating of the reactor in question should be calculated by using proper system constants for a single line-to-ground fault at the terminal of the machine being grounded. Equations in

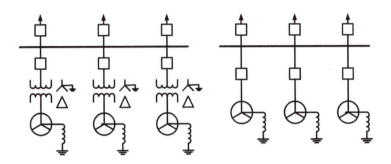

FIGURE 2.9.10 Two generator arrangements where a reactor can be used as a neutral-grounding device: unit system (left) and three-wire system (right).

chapter 19 of the Westinghouse Electrical Transmission and Distribution Reference Book [2] can be employed for this case.

3. The rated short-circuit duration for the grounding reactor shall also be specified. When reactors are used for a single isolated generator or in a unit system, a 10-sec rating is usually employed. When grounding reactors are used in systems having feeders at generator voltage, a 1-min rating is usually employed to accommodate for repetitive feeder faults.

4. The rated continuous current of the grounding reactor should be specified considering the allowable unbalance current and third-harmonic current. In the absence of this information, 3% of the reactor short-circuit rating can be specified as the continuous-current rating when the duration of the rated short-circuit current is 10 sec. In cases when the rated short-circuit duration is 1 min, 7% of rated short-circuit current is recommended as the rated continuous current. See IEEE Std. 32-1972 for more information. [3]

5. Insulation class and associated BIL (basic impulse insulation level) rating should be specified based on (1) the reactor voltage drop during a single line-to-ground fault and (2) the nominal system voltage. Refer to Table 4 of IEEE Std. 32-1972. [3]

Transient overvoltages are another important consideration in the application of generator neutral-grounding reactors. When the neutral of a generator is not solidly grounded, transient overvoltages can be expected. These overvoltages are usually caused by phase-to-ground arcing faults in air or by a switching operation followed by one or more restrikes in the breaker. When a grounding reactor is used for generator neutral grounding, the X_0/X_1 ratio at the generator terminals shall be less than three to keep transient overvoltages within an acceptable level. (X_0 and X_1 are the resultant of the generator and system zero- and positive-sequence reactances, respectively.) When a neutral-reactor installation is intended only for reduction of a single line-to-ground fault to the three-phase fault level, X_0/X_1 is equal to unity, which is a safe ratio in terms of imposed transient overvoltages.

2.9.2.2.4   Arc-Suppression Reactors (Petersen Coils)

An arc-suppression coil is a single-phase, variable-inductance, oil-immersed, iron-core reactor that is connected between the neutral of a transformer and ground for the purpose of achieving a resonant neutral ground. The zero-sequence impedance of the transformer is taken into consideration in rating the inductance of the arc-suppression coil. The adjustment of inductance is achieved in steps by means of taps on the winding, or inductance can be continuously adjusted by varying the reluctance of the magnetic circuit. The length of an air gap is adjusted by means of a central moveable portion of the core (usually motor driven). See Figure 2.9.11. The inductance is adjusted, in particular during nonground-fault conditions, to achieve cancellation of the capacitive ground-fault current, so that in the case of a single line-to-ground fault, cancellation of the capacitive fault current is achieved with an inductive current of equal magnitude. Current injection by an active component (power converter) into the neutral, usually through an auxiliary winding of the arc-suppression coil, can also provide cancellation of the resistance component of the fault current. See Figure 2.9.12 and Figure 2.9.13, which illustrate this

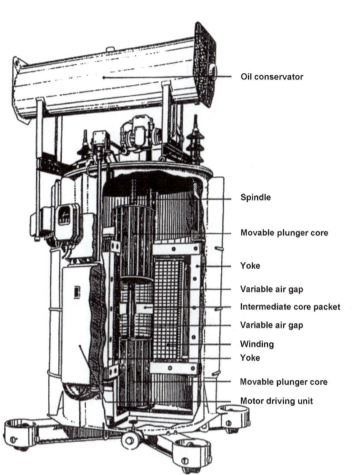

FIGURE 2.9.11 Arc-suppression reactor.

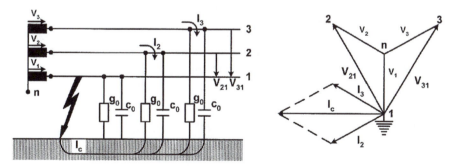

FIGURE 2.9.12 Single line-to-ground fault in nongrounded system.

principle for nongrounded and resonant-grounded systems, respectively. Resonant grounding is used in distribution systems in Europe, parts of Asia and in a few areas of the U.S. The type of system ground employed is a complex function of system design, safety considerations, contingency (fault) operating practices, and legislation. Arc-suppression reactors are typically used to the best advantage on distribution systems with overhead lines to reduce the intermittent arcing-type single line-to-ground faults that may otherwise occur on ungrounded systems.

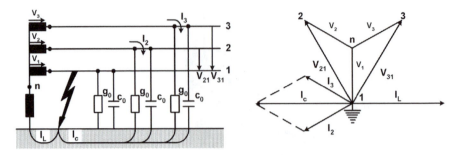

FIGURE 2.9.13 Single line-to-ground fault in resonant-grounded system.

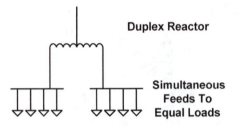

FIGURE 2.9.14 Typical duplex-reactor connection.

2.9.2.2.5 Duplex Reactors

Duplex reactors are usually installed at the point where a large source of power is split into two simultaneously and equally loaded buses (Figure 2.9.14). They are designed to provide low-rated reactance under normal operating conditions and full-rated or higher reactance under fault conditions. A duplex reactor consists of two magnetically coupled coils per phase. This magnetic coupling, which is dependent upon the geometric proximity of the two coils, determines the properties of a duplex reactor under steady-state and short-circuit operating conditions. During steady-state operation, the magnetic fields produced by the two windings are in opposition, and the effective reactance between the power source and each bus is a minimum. Under short-circuit condition, the linking magnetic flux between the two coils becomes unbalanced, resulting in higher impedance on the faulted bus, thus restricting the fault current. The voltage on the unfaulted bus is supported significantly until the fault is cleared, both by the effect of the reactor impedance between the faulted and unfaulted bus and also by the "voltage boosting" effect caused by the coupling of the faulted leg with the unfaulted leg of duplex reactors.

The impedance of a duplex reactor can be calculated using Equation 2.9.1 and Equation 2.9.2, the same as those used for phase reactors.

2.9.2.3 Capacitor Inrush/Outrush Reactors

Capacitor switching can cause significant transients at both the switched capacitor and remote locations. The most common transients are:

- Overvoltage on the switched capacitor during energization
- Voltage magnification at lower-voltage capacitors
- Transformer phase-to-phase overvoltages at line termination
- Inrush current from another capacitor during back-to-back switching
- Current outrush from a capacitor into a nearby fault
- Dynamic overvoltage when switching a capacitor and transformer simultaneously

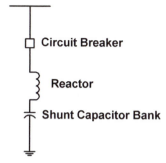

FIGURE 2.9.15 Typical capacitor inrush/outrush reactor connection.

Capacitor inrush/outrush reactors (Figure 2.9.15) are used to reduce the severity of some of the transients listed above in order to minimize dielectric stresses on breakers, capacitors, transformers, surge arresters, and associated station electrical equipment. High-frequency-transient interference in nearby control and communication equipment is also reduced.

Reactors are effective in reducing all transients associated with capacitor switching, since they limit the magnitude of the transient current (Equation 2.9.5), in kA, and significantly reduce the transient frequency (Equation 2.9.6), in Hz.

$$I_{peak} = V_{LL} * \sqrt{(C_{eq.} / L_{eq.})} \tag{2.9.5}$$

$$f = 1 / \left[2\pi \sqrt{\left(L_{eq.} C_{eq} \right)} \right] \tag{2.9.6}$$

where

C_{eq} = equivalent capacitance of the circuit, F
L_{eq} = equivalent inductance of the circuit, H
V_{LL} = system line-to-line voltage, kV

Therefore, reflecting the information presented in the preceding discussion, IEEE Std. 1036-1992, Guide for Application of Shunt Power Capacitors, calls for the installation of reactors in series with each capacitor bank, especially when switching back-to-back capacitor banks.

Figure 2.9.16 shows a typical EHV shunt-capacitor installation utilizing reactors rated at 550 kV/1550 kV BIL, 600 A, and 3.0 mH.

FIGURE 2.9.16 550-kV capacitor inrush/outrush reactors.

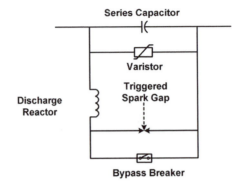

FIGURE 2.9.17 Typical discharge-current-limiting reactor connection.

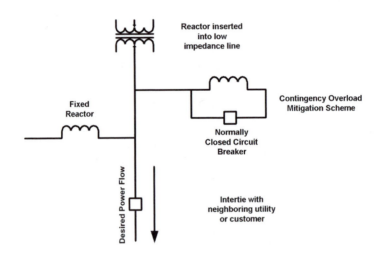

FIGURE 2.9.18 Typical high-voltage power-flow-control reactor connections.

2.9.2.4 Discharge-Current-Limiting Reactors

High-voltage series-capacitor banks are utilized in transmission systems to improve stability operating limits. Series-capacitor banks can be supplied with a number of discrete steps, insertion or bypass being achieved using a switching device. For contingencies, a bypass gap is also provided for fast bypass of the capacitors. In both cases, bypass switch closed or bypass gap activated, a discharge of the capacitor occurs, and the energy associated with the discharge must be limited by a damping circuit. A discharge-current-limiting reactor is an integral part of this damping circuit. Therefore, the discharge-current-limiting reactor must be designed to withstand the high-frequency discharge current superimposed on the system power-frequency current. The damping characteristic of this reactor is a critical parameter of the discharge circuit. Sufficient damping can be provided as an integral component of the reactor design (de-Q'ing), or it can be supplied as a separate element (resistor). See Figure 2.9.17.

2.9.2.5 Power-Flow-Control Reactors

A more recent application of series reactors in transmission systems is that of power-flow control (Figure 2.9.18) or its variant, overload mitigation. The flow of power through a transmission system is a function of the path impedance and the complex voltage (magnitude and phase) at the ends of the line. In interconnected systems, the control of power flow is a major concern for the utilities, because unscheduled power flow can give rise to a number of problems, such as:

- Overloading of lines
- Increased system losses
- Reduction in security margins
- Contractual violations concerning power import/export
- Increase in fault levels beyond equipment rating

Typical power-flow inefficiencies and limitations encountered in modern power systems may be the result of one or more of the following:

- Nonoptimized parallel line impedances resulting in one line reaching its thermal limit well before the other line, thereby limiting peak power transfer
- Parallel lines having different X/R ratios, where a significant reactive component flows in the opposite direction to that of the active power flow
- High-loss line more heavily loaded than lower-loss parallel line, resulting in higher power-transfer losses
- "Loop flow" (the difference between scheduled and actual power flow), although inherent to interconnected systems, can be so severe as to adversely affect the system reliability

Power-flow-control reactors are used to optimize power flow on transmission lines through a modification of the transfer impedance. As utility systems grow and the number of interties increases, parallel operation of ac transmission lines is becoming more common in order to provide adequate power to load centers. In addition, the complexity of contemporary power grids results in situations where the power flow experienced by a given line of one utility can be affected by switching, loading, and outage conditions occurring in another service area. Strategic placement of power-flow reactors can serve to increase peak power transfer, reduce power-transfer loss, and improve system reliability. The paper, "A Modern Alternative for Power Flow Control, [4] provides a good case study. The insertion of high-voltage power-flow-control reactors in a low-impedance circuit allows parallel lines to reach their thermal limits simultaneously and hence optimize peak power transfer at reduced overall losses. Optimum system performance can be achieved by insertion of one reactor rating to minimize line losses during periods of off-peak power transfer and one of an alternative rating to achieve simultaneous peak power transfer on parallel lines during peak load periods or contingency conditions.

Contingency overload-mitigation reactor schemes are used when the removal of generation sources and/or lines in one area affects the loading of other lines feeding the same load center. This contingency can overload one or more of the remaining lines. The insertion of series reactors, shunted by a normally closed breaker, in the potentially overloaded line(s) keeps the line current below thermal limits. The parallel breaker carries the line current under normal line-loading conditions, and the reactor is switched into the circuit only under contingency situations.

2.9.2.6 Shunt Reactors (Steady-State Reactive Compensation)

High-voltage transmission lines, particularly long ones, generate a substantial amount of leading reactive power when lightly loaded. Conversely, they absorb a large amount of lagging reactive power when heavily loaded. As a consequence, unless the transmission line is operating under reactive power balance, the voltage on the system cannot be maintained at rated values.

reactive power balance = total line-charging capacitive VARs − line inductive VARs

To achieve an acceptable reactive power balance, the line must be compensated for a given operational condition. For details of the definition of reactive power balance, refer to Section 2.9.3.3, Power-Line Balance. Under heavy load, the power balance is negative, and capacitive compensation (voltage support) is required. This is usually supplied by the use of shunt capacitors. Conversely, under light load, the power balance is positive, and inductive compensation is required. This is usually supplied by the use of shunt reactors.

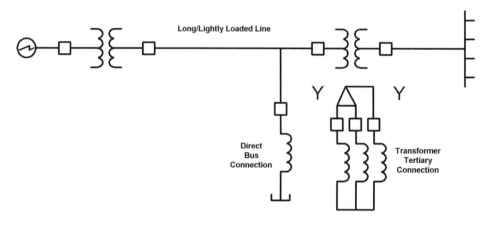

FIGURE 2.9.19 Typical shunt-reactor connections.

The large inherent capacitance of lightly loaded transmission systems can cause two types of overvoltage in the system that can be controlled by employing shunt reactors. The first type of overvoltage occurs when the leading capacitive charging current of a lightly loaded, long transmission line flows through the inductance of the line and the system. This is referred to as the Ferranti effect; operating voltage increases with distance along the transmission line. Lagging reactive current consumed by a shunt reactor reduces the leading capacitive charging current of the line and thus reduces the voltage rise.

Another type of overvoltage is caused by the interaction of line capacitance with any saturable portion of system inductive reactance, or ferroresonance. When switching a transformer-terminated line, the voltage at the end of the line can rise to a sufficient value to saturate the transformer inductance. Interaction between this inductance and the capacitance of the line can generate harmonics, causing overvoltages. Application of a shunt reactor on the tertiary of the transformer can mitigate this type of overvoltage by reducing the voltage to values below that at which saturation of the transformer core can occur, while also providing a low nonsaturable inductance in parallel with the transformer impedance.

Shunt reactors can be connected to the transmission system through a tertiary winding of a power transformer connected to the transmission line being compensated, typically 13.8, 34.5, and 69 kV (see Figure 2.9.20). Tertiary-connected shunt reactors (Figure 2.9.19) may be of dry-type, air-core, single-phase-per-unit construction, or oil-immersed three-phase construction, or oil-immersed single-phase-per-unit construction.

Alternatively, shunt reactors can be connected directly to the transmission line to be compensated (Figure 2.9.19). Connection can be at the end of a transmission line or at an intermediate point, depending on voltage-profile considerations. Directly connected shunt reactors are usually of oil-immersed construction; dry-type air-core shunt reactors are presently available only for voltages up to 235 kV.

For both tertiary and direct connected shunt reactors, protection is an important consideration. Details regarding protection practices can be found in the IEEE paper, "Shunt Reactor Protection Practices" [13] and also in the IEEE C37.109.[19]

Oil-immersed shunt reactors are available in two design configurations: coreless and iron core (and either self-cooled or force cooled). Coreless oil-immersed shunt-reactor designs utilize a magnetic circuit or shield, which surrounds the coil to contain the flux within the reactor tank. The steel core that normally provides a magnetic flux path through the primary and secondary windings of a power transformer is replaced by insulating support structures, resulting in an inductor that is nearly linear with respect to applied voltage.

Conversely, the magnetic circuit of an oil-immersed iron-core shunt reactor is constructed in a manner similar to that used for power transformers, with the exception that an air gap or distributed air gap is introduced to provide the desired reluctance. Because of the very high permeability of the core material, the reluctance of the magnetic circuit is dominated by the air gap, where magnetic energy is primarily

FIGURE 2.9.20 20-kV, 20-MVA (per phase) dry-type, air core, tertiary-connection shunt reactors.

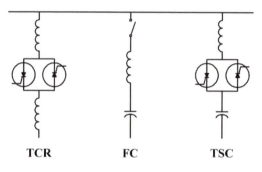

FIGURE 2.9.21 Static VAR compensator.

stored. Inductance is less dependent on core permeability, and core saturation does not occur in the normal steady-state-current operating range, resulting in a linear inductance. A distributed air gap is employed to minimize fringing flux effects, to reduce winding eddy losses (adjacent to the gap[s]), and improve ampere-turns efficiency. Both types of oil-immersed shunt reactors can be constructed as single-phase or three-phase units and are similar in appearance to conventional power transformers.

2.9.2.7 Thyristor-Controlled Reactors (Dynamic Reactive Compensation)

As the network operating characteristics approach system limits, such as dynamic or voltage stability, or in the case of large dynamic industrial loads, such as arc furnaces, the need for dynamic compensation arises. Typically, static VAR compensators (SVC) are used to provide dynamic compensation at a receiving end bus, through microprocessor control, for maintaining a dynamic reserve of capacitive support when there is a sudden need.

Figure 2.9.21 illustrates a typical configuration for an SVC. Figure 2.9.22a shows the voltage and current in one phase of a TCR when the firing angle (α) of the thyristor(s) is not zero. Figure 2.9.22b depicts the various harmonic current spectra, as a percent of fundamental current, generated by the TCR for various firing angles α. By varying the firing angle, α, of the thyristor-controlled reactor (TCR), the amount of current absorbed by the reactor can be continuously varied. The reactor then behaves as an infinitely variable inductance. Consequently, the capacitive support provided by the fixed capacitor (FC) and/or by the thyristor-switched capacitor (TSC) can be adjusted to the specific need of the system.

The efficiency, as well as voltage control and stability, of power systems is greatly enhanced with the installation of SVCs. The use of SVCs is also well established in industrial power systems. Demands for increased production and the presence of stricter regulations regarding both the consumption of reactive

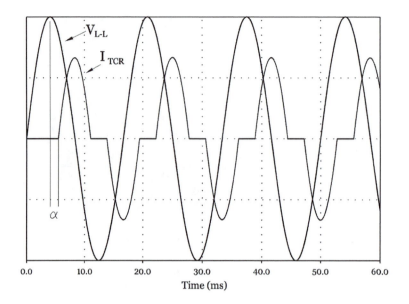

FIGURE 2.9.22A   TCR current and voltage waveforms.

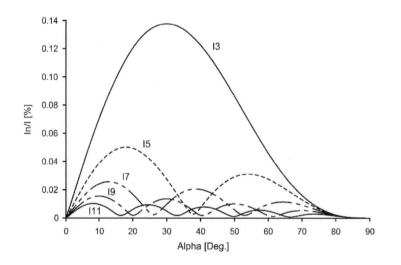

FIGURE 2.9.22B   TCR harmonic-current spectra as a percentage of fundamental current.

power and disturbance mitigation on the power system may require the installation of SVCs. A typical example of an industrial load that can cause annoyance to consumers, usually in the form of flicker, is the extreme load fluctuations of electrical arc furnaces in steel works. A typical installation at a steel mill is shown in Figure 2.9.23. The thyristor-controlled reactors are rated at 34 kV, 710 A, and 25 MVAr per phase.

2.9.2.8   Filter Reactors

The increasing presence of nonlinear loads and the widespread use of power electronic-switching devices in industrial power systems is causing an increase of harmonics in the power system. Major sources of harmonics include industrial arcing loads (arc furnaces, welding devices), power converters for variable-speed motor drives, distributed arc lighting for roads, fluorescent lighting, residential sources such as TV sets and home computers, etc.

FIGURE 2.9.23 34-kV, 25-MVAR (per phase) thyristor-controlled reactors.

Power electronic-switching devices are also applied in modern power-transmission systems and include HVDC converters as well as FACTS (flexible ac transmission system) devices such as SVCs. Harmonics can have detrimental effects on equipment such as transformers, motors, switchgear, capacitor banks, fuses, and protective relays. Transformers, motors, and switchgear can experience increased losses and excessive heating. Capacitors can fail prematurely from increased heating and higher dielectric stress. If distribution feeders and telephone lines have the same "right of way," harmonics can also cause telephone interference problems.

In order to minimize the propagation of harmonics into the connected power distribution or transmission system, shunt filters are often applied close to the origin of the harmonics. Such shunt filters, in their simplest embodiment, consist of a series inductance (filter reactor) and capacitance (filter capacitor). Figure 2.9.24 shows a typical filter-reactor connection. If more than one harmonic is to be filtered, several sets of filters of different rating are applied to the same bus. More-complex filters are also used to filter multiple harmonics. More background information can be found in the IEEE paper, "Selecting Ratings for Capacitors and Reactors in Applications Involving Multiple Single Tuned Filters." [5]

2.9.2.9 Reactors for HVDC Application

In an HVDC system, reactors are used for various functions, as shown, in principle, in Figure 2.9.25. The HVDC-smoothing reactors are connected in series with an HVDC transmission line or inserted in the intermediate dc circuit of a back-to-back link to reduce the harmonics on the dc side, to reduce the current rise caused by faults in the dc system, and to improve the dynamic stability of the HVDC transmission system.

Filter reactors are installed for harmonic filtering on the ac and on the dc side of the converters. The ac filters serve two purposes simultaneously: the supply of reactive power and the reduction of harmonic currents. The ac filter reactors are utilized in three types of filter configurations employing combinations of resistors and capacitors, namely single-tuned filters, double-tuned filters, and high-pass filters. A single-tuned filter is normally designed to filter the low-order harmonics on the ac side of the converter. A double-tuned filter is designed to filter multiple discrete frequencies using a single combined filter circuit. A high-pass filter is essentially a single-tuned damped filter. Damping flattens and extends the filter response to more effectively cover high-order harmonics. The dc filter reactors are installed in shunt with the dc line, on the line side of the smoothing reactors. The function of these dc filter banks is to further reduce the harmonic currents on the dc line (see Figure 2.9.24 and Figure 2.9.25).

PLC (power-line carrier) and RI (radio interference) filter reactors are employed on the ac or dc side of the HVDC converter to reduce high-frequency noise propagation in the lines.

2.9.2.10 Series Reactors for Electric-Arc-Furnace Application

Series reactors can be installed in the medium-voltage feeder (high-voltage side of the furnace transformer) of an ac electric-arc furnace (EAF) to improve efficiency, reduce furnace electrode consumption,

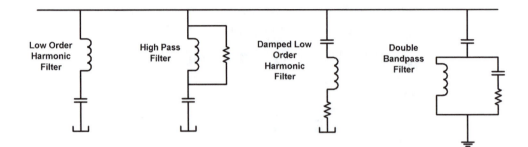

FIGURE 2.9.24 Typical filter-reactor connections.

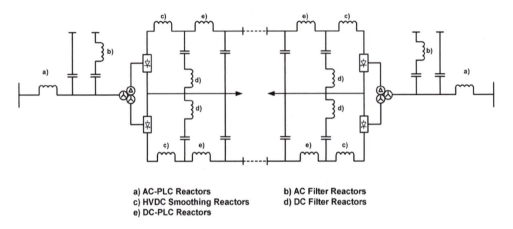

a) AC-PLC Reactors   b) AC Filter Reactors
c) HVDC Smoothing Reactors   d) DC Filter Reactors
e) DC-PLC Reactors

FIGURE 2.9.25 One-line diagram of a typical HVDC bipole link illustrating reactor applications.

and limit short-circuit current (thus reducing mechanical forces on the furnace electrodes). Such reactors can either be "built into" the furnace transformer, or they can be separate, stand-alone units of oil-immersed or dry-type air-core construction. Usually, the reactors are equipped with taps to facilitate optimization of the furnace performance. See Figure 2.9.26 and Figure 2.9.27.

2.9.2.11 Other Reactors

Reactors are also used in such diverse applications as motor starting (current limiting), test-laboratory circuits (current limiting, dv/dt control, di/dt control), and insertion impedance (circuit switchers). Design considerations, insulation system, conductor design, cooling-method-construction concept (dry type, oil immersed), and subcomponent/subassembly variants (mounting and installation consider-ations) are selected based on the application requirements.

2.9.3 Some Important Application Considerations

2.9.3.1 Short Circuit: Basic Concepts

Figure 2.9.28 represents a radial system in which the sending-end bus is connected to an infinite source. From inspection, the following equations can be established:

$$V_S = V_R + \Delta V \qquad (2.9.7)$$

$$I = (V_R + \Delta V)/(Z_S + Z_T + Z_L) \qquad (2.9.8)$$

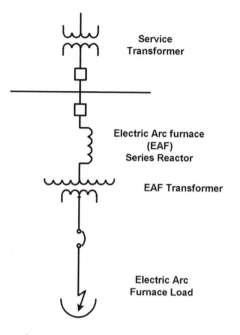

FIGURE 2.9.26 Typical electric-arc-furnace series-reactor connection.

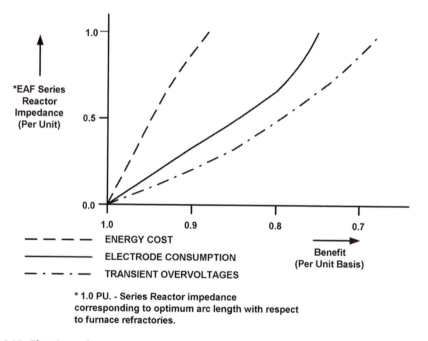

FIGURE 2.9.27 Electric-arc-furnace series-reactor benefits (reduced energy cost, reduced electrode consumption, reduced magnitude of transient over voltages.)

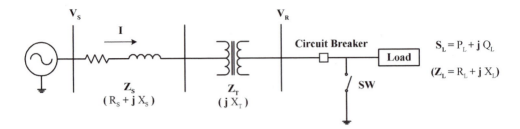

FIGURE 2.9.28 Radial system.

where

V_S = sending end voltage, p.u.

V_R = receiving end voltage, p.u.

ΔV = bus voltage drop, p.u.

I = bus current, p.u.

Z_S = source impedance, p.u.

Z_T = transformer impedance, p.u.

Z_L = load impedance, p.u.

Under steady state, the load impedance[4] Z_L essentially controls the current I, since both Z_S and Z_T are small. Also, typically $X_S \gg R_S$, $X_T \gg X_S$, and the load $P_L > Q_L$. Therefore,

$$\Delta V = V_S - V_R \cong [PR_S + Q(X_S + X_T)] \tag{2.9.9}$$

where

P = real power, p.u.

Q = reactive power, p.u.

R_S = source resistance, p.u.

X_S = source reactance, p.u.

X_T = transformer reactance, p.u.

ΔV = voltage drop, p.u.

and ΔV, per unit, is small

When a short circuit occurs (closing the switch SW), $V_R \to 0$ and $Z_L = 0$ (bolted fault), and Equation 2.9.8 can be rewritten as

$$I_{SC} = \Delta V / [R_S + j\ (X_S + X_T)] \tag{2.9.10}$$

---

[4]The load may, in fact, be a rotating machine, in which case the back electromotive force (EMF) generated by the motor is the controlling factor in limiting the current. Nevertheless, this EMF may be related to an impedance, which is essentially reactive in nature.

$$\cong \frac{\Delta V}{(X_s + X_T)} \angle - 90° \qquad (2.9.10a)$$

where, I_{SC} = short circuit current, p.u.

Since $|(X_S + X_T)|$ is small, the short-circuit current (I_{SC}) can become very large. The total transmitted power then equals the available power from the source (MVA_{SC}),

$$S = V_S I_{SC} \qquad (2.9.11)$$

$$S = \frac{V_s (\Delta V)}{(X_s + X_T)} \angle - 90° \qquad (2.9.11a)$$

where, S = transmitted power, p.u.

Therefore, as voltage drops, the system voltage is shared between the system impedance (transmission lines) and the transformer impedance:

$$V_S = \Delta V \cong Q_S (X_S + X_T) \qquad (2.9.12)$$

where, Q_S = transmitted reactive power, p.u.

Since X_T is typically much larger than X_S, the voltage drop across the transformer almost equals the system voltage. Two major concerns arise from this scenario:

1. Mechanical stresses in the transformer, with the windings experiencing a force proportional to the square of the current
2. Ability of the circuit breaker to successfully interrupt the fault current

Therefore, it is imperative to limit the short-circuit current so that it will not exceed the ratings of equipment exposed to it. Basic formulas are as follows:

Three-phase fault:

$$I_{3\Phi}, kA = 100 \, [MVA]/[\sqrt{3} \, V_{LL} \, Z_1] \qquad (2.9.13)$$

1Φ-to-ground fault:

$$I_{SLG}, kA = 3 \times 100 \, [MVA]/[\sqrt{3} \, V_{LL} \, Z_T] \qquad (2.9.14)$$

$$Z_T = Z_1 + Z_2 + Z_0 + 3Z_N \qquad (2.9.14a)$$

where

V_{LL} = line-to-line base voltage, kV

Z_T = total equivalent system impedance seen from the fault, p.u.

Z_1, Z_2, and Z_0 = equivalent system positive-, negative-, and zero-sequence impedance seen from the fault (in p.u. @ 100-MVA base)

Z_N = any impedance intentionally connected to ground in the path of the fault current, p.u.

2.9.3.2 Phase Reactors vs. Bus-Tie Reactors

A method to evaluate the merits of using phase reactors vs. bus-tie reactors is presented here. Refer to Figure 2.9.29 and Figure 2.9.30 and the following definitions:

I_{S1} and I_{S2} = available fault contribution from the sources, kA.

I_L = rated current of each feeder, kA.

n = total number of feeders in the complete bus (assuming all feeders are identical)

$I_b = \alpha I_{S2}$, the interrupting rating of the feeder circuit breakers, $\alpha > 1$, kA

$I_{S2} = K\ I_{S1}$ (I_{S2} assumed $< I_{S1}$, $K < 1$), kA
$I_{bt} = \beta\ I_L$, the rated current of the bus-tie reactor, kA
β = number of feeders on the section of the bus with the highest number of feeders

Therefore, the required reactor impedance, in Ohms, in each configuration, their ratio, and the ratio of the rated power of the reactors are given by Equations 2.9.15a–d:

$$X_{bt} = \frac{V_{LL}}{\sqrt{3}\ I_{S2}} \left[\frac{K(1-\alpha)+1}{\alpha K - 1} \right] \tag{2.9.15a}$$

$$X_{fd} = \frac{V_{LL}}{\sqrt{3}\ I_{S2}} \left[\frac{K(1-\alpha)+1}{\alpha(1+K)} \right] \tag{2.9.15b}$$

$$\frac{X_{bt}}{X_{fd}} = \frac{\alpha(1+K)}{\alpha K - 1} \tag{2.9.15c}$$

$$\frac{MVA_{bt}}{Total\ MVA_{fd}} = \frac{\beta^2}{n} \left[\frac{\alpha(1+K)}{\alpha K - 1} \right] \tag{2.9.15d}$$

From the above, it is apparent that bus-tie reactors are a good solution where a relatively small reduction in fault level is required on a number of downstream feeders. However, bus-tie reactors increase rapidly in size and cost when (1) the fault contributions on either side of the reactor are significantly different (i.e., as K moves away from 1.0) and (2) when the largest fault contribution (I_{S1}) approaches the breaker rating I_b. Conversely, bus-tie reactors decrease rapidly in size and cost when the reactor can be given a low continuous rating due to low normal power transfer across the tie.

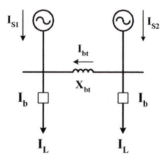

FIGURE 2.9.29 Bus-tie-reactor connection.

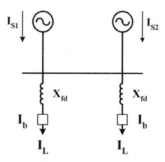

FIGURE 2.9.30 Phase-reactor connection.

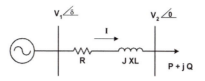

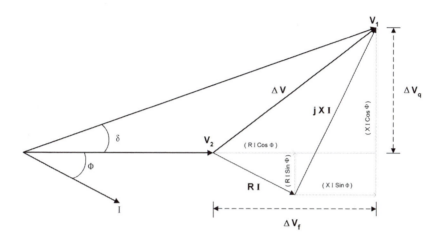

FIGURE 2.9.31 Simplified radial system.

FIGURE 2.9.32 Power-line-balance phasor diagram.

2.9.3.3 Power-Line Balance

Consider the radial system shown in Figure 2.9.31, in which the sending-end bus is fed from an infinite power source. By inspection of Figure 2.9.32, the following equations can be written:

$$\Delta V = V_1 - V_2 \tag{2.9.16}$$

$$\Delta V = Z\, I \tag{2.9.17}$$

$$\Delta V = [(R\, I\, \cos \phi) + (X\, I\, \sin \phi)] + j\, [(X\, I\, \cos \phi) - (RI\, \sin \phi)] \tag{2.9.18}$$

where

$\phi = \tan^{-1} (Q/P)$
V_1 = sending end voltage, kV
V_2 = receiving end voltage, kV
ΔV = line voltage drop, kV
R = line resistance, Ω
X = line reactance, Ω
I = line current, kA
P = real power, kW
Q = reactive power kVA
δ = transmission angle, deg.
ϕ = current phase angle, deg.

Since $P = V_2\, I \cos \phi$ and $Q = V_2\, I \sin \phi$, then

$$\Delta V = [(P\, R + Q\, X) + j\, (P\, X - Q\, R)]/V_2 \tag{2.9.19}$$

$$\Delta V = \Delta V_f + j\, \Delta V_q \tag{2.9.20}$$

where

ΔV_f = in phase component of ΔV, kV
ΔV_q = quadrature component of ΔV, kV

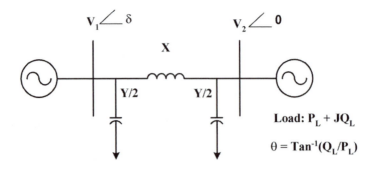

FIGURE 2.9.33 Transmission system π equivalent circuit.

The phasor diagram in Figure 2.9.32 illustrates the meaning of Equation 2.9.19 and Equation 2.9.20. By inspection of Figure 2.9.32, it is clear that $\sin \delta = \Delta V_q/V_1$. Therefore

$$P = [(V_1 V_2 \sin \delta)/X] + [Q (R/X)] \tag{2.9.21}$$

and

$$\delta = \tan^{-1}\{(P\ X - Q\ R)/[V_2 + (P\ R + Q\ X)]\} \tag{2.9.22}$$

In transmission systems, the X/R ratio is large, and the grid usually operates with power factor close to unity. Thus, assuming R = 0 and Q = 0,

$$\Delta V = j\ P\ (X/V_2) \tag{2.9.23}$$

$$P = (V_1 V_2 \sin \delta)/X \tag{2.9.24}$$

$$\delta = \tan^{-1}[P\ (X/V_2)] \tag{2.9.25}$$

Therefore, apart from the voltage magnitude, which must be kept within regulated limits, control of power flow can only be achieved by variation of the line reactance (X), the transmission angle (δ), or both.

2.9.3.4 Reactive Power Balance

Figure 2.9.33 shows a transmission system represented by its π equivalent. By inspection, the following expressions can be derived:

$$V_1 = V_2 + \Delta V \tag{2.9.26}$$

$$(V_1 \cos \delta) + j(V_1 \sin \delta) = V_2 + (X\ I \sin \theta) + j(X\ I \cos \theta) \tag{2.9.27}$$

$$X\ I \cos \theta = P_L\ (X/V_2) = V_1 \sin \delta \tag{2.9.28}$$

$$P_L = (V_1 V_2 \sin \delta)/X \tag{2.9.29}$$

$$X\ I \sin \theta = Q_L\ (X/V_2) = (V_1 \cos \delta) - V_2 \tag{2.9.30}$$

$$Q_L = V_1\ V_2 \cos \delta - (V_2^2/X) \tag{2.9.31}$$

$$S_i = \sqrt{(P_i^2 + Q_l^2)} \tag{2.9.32}$$

Therefore,

$$\text{Line reactive losses} = (S_i/V_i)^2\ X \qquad (2.9.33)$$

$$\text{Line charging (at each end)} = V_i^2\ (Y/2)\quad ,\ i = 1,\ 2 \qquad (2.9.34)$$

$$\text{Line - surge impedance, } Z_S = \sqrt{X\,/\,Y} = \sqrt{L\,/\,C} \qquad (2.9.35)$$

$$\text{Surge-impedance loading (SIL)} = V_{LL}^2/Z_S \qquad (2.9.36)$$

$$\text{Power balance} = \text{total line charging} - \text{line reactive losses}$$

2.9.4   Shunt Reactors Switching Transients

Since the amount of the reactive compensation needed in a power system varies with the loading of the transmission line, shunt reactors are typically switched daily. Shunt reactors will thus experience a large number of switching transients. Transient overvoltages occur, mainly while disconnecting the reactor from the circuit, due to the following two phenomena in the switching device.

1. Current chopping
2. Single or multiple restrikes

The behavior of these overvoltages depends on a number of factors, such as:

- Circuit connection (wye or delta)
- Method of neutral grounding (floating neutral, solidly grounded, or grounded through a reactor)
- Rated MVAR of the reactor
- Construction of the reactor (air core, iron core with three legs or five legs), which impacts the high-frequency characteristics
- Type of connection to the system (tertiary winding or direct connection)
- Type and ratings of circuit breaker
- Characteristics of neighboring equipment

2.9.4.1   Current Chopping

When the contacts of a breaker part, current in the circuit is not interrupted immediately. Current continues to flow through the arc established between the contacts right after the instant of contact parting. Normally the arc extinguishes when the ac current crosses zero. In some cases, however, due to arc instability caused by the circuit parameters and the breaker characteristics, the arc extinguishes abruptly and prematurely ahead of the natural zero crossing of the ac current. When this happens, the energy trapped in the magnetic field of the reactor transfers to the electric field of the stray capacitances in the circuit, thus initiating a resonant response. The resonance frequency is typically a few kHz, and its magnitude is directly proportional to the chopped current and the surge impedance of the circuit, and it may exceed the dielectric withstand of the reactor. Figure 2.9.34 shows a single-phase equivalent circuit that can be used to simulate the chopping overvoltages for a star-connected and solidly grounded shunt-reactor bank with negligible coupling between phases. The legend for Figure 2.9.34 is as follows:

L_L = reactor reactance (For gapped iron-core shunt reactors, the manufacturer should be consulted to obtain the core saturation level at high transient frequencies and thus the resultant inductance; for air-core shunt reactors, the inductance can always be considered a constant value equal to the 60-Hz rated value.)

C_L = equivalent capacitance at the reactor side of the circuit breaker, F

$R(f)$ = reactor frequency-dependent resistance, Ω

L_S = system equivalent reactance, H

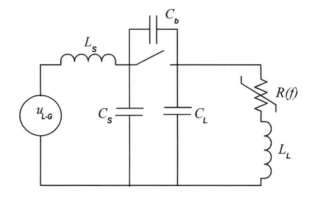

FIGURE 2.9.34 Single-phase equivalent circuit for a star-connected and solidly grounded shunt-reactor bank.

C_b = equivalent capacitance across the breaker terminals, F
C_S = system equivalent capacitance, F
u_{L-G} = system line-to-ground voltage, kV

The parameter k_a in Equation 2.9.37 provides a relative indication of the magnitude of the overvoltage for this type of reactor connection. Circuit damping effects are neglected.

$$k_a = \sqrt{1 + \frac{3.I_{ch}^2}{2.\omega.C_1.Q}} \qquad k_a = \frac{V_{reactor\_chopping}}{V_{L-G}} \qquad (2.9.37)$$

where

k_a = per-unit parameter that indicates the relative magnitude of the overvoltage due to current chopping
$V_{reactor\_chopping}$ = maximum peak of the overvoltage across the reactor terminals after current chopping, kV
V_{L-G} = maximum peak line-to-ground system voltage, kV
I_{ch} = magnitude of current chopped, A
C_1 = total capacitance on the reactor side, μF
ω = angular power frequency, rad/s
Q = shunt-reactor-bank three-phase MVAR

For SF$_6$, bulk-oil, and air-blast circuit breakers, I_{ch} is a function of the high-frequency stray and grading capacitances in parallel with the circuit-breaker terminals [12]. Equation 2.9.38 shows this relationship.

$$I_{ch} = \lambda.\sqrt{C_{CB}} \qquad (2.9.38)$$

where

λ = chopping number (AF$^{-0.5}$), which depends on the circuit breaker construction and arc-extinguishing media
C_{CB} = equivalent capacitance across the circuit breaker terminals for circuit represented in Figure 2.9.34, F

$$C_{CB} = C_b + \frac{C_S.C_L}{C_S + C_L} \qquad (2.9.39)$$

Since C_S in the case of tertiary-winding-connected shunt reactor is relatively smaller than C_S for a shunt reactor directly connected to the high-voltage bus, tertiary-connected reactors normally experience overvoltages of lower magnitude.

Vacuum circuit breakers have a constant I_{ch}, which is weakly affected by the capacitance seen across its terminal. To obtain current chopping characteristics of a vacuum breaker, the breaker manufacturer should be consulted.

IEEE C37.015-1993 provides helpful information for evaluating overvoltages caused by circuit breaker current chopping. [7]

Current chopping can be mitigated by employing an opening resistor with the circuit breaker. The shunt reactor can also be protected against current-chopping overvoltages by a surge arrester installed across the reactor terminals.

2.9.4.2 Restrike

When interrupting small inductive current just before the natural current zero, in a circuit with a critical combination of source-side and load-side capacitances and inductances, the voltage across the circuit breaker terminals may exceed its transient-recovery-voltage (TRV) capability and lead to circuit breaker restrikes. This process can repeat several times until the gap between the circuit breaker contacts becomes sufficiently large so that its dielectric withstand exceeds the voltage across the circuit breaker terminals. Each time the circuit breaker restrikes, a transient overvoltage is imposed on the reactor. This type of overvoltage has a very fast rate of rise of voltage that can distribute nonlinearly across the reactor winding turns. Current chopping can also increase the magnitude of transient overvoltages produced by restrike.

Multiple restrikes can excite a resonant oscillation in the reactor winding. This can lead to high-frequency overvoltages between some coil winding turns. Equation 2.9.40 approximates the relative magnitude of the first restrike overvoltage for the reactor shown in Figure 2.9.34.

$$k = 1 + \left(k_a + 1\right) \left(\frac{C_S - C_L}{C_S + C_L} \right) \qquad k = \frac{V_{reactor\_restrike}}{V_{L-G}} \qquad (2.9.40)$$

When reactors are not solidly grounded, higher restrike voltages can occur. IEEE C37.015-1993 [7] provides detailed information for evaluating this type of overvoltage for reactors with different types of grounding.

Restrike can be mitigated by the use of a surge arrester across the circuit-breaker terminals or by the use of a synchronous opening device with the circuit breaker. Overvoltages caused by restrike can be limited by adding a surge arrester across the reactor terminals. If multiple restrikes excite the natural frequency of the reactor winding, employment of an RC circuit can change the resonant frequency of the circuit and avert high-frequency overvoltages between coil winding turns.

2.9.5 Current-Limiting Reactors and Switching Transients

2.9.5.1 Definitions

Transient-recovery voltage (TRV) is the voltage that appears across the contacts of a circuit-breaker pole upon interruption of a fault current. The first time that the short-circuit current passes through zero after the circuit breaker contacts part, the arc extinguishes and the voltage across the circuit breaker contacts rapidly increases. If the dielectric strength between the circuit breaker contacts does not recover as fast as the recovery voltage across the contacts, the circuit breaker will restrike and continue to conduct.

There are various sources of the high rate of rise of transient-recovery voltage, which can potentially cause the circuit breaker to restrike, e.g., transformer impedance, short-line fault, or distant-fault reflected waves. In cases where the fault current is limited by the inductance of a current-limiting reactor, a high rate of rise of transient-recovery voltage can also occur. It is recommended that a TRV study be conducted prior to selection of a circuit breaker and/or prior to installation of a current-limiting reactor.

2.9.5.2 Circuit-Breaker TRV Capabilities

Rated circuit-breaker TRV capabilities can be obtained from the manufacturer or from various standards such as ANSI C37.06 [15] or IEC 62271-100. [16] Circuit breaker TRV capabilities are normally defined by two sets of parameters. One indicates the maximum voltage peak that the circuit breaker can withstand, and the other one represents the rate of rise of the voltage, i.e., the minimum time to the voltage peak.

Parameters introduced in ANSI C37.06 define the TRV capability of a circuit breaker by means of two types of envelopes. For circuit breakers rated 123 kV and above, the envelope has an exponential-cosine shape if the symmetrical short-circuit current is above 30% of the circuit breaker short-circuit capability (see Figure 2.9.35). At 30% and below, the envelope has a 1-cosine shape. For circuit breakers rated 72.5 kV and below, the envelope has a 1-cosine shape over the entire fault-current range. Circuit breakers are also required to withstand line-side TRV originating from a "kilometric" or short-line fault (SLF). This type of TRV normally reaches its peak during the delay time of the source-side TRV (see Figure 2.9.35). When the actual fault current is less than the circuit-breaker rated fault current, circuit breakers can withstand higher TRV voltages in a short rise time versus their rated value. As shown in Figure 2.9.35, when the actual fault current is 60% of rated value, the circuit breaker can withstand a TRV of 7% higher voltage for a duration 50% shorter than rated time. Therefore when using reactors to reduce the short-circuit level at a substation, it might be more beneficial, from the TRV point of view, to install current-limiting reactors at the upstream feeding substation, where circuit breakers are rated for higher short-circuit level.

2.9.5.3 TRV Evaluation

Computer simulation, or even manual calculation using the current-injection method, can be employed to evaluate the TRV across the circuit-breaker terminals. IEEE C37.011-1994 [17] also provides some guidance for evaluating the TRV. To conduct the study, the electrical characteristics of all the equipment

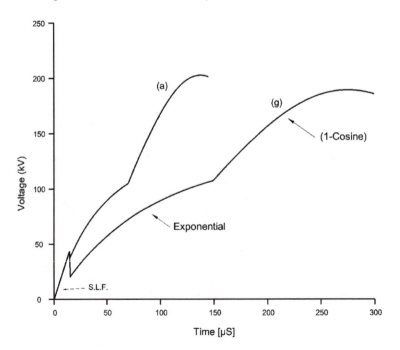

FIGURE 2.9.35 Typical TRV capability envelopes for 123-kV-class circuit breaker at (a) 60% and (b) 100% of rated fault current.

involved in the TRV circuit is required. Equipment characteristics can normally be obtained from the equipment nameplate. However, stray capacitances or inductances are not normally shown on the nameplates. To obtain this information, the equipment manufacturer should be consulted. When dealing with TRV associated with fault current limited by a current-limiting reactor, normally the first pole to open, during an ungrounded three-phase fault at the terminal, experiences the highest peak of TRV. (For certain conditions during a two-phase-to-ground fault, higher TRV can be experienced [11]). The shape of TRV when fault current is limited by a reactor is oscillatory (underdamped) and is very similar to the TRV generated when fault current is limited by a transformer impedance.

When modeling a reactor for TRV studies, the following items shall be considered:

- All types of reactors have stray capacitances, whose values can be obtained from the manufacturer. Generally, the stray capacitance of a dry-type reactor is on the order of a few hundred pF and that of an oil-filled reactor is on the order of a few ηF (see Figure 2.9.36).
- Reactor losses at power frequency are very small and insignificant. However, as a result of skin effect and eddy losses at high frequencies (in the range of kHz), the reactor losses are significant (see Figure 2.9.36).

This characteristic of the reactor can help to damp the transients and shall be reflected in the model by including a frequency-dependent resistance in series with the reactance.

2.9.5.4 TRV Mitigation Methods

Should the system TRV exceed the circuit breaker capability, one or a combination of the following mitigation methods can be employed:

1. Installation of a capacitor across the reactor terminals
2. Installation of a capacitor to ground from the reactor terminal connected to the circuit breaker. For bus-tie reactors, capacitors shall be installed to ground at both reactor terminals.

From a TRV point of view, either of the previous mitigation methods is acceptable. However, from an economic point of view, it is more cost-effective to install the capacitor between the reactor terminals, since the steady-state voltage drop across the reactor is significantly lower than the line-to-ground voltage. Consequently, a lower voltage capacitor can be used.

Figure 2.9.37 shows the result of a TRV study for a subtransmission substation. Curve b in this figure shows the system TRV prior to installation of a current-limiting reactor. It slightly exceeds the circuit breaker capability, as seen in Curve a. After installation of a current-limiting reactor, the system TRV exceeds the circuit breaker capability significantly, as seen in Curve c. To reduce the rate of rise of TRV, a capacitor was installed across the reactor terminals. As seen in Curve d, the system TRV has been modified to an acceptable level.

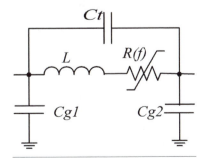

FIGURE 2.9.36 Reactor model for TRV studies, where: C_{g1} and C_{g2} = stray capacitance of reactor terminals to ground, C_t = terminal-to-terminal stray capacitance, L = reactor inductance, and $R_{(f)}$ = reactor-frequency-dependent resistance.

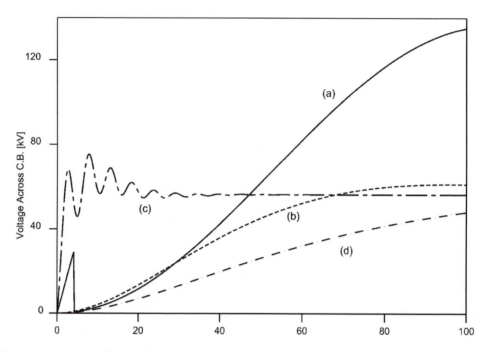

FIGURE 2.9.37 Results of a TRV study for a subtransmission substation: (a) circuit breaker TRV capability, (b) system TRV with no reactor installed, (c) system TRV with a current-limiting reactor installed, (d) system TRV with capacitor across reactor terminals.

2.9.6 Reactor Loss Evaluation

Loss evaluation is often given a high profile by utilities in their tender analysis when purchasing reactors. The consideration of the cost of losses can be a significant factor in the design of reactors, so that the purchaser needs to ensure that loss evaluation, if applied, is made under clearly defined conditions according to clearly defined procedures.

Reactor designs can readily be modified to provide the lowest-loss evaluated equipment total cost (the sum of capital cost and cost of losses as calculated from the loss evaluation).

2.9.6.1 Reactor Losses

The losses in a reactor loaded with fundamental current and, if present, with harmonic current can be determined by Equation 2.9.41,

$$P = \sum_{n=1}^{n=N} \frac{I_n^2 \cdot X_n}{Q_n} \tag{2.9.41}$$

where

P = reactor loss, kW
n = harmonic number (n = 1 ... fundamental current)
N = maximum harmonic order
I_n = current in the reactor at harmonic order n, A
X_n = reactor reactance at harmonic order n, Ω
Q_n = reactor q-factor at harmonic order n

Unlike transformers, reactors do not have no-load losses, and in most cases they have nil or negligible auxiliary losses (oil-immersed reactors only have auxiliary losses associated with cooling fans, oil pumps, etc.). If a reactor is off-load, then the losses are zero, and if the reactor is energized, then it exhibits power-loss consumption. Thus reactor losses can be treated in the same way as transformer load losses.

The main aspects that need to be considered when assessing reactor losses are:

- Anticipated load profile (load factor)
- Rated (operating) bus-bar voltage (for shunt and filter reactors)
- Ambient temperature
- Harmonics

Series reactors (current-limiting and load-balancing reactors) are load-cycled, and thus the loss will vary with current. Shunt reactors are constantly at full load when they are connected to the bus. Their losses depend only on the operating voltage of the bus bar, and the cost of losses is a function of connection time.

As the reactor losses vary with temperature it is important to specify an appropriate reference temperature at which the losses shall be declared by the reactor manufacturer. A standard reference temperature is 75°C. In case of highly loss-evaluated reactors, the temperature rise of the winding is usually low. For this condition, it is reasonable to specify a lower reference temperature. For example, if the average winding-temperature rise of a reactor is 35°C and the average ambient is 20°C, then 55°C instead of 75°C will better represent a reference temperature for loss evaluation.

It should be noted that harmonic currents can be decisive in declaring the losses for filter reactors, but the losses at harmonic frequencies of reactors other than filter reactors are usually almost negligible in comparison with those at fundamental frequency. Also, harmonic losses (Q factor) are not usually included in loss evaluations, since filter-quality factors at the tuning frequency are often dictated by other filter design considerations, such as filter bandwidth, damping requirements, and required level of harmonic filtering.

2.9.6.2 Basic Concepts for Loss Evaluation

Each supply authority determines the cost of losses for its own particular operating conditions. Because these differ from authority to authority, the cost of capitalized losses will also vary. The cost of losses is generally the sum of two components:

1. The cost reflecting the generating capacity (installation cost)

   The cost of installing the generating capacity to supply the peak losses in the supply system so that it, in turn, can supply the power losses in the reactors. The cost of capacity is generally assumed to be the lowest-cost means of providing the peaking capacity, e.g., gas-turbine peaking units, or alternatively it can be based on the average cost of new generating capacity.

2. The energy cost (kilowatt-hour cost) of reactor losses

   The energy cost to reflect the actual cost of supplying the energy consumed by the reactor. Because energy is consumed throughout the entire life of the reactor, further factors need to be considered to determine a present-value worth of the future losses. These include:

   Load factor or connection time (shunt reactors)

   Efficiency of supplying the energy consumed by the reactor

   Cost escalation rate for energy

   Discount rate for present-valuing future costs

   Anticipated operating life of the reactor

Typical loss-evaluation figures (load-power-loss cost rates, LLCR), range from a few hundred dollars per kilowatt to a few thousands of dollars per kilowatt. Loss penalties for reactors with measured losses exceeding guaranteed values are often applied at the same rate as the evaluation. A method of calculating the LLCR is provided in IEEE Std. C57.120-1991, IEEE Loss Evaluation Guide for Power Transformers and Reactors.[18]

2.9.6.3 Operational Benefits of Loss Evaluation

Where loss evaluations are significant and result in lower-loss reactor designs, further operational advantages may ensue, including:

- Lower reactor operating temperature and hence increased service life
- Increased reactor-overload capability
- Less cooling required for indoor units

2.9.7 De-Q'ing

Various levels of electrical damping are required in a number of reactor applications, including harmonic filters, shunt-capacitor banks, and series-capacitor banks. All are inductive/capacitive circuits, and damping is usually governed by the resistive component of the reactor impedance. If this is insufficient, then other means of providing damping must be employed. The required level of damping in a harmonic filter depends on system parameters. In the case of harmonic-filter, shunt-capacitor-bank, and series-capacitor-bank applications, the required level of damping is driven by system design. Damping is usually required at a specific frequency. The Q factor is a measure of the damping, with a lower Q indicating higher damping. The Q factor is the ratio of reactive power to active power in the reactor at a specific frequency. In cases requiring high damping, the natural Q factor of the reactor is usually too high. However, there are methods available to reduce the Q of a reactor by increasing the stray losses through special design approaches, including increasing conductor eddy loss and mechanical clamping-structure eddy loss. In the case of reactors for shunt-capacitor-bank and series-capacitor-bank applications, this method is usually sufficient. In the case of reactors for harmonic-filter applications, other more-stringent approaches may be necessary. One traditional method involves the use of resistors, which, depending on their rating, can be mounted in the interior or on top of the reactor or separately mounted. Resistors are usually connected in parallel with the reactor. Figure 2.9.38 shows a tapped-filter reactor with a separately mounted resistor for an ac filter on an HVDC project.

Another more-innovative and patented approach involves the use of de-Q'ing rings, which can reduce the Q factor of the reactor by a factor of as much as 10. A filter reactor with de-Q'ing rings is illustrated in Figure 2.9.39. The de-Q'ing system comprises a single or several coaxially arranged closed rings that couple with the main field of the reactor. The induced currents in the closed rings dissipate energy in the rings, which lowers the Q-factor of the reactor.

Because of the large energy dissipated in the rings, they must be constructed to have a very large surface-to-volume ratio in order to dissipate the heat. Therefore, they are usually constructed of thin, tall sheets of stainless steel. Cooling is provided by thermal radiation and by natural convection of the surrounding air, which enters between the sheets at the bottom end of the de-Q'ing system and exits at its top end. The stainless-steel material used for the rings can be operated up to about 300°C without altering the physical characteristics/parameters. Especially important is that the variation of resistance with temperature be negligible. The physical dimensions of the rings, their number, and their location with respect to the winding are chosen to give the desired Q-factor at the appropriate frequency.

The Q characteristic of a reactor with de-Q'ing rings is very similar to that of a reactor shunted by a resistor. The basic theory for both approaches is described in the following paragraphs.

2.9.7.1 Paralleled Reactor and Resistor

Figure 2.9.40a shows a circuit diagram corresponding to a resistor in parallel with a fully modeled reactor, i.e., one having series resistance. For simplicity, it is assumed that the series resistance is not frequency dependent. For this nonideal case, the expression for Q is given by Equation 2.9.42.

FIGURE 2.9.38 AC filter for an HVDC project; capacitors, tapped filter reactors, and separately mounted resistors.

$$Q = \frac{\omega L_1 R_p}{R_s R_p + R_s^2 + \omega^2 L_1^2}$$
(2.9.42)

where

L_1 = self inductance of reactor, H
R_p = resistance of parallel resistor, Ω
R_s = series resistance of reactor, Ω

The Q-vs.-frequency characteristic is shown in Figure 2.9.40b. For very low frequencies, the system behaves like a series R-L circuit, i.e., Q is approximately equal to $\omega L_1/R_s$, and Q equals zero when ω equals zero. For very high frequencies, the system behaves like a paralleled R-L circuit, with Q approximately equal to $R_p/\omega L_1$ and approaching zero as ω approaches infinity. It can be shown that Q reaches a maximum at a frequency given by Equation 2.9.43.

$$\omega = \frac{1}{L_1} \sqrt{R_s R_p + R_s^2}$$
(2.9.43)

The maximum value of the Q is given by Equation 2.9.44.

$$Q\max = \frac{R_p}{2\sqrt{R_s R_p + R_s^2}}$$
(2.9.44)

2.9.7.2 Reactor with a De-Q'ing Ring

The circuit diagram for a reactor with a de-Q'ing ring is shown in Figure 2.9.41a. For simplicity, it is assumed that the series resistance is not a function of frequency. The expression for Q as a function of frequency is given by Equation 2.9.45,

FIGURE 2.9.39 Tapped filter reactor with de-Q'ing rings.

$$Q = \frac{\omega^2 \, (L_1 L_2^2 - M^2 L_2) + \omega \, R_2^2 L_2}{\omega^2 (R_2 M^2 + R_1 L_2^2 + R_1 R_2^2)} \tag{2.9.45}$$

where

L_1 = self inductance of reactor, H
L_2 = self inductance of de-Q'ing ring system, H
M = mutual inductance between reactor and de-Q'ing ring system, H
R_1 = series resistance of reactor, Ω
R_2 = resistance of de-Q'ing ring system, Ω

The shape of the Q-vs.-frequency characteristic depends on the value of the coupling factor and also on the values of the series resistance and load resistance. Figure 2.9.41b is a typical characteristic for a practical reactor plus d-Q'ing ring system. It is seen that the Q curve has both a maximum and a minimum. In general, the maximum depends mostly on the series resistance, while the minimum depends primarily on the coupling factor and the ring resistance.

2.9.7.3 Summary

Figure 2.9.42 contains a graph of Q vs. frequency for a natural Q-filter reactor, a filter reactor plus parallel resistor, and a filter reactor with a de-Q'ing ring system. Note the virtually identical Q-vs.-f characteristic for the parallel resistor-vs.-de-Q'ing-ring option.

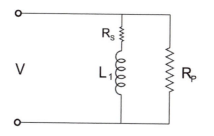

FIGURE 2.9.40A Circuit diagram corresponding to a resistor in parallel with a fully modeled reactor, i.e., one having series resistance.

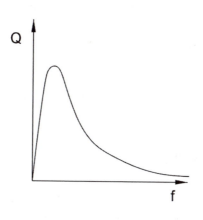

FIGURE 2.9.40B Q-vs.-frequency characteristic of a reactor with parallel resistor.

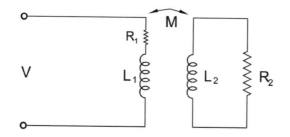

FIGURE 2.9.41A Circuit diagram for a reactor with a de-Q'ing ring.

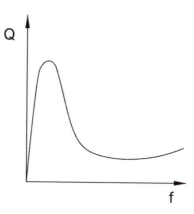

FIGURE 2.9.41B Typical graph of Q vs. frequency for a reactor plus de-Q'ing rings.

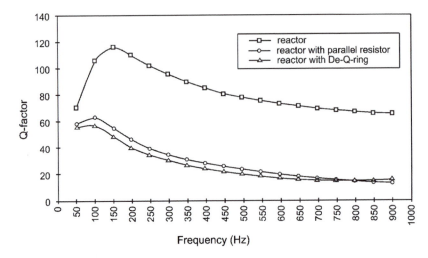

FIGURE 2.9.42 Graph of Q vs. frequency for a natural Q-filter reactor, a filter reactor plus parallel resistor, and a filter reactor with a de-Q'ing ring system.

2.9.8   Sound Level and Mitigation

2.9.8.1   General

The primary source of sound from dry-type and oil-immersed reactors is related to electromagnetic forces generated at fundamental power-frequency current and, where applicable, harmonic frequency. The source of sound in dry-type air-core reactors is the "breathing mode" (expansion/relaxation) vibrations of the windings resulting from the interaction of the winding currents and the "global" magnetic field of the reactor. In the case of oil-immersed reactors, the sound sources are more complex and, depending on design approach, include combinations of and contributions from winding, core (including air gap), magnetic shields, nonmagnetic shielding, and ancillary equipment such as cooling fans. The basic mechanism involves the magnetic field at the iron/air interfaces and the resultant "pulling" forces on the magnetic core material.

2.9.8.2   Oil-Immersed Reactors

Oil-immersed shunt reactors utilize two basic design approaches: air-core magnetically shielded and distributed air-gap iron core. Unlike power transformers, where magnetostriction in the core material is the primary source of noise in an unloaded transformer, the major source of noise in a shunt reactor is vibrational forces resulting from magnetic "pull" effects at iron/air interfaces, primarily at the air gaps. On a secondary-order, leakage flux penetrates structural components of the reactor, and the resultant electromagnetic forces generate vibrational movement and audible noise at twice the power frequency. In the case of oil-immersed magnetically shielded air-core designs, the forces primarily act on the end shield, producing bending forces in the laminations. The resulting vibrations depend on the geometry of the laminated iron-core shields and the mechanical clamping structure. Additional forces/vibrations result from leakage–flux interaction with the tank walls and any ancillary laminated magnetic core material used to shield the tank walls. Distributed air-gap iron-core reactors produce a major portion of their noise as a result of the large magnetic attraction forces in the gaps and also, due to similar forces, at the end-yoke/core-leg interfaces. The avoidance of mechanical resonance is key to minimizing sound levels. It should be noted that gapped iron-core technology was used in the past for the design of high-voltage filter reactors for HVDC application, and the issues described above were exacerbated by the presence of harmonic currents.

Another design approach that can be used for oil-immersed reactors is essentially an air-core reactor design that is placed in a conducting tank (usually aluminum) or in a steel tank with continuous aluminum shielding, i.e., an oil-immersed air-core reactor with nonmagnetic conductive shielding. In this case, noise sources are the windings and the shielding.

Mitigation of sound begins at the design stage by ensuring that critical mechanical resonances are avoided in clamping structure, tank walls, etc., and, where necessary, by utilizing mechanical damping techniques such as vibration-isolation mounting of core material. Site mitigation measures include the use of large, tank-supported, external screens with thick acoustic absorbent material, walled compounds, and barriers made of, for instance, acoustic (resonator) masonry blocks and full acoustic enclosures.

2.9.8.3 Air-Core Reactors

For air-core reactors, the primary source of acoustic noise is the radial vibration of the winding due to the interaction of the current flowing through the winding and the "global" magnetic field. Air-core reactors carrying only power-frequency current, such as series reactors, shunt reactors, etc., produce noise at twice the fundamental power-frequency current. In the case of filter reactors and TCRs, the noise generated by harmonic currents contributes more to the total sound level than the noise resulting from the fundamental power-frequency current because of the A -frequency weighting of the sound. Because there are multiple harmonic currents present in reactors employed on HVDC systems, the design for low operating sound level is a challenge; avoidance of mechanical resonances at the numerous forcing functions requires excellent design-linked modeling tools.

The winding can be regarded in simplified modeling as a cylinder radiating sound from the surface due to radial pulsation. A reactor winding has several mechanical self-resonance frequencies. However, predominately one mode shape — the first tension mode or the so-called breathing mode — will be excited, since this mode shape coincides with the distribution of the electromagnetic forces. The breathing-mode mechanical frequency is inversely proportional to the winding diameter. For example, a cylindrical aluminum winding with a diameter of 1400 mm has a natural breathing-mode mechanical frequency of approximately 1000 Hz. To avoid dynamic resonance amplification, the model frequency should be designed so that it is not near the forcing frequency.

The exciting electromagnetic forces are proportional to the square of the current and oscillate with twice the frequency of the current. If, however, the reactor is simultaneously loaded by several currents of different frequencies, in addition to vibration modes at double the electrical frequencies, additional vibration frequencies occur as shown below.

- Loading with two ac currents with frequencies f_1 and f_2 generates acoustic sound with frequencies of $2f_1$, $2f_2$, $f_1 + f_2$, $f_1 - f_2$.
- Loading with dc current and one ac current with a frequency f_1 generates acoustic frequencies f_1, $2f_1$.

The acoustic-frequency spectrum substantially increases if the reactor's current spectrum includes multiple harmonics; "n" harmonic currents can generate at most n^2 forcing frequencies, but the practical number is usually less because some overlapping occurs.

With the increasing concern for the environment, there are now often stringent sound-level requirements for many sites. Extensive sound-modeling software and mitigation techniques have been developed for dry-type air-core reactors. The predictive software allows the design of air-core reactors with mechanical resonances distant from any major exciting frequency and facilitates the optimum use of component materials to reduce sound level.

Where extremely low sound levels are required, mitigating methods such as acoustic-foam-lined sound shields are also available and have been used with great success. Figure 2.9.43 shows an installation of harmonic filter reactors for an HVDC project with a sound-shield enclosure. Sound-shield enclosures can typically reduce the sound pressure level by up to 10 to 12 dB. Very low sound levels were required, as people were living in houses on the hillsides overlooking the HVDC site.

Figure 2.9.44 shows three curves of the sound-pressure level (SPL) plotted as a function of the frequency of the filter-reactor current for one of the filter reactors shown in Figure 2.9.43. One curve shows the sound characteristic of the original or natural design. The second curve shows the sound characteristic of

FIGURE 2.9.43 Installation of harmonic filter reactors for an HVDC project with a sound-shield enclosure.

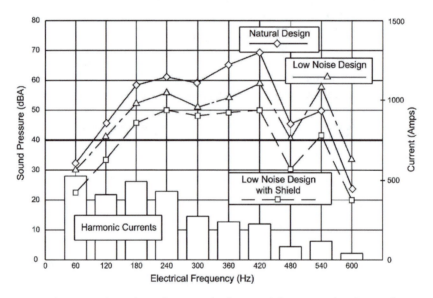

FIGURE 2.9.44 Design comparison of sound-pressure levels; natural design vs. reduced noise design vs. shielded reduced noise design.

the reactor redesigned to avoid mechanical resonances and with increased damping material, thus resulting in a lower amplitude of the sound pressure level at the dominating frequencies of the forcing function. The third curve shows the effect of sound reduction due to a sound-shield enclosure.

References

1. NEMA, Motors and Generators, Part 22, NEMA MG 1-1987, National Electrical Manufacturers Association, Rosslyn, VA, 1987.
2. Westinghouse Electric Corp., *Electrical Transmission and Distribution Reference Book*, 4th ed., Westinghouse Electric Corp., East Pittsburgh, PA, 1964
3. IEEE, Standard Requirements, Terminology, and Test Procedure for Neutral Grounding Devices, IEEE Std. 32-1972, Institute of Electrical and Electronics Engineers, Piscataway, NJ, 1972.

4. Bonheimer, D., Lim, E., Dudley, R.F., and Castanheira, A., A Modern Alternative for Power Flow Control, IEEE/PES Transmission and Distribution Conference, Sept. 22–27, 1991, Dallas, TX.

5. Bonner, J.A., Hurst, W., Rocamora, R.G., Dudley, R.F., Sharp, M.R., and Twiss, J.A., Selecting Ratings for Capacitors and Reactors in Applications Involving Multiple Single Tuned Filters, *IEEE Trans. Power Delivery*, 10, 1, 547–555, January 1995.

6. IEEE, Standard Requirements, Terminology and Test Code for Shunt Reactors Rated Over 500 kVA, IEEE C57.21-1990 (R 1995), Institute of Electrical and Electronics Engineers, Piscataway, NJ, 1995.

7. IEEE, Application Guide for Shunt Reactor Switching, IEEE Std. C37.015-1993, Institute of Electrical and Electronics Engineers, Piscataway, NJ, 1993.

8. IEEE, Standard Requirements, Terminology and Test Code for Dry-Type Air Core Series Connected Reactors, IEEE Std. C57.16-1996, Institute of Electrical and Electronics Engineers, Piscataway, NJ, 1996.

9. IEEE, General Requirements and Test Code for Dry Type and Oil-Immersed Smoothing Reactors for DC Power Transmission, IEEE Std. 1277-2002, Institute of Electrical and Electronics Engineers, Piscataway, NJ, 2002.

10. Peelo, D.F. and Ross, E.M., A New IEEE Application Guide for Shunt Reactor Switching, *IEEE Trans. Power Delivery*, 11, 881–887, 1996.

11. Skeats, W.F., Short-Circuit Currents and Circuit Breaker Recovery Voltages Associated with Two-Phase-to-Ground Short Circuits, *AIEE Trans. Power Appar. Syst.*, 74, 688–693, 1955.

12. CIGRE Working Group 13.02, Interruption of Small Inductive Currents, chap. 4, part A, *Electra*, 101, 13–39, 1985.

13. Power System Relaying Committee Report prepared by the Shunt Reactor Protection Working Group, Shunt Reactor Protection Practices, IEEE/PES 1984 Meeting, Dallas, TX, Jan. 19–Feb. 3, 1984.

14. IEEE Green Book, IEEE Std. 142-1991, Institute of Electrical and Electronics Engineers, Piscataway, NJ, 1991.

15. ANSI, AC High-Voltage Circuit Breakers Rated on a Symmetrical Current Basis-Preferred Ratings and Related Required Capabilities, ANSI C37.06-2000, American National Standards Institute, 2000.

16. IEC, High-Voltage Switchgear and Controlgear. Part 100: High-Voltage Alternating Current Circuit Breakers. IEC 62271-100-2003. International Electrotechnical Commission, Geneva, Switzerland, 2003.

17. IEEE, Application Guide for Transient Recovery Voltage for AC High-Voltage Circuit Breakers Rated on a Symmetrical Current Basis, IEEE Std. C37.011-1994, Institute of Electrical and Electronics Engineers, Piscataway, NJ, 1994.

18. IEEE, Loss Evaluation Guide for Power Transformers and Reactors, IEEE Std. C57.120.1991, Institute of Electrical and Electronics Engineers, Piscataway, NJ, 1991.

19. IEEE, Guide for the Protection of Shunt Reactors, IEEE C37.109-1988, Institute of Electrical and Electronics Engineers, Piscataway, NJ, 1988.

3

Ancillary Topics

Leo J. Savio
ADAPT Corporation

Ted Haupert
TJ/H2b Analytical Services

Loren B. Wagenaar
America Electric Power

Dieter Dohnal
Maschinenfabrik Reinhausen GmbH

Robert F. Tillman, Jr.
Alabama Power Company

Dan D. Perco
Perco Transformer Engineering

Shirish P. Mehta
William R. Henning
Waukesha Electric Systems

James H. Harlow
Harlow Engineering Associates

Armando Guzmán
Hector J. Altuve
Gabriel Benmouyal
Schweitzer Engineering Laboratories

Jeewan Puri
Transformer Solutions

Robert C. Degeneff
Rensselaer Polytechnic Institute

Alan Oswalt
Consultant

Wallace Binder
Consultant

Harold Moore
H. Moore & Associates

Andre Lux
KEMA T&D Consulting

Philip J. Hopkinson
HVOLT, Inc.

3.1   Insulating Media

Leo J. Savio and Ted Haupert

Insulating media in high voltage transformers consists of paper wrapped around the conductors in the transformer coils plus mineral oil and pressboard to insulate the coils from ground. From the moment a transformer is placed in service, both the solid and liquid insulation begin a slow but irreversible process of degradation.

3.1.1   Solid Insulation — Paper

3.1.1.1   Composition of Paper — Cellulose

Paper and pressboard are composed primarily of cellulose, which is a naturally occurring polymer of plant origin. From a chemical perspective, cellulose is a naturally occurring polymer. Each cellulose molecule is initially composed of approximately 1000 repeating units of a monomer that is very similar to glucose. As the cellulose molecule degrades, the polymer chain ruptures and the average number of repeating units in each cellulose molecule decreases. With this reduction in the degree of polymerization of cellulose, there is a decrease in the mechanical strength of the cellulose as well as a change in brittleness and color. As a consequence of this degradation, cellulose will reach a point at which it will no longer properly function as an insulator separating conductors. When cellulose will reach its end of life as an insulator depends greatly on the rate at which it degrades.

3.1.1.2 Parameters that Affect Degradation of Cellulose

3.1.1.2.1 *Heat*

Several chemical reactions contribute to the degradation of cellulose. Oxidation and hydrolysis are the most significant reactions that occur in oil-filled electrical equipment. These reactions are dependent on the amounts of oxygen, water, and acids that are in contact with the cellulose. In general, the greater the level of these components, the faster are the degradation reactions. Also, the rates of the degradation reactions are greatly dependent on temperature. As the temperature rises, the rates of chemical reactions increase.

For every 10° (Celsius) rise in temperature, reaction rates double.

Consequently, the useful life of cellulose and oil is markedly reduced at higher temperatures. Paper and oil subjected to an increased temperature of 10°C will have their lives reduced by a factor of 50%. Elevations in temperature can result from voluntary events such as increased loading, or they can result from a large number of involuntary events, such as the occurrence of fault processes (partial discharge and arcing).

3.1.1.2.2 *Oxygen*

The cellulose that is present in paper, pressboard, and wood oxidizes readily and directly to carbon oxides. The carbon oxides (carbon dioxide and carbon monoxide) that are found in oil-filled electrical equipment result primarily from the cellulose material. This has very important consequences, since the useful life of major electrical devices such as power transformers is generally limited by the integrity of the solid insulation — the paper. It is now possible to determine more closely the extent and the rate of degradation of the cellulose by observing the levels of the carbon oxides in the oil as a function of time.

As cellulose reacts with oxygen, carbon dioxide, water, and possibly carbon monoxide are produced. Carbon monoxide is produced if there is an insufficient supply of oxygen to meet the demands of the oxidation reaction. The levels of these products in the oil continue to increase as oxidation continues. However, they never exceed concentrations in the oil that are referred to as their solubility limits, which are temperature and pressure dependent. After the solubility limit of each has been reached, further production cannot increase their concentration in the oil. If carbon monoxide and carbon dioxide were to ever exceed their solubility limits, they would form bubbles that would be lost to the atmosphere or to a gas blanket; this rarely happens. Any water that forms will fall to the bottom of the tank or be adsorbed into the solid insulation (the cellulose).

3.1.1.2.3 *Moisture*

Cellulose has a great affinity for holding water (notice how well paper towels work). Water that is held in the paper can migrate into the oil as the temperature of the system increases, or the reverse can happen as the temperature of the system decreases. In a typical large power transformer, the quantity of cellulose in the solid insulation can be several thousand pounds.

For new transformers, the moisture content of the cellulose is generally recommended to be no more than 0.5%. Water distributes between the oil and the paper in a constant ratio, depending on the temperature of the system. As the temperature increases, water moves from the paper into the oil until the distribution ratio for the new temperature is achieved. Likewise, as the temperature decreases, water moves in the opposite direction.

In addition to the water that is in the paper and the oil at the time a transformer is put into service, there is also water introduced into the system because of the ongoing oxidation of the cellulose. Water is a product of the oxidation of cellulose, and it is therefore always increasing in concentration with time. Even if the transformer were perfectly sealed, the moisture concentration of the paper would continue to increase. The rate of generation of water is determined primarily by the oxygen content of the oil and the temperature of the system. An increase in either of these factors increases the rate of water generation.

3.1.1.2.4 *Acid*

Cellulose can degrade by a chemical process referred to as hydrolysis. During hydrolysis, water is consumed in the breaking of the polymeric chains in the cellulose molecules. The process is catalyzed by acids. Acids are present in the oil that is in contact with the cellulose. Carboxylic acids are produced from the oil as a result of oxidation. The acid content of the oil increases as the oil oxidizes. With an increase in acidity, the degradation of the cellulose increases.

3.1.2 Liquid Insulation — Oil

The insulating fluid that has the greatest use in electrical equipment is mineral oil. There are insulating materials that may be superior to mineral oil with respect to both dielectric and thermal properties; however, to date, none has achieved the requisite combination of equal or better performance at an equal or better price. Consequently, mineral oil continues to serve as the major type of liquid insulation used in electrical equipment.

3.1.2.1 Composition of Oil

3.1.2.1.1 *Types of Hydrocarbons and Properties of Each*

Mineral oil can vary greatly in its composition. All mineral oils are mixtures of hydrocarbon compounds with about 25 carbon atoms per molecule. The blend of compounds that is present in a particular oil is dependent on several factors, such as the source of the crude oil and the refining process. Crude oils from different geographical areas will have different chemical structures (arrangement of the carbon atoms within the molecules). Crude oils from some sources are higher in *paraffinic* compounds, whereas others are higher in *naphthenic* compounds. Crude oils also contain significant amounts of aromatic and polyaromatic compounds. Some of the polyaromatic compounds are termed "heterocyclics" because, besides carbon and hydrogen, they contain other atoms such as nitrogen, sulfur, and oxygen. Some heterocyclics are beneficial (e.g., oxidation inhibitors), but most are detrimental (e.g., oxidation initiators, electrical charge carriers).

The refining of crude oil for the production of dielectric fluids reduces the aromatic and polyaromatic content to enhance the dielectric properties and stability of the oil.

The terms *paraffinic* and *naphthenic* refer to the arrangement of carbon atoms in the oil molecule. Carbon atoms that are arranged in straight or branched chains, i.e., carbon atoms bonded to one another in straight or branched lines, are referred to as being *paraffinic*. Carbon atoms that are bonded to one another to form rings of generally five, six, or seven carbons are referred to as being *naphthenic*. Carbon atoms that are bonded as rings of benzene are referred to as being *aromatic*. Carbon atoms that are contained in "fused" benzene rings are referred to as being *polyaromatic*. These forms of bonded carbon atoms are depicted in Figure 3.1.1. The straight lines represent the chemical bonds between carbon atoms that are present (but not depicted) at the ends and vertices of the straight lines.

Figure 3.1.2 illustrates a typical oil molecule. Remember that a particular oil will contain a mixture of many different molecular species and types of carbon atoms. Whether a particular oil is considered paraffinic or naphthenic is a question of degree. If the oil contains more paraffinic carbon atoms than naphthenic carbons, it is considered a paraffinic oil. If it contains more naphthenic carbons, it is considered a naphthenic oil.

The differences in the chemical composition will result in differences in physical properties and in the chemical behavior of the oils after they are put in service. The chemical composition has profound effects on the physical characteristics of the oil.

For electrical equipment, the main concerns are:

- Paraffinic oils tend to form waxes (solid compounds) at low temperature.
- Paraffinic oils have a lower thermal stability than that of naphthenic and aromatic oils.
- Paraffinic oils have a higher viscosity at low temperature than that of naphthenic and aromatic oils.

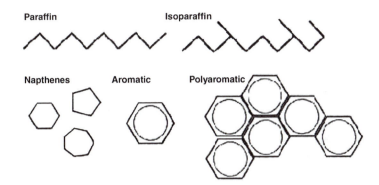

FIGURE 3.1.1 Carbon configurations in oil molecules.

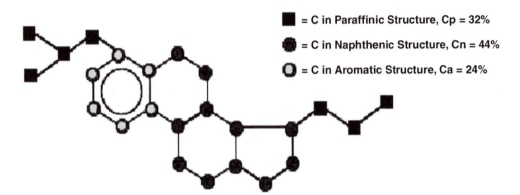

FIGURE 3.1.2 Typical oil molecule.

These factors can impair the performance of high-voltage electrical equipment. The first two factors have an unfavorable effect on the dielectric characteristics of the oil. The third factor unfavorably affects the heat/dissipation ability of the oil. Unfortunately, the availability of insulating oil is limited. Therefore, electrical equipment owners have a choice of only a few producers, who produce only a very few different products.

3.1.2.1.2 Oxidation Inhibitors

Oxidation inhibitors, such as DBPC (di-tertiary butyl paracresol) and DBP (di-tertiary butylphenol), are often added to oil to retard the oxidation process. These compounds work by attracting oxygen molecules to themselves rather than allowing oxygen to bind with oil molecules. With time, the inhibitor gets consumed because of its preferential reaction with oxygen. As a result, the oil will then oxidize at a more rapid rate. The remedy is to add inhibitor to oil that has lost its antioxidant capabilities.

3.1.2.2 Functions of Oil

3.1.2.2.1 Electrical Insulation

The primary function of insulating oil is to provide a dielectric medium that acts as insulation surrounding various energized conductors. Another function of the insulating oil is to provide a protective coating to the metal surfaces within the device. This coating protects against chemical reactions, such as oxidation, that can influence the integrity of connections, affect the formation of rust, and contribute to the consequent contamination of the system.

Insulating oil, however, is *not* a good lubricant. Despite this fact, it is widely used in load tap changers, circuit breakers, and transformers. Therefore, Its use in these devices presents a challenge to the mechanical design of the system.

3.1.2.2.2   *Heat Dissipation*

A secondary function of the insulating fluid is to serve as a dissipater of heat. This is of particular importance in transformers where localized heating of the windings and core can be severe. The oil aids in the removal of heat from these areas and distributes the thermal energy over a generally large mass of oil and the tank of the device. Heat from the oil can then be transferred by means of conduction, convection, and radiation to the surrounding environment.

All mineral oils are comparable in their ability to conduct and dissipate heat. To ensure that a given oil will perform satisfactorily with respect to heat dissipation, several specifications are placed on the oil. These specifications are based upon certain factors that influence the oil's ability to dissipate heat over a wide range of possible operating conditions. These factors include such properties as viscosity, pour point, and flash point.

3.1.2.2.3   *Diagnostic Purposes*

The third function of the insulating fluid is to serve as an indicator of the operational condition of the liquid-filled equipment. The condition (both chemical and electrical) of the insulating fluid reflects the operational condition of the electrical device. In a sense, the fluid can provide diagnostic information about the electrical device much like blood can provide diagnostic information about the human body. The condition of the blood is important as it relates to its primary function of transporting oxygen and other chemical substances to the various parts of the body. Indeed the condition of the blood is symptomatic of the overall health of the body. For example, the analysis of the blood can be used to diagnose a wide variety of health problems related to abnormal organ function.

In much the same way, insulating fluid can be viewed as serving its primary functions as an insulator and heat dissipater. It can also be viewed as serving another (and perhaps equally important) function as a diagnostic indicator of the operational health of liquid-filled equipment. This is possible because when faults develop in liquid-filled equipment, they cause energy to be dissipated through the liquid. This energy can cause a chemical degradation of the liquid. An analysis for these degradation products can provide information about the type of fault that is present.

3.1.2.3   Parameters that Affect Oil Degradation

3.1.2.3.1   *Heat*

Just as temperature influences the rate of degradation of the solid insulation, so does it affect the rate of oil degradation. Although the rates of both processes are different, both are influenced by temperature in the same way. As the temperature rises, the rates of degradation reactions increase.

For every 10° (Celsius) rise in temperature, reaction rates double!

3.1.2.3.2   *Oxygen*

Hydrocarbon-based insulating oil, like all products of nature, is subject to the ongoing, relentless process of oxidation. Oxidation is often referred to as aging. The abundance of oxygen in the atmosphere provides the reactant for this most common degradation reaction. The ultimate products of oxidation of hydrocarbon materials are carbon dioxide and water. However, the process of oxidation can involve the production of other compounds that are formed by intermediate reactions, such as alcohols, aldehydes, ketones, peroxides, and acids.

3.1.2.3.3   *Partial Discharge and Thermal Faulting*

Of all the oil degradation processes, hydrogen gas requires the lowest amount of energy to be produced. Hydrogen gas results from the breaking of carbon–hydrogen bonds in the oil molecules. All of the three fault processes (partial discharge, thermal faulting, and arcing) will produce hydrogen, but it is only with partial discharge or corona that hydrogen will be the only gas produced in significant quantity. In the presence of thermal faults, along with hydrogen will be the production of methane together with ethane and ethylene. The ratio of ethylene to ethane increases as the temperature of the fault increases.

3.1.2.3.4 Arcing

With arcing, acetylene is produced along with the other fault gases. Acetylene is characteristic of arcing. Because arcing can generally lead to failure over a much shorter time interval than faults of other types, even trace levels of acetylene (a few parts per million) must be taken seriously as a cause for concern.

3.1.2.3.5 Acid

High levels of acid (generally acid levels greater than 0.6 mg KOH/g of oil) cause sludge formation in the oil. Sludge is a solid product of complex chemical composition that can deposit throughout the transformer. The deposition of sludge can seriously and adversely affect heat dissipation and ultimately result in equipment failure.

3.1.3 Sources of Contamination

3.1.3.1 External

External sources of contamination can generally be minimized by maintaining a sealed system, but on some types of equipment (e.g., free-breathing devices) this is not possible. Examples of external sources of contamination are moisture, oxygen, and solid debris introduced during maintenance of the equipment or during oil processing.

3.1.3.2 Internal

Internal sources of contamination can be controlled only to a limited extent because these sources of contamination are generally chemical reactions (like the oxidation of cellulose and the oxidation of oil) that are constantly ongoing. They cannot be stopped, but their rates are determined by factors that are well understood and often controllable. Examples of these factors are temperature and the oxygen content of the system.

Internal sources of contamination are:

- Nonmetallic particles such as cellulose particles from the paper and pressboard
- Metal particles from mechanical or electrical wear
- Moisture from the chemical degradation of cellulose (paper insulation and pressboard)
- Chemical degradation products of the oil that result from its oxidation (e.g., acids, aldehydes, ketones)

3.2 Electrical Bushings

Loren B. Wagenaar

3.2.1 Purpose of Electrical Bushings

ANSI/IEEE Std. C57.19.00 [1] defines an electrical bushing as "an insulating structure, including a through conductor or providing a central passage for such a conductor, with provision for mounting a barrier, conducting or otherwise, for the purpose of insulating the conductor from the barrier and conducting current from one side of the barrier to the other." As a less formal explanation, the purpose of an electrical bushing is simply to transmit electrical power in or out of enclosures, i.e., barriers, of an electrical apparatus such as transformers, circuit breakers, shunt reactors, and power capacitors. The bushing conductor may take the form of a conductor built directly as a part of the bushing or, alternatively, as a separate conductor that is drawn through, usually through the center of, the bushing.

Since electrical power is the product of voltage and current, insulation in a bushing must be capable of withstanding the voltage at which it is applied, and its current-carrying conductor must be capable of carrying rated current without overheating the adjacent insulation. For practical reasons, bushings are not rated by the power transmitted through them; rather, they are rated by the maximum voltage and current for which they are designed.

3.2.2 Types of Bushings

There are many methods to classify the types of bushings. These classifications are based on practical reasons, which will become apparent in the following discussion in three broad areas. Bushings can be classified:

1. According to insulating media on ends
2. According to construction
3. According to insulation inside bushing

3.2.2.1 According to Insulating Media on Ends

One method is to designate the types of insulating media at the ends of the bushing. This classification depends primarily on the final application of the bushing.

An air-to-oil bushing has air insulation at one end of the bushing and oil insulation at the other. Since oil is more than twice as strong dielectrically as air at atmospheric pressure, the oil end is approximately half as long (or less) than the air end. This type of bushing is commonly used between atmospheric air and any oil-filled apparatus.

An air-to-air bushing has air insulation on both ends and is normally used in building applications where one end is exposed to outdoor atmospheric conditions and the other end is exposed to indoor conditions. The outer end may have higher creep distances to withstand higher-pollution environments, and it may also have higher strike distances to withstand transient voltages during adverse weather conditions such as rainstorms.

Special application bushings have limited usage and include: air-to-SF_6 bushings, usually used in SF_6-insulated circuit breakers; SF_6-to-oil bushings, used as transitions between SF_6 bus ducts and oil-filled apparatus; and oil-to-oil bushings, used between oil bus ducts and oil-filled apparatus.

3.2.2.2 According to Construction

There are basically two types of construction, the solid or bulk type and the capacitance-graded or condenser type.

3.2.2.2.1 Solid Bushing

The solid-type bushing, depicted in Figure 3.2.1, is typically made with a central conductor and porcelain or epoxy insulators at either end and is used primarily at the lower voltages through 25 kV. Generally, this construction is relatively simple compared with the capacitance-graded type. This was the construction method used for the original bushings, and its current usage is quite versatile with respect to size. Solid bushings are commonly used in applications ranging from small distribution transformers and circuit switchers to large generator step-up transformers and hydrogen-cooled power generators.

At the lower end of the applicable voltage range, the central conductor can be a small-diameter lead connected directly to the transformer winding, and such a lead typically passes through an arbitrarily shaped bore of an outer and inner porcelain or epoxy insulator(s). Between the two insulators there is typically a mounting flange for mounting the bushing to the transformer or other apparatus. In one unique design, only one porcelain insulator was used, and the flange was assembled onto the porcelain after the porcelain had been fired. At higher voltages, particularly at 25 kV, more care is taken to make certain that the lead and bore of the insulator(s) are circular and concentric, thus ensuring that the electric stresses in the gap between these two items are more predictable and uniform. For higher-current bushings, typically up to 20 kA, large-diameter circular copper leads or several copper bars arranged in a circle and brazed to copper end plates can be used.

The space between the lead and the insulator may consist of only air on lower-voltage solid-type bushings, or this space may be filled with electric-grade mineral oil or some other special compound on higher-voltage bushings. The oil may be self-contained within the bushing, or it may be oil from the apparatus in which the bushing is installed. Special compounds are typically self-contained. Oil and compounds are used for three reasons: First, they enable better cooling of the conductor than does air. Second, they have higher dielectric constants (about 2.2 for oil) than air, and therefore, when used with

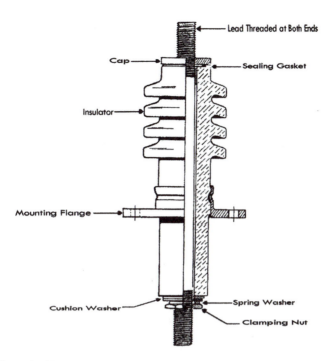

FIGURE 3.2.1 Solid-type bushing.

materials with higher dielectric constants, such as porcelain or epoxy, they endure a smaller share of the voltage than an equally sized gap occupied by air. The result is that oil and compounds withstand higher voltages than air alone. Third, oil and other compounds display higher breakdown strengths than air.

The primary limitation of the solid bushing is its ability to withstand 60-Hz voltages above 90 kV. Hence, its applications are limited to 25-kV equipment ratings, which have test voltages of 70 kV. Recent applications require low partial-discharge limits on the 25-kV terminals during transformer test and have caused further restrictions on the use of this type of bushing. In these cases, either a specially designed solid bushing, with unique grading shielding that enables low inherent partial-discharge levels, or a more expensive capacitance-graded bushing must be used.

3.2.2.2.2 Capacitance-Graded Bushings
Technical literature dating back to the early twentieth century describes the principles of the capacitance-graded bushing [2]. R. Nagel of Siemens published a German paper [3] in 1906 describing an analysis and general principles of condenser bushings, and A.B. Reynders of Westinghouse published a U.S. paper [4] that described the principles of the capacitance-graded bushing and compared the characteristics of these bushings with those of solid-type construction. Thereafter, several additional papers were published, including those by individuals from Micafil of Switzerland and ASEA of Sweden.

The value of the capacitance-graded bushing was quickly demonstrated, and this bushing type was produced extensively by those companies possessing the required patents. Currently, this construction is used for virtually all voltage ratings above 25-kV system voltage and has been used for bushings through 1500-kV system voltage. This construction uses conducting layers at predetermined radial intervals within oil-impregnated paper or some other insulation material that is located in the space between the central conductor and the insulator. Different manufacturers have used a variety of materials and methods for making capacitance-graded bushings. Early methods were to insert concentric porcelain cylinders with metallized surfaces or laminated pressboard tubes with embedded conductive layers. Later designs used conductive foils, typically aluminum or copper, in oil-impregnated kraft paper. An alternative method is to print semiconductive ink (different manufacturers have used different conductivities) on all or some of the oil-impregnated kraft-paper wraps.

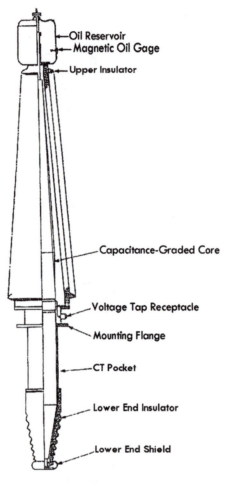

FIGURE 3.2.2 Oil-filled, capacitance-graded bushing.

Figure 3.2.2 shows the general construction of an oil-filled, capacitance-graded bushing. The principal elements are the central circular conductor, onto which the capacitance-graded core is wound; the top and lower insulators; the mounting flange; the oil and an oil-expansion cap; and the top and bottom terminals. Figure 3.2.3 is a representation of the equipotential lines in a simplified capacitance-graded bushing in which neither the expansion cap nor the sheds on either insulator are shown. The bold lines within the capacitance-graded core depict the voltage-grading elements. The contours of the equipotential lines show the influence of the grading elements, both radially within the core and axially along the length of the insulators.

The mathematical equation for the radial voltage distribution as a function of diameter between two concentric conducting cylinders is:

$$V(d) = V \ [\ln (D_2/d)]/[\ln (D_2/D_1)] \tag{3.2.1}$$

where

V = voltage between the two cylinders
d = position (diameter) at which the voltage is to be calculated
D_1 and D_2 = diameters of the inner and outer cylinders, respectively

Since this is a logarithmic function, the voltage is nonlinear, concentrating around the central conductor and decreasing near the outer cylinder. Likewise, the associated radial electric stress, calculated by

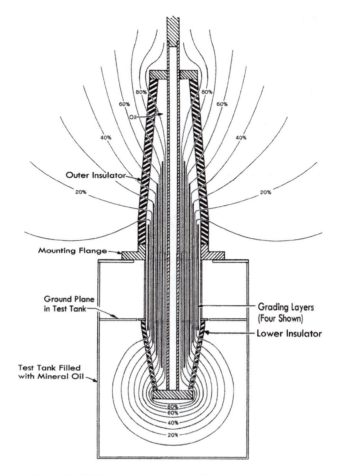

FIGURE 3.2.3 Exponential plot of oil-filled, capacitance-graded bushing.

$$E(d) = 2V/[d \ln (D_2/D_1)] \tag{3.2.2}$$

will be the greatest at $d = D_1$. The lengths of grading elements and the diameters at which they are positioned are such as to create a more uniform radial-voltage distribution than found in a solid-type bushing.

As seen from Figure 3.2.3, the axial voltage distribution along the inner and outer insulators is almost linear when the proper capacitance grading is employed. Thus both insulators on capacitance-graded bushings can be shorter than their solid-bushing counterparts.

Capacitance-graded bushings involve many more technical and manufacturing details than solid bushings and are therefore more expensive. These details include the insulation/conducting layer system, equipment to wind the capacitor core, and the oil to impregnate the paper insulation. However, it should be noted that the radial dimension required for the capacitance-graded bushing is much less than the solid construction, and this saves on material within the bushing as well in the apparatus in which the bushing is used. Also, from a practical standpoint, higher-voltage bushings could not possibly be manufactured with a solid construction.

3.2.2.3 According to Insulation inside Bushing

Still another classification relates to the insulating material used inside the bushing. In general, these materials can be used in either the solid- or capacitance-graded construction, and in several types, more than one of these insulating materials can be used in conjunction. The following text gives a brief description of these types:

3.2.2.3.1 Air-Insulated Bushings
Air-insulated bushings generally are used only with air-insulated apparatus and are of the solid construction that employs air at atmospheric pressure between the conductor and the insulators.

3.2.2.3.2 Oil-Insulated or Oil-Filled Bushings
Oil-insulated or oil-filled bushings have electrical-grade mineral oil between the conductor and the insulators in solid-type bushings. This oil can be contained within the bushing, or it can be shared with the apparatus in which the bushing is used. Capacitance-graded bushings also use mineral oil, usually contained within the bushing, between the insulating material and the insulators for the purposes of impregnating the kraft paper and transferring heat from the conducting lead.

3.2.2.3.3 Oil-Impregnated Paper-Insulated Bushings
Oil-impregnated paper-insulated bushings use the dielectric synergy of mineral oil and electric grades of kraft paper to produce a composite material with superior dielectric-withstand characteristics. This material has been used extensively as the insulating material in capacitance-graded cores for approximately the last 50 years.

3.2.2.3.4 Resin-Bonded or -Impregnated Paper-Insulated Bushings
Resin-bonded paper-insulated bushings use a resin-coated kraft paper to fabricate the capacitance-graded core, whereas resin-impregnated paper-insulated bushings use papers impregnated with resin, which are then used to fabricate the capacitance-graded core. The latter type of bushing has superior dielectric characteristics, comparable with oil-impregnated paper-insulated bushings.

3.2.2.3.5 Cast-Insulation Bushings
Cast-insulation bushings are constructed of a solid-cast material with or without an inorganic filler. These bushings can be either of the solid or capacitance-graded types, although the former type is more representative of present technology.

3.2.2.3.6 Gas-Insulated Bushings
Gas-insulated bushings [5] use pressurized gas, such as SF_6 gas, to insulate between the central conductor and the flange. The bushing shown in Figure 3.2.4 is one of the simpler designs and is typically used with circuit breakers. It uses the same pressurized gas as the circuit breaker, has no capacitance grading, and uses the dimensions and placement of the ground shield to control the electric fields. Other designs use a lower insulator to enclose the bushing, which permits the gas pressure to be different than within the circuit breaker. Still other designs use capacitance-graded cores made of plastic-film material that is compatible with SF_6 gas.

3.2.3 Bushing Standards

Several bushing standards exist in the various countries around the world. The major standards have been established by the Transformers Committee within the IEEE Power Engineering Society and by IEC Committee 37. Five important standards established by these committees include the following:

1. ANSI/IEEE Std. C57.19.00, Standard Performance Characteristics and Test Procedure for Outdoor Power Apparatus Bushings [1]. This is the general standard that is widely used by countries in the Western Hemisphere and contains definitions, service conditions, ratings, general electrical and mechanical requirements, and detailed descriptions of routine and design test procedures for outdoor-power-apparatus bushings.
2. IEEE Std. C57.19.01, Standard Performance Characteristics and Dimensions for Outdoor Power Apparatus Bushings [6]. This standard lists the electrical-insulation and test-voltage requirements for power-apparatus bushings rated from 15 through 800-kV maximum system voltages. It also lists dimensions for standard-dimensioned bushings, cantilever-test requirements for bushings

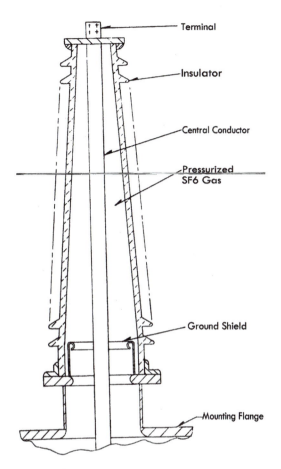

FIGURE 3.2.4 Pressurized SF$_6$ gas bushing.

rated through 345-kV system voltage, and partial-discharge limits as well as limits for power factor and capacitance change from before to after the standard electrical tests.

3. IEEE Std. C57.19.03, Standard Requirements, Terminology and Test Procedures for Bushings for DC Applications [7]. This standard gives the same type of information as ANSI/IEEE Std. C57.19.00 for bushings for direct-current equipment, including oil-filled converter transformers and smoothing reactors. It also covers air-to-air dc bushings.

4. IEEE Std. C57.19.100, Guide for Application of Power Apparatus Bushings [8]. This guide recommends practices to be used (1) for thermal loading above nameplate rating for bushings applied on power transformers and circuit breakers and (2) for bushings connected to isolated-phase bus. It also recommends practices for allowable cantilever loading caused by the pull of the line connected to the bushing, applications for contaminated environments and high altitudes, and maintenance practices.

5. IEC Publication 137 [9], Bushings for Alternating Voltages above 1000 V. This standard is the IEC equivalent to the first standard listed above and is used widely in European and Asian countries.

3.2.4 Important Design Parameters

3.2.4.1 Conductor Size and Material

The conductor diameter is determined primarily by the current rating. There are two factors at work here. First, the skin depth of copper material at 60 Hz is about 1.3 cm and that of aluminum is about

1.6 cm. This means that most of the current will flow in the region from the outer portion of the conductor and radially inward to a depth of the skin depth δ. Second, the losses generated within a conductor will be:

$$P_{loss} = I^2R = I^2\rho L/A = 4I^2\rho L/\pi(D_1^2 - D_0^2) \tag{3.2.3}$$

where

I = rated current
ρ = resistivity of the conductor material, ohm·m
ρ = 1.7241×10^{-3} ohm·m for copper with 100% IACS (international annealed copper standard)
ρ = 3.1347×10^{-3} ohm·m for electrical aluminum alloy with 55% IACS
L = length of conductor, m
A = cross section of conductor = $\pi(D_1^2 - D_0^2)/4$
D_1 = outside diameter of conductor, m
D_0 = $D_1 - \delta$, m
δ = skin depth of the conductor in the case that a tubular conductor is used
δ ≈ 0.0127 m for copper, 0.0159 m for aluminum, both at 60-Hz frequency

It can be seen from Equation 3.2.3 that P_{loss} decreases as D_1 increases. Hence, design practice is to increase the outside diameter of the conductor for higher current ratings and to limit the wall thickness to near the skin depth. There are other technical advantages to increasing the outside conductor diameter: First, from Equation 3.2.2, observe that electric-field stress reduces as $d = D_1$ increases. Therefore, a larger diameter conductor will have higher partial-discharge inception and withstand voltages. Second, the mechanical strength of the conductor is dependent on the total cross-sectional area of the conductor, so that a larger diameter is sometimes used to achieve higher withstand forces in the conductor.

3.2.4.2 Insulators

Insulators must have sufficient length to withstand the steady-state and transient voltages that the bushing will experience. Adequate lengths depend on the insulating media in which the insulator is used and on whether the bushing is capacitively graded. In cases where there are two different insulating media on either side of an insulator, the medium with the inferior dielectric characteristics determines the length of the insulator.

3.2.4.2.1 Air Insulators

Primary factors that determine the required length of insulators used in air at atmospheric pressure are lightning-impulse voltage under dry conditions and power-frequency and switching-impulse voltages under wet conditions. Standard dry conditions are based on 760-mm (Hg) atmospheric pressure and 20°C, and wet conditions are discussed in Section 3.2.6.1, High-Altitude Applications.

Bushings are normally designed to be adequate for altitudes up to 1000 m. Beyond 1000 m, longer insulators must be used to accommodate the lower air density at higher altitudes. Under clean conditions, air insulators for capacitively graded bushings are normally shorter in length than insulator housings without grading elements within them. However, as the insulator becomes more contaminated, the effects of grading elements disappear, and the withstand characteristics of graded and nongraded insulators become the same over the long term (15 to 30 min) [10]. Further guidance on this subject is given in Section 3.2.6.2, Highly Contaminated Environments.

3.2.4.2.2 Oil Insulators

Since mineral oil is dielectrically stronger than air, the length of insulators immersed in oil is typically 30 to 40% the length of air insulators. In equipment having oil with low contamination levels, no sheds are required on oil-immersed insulators. In situations where some contamination exists in the oil, such as carbon particles in oil-insulated circuit breakers, small ripples are generally cast on the outer insulator surface exposed to the oil.

3.2.4.2.3 *Pressurized SF $_6$ Gas Insulators*

Since various pressures can be used for this application, the length of the insulator can be equal to or less than an insulator immersed in oil. Since particles are harmful to the dielectric strength of any pressurized gas, precautions are generally taken to keep the SF$_6$ gas free of particles. In such cases, no sheds are required on the insulators.

3.2.4.3 Flange

The flange has two purposes: first, to mount the bushing to the apparatus on which it is utilized, and second, to contain the gaskets or other means of holding the insulators in place located on the extreme ends of the flange, as described in Section 3.2.4.5, Clamping System. Flange material can be cast aluminum for high-activity bushings, where the casting mold can be economically justified. In cases where production activities are not so high, flanges can be fabricated from steel or aluminum plate material. A further consideration for high-current bushings is that aluminum, or some other nonmagnetic material, is used in order to eliminate magnetic losses caused by currents induced in the flange by the central conductor.

3.2.4.4 Oil Reservoir

An oil reservoir, often called the expansion cap, is required on larger bushings with self-contained oil for at least one and often two related reasons: First, mineral oil expands and contracts with temperature, and the oil reservoir is required to contain the oil expansion at high oil temperatures. Second, oil-impregnated insulating paper must be totally submerged in oil in order to retain its insulating qualities. Hence, the reservoir must have sufficient oil in it to maintain oil over the insulating paper at the lowest anticipated temperatures. Since oil is an incompressible fluid, the reservoir must also contain a sufficient volume of gas, such as nitrogen, so that excessive pressures are not created within the bushing at high temperatures. Excessive pressures within a bushing can cause oil leakage.

On bushings for mounting at angles up to about 30° from vertical, the reservoir is mounted on the top end of the bushing. On smaller, lower-voltage bushings, the reservoir can be within the top end of the upper insulator. Oil-filled bushings that are horizontally mounted usually have an oil reservoir mounted on the flange, but some have bellows, either inside or outside the bushing, which expand and contract with the temperature of the oil.

For the purpose of checking the oil level in the bushing, an oil-level gauge is often incorporated into the reservoir. There are two basic types of oil gauges, the clear-glass type and the magnetic type. The former type is cast from colored or clear glass such that the oil level can be seen from any angle of rotation around the bushing. The second type is a two-piece gauge, the part inside the reservoir being a float attached to a magnet that rotates on an axis perpendicular to the reservoir wall. The part outside the reservoir is then a gauge dial attached to a magnet that follows the rotation of the magnet mounted inside the reservoir. This type of gauge suffers a disadvantage in that it can only be viewed at an angle of approximately 120° around the bushing. For this reason, bushings with this type of gauge are normally rotated on the apparatus such that the gauge can be seen from ground level.

3.2.4.5 Clamping System

The clamping system used on bushings is very important because it provides the mechanical integrity of the bushing. A thorough discussion and excellent illustration of different types of clamping systems used for all insulators, including those used on bushings, is given in Section Q.2.2 of Appendix Q of the IEEE 693-1997, Recommended Practice for Seismic Design of Substations [11].

Two types of clamping systems are generally used on bushings, and a third type is used less frequently. The first, the mechanically clamped type, uses an external flange on the end of each insulator, and bolts are used to fasten them to mating parts, i.e., the mounting flange and the top and bottom terminals. A grading ring is often placed over this area on higher-voltage designs to shield the bolts from electric fields. The mechanically clamped type is economical and compact, but it has an increased potential for breakage due to stress concentration present at the bolted clamps.

The second type, the center-clamped type, involves the use of a compression-type spring assembly in the reservoir located at the top of the bushing, thereby placing the central conductor in tension when the spring assembly is released. This action simultaneously places the insulators, the flange and gaskets between these members, and the terminals at the extreme ends of the insulators in compression, thereby sealing the gaskets. The center-clamped type is also an economical, compact design, but it has the potential of oil leaks due to cantilever or seismic forces placed on the insulator. The capacitance-graded bushing shown in Figure 3.2.2 uses a center-clamped type of clamping system.

The third type, the cemented type, uses a metal flange to encircle of the ends of the insulator. A small radial gap is left between the outer diameter of the insulator and the inside diameter of the flange. This gap is filled with grout material rigid enough to transfer the compressive loads but pliable enough to prevent load concentrations on the porcelain. As with the mechanically clamped type, bolts are used to fasten them to mating parts, and grading rings are used at the higher voltages. This type of clamping system minimizes the potential for oil leakage or breakage due to mechanical stress concentrations, but the overall length of the insulator must be increased slightly in order to maintain electrical metal–metal clearances. The pressurized gas bushing shown in Figure 3.2.4 uses the cemented type of clamping system.

Whatever method is used for the clamping system, the clamping force must be adequate to withstand the cantilever forces that will be exerted on the ends of a bushing during its service life. The major mechanical force to which the top end of an outdoor bushing is subjected during service is the cantilever force applied to the top terminal by the line pull of the connecting lead. This force comprises the static force exerted during normal conditions plus the forces exerted due to wind loading and/or icing on the connecting lead. In addition, bushings mounted at an angle from vertical exert a force equivalent to a static cantilever force at the top of the bushing, and this force must be accounted for in the design.

In addition to the static forces, bushings must also withstand short-time dynamic forces created by short-circuit currents and seismic shocks. In particular, the lower end of bushings mounted in circuit breakers must also withstand the forces created by the interruption devices within the breaker.

Users can obtain guidance for allowable line pull from IEEE Std. C57.19.100-1995 [8], which recommends permissible loading levels. According to the standard, the static line loading should not exceed 50% of the test loading, as defined later in Section 3.2.8.3, and the short-time, dynamic loading should not exceed 85% of the same test loading.

3.2.4.6 Temperature Limits

Temperature limits within bushings depend on the type of bushing and the materials used in them. Solid-type bushings are made of only the central conductor, the porcelain or epoxy insulator(s), and the sealing gaskets. These bushings are therefore limited to the maximum allowable temperatures of the sealing gaskets and possibly the epoxy insulators, if used.

The kraft-paper insulation typically used to provide electrical insulation and mechanical support for the grading elements in a capacitance-graded bushing is severely limited by temperature. The maximum temperature that this paper can endure without accelerated loss of life is 105°C. Standards [1] have therefore established the following maximum temperatures for this type of bushing:

Temperature of immersion oil: 95°C average over a 24-hr period, with a maximum of 105°C
Ambient air temperature: 40°C
Top terminal temperature: 70°C (30°C rise over ambient air)
Bushing hottest-spot temperature: 105°C

IEEE Guide C57.19.100 [8] gives a detailed procedure for establishing thermal constants for conductor hottest spot of bottom-connected bushings with no significant dielectric losses and no cooling ducts. After the tests have been performed, an estimate for the steady-state temperature rise at any current can be made with the following equation [12]:

$$\Delta\Theta_{HS} = k_1 I^n + k_2 \Delta\Theta_o \qquad (3.2.4)$$

where

$\Delta\Theta_{HS}$ = steady-state bushing hottest-spot rise over ambient air, °C
$\Delta\Theta_o$ = steady-state immersion-oil rise over ambient air, i.e., transformer top-oil rise, °C
I = per unit load current based on bushing rating
$k_2 = \Delta\Theta_{HS}/\Delta\Theta_o$, with both quantities being obtained by measurement when I = 0
$k_1 = \Delta\Theta_{HS} - k_2 \Delta\Theta_o$, with both quantities being obtained by measurement when I = 1.0

$$n = \{\ln[\Delta\Theta_{HS} (I = X \text{ pu}) - k_2 \Delta\Theta_o (I = X \text{ pu})]\}/[k_1 \ln I (I = X \text{ pu})] \qquad (3.2.5)$$

Typical values of k_1, k_2, and n generally range from 15 to 32, 0.6 to 0.8, and 1.6 to 2.0, respectively.

A transformer's top-oil temperature sometimes exceeds 95°C (55°C rise) at rated load. In such cases, the rating of the bushing must be derated such that the kraft-paper insulation does not exceed 105°C (65°C rise). The following equations can be used to establish a derating factor, based on the transformer's top-oil temperature, to be multiplied by the bushing's current rating in order to determine the maximum current rating for that particular application:

$$I_d = d \, I_r \qquad (3.2.6)$$

where

I_d = derated current at transformer top-oil temperature rise $\Delta\Theta_o$

$$d = [(65 - k_2 \Delta\Theta_o)/k_1]^{1/n} \qquad (3.2.7)$$

I_r = rated current
$\Delta\Theta_o$ = transformer top-oil rise, °C
Constants k_1, k_2, and n are as defined above

Note from Equation 3.2.7 that d = 1 when $\Delta\Theta_o$ = 55°C. This consequently leads to the following dependence between k_1 and k_2:

$$k_1 = 65 - 55 k_2 \qquad (3.2.8)$$

Figure 3.2.5 shows the bushing derating factors for transformer top-oil rises between 55 and 65°C. The curves in Figure 3.2.5 are defined in Table 3.2.1.

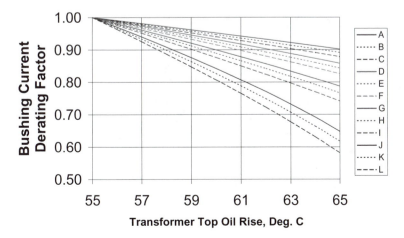

FIGURE 3.2.5 Bushing derating factors for transformer top-oil rises between 55 and 65°C.

TABLE 3.2.1 Definition of Curves Depicted in Figure 3.2.5

Constants k_1 and k_2	Exponent, n		
	1.6	1.8	2.0
$k_1 = 15.5$	L	K	J
$k_2 = 0.9$			
$k_1 = 21.0$	I	H	G
$k_2 = 0.8$			
$k_1 = 26.5$	F	E	D
$k_2 = 0.7$			
$k_1 = 32.0$	C	B	A
$k_2 = 0.6$			

3.2.5   Other Features of Bushings

3.2.5.1   Voltage Taps

It is possible within capacitance-graded bushings to create a capacitance divider arrangement wherein a small voltage, on the order of 5 to 10 kV, appears at the "voltage tap" when the bushing is operated at normal voltage. The voltage tap is created by attaching to one of the grading elements just to the inside of the grounded element. This tap is normally located in the flange, as shown in Figure 3.2.2 and in more detail as an example in Figure 3.2.6. Two sets of standard dimensions were used for voltage taps in the past, but modern bushings use only one set of standard dimensions [6]. The center conductor of the tap is grounded during normal operation unless voltage is required to power some measuring equipment.

The voltage tap can be used during the testing operation of the bushing and the apparatus into which it is installed, as well as during field operation. In the former application, it is used to perform power-factor and capacitance measurements on the bushing starting at the factory and throughout its life, as well as partial-discharge measurements within the bushing tested by itself or within the transformer. It is used during field operation to provide voltage to relays, which monitor phase voltages and instruct the circuit breakers to operate under certain conditions, in conjunction with the bushing potential device, discussed in Section 3.2.7.1, Bushing Potential Device.

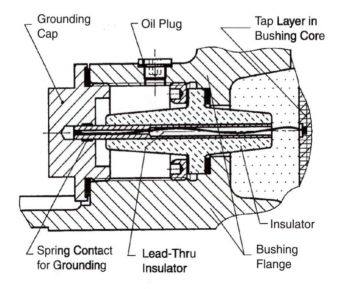

FIGURE 3.2.6 Voltage tap.

3.2.5.2 Bushing Current Transformer Pockets

The bushing flange creates a very convenient site to locate bushing current transformers (BCTs). The flange is extended on its inner end, and the BCTs, having 500 to 5000 turns in the windings, are placed around the flange. This location is called the BCT pocket and is shown in Figure 3.2.2. In this case, the bushing central conductor forms the single-turn primary of the BCT, and the turns in the windings form the secondary. Bushings built with standard dimensions have standard lengths for the BCT pocket [6].

3.2.5.3 Lower Support/Lower Terminal

Bushings that do not use draw-through leads, described in Section 3.2.7.3, must have a lower terminal in order to connect to the transformer winding or the circuit-breaker internal mechanism. This terminal can be one of any number of shapes, e.g., a smooth stud, threaded stud, spade, tang, or simply a flat surface with tapped holes for an additional terminal to be attached. Standards [6] prescribe several of these for various sizes of bushings.

One lower terminal specified by standards for bushings with voltage ratings 115 through 230 kV incorporates the lower-support function of the bushing with the lower terminal. The lower support is an integral part of the second type of clamping system (compression) described above, and in this function, it helps create the required forces for compressing the seals. The lower surface of the flat support has tapped holes in it so that the desired lower terminal can be attached. Two lower terminals specified by standards have a spherical radius of 102 mm (4.0 in.) machined into their bottom surfaces. This spherical radius enables the use of an additional lower terminal with a suitably shaped matching surface to be attached via the tapped holes, but at a small angle with the bottom of the bushing. This feature is useful for attaching rigid leads that are not always perfectly true with respect to the placement of the bushing, or when bushings are mounted at a small angle from vertical.

3.2.5.4 Lower-End Shield

It can be seen from Figure 3.2.3 that all regions of the lower end of air-to-oil bushings experience high dielectric stresses. In particular, the areas near the corners of the lower support and terminal are very highly stressed. Therefore, electrostatic shields with large radii, such as the one shown in Figure 3.2.2 and Figure 3.2.7, are attached to the lower end of these bushings in order to reduce the electric fields

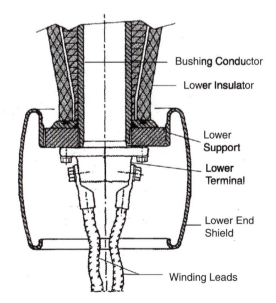

FIGURE 3.2.7 Lower-end shield.

that appear in this area. Shields also serve the purpose of shielding the bolted connections used to connect the leads to the bushing. Since shields with a thin dielectric barrier are somewhat stronger dielectrically, crepe paper is wrapped, or molded pressboard is placed on, the outer surfaces of the shield.

3.2.6 Bushings for Special Applications

3.2.6.1 High-Altitude Applications

Bushings intended for application at altitudes higher than 1000 m suffer from lower air density along the outer insulator. Standards [1] specify that, when indicated, the minimum insulation necessary at the required altitude can be determined by dividing the standard insulation length at 1000 m by the correction factor given in Table 3.2.2. For instance, suppose that the required length of the air insulator on a bushing is 2.5 m at 1000-m altitude. Further, suppose that this bushing is to be applied at 3000 m. Hence, the air insulator must be at least 2.5/0.8 = 3.125 m in length. The air insulator on the bushing designed for 1000 m must be replaced with a 3.125-m-long insulator, but the remainder of the bushing, i.e., the central core and the oil insulator, will remain the same as the standard bushing because these parts are not affected by air insulation. These rules do not apply to altitudes higher than 4500 m.

3.2.6.2 Highly Contaminated Environments

Insulators exposed to pollution must have adequate creep distance, measured along the external contour of the insulator, to withstand the detrimental insulating effects of contamination on the insulator surface. Figure 3.2.2 shows the undulations on the weather sheds, and additional creep distance is obtained by adding undulations or increasing their depth. Recommendations for creep distance [8] are shown in Table 3.2.3 according to four different classifications of contamination.

For example, a 345-kV bushing has a maximum line-to-ground voltage of 220 kV, so that the minimum creep is $220 \times 28 = 6160$ mm for a light contamination level and $220 \times 44 = 9680$ mm for a heavy contamination level. The term ESDD (equivalent salt-density deposit) used in Table 3.2.3 is

TABLE 3.2.2 Dielectric-Strength Correction Factors for Altitudes Greater than 1000 m

Altitude, m	Altitude Correction Factor for Dielectric Strength
1000	1.00
1200	0.98
1500	0.95
1800	0.92
2100	0.89
2400	0.86
2700	0.83
3000	0.80
3600	0.75
4200	0.70
4500	0.67

Source: ANSI/IEEE, 1997 [1]. With permission.

TABLE 3.2.3 Recommended Creep Distances for Four Contamination Levels

Contamination Level	Equivalent Salt-Deposit Density (ESDD), mg/cm^2	Recommended Minimum Creep Distance, mm/kV
Light	0.03–0.08	28
Medium	0.08–0.25	35
Heavy	0.25–0.6	44
Extra heavy	above 0.6	54

Source: IEEE Std. C57.19.100-1995 (R1997) [8]. With permission.

the conductivity of the water-soluble deposits on the insulator surface. It is expressed in terms of the density of sodium chloride deposited on the insulator surface that will produce the same conductivity. Following are typical environments for the four contamination levels listed [8]:

Light-contamination areas include areas without industry and with low-density emission-producing residential heating systems, and areas with some industrial areas or residential density but with frequent winds and/or precipitation. These areas are not exposed to sea winds or located near the sea.

Medium-contamination areas include areas with industries not producing highly polluted smoke and/ or with average density of emission-producing residential heating systems, areas with high industrial and/or residential density but subject to frequent winds and/or precipitation, and areas exposed to sea winds but not located near the sea coast.

Heavy-contamination areas include those areas with high industrial density and large city suburbs with high-density emission-producing residential heating systems, and areas close to the sea or exposed to strong sea winds.

Extra-heavy-contamination areas include those areas subject to industrial smoke producing thick, conductive deposits and small coastal areas exposed to very strong and polluting sea winds.

3.2.6.3 High-Current Bushings within Isolated-Phase Bus Ducts

As already noted, there are applications where temperatures can exceed the thermal capabilities of kraft-paper insulation used within bushings. One such application is in high-current bushings that connect between generator step-up transformers (GSUT) and isolated-phase bus duct. Typically, forced air is used to cool the central conductors in the isolated-phase bus duct, air forced toward the GSUT in the two outer phases and returned at twice the speed in the center phase. Air temperatures at the outer ends of the bushings typically range from 80 to 100°C, well above the standard limit of 40°C. This means that either a derating factor, sometimes quite severe, must be applied to the bushing's current rating, or materials with higher temperature limits must be used.

In older bushings, which were of the solid type, the only materials that were temperature limited were the gaskets, typically cork neoprene or nitrile. In this case, these gasket materials were changed to higher-temperature, oil-compatible fluorosilicon or fluorocarbon materials. However, solid-type bushings do not have low-partial-discharge characteristics. Therefore, as requirements for low-partial-discharge characteristics arose for GSUTs, a capacitance-grade core was used. As has already been explained, kraft-paper insulation is limited to 105°C, so that higher-temperature materials have been adapted for this purpose. This material is a synthetic insulation called aramid, i.e., Nomex®, and it has a limiting temperature in the order of 200°C. The material with the next-highest limiting temperature is the mineral oil, and to date, its temperature limits have been adequate for the high-temperature, high-current bushing application.

3.2.7 Accessories Commonly Used with Bushings

3.2.7.1 Bushing Potential Device

It is often desirable to obtain low-magnitude voltage and moderate wattage at power frequency for purposes of supplying voltage to synchroscopes, voltmeters, voltage-responsive relays, or other devices. This can be accomplished by connecting a bushing potential device (BPD) [13] to the voltage tap of a condenser-type bushing. Output voltages of a BPD are commonly in the 110 to 120-V range, or these values divided by 1.732, and output power typically ranges from about 25 W for 115-kV bushings to 200 W for 765-kV bushings.

A simple schematic of the BPD and bushing voltage tap is shown in Figure 3.2.8. The BPD typically consists of several components: a special fitting on the end of a shielded, weatherproof cable that fits into the voltage tap of the bushing; a padding capacitor that reduces the voltage seen by the BPD; a main transformer having an adjustable reactance; an adjustable-ratio auxiliary transformer; a tapped capacitor used to correct the power factor of the burden; a protective spark gap in case a transient voltage appears on the bushing; and a grounding switch that enables de-energization of the device. All items except the

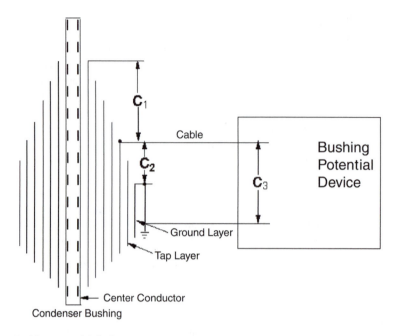

FIGURE 3.2.8 Bushing potential device.

first are housed in a separate cabinet, typically mounted to the side of the transformer or circuit-breaker tank. Since the BPD is essentially a series-tuned device, output phase shift is sensitive to output frequency. The greatest phase shift is experienced when the BPD is loaded to its rating and system voltage is low relative to the bushing rating.

If the BPD is called upon to carry a burden beyond its capacity, the voltage appearing on the tap rises. If it rises enough, it will cause the protective gap to operate. This phenomenon is also a consequence of the series-resonant circuit in the BPD.

3.2.7.2 Upper Test Terminals

In order to perform periodic maintenance tests on bushings, transformers, and other electrical equipment, it is necessary to disconnect the line leads from the bushing terminals. This often requires the use of bucket trucks and/or lifting cranes to loosen the connections and lower the leads, particularly on the higher voltage ratings. This operation therefore requires several people and a substantial amount of time.

A device known as the Lapp test terminal [14] is used to simplify this operation. This device, shown with the shunting bars opened and closed in Figure 3.2.9, is made of a short length of porcelain, mounting terminals on both ends, and some shunting bars that connect both terminals during normal operation. Its bottom terminal connects to the top terminal of the bushing, and the line leads are connected to its top terminal. When maintenance tests are required, one end of each shunt is loosened, and the shunts are swung away so that there is no connection from top to bottom. This enables the line to be isolated from the bushing without actually removing the line, and the testing on the transformer or other equipment can proceed, saving both time and manpower.

The bushing and outer-terminal design for the bushing must be adequate for the use of the Lapp test terminal. The outer terminal must be capable of withstanding the moment placed on the top of the test terminal without permanently bending the bushing's top terminal or upper part of the central conductor, and the bushing must be capable of withstanding the extra moment placed on it. The primary location of concern is at the bottom of the upper insulator, which can be lifted up off the sealing gasket, if one is used, and allow leakage of the internal insulating oil.

FIGURE 3.2.9 Lapp test terminal. (Photo courtesy of Lapp Insulator Company, Leroy, NY.)

At higher voltages, excessive amounts of corona may occur due to the relatively sharp edges present on the top or both ends of the Lapp test terminal. Large-diameter corona rings are therefore placed over one or both ends of the terminal.

3.2.7.3 Draw-Lead Conductors

Normally, bushings have current ratings of 1200 A or higher. Some applications of bushings mounted in transformers with lower MVA ratings do not require these current-carrying capacities. In these cases, it is practical to run a smaller-diameter cable inside the hollow central tube in the bushing and connect it directly to the transformer winding. If the bushing must be removed for some reason, the transformer oil can be lowered, as necessary, to a level below the top of the transformer tank or turret. Then, the top of the draw-lead is unfastened from the top terminal of the bushing. The bushing can then be lifted out of the transformer and replaced with a new one, and the top terminal is reinstalled to the draw-lead. Finally, the oil can be adjusted to the proper level in the transformer. The use of a draw-lead therefore enables much faster replacement of bushings and eliminates the need for time-consuming processing of oil for the transformer.

Current-carrying capabilities of draw-leads are established by transformer manufacturers and are not standardized at this time. In general, the capability will increase with the cross-sectional area of the cable and decrease with the length of the cable. For these reasons, larger-sized holes are placed in the central tubes of bushings with higher voltage ratings. Table 3.2.4 gives the maximum current ratings of draw-lead applications, the current rating of the same bushing for bottom-connected applications, and the minimum hole size in the central conductor. [6]

3.2.8 Tests on Bushings

3.2.8.1 Categories of Tests

Standards [1] designate three types of tests to be applied to bushings:

TABLE 3.2.4 Current Ratings of Bushings Capable of Being Used with Draw-Leads

Nominal System Voltage, kV	Maximum Draw-Lead Current Rating, A	Bottom-Connected Current Rating, A	Minimum Diameter inside Tube, mm
34.5–69	400	1200	22
138–230	800	1200	41
345–765	800	1200	51

1. Design tests
2. Routine tests
3. Special tests

3.2.8.1.1 Design Tests

Design, or type, tests are only made on prototype bushings, i.e., the first of a design. The purpose of design tests is to ascertain that the bushing design is adequate to meet its assigned ratings, to ensure that the bushing can operate satisfactorily under usual or special service conditions, and to demonstrate compliance with industry standards. These tests need not be repeated unless the customer deems it necessary to have them performed on a routine basis.

Test levels at which bushings are tested during design tests are higher than the levels encountered during normal service so as to establish margins that take into account dielectric aging of insulation as well as material and manufacturing variations in successive bushings. Bushings must withstand these tests without evidence of partial or full failure, and incipient damage that initiates during the dielectric tests is usually detected by comparing values of power factor, capacitance, and partial discharge before and after the testing program.

Standards [1] prescribe the following design tests:

Low-frequency wet-withstand voltage on bushings rated 242 kV maximum system voltage and less
Full-wave lightning-impulse-withstand voltage
Chopped-wave lightning-impulse-withstand voltage
Wet-switching-impulse-withstand voltage on bushings rated 345 kV maximum system voltage and greater
Draw-lead bushing-cap pressure test
Cantilever-withstand test
Temperature test at rated current

3.2.8.1.2 Routine Tests

Routine, or production, tests are made on every bushing produced, and their purpose is to check the quality of the workmanship and the materials used in the manufacture. Standards [1] prescribe the following routine tests:

Capacitance and power-factor measurements at 10 kV
Low-frequency dry-withstand test with partial-discharge measurements
Tap-withstand voltage test
Internal hydraulic pressure test

3.2.8.1.3 Special Tests

Special tests are for establishing the characteristics of a design practice and are not part of routine or design tests. The only special test currently included in standards [1] is the thermal-stability test, only applicable to extra high voltage (EHV) bushings, but other tests could be added in the future. These include short-time, short-circuit withstand and seismic capabilities.

3.2.8.2 Dielectric Tests

3.2.8.2.1 Low-Frequency Tests

There are two low-frequency tests:

1. Low-frequency wet-withstand voltage test
2. Low-frequency dry-withstand voltage test

3.2.8.2.1.1 Low-Frequency Wet-Withstand Voltage Test — The low-frequency wet-withstand voltage test is applied on bushings rated 242 kV and below while a waterfall at a particular precipitation rate and conductivity is applied. The values of precipitation rate, water resistivity, and the time of application vary in different countries. American standard practice is a precipitation rate of 5 mm/min, a resistivity of 178 ohm-m, and a test duration of 10 sec, whereas European practice is 3 mm/min, 100 ohm·m, and 60 sec, respectively [15]. If the bushing flashes over externally during the test, it is allowed that the test be applied one additional time. If this attempt also flashes over, then the test fails and something must be done to modify the bushing design or test setup so that the capability can be established.

3.2.8.2.1.2 Low-Frequency Dry-Withstand Voltage Test — The low-frequency dry-withstand test was, until recently, made for a 1-min duration without the aid of partial-discharge measurements to detect incipient failures, but standards [1] currently specify a one-hour duration for the design test, in addition to partial-discharge measurements. The present test procedure is:

Partial discharge (either radio-influence voltage or apparent charge) shall be measured at 1.5 times the maximum line-ground voltage. Maximum limits for partial discharge vary for different bushing constructions and range from 10 to 100 µV or pC.

A 1-min test at the dry-withstand level, approximately 1.7 times the maximum line-ground voltage, is applied. If an external flashover occurs, it is allowed to make another attempt, but if this one also fails, the bushing fails the test. No partial-discharge tests are required for this test.

Partial-discharge measurements are repeated every 5 min during the one-hour test duration at 1.5 times maximum line-ground voltage required for the design test. Routine tests specify only a measurement of partial discharge at 1.5 times maximum line-ground voltage, after which the test is considered complete.

Bushing standards were changed in the early 1990s to align with the transformer practice, which started to use the one-hour test with partial-discharge measurements in the late 1970s. Experience with this new approach has been good in that incipient failures were uncovered in the factory test laboratory, rather than in service, and it was decided to add this procedure to the bushing test procedure. Also from a more practical standpoint, bushings are applied to every transformer, and transformer manufacturers require that these tests be applied to the bushings prior to application so as to reduce the number of bushing failures during the transformer tests.

3.2.8.2.2   Wet-Switching-Impulse-Withstand Voltage

This test is required on bushings rated for 345-kV systems and above. The test waveshape is 250 µs time-to-crest and 2500 µs time-to-half-value with tolerance of ±30% on the time to crest and ±20% on the time-to-half-value. This is the standard waveshape for testing insulation systems without magnetic-core steel present in the test object and is different than the waveshape for transformers.

Three different standard test procedures are commonly used to establish the wet-switching-impulse-withstand voltage of the external insulation:

Fifteen impulses of each polarity are applied, with no more than two flashovers allowed.

Three impulses of each polarity are applied. If a flashover occurs, then it is permitted to apply three additional impulses. If no flashovers occur at either polarity, then the bushing passes the test. Otherwise, the bushing fails the test.

The 90% (1.3 σ) level is established from the 50% flashover tests.

3.2.8.2.3   Lightning-Impulse Tests

The same waveshapes are used to establish the lightning-impulse capability of bushings and transformers. The waveshape for the full wave is 1.2 µs for the wavefront and 50 µs for the time-to-half-value, and the chopped wave flashes over at a minimum of 3.0 µs. One of the same procedures as described above for the wet-switching-impulse tests is followed to establish the full-wave capability for both polarities. The chopped-wave capability is established by applying a minimum of three chopped impulses at each polarity.

3.2.8.3   Mechanical Tests

IEEE Std. C57.19.01 [6] specifies the static-cantilever-withstand forces to be applied separately to the top and bottom ends of outdoor-apparatus bushings. The forces applied to the top end range from 68 kg

(150 lb) for the smaller, lower-voltage bushings to 545 kg (1200 lb) for the larger, higher-voltage or current bushings, and the forces applied to the lower end are generally about twice the top-end forces.

The test procedure is to apply the specified forces perpendicular to the bushing axis, first at one end then at the other, each application of force lasting 1 min. Permanent deflection, measured at the bottom end, shall not exceed 0.76 mm, and there shall be no oil leakage at either end at any time during the test or within 10 min after removing the force.

3.2.8.4 Thermal Tests

There are two thermal tests. The first is the thermal test at rated current, and it is applied to all bushing designs. The second test is the thermal-stability test [11], and it is applied for only EHV bushings.

3.2.8.4.1 Thermal Test at Rated Current

This test demonstrates a bottom-connected bushing's ability to carry rated current. The bushing is first equipped with a sufficient number of thermocouples, usually placed inside the inner diameter of the hollow-tube conductor, to measure the hottest-spot temperature of the conductor. The bushing is then placed in an oil-filled tank, the oil is heated to a temperature rise above ambient air of 55°C for transformers and 40°C for circuit breakers, and rated current is passed through the central conductor until thermal equilibrium is reached. The bushing passes the test if the hottest-spot temperature rise above ambient air does not exceed 65°C.

3.2.8.4.2 Thermal-Stability Test [16]

Capacitive leakage currents in the insulating material within bushings cause dielectric losses. Dielectric losses within a bushing can be calculated by the following equation using data directly from the nameplate or test report:

$$P_d = 2 \pi f C V^2 \tan \delta \tag{3.2.9}$$

where

P_d = dielectric losses, W
f = applied frequency, Hz
C = capacitance of bushing (C_1), F
V = operating voltage, rms V
$\tan \delta$ = dissipation factor, p.u.

A bushing operating at rated voltage and current generates both ohmic and dielectric losses within the conductor and insulation, respectively. Since these losses, which both appear in the form of heat, are generated at different locations within the bushing, they are not directly additive. However, heat generated in the conductor influences the quantity of heat that escapes from within the core. A significant amount of heat generated in the conductor will raise the conductor temperature and prevent losses from escaping from the inner surface of the core. This causes the dielectric losses to escape from only the outer surface of the core, consequently raising the hottest-spot temperature within the core. Most insulating materials display an increasing dissipation factor, $\tan \delta$, with higher temperatures, such that as the temperature rises, $\tan \delta$ also rises, which in turn raises the temperature even more. If this cycle does not stabilize, then $\tan \delta$ increases rapidly, and total failure of the insulation system ensues.

Bushing failures due to thermal instability have occurred both on the test floor and in service. One of the classic symptoms of a thermal-stability failure is the high internal pressure caused by the gases generated from the deteriorating insulation. These high pressures cause an insulator, usually the outer one because of its larger size, either to lift off the flange or to explode. If the latter event occurs with a porcelain insulator, shards of porcelain saturated with oil become flaming projectiles, endangering the lives of personnel and causing damage to nearby substation equipment.

Note from Equation 3.2.9 that the operating voltage, V, particularly influences the losses generated within the insulating material. It has been found from testing experience that thermal stability only becomes a factor at operating voltages 500 kV and above.

The test procedure given in ANSI/IEEE Std. C57.19.00-1991 [1] is to first immerse the lower end of the bushing in oil at a temperature of 95°C and then pass rated current through the bushing. When the bushing comes to thermal equilibrium, a test voltage equal to 1.2 times the maximum line-ground voltage is applied, and tan δ is measured at regular, normally hourly, intervals. These conditions are maintained until tan δ rises no more than 0.02% (0.0002 p.u.) over a period of five hours. The bushing is considered to have passed the test if it has reached thermal stability at this time and it withstands all of the routine dielectric tests without significant change from the previous results.

3.2.9   Maintenance and Troubleshooting

Once a bushing has been successfully installed in the apparatus, without breakage or other damage, it will normally continue to operate without problems for as long as its parent apparatus is in service unless it is thermally, mechanically, or electrically overloaded. However, such overloads do occur, plus some designs have design flaws, and bushings can sustain incipient damage during shipment or installation. All of these potential problems must be watched by periodic inspections and maintenance. The type and frequency of inspection and maintenance performed on bushings depend on the type of bushing plus the relative importance and cost of the apparatus in which the bushing is installed. The following gives a brief description of some common problems and maintenance that should be performed [8, 17, 18]:

3.2.9.1   Oil Level

Proper oil level is very important in the operation of a bushing, and abnormal change in oil level can indicate problems within the bushing. Loss of oil can indicate that the bushing has developed a leak, possibly through a gasket, a soldered or welded seal, or an insulator that has been cracked or broken. A leak on the air-end side may be an indication that water has entered the bushing, and the water content and dielectric properties of the bushing oil should be checked as soon as possible to determine whether this has occurred. Excessive water can cause deterioration of the dielectric integrity along the inside of the insulators, particularly the oil-end insulator and, if used, the bushing core. Furthermore, excessive loss of oil in a capacitance-graded bushing can cause oil level to drop below the top of the core. Over time, this will cause the insulating paper to become unimpregnated, and the bushing can suffer dielectric failure, possibly an explosive one.

An abnormal increase in oil level on a bushing installed in a transformer with a conservator oil-preservation system may indicate a leak in an oil-end gasket or seal, or the oil-end insulator may be cracked or broken.

3.2.9.2   Power-Factor/Capacitance Measurements

Two methods are used to make power (dissipation) factor and capacitance measurements [17, 18]. The first is the grounded specimen test (GST), where current, watts, and capacitance of all leakage paths between the energized central conductor and all grounded parts are measured. Measurements include the internal core insulation and oil as well as leakage paths over the insulator surfaces. The use of a guard-circuit connection can be used to minimize the effects of the latter.

The second method is the ungrounded specimen test (UST), where the above quantities are measured between the energized center conductor and a designated ungrounded test electrode, usually the voltage or test tap. The two advantages of the UST method are that the effects of unwanted leakage paths, for instance across the insulators, are minimized, and separate tests are possible while bushings are mounted in apparatus.

Standards [8] recommend that power factor and capacitance measurements be made at the time of installation, a year after installation, and every three to five years thereafter. A significant increase in a bushing's power factor indicates deterioration of some part of the insulating system. It may mean that one of the insulators, most likely the air-end insulator, is dirty or wet, and excessive leakage currents are flowing along the insulator. A proper reading can be obtained by cleaning the insulator. On the other hand, a significant increase of the power factor may also indicate deterioration within the bushing. An increase in the power factor across the C_1 portion, i.e., from conductor to tap, typically indicates deterioration within the core. An increase across the C_2 portion of a bushing using a core, i.e., from tap to flange, typically indicates deterioration of that part of the core or the bushing oil. If power factor doubles from the reading immediately after initial installation, the rate of change of the increase should be monitored at more frequent intervals. If it triples, then the bushing should be removed from service [8].

An increase of bushing capacitance is also a very important indicator that something is wrong inside the bushing. An excessive change, on the order of 2 to 5%, depending on the voltage class of the bushing, over its initial reading probably indicates that insulation between two or more grading elements has shorted out. Such a change in capacitance is indication that the bushing should be removed from service as soon as possible.

3.2.9.3 Damage or Contamination of Air-End Insulator

Insulators, particularly those that are porcelain, are susceptible to damage during handling, shipping, or from flying parts of other failed equipment. Chips can be broken out of the sheds, and these can generally be repaired by grinding off the rough edges with a file and painting over the break with a suitable paint. In some cases, a composite material can be used to fill the void caused by the break [17]. On the other hand, some damage may extend into the body of the insulator. These areas should be watched closely for oil leakage for a period after the damage is noticed, since the impact may have cause the insulator to crack.

Bushings used in highly polluted environments should be washed periodically. Washing involves either de-energizing the apparatus and cleaning the insulators by hand with a cleaning agent or using a suitable jet of low-conductivity water while energized [17].

3.2.9.4 Improper Installation of Terminals

Improper installation of either the top or bottom terminals can cause thermal problems. A common problem with threaded terminals is that force of some kind must be placed on the mating threads so that a good current path is formed; in the absence of such force on the joint, the terminal is not suitable for carrying large currents. These problems can be found by using thermography methods [17].

Some manufacturers require that threaded terminals be torqued to a certain minimum value. Another method commonly used is to cut the member with the female threads in half. Then, it is first threaded onto the bushing top terminal and finally clamped such that it compresses onto the male threads of the top terminal.

3.2.9.5 Misaligned or Broken Voltage Tap Connections

In some cases, usually during rough shipment of bushings, the core will rotate within the insulators. This can cause the connection to the tap layer to break. In some older bushings, a spring connection was used to make the tap connection, and this connection sometimes shifted during handling or shipment. Both of these problems cause the tap to become inoperative, and the power-factor/capacitance measurements involving the tap cannot be made.

3.2.9.6 Dissolved-Gas-in-Oil Analysis

It is not recommended that dissolved-gas-in-oil analysis (DGA) samples be made on a routine basis because bushings have only a limited supply of oil, and the oil will have to be replenished after several such oil samples have been taken. However, if power-factor/capacitance measurements indicate that something is wrong with the bushing, DGA samples are indicated. Large amounts of CO and CO_2 gases indicate deterioration of paper insulation within the bushing, whereas other DGA gases indicate by-products of arcing or thermal overheating, just as they do in transformer insulation.

References

1. ANSI/IEEE, Standard Performance Characteristics and Test Procedure for Outdoor Power Apparatus Bushings, IEEE Std. C57.19.00-1991 (R1997), Institute of Electrical and Electronics Engineers, Piscataway, NJ, 1997.

2. Easley, J.K. and Stockum, F.R., Bushings, IEEE Tutorial on Transformers, IEEE EH0209-7/84/0000-0032, Institute of Electrical and Electronics Engineers, Piscataway, NJ, 1983.

3. Nagel, R., Uber Eine Neuerung An Hochspannungstransformer Der Siemens- Schuckertwerke, *Elektrische Bahnen Betriebe*, 4, 275–278, May 23, 1906.

4. Reynders, A.B., Condenser Type of Insulation for High-Tension Terminals, *AIEE Trans.*, 23, Part I, 209, 1909.

5. Spindle, H.E., Evaluation, Design and Development of a 1200 kV Prototype Termination, USDOE Report DOE/ET/29068-T8 (DE86005473), U.S. Department of Energy, Washington, DC.

6. IEEE, Standard Performance Characteristics and Dimensions for Outdoor Power Apparatus Bushings, IEEE Std. C57.19.01-2000, Institute of Electrical and Electronics Engineers, Piscataway, NJ, 2000.

7. IEEE, Standard Requirements, Terminology and Test Procedures for Bushings for DC Applications, IEEE Std. C57.19.03-1996, Institute of Electrical and Electronics Engineers, Piscataway, NJ, 1996.

8. IEEE, Guide for Application of Power Apparatus Bushings, IEEE Std. C57.19.100-1995 (R1997), Institute of Electrical and Electronics Engineers, Piscataway, NJ, 1997.

9. IEC, Bushings for Alternating Voltages above 1000 V, IEC 137, International Electrotechnical Commission, Geneva, 1995.

10. Ueda, M., Honda, M., Hosokawa, M., Takahashi, K., and Naito, K., Performance of Contaminated Bushing of UHV Transmission Systems, *IEEE Trans.*, PAS-104, 891–899, 1985.

11. IEEE, Recommended Practice for Seismic Design of Substations, IEEE Std. 693-1997, Institute of Electrical and Electronics Engineers, Piscataway, NJ, 1997.

12. McNutt, W.J. and Easley, J.K., Mathematical Modeling — A Basis for Bushing Loading Guides, *IEEE Trans.*, PAS-97, 2395–2404, 1978.

13. ABB, Inc., Instructions for Bushing Potential Device Type PBA2, instruction leaflet IB 33-357-1F, ABB, Inc., Alamo, TN, September 1993.

14. Lapp Insulator Co., Extend Substation Equipment Life with Lapp-Doble Test Terminal, Bulletin 600, Lapp Insulator Co., LeRoy, NY, 2000.

15. IEEE, Standard Techniques for High Voltage Testing, IEEE Std. 4-1995, Institute of Electrical and Electronics Engineers, Piscataway, NJ, 1995.

16. Wagenaar, L.B., The Significance of Thermal Stability Tests in EHV Bushings and Current Transformers, Paper 4-6, presented at 1994 Doble Conference, Boston, MA.

17. EPRI, Guidelines for the Life Extension of Substations, EPRI TR-105070-R1CD, Electric Power Research Institute, Palo Alto, CA, November 2002.

18. Doble Engineering Co., A-C Dielectric-Loss, Power Factor and Capacitance Measurements as Applied to Insulation Systems of High Voltage Power Apparatus in the Field (A Review), Part I, Dielectric Theory, and Part II, Practical Application, Report ACDL-I and II-291, Doble Engineering Co., Boston, MA, Feb. 1991.

3.3 Load Tap Changers

Dieter Dohnal

For many decades power transformers equipped with load tap changers (LTC) have been the main components of electrical networks and industry. The LTC allows voltage regulation and/or phase shifting by varying the transformer ratio under load without interruption.

From the beginning of LTC development, two switching principles have been used for the load-transfer operation, the high-speed-resistance type and the reactance type. Over the decades, both principles have been developed into reliable transformer components available in a broad range of current and voltage applications to cover the needs of today's network and industrial-process transformers as well as ensuring optimum system and process control (Goosen, 1996).

This section refers to LTCs immersed in transformer mineral oil. The use of other insulating fluids or gas insulation requires the approval of the LTC's manufacturer and may lead to a different LTC design.

3.3.1 Design Principle

The LTC changes the ratio of a transformer by adding turns to or subtracting turns from either the primary or the secondary winding. The main components of an LTC are contact systems for make-and-break currents as well as carrying currents, transition impedances, gearings, spring energy accumulators, and a drive mechanism. In new LTC-designs, the contacts for make-and-break-currents will be replaced more and more by vacuum interrupters.

The transition impedance in the form of a resistor or reactor consists of one or more units that are bridging adjacent taps for the purpose of transferring load from one tap to the other without interruption or appreciable change in the load current. At the same, time they are limiting the circulating current for the period when both taps are used. Normally, reactance-type LTCs use the bridging position as a service position and, therefore, the reactor is designed for continuous loading.

The voltage between the mentioned taps is the step voltage. It normally lies between 0.8% and 2.5% of the rated voltage of the transformer.

The majority of resistance-type LTCs are installed inside the transformer tank (in-tank LTCs), whereas the reactance-type LTCs are in a separate compartment that is normally welded to the transformer tank.

3.3.1.1 Resistance-Type Load Tap Changer

The LTC design that is normally applied to larger powers and higher voltages comprises an arcing switch and a tap selector. For lower ratings, LTC designs are used where the functions of the arcing switch and the tap selector are combined in a so-called arcing tap switch.

With an LTC comprising an arcing switch and a tap selector (Figure 3.3.1), the tap change takes place in two steps (Figure 3.3.2). First, the next tap is preselected by the tap selector at no load (Figure 3.3.2,

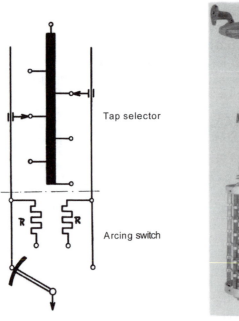

Tap selector

Arcing switch

Arcing switch

Tap selector

Switching principle   Design

FIGURE 3.3.1   Design principle — arcing switch with tap selector.

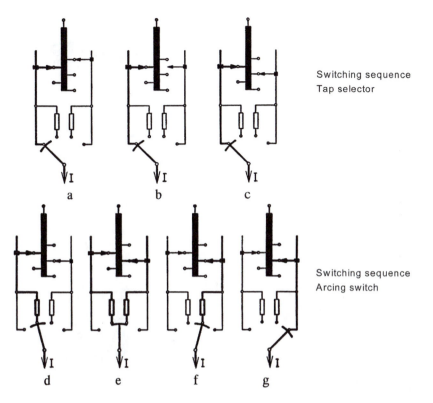

FIGURE 3.3.2   Switching sequence of tap selector — arcing switch.

positions a–c). Then the arcing switch transfers the load current from the tap in operation to the preselected tap (Figure 3.3.2, positions c–g). The LTC is operated by means of a drive mechanism. The tap selector is operated by a gearing directly from the drive mechanism. At the same time, a spring energy accumulator is tensioned. This operates the arcing switch — after releasing in a very short time — independently of the motion of the drive mechanism. The gearing ensures that this arcing switch operation always takes place after the tap preselection operation has been finished. With today's designs, the switching time of an arcing switch lies between 40 and 60 ms.

During the arcing switch operation, transition resistors are inserted (Figure 3.3.2, positions d–f), which are loaded for 20 to 30 ms, i.e., the resistors can be designed for short-term loading. The amount of resistor material required is therefore relatively small. The total operation time of an LTC is between 3 and 10 sec, depending on the respective design.

An arcing tap switch (Figure 3.3.3) carries out the tap change in one step from the tap in service to the adjacent tap (Figure 3.3.4). The spring energy accumulator, wound up by the drive mechanism actuates the arcing tap switch sharply after releasing. For switching time and resistor loading (Figure 3.3.4, positions b–d), the above statements are valid.

The details of switching duty, including phasor diagrams, are described by IEEE (Annex A [IEEE, 1995]) and IEC (Annex A [IEC, 2003]).

3.3.1.2   Reactance-Type Load Tap Changer

For reactance-type LTCs, the following types of switching are used (Annex B of [IEEE, 1995 and IEC, 2003]):

- Arcing tap switch
- Arcing switch with tap selector
- Vacuum interrupter with tap selector

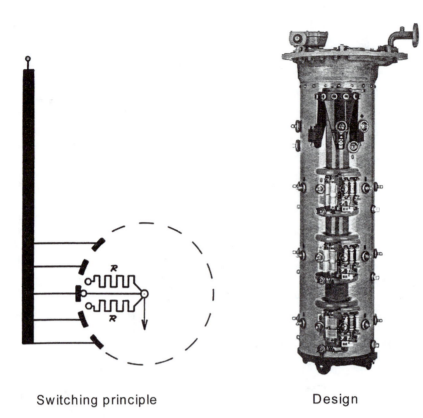

Switching principle   Design

FIGURE 3.3.3   Design principle — arcing tap switch.

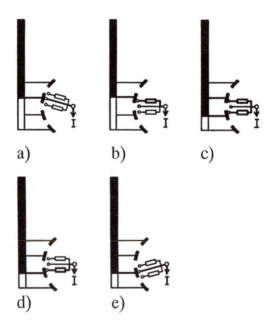

a)   b)   c)

d)   e)

FIGURE 3.3.4   Switching sequence of arcing tap switch.

FIGURE 3.3.5 Reactance-type LTC, inside view showing vacuum interrupter assembly.

Today the greater part of arcing tap switches is produced for voltage regulators, whereas the vacuum-type LTC is going to be the state of the art in the field of power transformers.

Figure 3.3.5 shows a three-phase vacuum-type LTC with full insulation between phases and to ground (nominal voltage level 69 kV). It consists of an oil compartment containing tap selector and reversing/coarse change-over selector, vacuum interrupters, and bypass switches.

A typical winding layout and the operating sequence of the said LTC is shown in Figure 3.3.6. The operating sequence is divided into three major functions:

1. Current transfer from the tap selector part preselecting the next tap to the part remaining in position by means of the vacuum interrupter in conjunction with the associated bypass switch (Figure 3.3.6b, positions A–C).
2. Selection of the next tap position by the tap selector in proper sequence, with the reclosing of the vacuum interrupter and bypass switch (Figure 3.3.6b, positions C–F). Contrary to resistance-type LTCs, the bridging position—in which the moving selector contacts p1 and p4 are on neighboring fixed selector contacts (in Figure 3.3.6b, contacts 4 and 5, position F)—is a service position, and therefore the preventive autotransformer/reactor (normally produced by the transformer manufacturer) is designed for continuous loading; i.e., the number of tap positions is twice the number of steps of the tap winding. In other words, the preventive autotransformer works as a voltage divider for step voltage of the tap winding in the bridging position. In comparison with the resistance-type LTC, the reactance-type LTC requires only half the number of taps of the tap winding for the equivalent number of service tap positions.
3. Operation of reversing or coarse changeover selector in order to double the number of positions; for this operation, the moving selector contacts p1 and p4 have to be on the fixed selector contact M (Figure 3.3.6a).

For more detailed information about switching duty and phasor diagrams, see Annex B (IEEE, 1995 and IEC, 2003).

3.3.1.3 Tap Position Indication

There are no general rules for defining the numerals on the tap-position indicator dial. This is a question of the user's specifications or national standards. Some users are accustomed to designations such as 1 through 33 (or 0 through 32), while other have traditionally known 16L (lower), 15L, 14L, ... N (neutral); 1R (raise), 2R ... 16R. An additional point of confusion comes about with the selection of the placement of the tap changer on the primary or secondary winding of the transformer. A tap changer on the primary

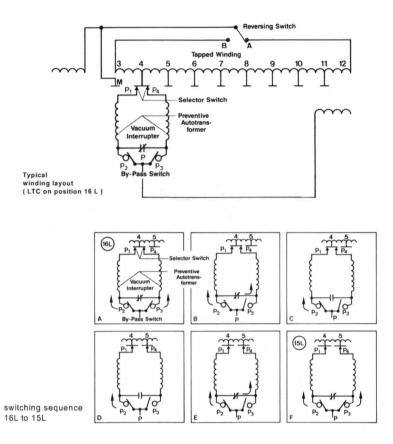

FIGURE 3.3.6 Reactance-type LTC. Top) Typical winding layout, LTC in position 16L; bottom) Switching sequence, position 16L to 15L.

is sometimes designated such as 1 through 33, but position #1 may indicate the greatest degree of voltage boost or buck, depending upon the transformer designer.

3.3.2 Applications of Load Tap Changers

3.3.2.1 Basic Arrangements of Regulating Transformers

The following basic arrangements of tap windings are used (Figure 3.3.7).

Linear arrangement (Figure 3.3.7a), is generally used on power transformers with moderate regulating ranges up to a maximum of 20%.

With a **reversing changeover selector** (Figure 3.3.7b), the tap winding is added to or subtracted from the main winding so that the regulating range can be doubled or the number of taps reduced. During this operation the tap winding is disconnected from the main winding (for a discussion on problems arising from this disconnection, see Section 3.3.5.1 entitled "Voltage Connection of Tap Winding during Changeover Operation"). The greatest copper losses occur, however, in the position with the minimum number of effective turns. This reversing operation is realized with the help of a changeover selector, which is part of the tap selector or of the arcing tap switch. The **double reversing changeover selector** (Figure 3.3.7c) avoids the disconnection of tap winding during the changeover operation. In phase-shifting transformers (PST), this apparatus is called an **advance-retard switch (ARS).**

By means of a **coarse changeover selector** (Figure 3.3.7d), the tap winding is either connected to the plus or minus tapping of the coarse winding. Also during coarse-selector operation, the tap winding is disconnected from the main winding. (Special winding arrangements can cause the same disconnection

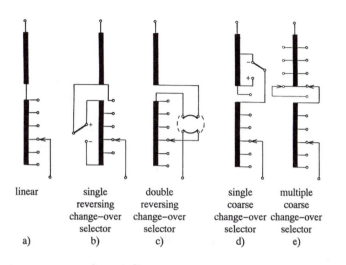

linear | single reversing change–over selector | double reversing change–over selector | single coarse change–over selector | multiple coarse change–over selector
a) | b) | c) | d) | e)

FIGURE 3.3.7   Basic arrangements of tap windings.

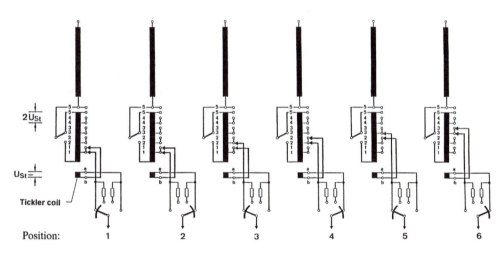

Position: 1   2   3   4   5   6

FIGURE 3.3.8   LTC for tap windings with tickler coil.

problems as above. The series impedance of coarse winding/tap winding has to be checked, see Section 3.3.5.2 entitled "Effects of the Leakage Impedance of Tap Winding/Coarse Winding during Operation of the Arcing Switch When Passing the Mid-Position of the Resistance-Type LTC".) In this case, the copper losses are lowest in the position of the lowest effective number of turns. This advantage, however, puts higher demands on insulation material and requires a larger number of windings. The **multiple coarse change over selector** (Figure 3.3.7e) allows a multiplication of the regulating range. It is mainly applied for industrial process transformers (rectifier/furnace transformers). The coarse change-over selector is also part of the LTC.

An arrangement of **LTC for tap winding with tickler coil** is applied when the tap winding has, for example, 8 steps and where the regulation should be carried out in ±16 steps. In this case, a tickler coil, whose voltage is half the step voltage of the tap winding, is used. The tickler coil is electrically separated from the tap winding and is looped into the arcing switch/tap selector connection. Figure 3.3.8 shows the switching sequence in positions 1,3,5,…, the tickler coil is out of circuit; in positions 2,4,6,…, the tickler coil is inserted.

Which of these basic winding arrangements is used in the individual case depends on the system and the operating requirements. These arrangements are applicable to two-winding transformers as well as to voltage and phase-shifting transformers (PST).

The position in which the tap winding and therefore the LTC is inserted in the windings depends on the transformer design.

3.3.2.2 Examples of Commonly Used Winding Arrangements

Two-winding transformers with **wye-connected windings** have the regulation applied to the neutral end as shown in Figure 3.3.9. This results in relatively simple and compact solutions for LTCs and tap windings.

Regulation of **delta-connected windings** (Figure 3.3.10) requires a three-phase LTC whose three phases are insulated according to the highest system voltage applied (Figure 3.3.10a), or 3 single-phase LTCs, or 1 single-phase and 1 two-phase LTC (Figure 3.3.10b). Today, the design limit for three-phase LTCs with phase-to-phase insulation is a nominal voltage level of 138 kV (BIL [basic impulse insulation level] 650 kV). To reduce the phase-to-phase stresses on the delta-LTC, the three-pole mid-winding arrangement (Figure 3.3.10c) can be used.

For regulated **autotransformers**, Figure 3.3.11 shows various schemes. Depending on their regulating range, system conditions and/or requirements, weight, and size restrictions during transportation, the most appropriate scheme is chosen. Autotransformers are always wye-connected.

- Neutral-end regulation (Figure 3.3.11a) may be applied with a ratio above 1:2 and a moderate regulating range up to 15%. It operates with variable flux.
- A scheme shown in Figure 3.3.11c is used for regulation of high voltage, U_1.

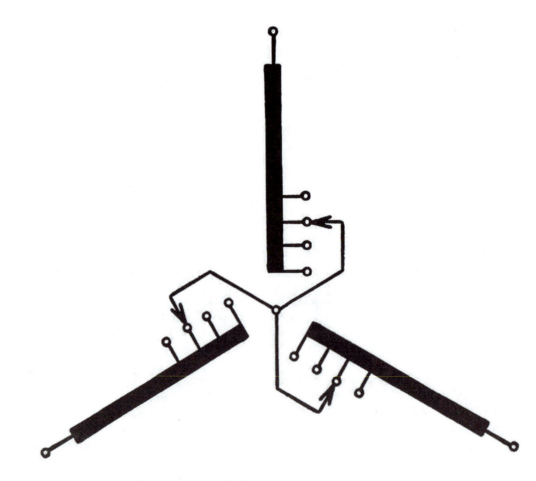

FIGURE 3.3.9 LTC with wye-connected windings.

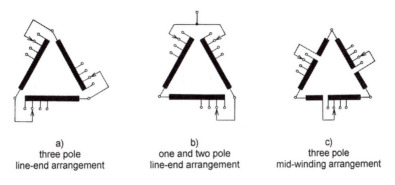

FIGURE 3.3.10   LTC with delta connected windings.

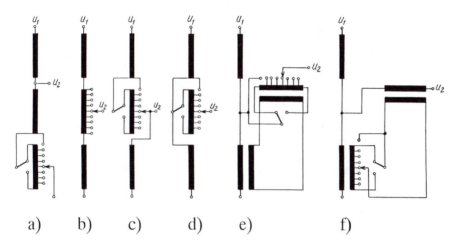

FIGURE 3.3.11   LTC in autotransformers.

- For regulation of low voltage U_2 the circuits in Figure 3.3.11b,d,e,f are applicable. The arrangements in Figure 3.3.11e and 3.3.11f are two core solutions. Circuit Figure 3.3.11f is operating with variable flux in the series transformer, but it has the advantage that a neutral-end LTC can be used. In case of arrangement according to Figure 3.3.11e, main and regulating transformers are often placed in separate tanks to reduce transport weight. At the same time, this solution allows some degree of phase shifting by changing the excitation connections within the intermediate circuit.

3.3.3   Phase-Shifting Transformers (PST)

In recent years, the importance of PSTs used to control the power flow on transmission lines in meshed networks has steadily been increasing (Krämer and Ruff, 1998). The fact that IEEE has prepared a "Guide for the Application, Specification and Testing of Phase-Shifting Transformers" proves the demand for PSTs (IEEE C57.135-2001). These transformers often require regulating ranges that exceed those normally used. To reach such regulating ranges, special circuit arrangements are necessary. Two examples are given in Figs. 3.3.12 and 3.3.13. Figure 3.3.12 shows a circuit with direct line-end regulation (single-core design). Figure 3.3.13 shows an intermediate circuit arrangement (two-core design).

Figure 3.3.12 illustrates very clearly how the phase-angle between the voltages of the source- and load-system can be varied by the LTC position. Various other circuit arrangements have been realized.

The number of LTC operations of PSTs is much higher than that of other regulating transformers in networks (10 to 15 times higher). In some cases, according to regulating ranges — especially for line-end

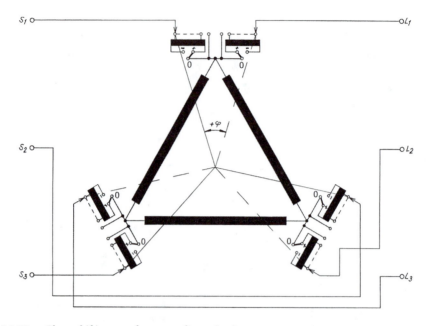

FIGURE 3.3.12   Phase-shifting transformer — direct circuit arrangement.

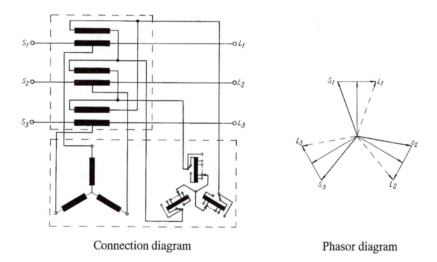

Connection diagram   Phasor diagram

FIGURE 3.3.13   Phase-shifting transformer — intermediate circuit arrangement.

arrangements (Figure 3.3.12) — the transient overvoltage stresses over tapping ranges have to be limited by the application of non-linear resistors. Furthermore, the short-circuit-current ability of the LTC must be checked, as the short-circuit power of the network determines the said current. The remaining features of LTCs for such transformers can be selected according to the usual rules (see Section 3.3.5 entitled "Selection of Load Tap Changers").

Significant benefits resulting from the use of a PST are:

- Reduction of overall system losses by elimination of circulating currents
- Improvement of circuit capability by proper load management
- Improvement of circuit power factor
- Control of power flow to meet contractual requirements

3.3.4 Rated Characteristics and Requirements for Load Tap Changers

The rated characteristics of an LTC are as follows:

- Rated through-current[1]
- Maximum rated through-current[1]
- Rated step voltage[1]
- Maximum rated step voltage[1]
- Rated frequency
- Rated insulation level

The basic requirements for LTCs are laid down in various standards (IEEE Std. C57.131-1995 [1995]; IEC 60214 [2003]; and IEC 60542 [1988]).

The main features to be tested during design tests are:

- Contact life: IEEE Std. C57.131-1995-6.2.1.1; IEC 60214-1-5.2.2.1
  50,000 operations at the max rated through-current and the relevant rated step voltage shall be performed. The result of these tests may be used by the manufacturer to demonstrate that the contacts used for making and breaking current are capable of performing, without replacement of the contacts, the number of operations guaranteed by the manufacturer at the rated through-current and the relevant rated step voltage.
- Temporary overload: IEEE Std. C57.131-1995-6.1.3 and 6.2.2; IEC 60214-1-4.3, 5.2.1, and 5.2.2.2

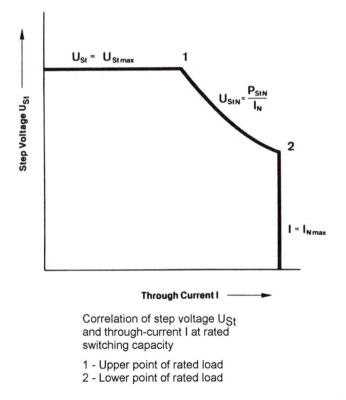

Correlation of step voltage U_{St} and through-current I at rated switching capacity

1 - Upper point of rated load
2 - Lower point of rated load

FIGURE 3.3.14 Characteristics of the arcing switch — through-current, step voltage, switching capacity.

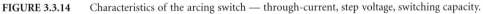

[1]Within the maximum rated through-current of the LTC, there may be different combinations of values of rated through-current and corresponding rated step voltage. Figure 3.3.14 shows that relationship. When a value of rated step voltage is referred to as a specific value of rated through-current, it is called the relevant rated step voltage.

At 1.2 times maximum rated through-current, temperature-rise tests of each type of contact carrying current continuously shall be performed to verify that the steady-state temperature rise does not exceed 20°C above the temperature of the insulating fluid, surrounding the contacts. In addition, breaking-capacity tests shall be performed with 40 operations at a current up to twice the maximum rated through-current and at the relevant rated step voltage.

LTCs that comply with the said design tests, and when installed and properly applied to the transformer, can be loaded in accordance with the applicable IEEE or IEC loading guides.

- Mechanical life: IEEE Std. C57.131-1995-6.5.1; IEC 60214-1-5.2.5.1

A mechanical endurance test of 500,000 tap-change operations without load has to be performed. During this test the LTC shall be assembled and filled with insulating fluid or immersed in a test tank filled with clean insulating fluid, and operated as for normal service conditions.

Compared with the actual number of tap-change operations in various fields of application (Table 3.3.1) it can be seen that the mechanical endurance test covers the service requirements.

- Short-circuit-current strength: IEEE Std. C57.131-1995-6.3; IEC 60214-1-5.2.3

All contacts of different design that carry current continuously shall be subjected to three short-circuit currents of 10 times maximum rated through-current (valid for I > 400 A), with an initial peak current of 2.5 times the rms value of the short-circuit current, each current application of at least 2-sec duration.

- Dielectric requirements: IEEE Std. C57.131-1995-6.6; IEC 60214-1-5.2.6

The dielectric requirements of an LTC depend on the transformer winding to which it is to be connected.

The transformer manufacturer shall be responsible not only for selecting an LTC of the appropriate insulation level, but also for the insulation level of the connecting leads between the LTC and the winding of the transformer.

The insulation level of the LTC is demonstrated by dielectric tests — in accordance with the standards (applied voltage, basic lightning impulse, switching impulse, partial discharge, if applicable) — on all relevant insulation spaces of the LTC.

The test and service voltages of the insulation between phases and to ground shall be in accordance with the standards. The values of the withstand voltages of all other relevant insulation spaces of an LTC shall be declared by the manufacturer of the LTC.

- Oil tightness of arcing-switch oil compartment: IEC 60214-1-5.2.5.4

The analysis of gases dissolved in the transformer oil is an important, very sensitive, and commonly used indication for the operational behavior of a power transformer. To avoid any influence of the switching gases produced by each operation of the arcing switch, on the results of the said gas-in-oil analysis, the arcing-switch oil compartment has to be oil tight. Furthermore, the arcing switch conservator tank must be completely separated from the transformer conservator tank on the oil and on the gas side.

Vacuum- and pressure-withstand values of the oil compartment shall be declared by the manufacturer of the LTC.

TABLE 3.3.1 Number of LTC Switching Operations in Various Fields of Application

Transformer	Transformer Data			Number of On-Load Tap Changer Operations Per Year		
	Power, MVA	Voltage, kV	Current, A	Min.	Medium	Max.
Power station	100–1300	110–765	100–2000	500	3,000	10,000
Interconnected	200–1500	110–765	300–3000	300	5,000	25,000
Network	15–400	60–525	50–1600	2,000	7,000	20,000
Electrolysis	10–100	20–110	50–3000	10,000	30,000	150,000
Chemistry	1.5–80	20–110	50–1000	1,000	20,000	70,000
Arc furnace	2.5–150	20–230	50–1000	20,000	50,000	300,000

3.3.5 Selection of Load Tap Changers

The selection of a particular LTC will render optimum technical and economical efficiency if requirements due to operation and testing of all conditions of the associated transformer windings are met. In general, usual safety margins may be neglected as LTCs designed, tested, selected, and operated in accordance with IEEE standards (1995) and IEC standards (1976; 2003) are most reliable. The details for the selection of LTCs are described in Krämer, 2000.

To select the appropriate LTC, the following important data of associated transformer windings should be known:

- MVA rating
- Connection of tap winding (for wye, delta, or single-phase connection)
- Rated voltage and regulating range
- Number of service tap positions
- Insulation level to ground
- Lightning-impulse and power-frequency voltage of the internal insulation

The following LTC operating data may be derived from this information:

- Maximum through-current: Imax
- Step voltage: Ust
- Switching capacity: Pst = Ust × Imax

and the appropriate tap changer can be determined:

- LTC type
- Number of poles
- Nominal voltage level of LTC
- Tap selector size/insulation level
- Basis connection diagram

If necessary, the following characteristics of the tap changer should be checked:

- Breaking capacity
- Overload capability
- Short-circuit current (especially to be checked in case of Figure 3.3.12 applications)
- Contact life

In addition to that, the following two important LTC stresses resulting from the arrangement and application of the transformer design have to be checked.

3.3.5.1 Voltage Connection of Tap Winding during Changeover Operation

During the operation of the reversing or coarse changeover selector, the tap winding is disconnected momentarily from the main winding. It thereby takes a voltage that is determined by the voltages of the adjacent windings as well as by the coupling capacities to these windings and to grounded parts. In general, this voltage is different from the voltage of the tap winding before the changeover selector operation. The differential voltages are the recovering voltages at the opening contacts of the changeover selector and, when reaching a critical level, they are liable to cause inadmissible discharges on the changeover selector. If these voltages exceed a certain limit value (for special product series, said limit voltages are in the range of 15 to 35 kV), measures regarding voltage control of the tap winding must be taken.

Especially in case of PSTs with regulation at the line end (e.g., Figure 3.3.12), high recovery voltages can occur due to the winding arrangement. Figure 3.3.15a illustrates a typical winding arrangement of PST according to Figure 3.3.12. Figure 3.3.15b gives the phasor diagram of that arrangement without limiting measures. As it can be seen, the recovery voltages appearing at the changeover selector contacts are in the range of the system voltages on the source and the load side.

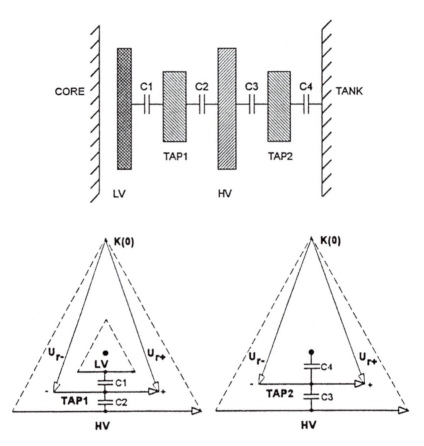

FIGURE 3.3.15 Phase-shifting transformer, circuit (as shown in Figure 3.3.12). Top) typical winding arrangement with two tap windings; bottom) recovery voltages (U_{r+}, U_{r-}) for tap windings 1 and 2 (phasor diagram).

It is sure that an LTC cannot be operated under such conditions. This fact has already been taken into account during the planning stage of the PST design. There are three ways to solve the above-mentioned problem:

1. Install screens between the windings. These screens must have the voltage of the movable changeover selector contact 0 (Figure 3.3.12). See Figs. 3.3.16a and b.
2. Connect the tap winding to a fixed voltage by a fixed ohmic resistor (tie-in resistor) or by an ohmic resistor, which is only inserted during changeover selector operation by means of a voltage switch. This ohmic resistor is usually connected to the middle of the tap winding and to the current take-off terminal of the LTC (Figure 3.3.17).
3. Use an advance–retard switch (ARS) as changeover selector (Figure 3.3.18). This additional unit allows the changeover operation to be carried out in two steps without interruption. With this arrangement, the tap winding is connected to the desired voltage during the whole changeover operation. As this method is relatively complicated, it is only used for high-power PSTs.

The common method for the voltage connection of tap windings is to use tie-in resistors. The following information is required to dimension tie-in resistors:

- All characteristic data of the transformer such as power, high and low voltages with regulating range, winding connection, insulation levels
- Design of the winding, i.e., location of the tap winding in relation to the adjacent windings or winding parts (in case of layer windings)

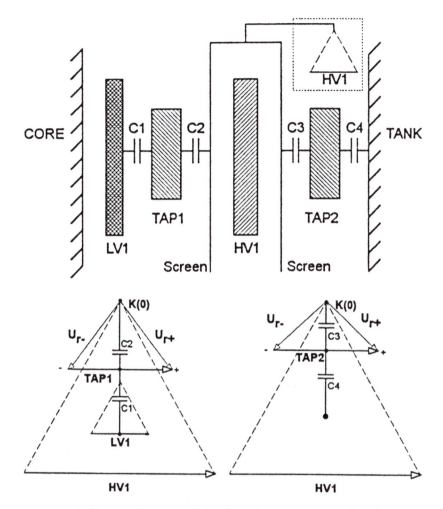

FIGURE 3.3.16 Phase-shifting transformer, circuit (as shown in Figure 3.3.12): top) winding arrangement with two windings and screens; bottom) recovery voltages (U_{r+}, U_{r-}) for tap windings 1 and 2 (phasor diagram).

- Voltages across the windings and electrical position of the windings within the winding arrangement of the transformer that is adjacent to the tap winding
- Capacity between tap winding and adjacent windings or winding parts
- Capacity between tap winding and ground or, if existing, grounded adjacent windings
- Surge stress across half of tap winding
- Service and test power-frequency voltages across half of the tap winding

3.3.5.2 Effects of the Leakage Impedance of Tap Winding/Coarse Winding during the Operation of the Arcing Switch When Passing the Mid-Position of the Resistance-Type LTC

During the operation of the arcing switch from the end of the tap winding to the end of the coarse winding and vice versa (passing mid-position, see Figure 3.3.19a), all turns of the whole tap winding and coarse winding are inserted in the circuit.

This results in a leakage-impedance value that is substantially higher than during operation within the tap winding, where only negligible leakage impedance of one step is relevant (Figure 3.3.19b). The higher impedance value in series with the transition resistors has an effect on the circulating current,

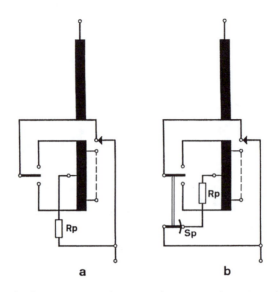

FIGURE 3.3.17 Methods of voltage connection (reversing changeover selector in mid-position): (a) fixed tie-in resistor Rp; (b) with voltage switch Sp and tie-in resistor Rp.

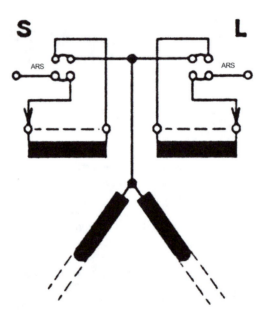

FIGURE 3.3.18 Phase-shifting transformer — changeover operation by means of an advance–retard switch (ARS).

which is flowing in the opposite direction through coarse winding and tap winding during the arcing switch operation.

Consequently, a phase shift between switched current and recovery voltage takes place at the transition contacts of the arcing switch and may result in an extended arcing time.

In order to ensure optimum selection and adaptation of the LTC to these operating conditions, it is necessary to specify the leakage impedance of coarse winding and tap winding connected in series.

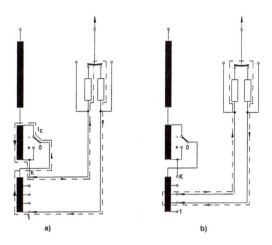

FIGURE 3.3.19   Effect of the leakage impedance of coarse winding/tap winding arrangement: (a) operation through mid-position; (b) operation through any tap position beside mid-position.

3.3.6   Protective Devices for Load Tap Changers

The protective devices for LTCs are designed to limit or prevent the effect of the following stresses: inadmissible increase of pressure within the arcing-switch compartment or the separate compartment of the reactance-type LTC, respectively; operation of LTCs with overcurrents above certain values; operation of LTCs at oil temperatures below the limit laid down in the standards (–25°C) (IEEE, 1995; IEC, 2003), and inadmissible voltage stresses of the insulation in the arcing switch caused by transient overvoltages. The following control and protective equipment are in use:

- Oil-flow relays inserted into the pipe between LTC head and conservator are mostly used for in-tank LTCs (Figure 3.3.20). They respond to disturbances in the arcing-switch compartment of relatively low-energy up to high-energy dissipation within a reasonable time, avoiding damages to the LTC and the transformer. The oil-flow relay has to disconnect the transformer. To give alarm only, as it is practiced in some users' systems, is not allowed because it is dangerous and could lead to severe faults.
- Pressure-sensing and/or pressure-releasing relays are also often used parallel to the oil-flow relay or alone. Their response time is a little bit shorter than that of the oil-flow relay. But decrease in the response time is of minor importance because the complete disconnecting time of the transformer is determined by the total response time of the control circuit that trips the circuit breakers of the transformer and that is much longer than the response time of the oil-flow relay.
- At oil temperatures below –25°C, it may be necessary to provide special devices, e.g., blocking the drive mechanism to obtain reliable service behavior of the LTC.
- An overcurrent blocking device that stops the LTC's drive mechanism during an overload is used in many utilities as a standard device. It is normally set at 1.5 times the rated current of the transformer.
- In transformers with regulation on the high-voltage side and coarse winding arrangements, extremely high voltage stresses can occur at the inner insulation of the arcing switch of the resistance-type LTC during impulse testing when the LTC is in mid-position (Figure 3.3.21). Up to 25% of the incoming wave for BIL-tests or 40% for chopped-wave tests can appear over the said distance. Critical values could be reached above a BIL of 550 kV.
- There are two principles to protect the arcing switch from undue overvoltages. Spark gaps or non-linear resistors could be installed in series to the transition resistors, as shown in Figure 3.3.22. The spark gap is a safe overvoltage protection for applications in medium-size power transformers.

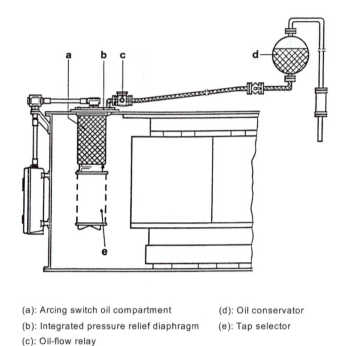

(a): Arcing switch oil compartment   (d): Oil conservator

(b): Integrated pressure relief diaphragm   (e): Tap selector

(c): Oil-flow relay

FIGURE 3.3.20   Arrangements of overpressure protection of LTCs in the transformer tank.

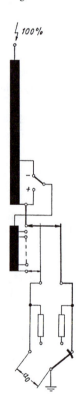

Voltage stress at "a_0":

wave shape	stress in % due to imput impulse
1.2/50 µs	10% – 20% max. 25%
chopped wave	15% – 30% max 40%

Coarse/fine winding arrangement
OLTC in mid–position

FIGURE 3.3.21   Voltage stress between selected and preselected tap of coarse-tap winding arrangement, LTC in mid-position.

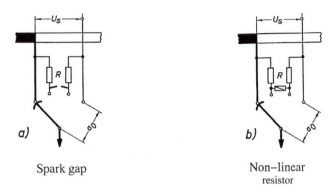

a) Spark gap

b) Non–linear resistor

FIGURE 3.3.22 Overvoltage protection devices arranged within the arcing switch.

Nonlinear resistors are solutions for high-power transformers and for all transformers where the service conditions would cause the spark gaps to respond frequently.

In the early stages, silicon-carbide (SiC) elements were installed. The specific characteristics of this material did not allow full-range application. When high-power zinc-oxide (ZnO) varistors came on the market, the application of these elements for overvoltage protection was studied in detail with good results. For more than 20 years, ZnO varistors have solely been in use.

3.3.7 Maintenance of Load Tap Changers

LTC maintenance is the basis for the regulating transformer's high level of reliability. The background for maintenance recommendations is as follows:

For LTCs where oil is used for arc-quenching, the arcing at the arcing switch or arcing tap switch contacts causes contact erosion and carbonization of the arcing switch oil. The degree of contamination depends upon the operating current of the LTC, the number of operations, and to some degree the quality of the insulating oil. For LTCs using vacuum interrupters for arc-quenching, contact life of the vacuum interrupters and the mechanically stressed parts of the device are the key indicators for the maintenance recommendations. The overall performance of vacuum-type LTCs leads more and more toward maintenance-free LTC designs.

Maintenance and inspection intervals depend on the type of LTC, the LTC rated through-current, the field experience, and the individual operating conditions. They are suggested as periodical measures with respect to a certain number of operations or after a certain operating time, whichever comes first.

The recommended maintenance intervals for an individual LTC type are given in the operating and inspection manuals available for each LTC type.

Normally, maintenance of an LTC can be performed within a few hours by qualified and experienced personnel, provided that it has been properly planned and organized. In countries with tropical or subtropical climate, the humidity must also be taken into consideration. In some countries, customers decide to start maintenance work only if the relative humidity is less than 75%.

Economical factors are taken more and more into consideration by users of large power transformers in distribution networks when assessing the operating parameters for cost-intensive operating equipment. While users are aiming at cost reduction for transformer maintenance, they are also demanding higher system reliability. Besides the new generation of LTCs with vacuum switching technology, modern supervisory concepts on LTCs (LTC monitoring) offer a solution for the control of these divergent development tendencies.

Today a few products are on the market that differ significantly in their performance.

A state-of-the-art LTC on-line monitoring system should include an early-fault-detection function and information on condition-based maintenance, which requires an expert-system of the LTC manufacturer. The data processing and visualization should provide information about status-signal messages, trend analyses, and prognoses. Monitoring application is a judgment of transformer size and importance and of maintenance and equipment costs.

3.3.8 Refurbishment/Replacement of Old LTC Types

With regard to system planning of power utilities, the lifetime of regulating transformers is normally assumed to be 25 to 30 years. The actual lifetime is, however, much longer. Due to economic aspects and aging networks, as well as the requirement to improve reliability, refurbishment/replacement is becoming a major policy issue for utility companies.

Refurbishment includes a complete overhaul of the regulating transformer plus other improvements regarding loading capability, an increase in insulation levels, a decrease in noise levels, and the possible replacement of the bushings and of the LTC or a complete overhaul of the LTC. This overhaul should be performed by specialists from the LTC manufacturer in order to avoid any risk when judging the condition of the LTC components, when deciding which components have to be replaced, and with regard to the disassembly and the reassembly as well as the cleaning of insulation material.

The replacement of an old risky LTC (for which neither maintenance work nor spare parts are available) by a new LTC may economically be justified, compared with the expenses for a new regulating transformer, even if the transformer design has to be modified for that reason. The manufacturer of the new LTC must, of course, guarantee maintenance work and spare parts for the foreseeable future.

3.3.9 Future Aspects

For the time being, no alternative to regulating transformers is expected. The LTC will therefore continue to play an essential part in the optimum operation of electrical networks and industrial processes in the foreseeable future.

With regard to the future of LTC systems, one can say that a static LTC, without any mechanical system and consisting only of power electronics, leads to extremely uneconomical solutions, and this will not change in the near future. Therefore, the mechanical LTC will still be used.

Conventional LTC technology has reached a very high level and is capable of meeting most requirements of the transformer manufacturer. This applies to the complete voltage and power fields of today, which will probably remain unchanged in the foreseeable future. It is very unlikely that, due to new impulses given to development, greater power and higher voltages will be required.

Today the main concern goes to service behavior as well as reliability of LTCs and how to keep this reliability at a consistently high level during the regulating transformer's life cycle.

At the present time and for the foreseeable future, the proper implementation of the vacuum switching technology in LTCs provides the best formula of quality, reliability, and economy achievable toward a maintenance-free design. The vacuum switching technology entirely eliminates the need for an on-line filtration system and offers reduced down-times with increased availability of the transformer and simplified maintenance logistics. All this translates into substantial savings for the end-user. Consequently, today's design concepts of LTCs — resistance- and reactance-type LTCs — are based more and more on vacuum interrupters. The vacuum switching technology — used in OLTCs — is in fact "state of the art" design today and tomorrow.

Another target of development is the insulation and cooling media with low or no flammability for regulating transformers, mainly relevant in the field of medium-power transformers (<100 MVA). Such media include: several synthetic/organic fluids provided as PCB-replacement, SF_6 gas, and air. The application of these transformers is preferred in areas where mineral oil would be too dangerous, considering pollution and fire. The application of these media may require a different LTC design. For many years SF_6-regulating transformers have been in service in Japan and a few other countries.

As an alternative, dry-type distribution transformers with regulation have been available for several years. The LTC is operating in air with vacuum interrupters. These transformers are used for indoor application with extreme fire hazard and/or pollution requirements, existing in metropolitan and special industrial areas.

References

Goosen, P.V. , Transformer accessories, (on behalf of Study Committee 12), *CIGRE*, Volume No., 12-104, 1996.

IEC International Standard 60214-1:2003, Tap-Changers, Part I: Performance requirements and test methods, 2003.

IEC Std. Publication 60542, Application Guide for On-Load Tap Changers, 1976, first amendment 1988.

IEEE Std. C57.135-2001, IEEE Guide for the Application, Specification, and Testing of Phase-Shifting Transformers, 2001.

IEEE Std. C57.131-1995, IEEE Standard Requirements for Load Tap Changers, 1995.

Krämer, A., *On-Load Tap-Changer for Power Transformers: Operation, Principles, Applications and Selection*, MR Publications, 2000.

Krämer, A. and Ruff, J., Transformers for phase angle regulation, considering the selection of on-load tap changers, *IEEE Trans. Power Delivery*, 13, 1998.

3.4 Loading and Thermal Performance

Robert F. Tillman, Jr.

3.4.1 Design Criteria

ANSI standards collection for the C57 series sets forth the general requirements for the design of power transformers. With respect to loading, the requirements of concern are those that define the transformer's thermal characteristics. The average ambient temperature of the air in contact with the cooling equipment for a 24-hr period shall not exceed 30°C, and the maximum shall not exceed 40°C (IEEE, 1994). For example, a day with a high temperature of 35°C and a low of 25°C has an approximate average ambient temperature of 30°C.

The average winding-temperature rise above ambient shall not exceed 65°C for a specified continuous-load current (IEEE, 1994). The industry uses this criterion rather than the more relevant hot-spot temperature rise because manufacturers can obtain it by simply measuring the resistance of the windings during temperature-rise tests. The manufacturer must guarantee to meet this requirement.

The hottest-conductor temperature rise (hot spot) in the winding shall not exceed 80°C under these same conditions (IEEE, 1994). This 80°C limit is no assurance that the hot-spot rise is 15°C higher than the average winding-temperature rise. In the past, using the 15°C adder to determine the hot-spot rise produced very conservative test results. The greatest thermal degradation of cellulose insulation occurs at the location of the hottest spot. The hot-spot location is near the top of the high- or low-voltage winding. The most common reason for hot spots to occur is that these regions have higher localized eddy losses because the leakage flux fringes radially at the winding ends.

Standards specify requirements for operation of the transformer above rated voltage. At no load, the voltage shall not exceed 110% (IEEE, 1994). At full load, the voltage shall not exceed 105% (IEEE, 1994). Modern transformers are usually capable of excitation beyond these limits without causing undue saturation of the core. However, this causes the core to contribute greater than predicted heating. Consequently, all temperature rises will increase. While not critical at rated load, this becomes important when loading transformers above the nameplate rating.

3.4.2 Nameplate Ratings

The transformer nameplate provides the voltage (excitation) rating for all tap positions. If a transformer has a four-step de-energized high-voltage tap changer (DETC) and a 32-step load tap changer (LTC) on the low-voltage winding, then it has five primary voltage ratings and 33 secondary voltage ratings. The excitation limits apply to all ratings.

The cooling class affects the design of the cooling package. Different cooling classes have different thermal profiles. Therefore, gauge readings can be different for equivalently rated transformers under the same loading conditions. Consequently, the operator must understand the cooling classes and their thermal profiles in order to confirm that a given transformer is responding properly.

ONAN cooling uses natural air flow through the radiators to dissipate heat. (See Section 2.1, Power Transformers, for description of cooling classes). ONAF cooling uses fans to force air through radiators in order to substantially increase the rate of cooling. There is natural flow by convection of the oil in the radiators from top to bottom with both ONAN and ONAF cooling. There may additionally be pumps used to draw the hot oil out of the top of the transformer and force it through the radiators into the bottom of the tank. If the design incorporates a cooling duct that directs the oil from the radiator outlet directly through the winding from bottom to top, it is classified as ODxx (Pierce, 1993).

Medium–size power transformers, 12 MVA and larger, have three stages of cooling, i.e., natural convection plus two stages of forced air through the radiators. Large power transformers usually employ directed-flow cooling to the main windings. Some large power transformers, especially generator step-up (GSU) transformers, are rated for only forced oil and air cooling; the nameplate reveals no rating unless the auxiliary cooling equipment is operating.

The nameplate kVA (or MVA) rating is the continuous load at rated voltage that produces an average winding-temperature rise within the 65°C limit and hot-spot temperature rise within the 80°C limit (IEEE, 1994). The rating of each cooling stage is given. The transformer design must meet temperature-rise guarantees for each cooling stage. Operators and planners generally express loading capability as a percent (or per unit) of the maximum MVA rating.

The windings contribute most of the heat that produces the temperature rises. The winding heat results mostly from the I^2R loss in the conductor. When users express loading as a percent on an MVA basis, they are assuming that the transformer is operating at rated voltage, both primary and secondary. Operators should express the load for thermal calculations as a percent (or per unit) of the full load current for a given set of primary and secondary tap positions (Tillman, 1998).

3.4.3 Other Thermal Characteristics

Other thermal characteristics are dependent on the design, loss optimization, and cooling class. Top-oil temperature is the temperature of the bulk oil at the top of the tank. Users can directly measure this with a gauge mounted on the transformer tank. The temperature of the oil exiting the oil duct in the top of the winding is actually the important characteristic (Chew, 1978; Pierce, 1993). Users normally do not have a direct measurement of this in practice. Manufacturers can measure this during heat run tests, but that is not common practice.

Bottom-oil rise is the temperature rise above ambient of the oil entering the bottom of the core and coil assembly. In all cooling classes, the bottom-oil temperature is approximately equal to the temperature of the oil exiting the radiators (Chew, 1978; Pierce, 1993). Manufacturers can easily measure it during heat run tests and provide the information on certified test reports. Users should require manufacturers to provide this information. In service, users can glean a measurement of the bottom-oil temperature with an infrared camera.

Average oil rise is the key oil thermal characteristic needed to calculate the winding gradient (Chew, 1978). It is approximately equal to the average of the temperature rise of oil entering the bottom of the core and coil assembly and the oil exiting the top. The location of the average oil is approximately equal to the location of the mid-winding.

The winding gradient is the difference between the average winding- and average oil-temperature rises (Chew, 1978). Designers assume that this gradient is the same from the bottom to the top of the winding except at the ends, where the hot spot exists. The hot-spot gradient is somewhat higher than the winding gradient. Designers calculate this gradient by determining the effects of localized eddy losses at the end of the winding. Computers allow designers to calculate this very accurately. An engineer can obtain an estimate of the hot-spot gradient by multiplying the winding gradient by 1.1 (Chew, 1978).

3.4.4 Thermal Profiles

Thermal profiles show the relationships between different temperatures inside a transformer. Operators must understand these thermal profiles for given cooling classes in order to understand a transformer's thermal response to a given load. A plot of winding temperature vs. its axial position in the tank is a thermal profile.

For ONxx cooling, the oil temperature outside the winding is approximately equal to the oil temperature inside the winding from bottom to top (Chew, 1978). This is due to the oil movement by natural convection both inside and outside the core and coil assembly. The oil temperature exiting the bottom of the radiators approximately equals that entering the bottom of the core and coil assembly (Chew, 1978; Pierce, 1993). The top bulk-oil temperature approximately equals the oil temperature at the top of the winding (IEEE, 1995). It is relatively easy for engineers to determine the gradients for this cooling class. They can verify that a transformer is responding properly to a given load by obtaining load information, reading the top-oil temperature gauge, measuring the bottom-oil temperature by use of an infrared camera, and comparing with calculations and data from the manufacturer's test reports.

For OFxx cooling, the oil pumps force oil into the bulk oil at the bottom of the tank. The higher rate of flow produces a small bottom-to-top temperature differential in the coolers. The oil flow in the winding is by natural convection. The pumps circulate cool oil throughout the tank and around the core and coil assembly while only a small portion passes through the windings. Therefore, the temperature profile of the oil internal to the windings does not match that of the external profile. The top winding-duct oil, the key temperature, is much hotter than the top bulk oil (Pierce, 1992; 1993).

For ODxx directed-flow cooling, the oil pumps force oil through a sealed oil duct directly into the bottom of the core and coil assembly. The oil flows within the winding and exits the top of the winding into the bulk oil. The temperature profile of the oil external to the winding approximately equals that of the oil internal to the winding. The bottom-to-top oil-temperature rise is very small. Top-oil temperature rises for this design are lower than equivalent ONxx designs. The hot-spot temperature is not lower because the hot-spot gradient is much larger (Pierce, 1992; 1993).

3.4.5 Temperature Measurements

The user specifies the gauges and monitors installed by the manufacturer. The top-oil temperature gauge directly measures the temperature of the top bulk oil. Most oil gauges in service are mechanical and provide instantaneous and maximum temperature readings. They also have contacts for controlling cooling and providing alarms. Some of these gauges are electronic and provide analog outputs for communication by supervisory control and data acquisition (SCADA). This gauge reading is approximately equal to the temperature of the top winding-duct oil for ONxx and ODxx cooling and is much lower for OFxx nondirected-flow cooling (IEEE, 1995; Chew, 1978; Pierce, 1992; 1993).

In most cases the winding temperature gauge is actually an assembly that simulates the hot-spot winding temperature. This hot-spot temperature simulator measures top bulk oil and adds to it another temperature increment (hot-spot gradient) proportional to the square of the load current. This incremental temperature is actually a simulation of the hot-spot gradient. An oil-temperature probe inserted in a resistive well and a current obtained from a BTCT in the main winding is a method used for simulating the hot-spot temperature. Designers calibrate the resistor and the current to the hot-spot gradient. For ODxx directed flow and ONxx cooling, this simulation is valid because the measured top-oil temperature is approximately equal to the top winding-duct oil temperature (IEEE, 1995; Chew, 1978). However, this simulation is not valid for OFxx nondirected-flow cooling because the measured top-oil temperature is less than the top winding-duct oil temperature (Pierce, 1992). It is more difficult to predict and verify the thermal response of a transformer with OFxx nondirected-flow cooling.

The hot-spot simulator exhibits its greatest accuracy for constant loads because there is a thermal lag between the top-oil and hot-spot winding temperatures. It takes several hours for the bulk oil to attain its ultimate temperature after a change of the load. On the other hand, the same load change takes only a few minutes to heat up the winding conductors to their ultimate temperature. Consequently, a transformer must have an applied constant load for several hours for the hot-spot simulator to indicate an accurate measurement. For transient loading, the instantaneous indication is meaningless to an operator. The actual hot-spot temperature is greater for increasing loads than the simulator indicates and lower for decreasing loads (IEEE, 1995).

Users also specify electronic gauges. These have many features built into one box. They include many adjustable alarm and control contacts, both top-oil and simulated hot-spot temperatures, and analog outputs for SCADA. This monitor creates the hot-spot simulation by routing the current directly to the monitor. Designers calibrate the microprocessor in the monitor to calculate the hot-spot gradient. This provides a more reliable indication than mechanical gauges, but it is still a simulation with the same drawbacks discussed above.

Direct measurements of the hot-spot temperature are possible through application of fiber optics. A sensor inserted in the oil duct between two winding discs replaces a key-spacer. Design engineers determine the locations to install the sensors. Fiber-optic probes transmit the temperature from the sensors to a storage device, often a personal computer. These applications are relatively unreliable. Designers typically call for many probes in order to have a few that work properly. The application of fiber optics also produces additional risk of internal dielectric failure in the transformer. While fiber optics is a good research tool and appropriate for special applications, its general application is usually not worth the added risk and cost (Pierce, 1993).

3.4.6 Predicting Thermal Response

The IEEE Std. C57.91-1995, Guide for Loading Mineral-Oil-Immersed Transformers, gives detailed formulas for calculating oil and winding temperatures. It gives formulas for constant steady-state and transient loading. Users apply the transient-loading equations in computer programs. Computer simulations are good for gaining understanding but are not necessary for most situations.

Users must calculate oil temperatures in order to predict a transformer's thermal response. The total losses in a transformer cause the oil temperature to rise for a given load (IEEE, 1995). Total losses include core, load, stray, and eddy losses. The oil temperatures vary directly with the ratio of the total losses raised to an exponent. The industry designation for the oil-rise exponent is n (IEEE, 1995; Chew, 1978; Pierce, 1993).

$$TO_2 = TO_1(TL_2/TL_1)^n \qquad (3.4.1)$$

where

TO_2 = ultimate top-oil rise
TO_1 = initial or known top-oil rise
TL_2 = total losses at ultimate load
TL_1 = total losses at known load

Users can calculate a conservative approximation by neglecting the core, stray, and eddy losses and considering only the i^2r load losses:

$$TO_2 = TO_1(i_2^2r/i_1^2r)^n = TO_1(i_2^2/i_1^2)^n = TO_1(i_2/i_1)^{2n} \qquad (3.4.2)$$

The value of n varies with different transformer designs. The industry generally accepts an approximate value of 0.9 for ONAF-class transformers (IEEE, 1995; Chew, 1978; Pierce, 1993). This is a conservative value.

$$TO_2 = TO_1(i_2/i_1)^{1.8} \qquad (3.4.3)$$

If i_1 is the full-load current rating, then i_2/i_1 is the per-unit current loading. As an example, consider a transformer with ONAF cooling that has a top-oil temperature rise of 55°C at full-load current. Calculate the ultimate steady-state top-oil temperature rise due to a constant 120% load:

$$TO_2 = 55(1.2)^{1.8} = 76°C$$

If the maximum ambient temperature is 35°C, then the ultimate top-oil temperature is 111°C. The user can find detailed transient equations in IEEE C57.91-1995. They appear complex but are only expanded forms of the above equations. The same basic principles apply. However, their use requires many incremental calculations summed over given time periods. There are computer programs available that apply these equations.

Transient analysis has the most value when actual temperature profiles are available. Except for special applications, this is often not the case. Usually, the temperature available is the maximum for a given time period. Furthermore, engineers do not always have time available to analyze and compare computer outputs vs. field data. The key issue is that the engineer must have the ability to predict a transformer's thermal response to a given load (steady state or transient) in order to ensure that the transformer is responding properly. They can use the above simplified steady-state equations to calculate maximum temperatures reached for a given load cycle. Using a load equal to 90% of a peak cyclical load in the steady-state equations yields a good approximation of the peak temperature for an operator to expect. In the above example, the resultant expected top-oil temperature for a cyclical 120% load is 98°C.

The relationships for winding-temperature equations are similar to the oil-temperature equations. The key element is the hot-spot gradient. Once an engineer calculates this gradient, it is added to the calculated top-oil temperature to obtain the hot-spot temperature. Similar to Equation 3.4.2, the hot-spot gradient varies directly with the ratio of the current squared raised to an exponent m (IEEE, 1995; Chew, 1978; Pierce, 1993).

$$HSG_2 = HSG_1(i_2/i_1)^{2m} \tag{3.4.4}$$

where

HSG_2 = ultimate hot-spot gradient
HSG_1 = initial or known hot-spot gradient
i_2 = current at the ultimate load
i_1 = current at the known or initial load

As with n, the values of m vary between transformers. However, the variance for m is much greater. It is very dependent on the current density and loss optimization used to design a particular transformer. High losses and current densities produce large values of m. Low losses and current densities produce small values for m. For example, a high-loss transformer design may have a lower hot-spot temperature at rated load than a similar low-loss transformer design. However, the low-loss transformer is likely more suitable for loading past its nameplate criteria. The industry generally accepts an approximate value of 0.8 for ONAF-class transformers (IEEE, 1995; Chew, 1978; Pierce, 1993). Again, this is a conservative value:

$$HSG_2 = HSG_1(i_2/i_1)^{1.6} \tag{3.4.5}$$

Engineers can calculate the m and n constants by requiring manufacturers to perform overload temperature tests (e.g., 125% load). This, along with temperature data from standard temperature tests, provides two points for calculating these values.

There are limiting elements in a transformer other than the active part. These include bushings, leads, tap changers, and bushing-type current transformers (BTCTs). Sometimes it is difficult to determine the bushing rating in older transformers. Infrared scanning can help determine if a bushing is heating excessively. In modern transformers, manufacturers choose bushings with ample margin for overloading the transformer. Insulated leads are also subject to overheating, especially if the manufacturer applies too much insulation. LTC contact life can accelerate. For arcing-under-oil tap changers, oil contamination increases. It is also important to know if a transformer includes a series transformer to reduce the LTC current to a manageable value. The user will probably not know the ratio of the series transformer for older transformers.

3.4.7 Load Cyclicality

Past practices for many users called for loading power transformers to a stated percentage of their nameplate kVA or MVA rating. A typical loading criteria was 100%. The measured load used to determine the percent loading was the annual peak 15-min integrated output demand in kVA or MVA. Few users considered the cyclical nature of the load or the ambient temperature. These thermal cycles allow operators to load transformers beyond their nameplate rating criteria for temporary peak loads with little or no increased loss of life or probability of failure.

To completely define the load profile, the user must know the current for thermal calculations and the power factor for voltage-regulation calculations as a function of time, usually in hourly increments. Cyclical loads generally vary with the ambient temperature due to air conditioning and heating. However, summer and winter profiles are very different. Summer peak loads occur when peak temperatures occur. The reverse occurs in winter. In fall and spring, not only is the ambient temperature mild, but there is little or no air conditioning or heating load. The thermal stress on a transformer with a cyclical normal load during fall and spring is negligible. It is also usually negligible during winter peak loading.

If ambient conditions differ from the nameplate criteria, then the user must adjust the transformer capability accordingly. IEEE C57.91-1995 provides tables and equations for making these adjustments. A good approximation is an adjustment of 1% of the maximum nameplate rating for every degree C above or below the nameplate rating (IEEE, 1995). If the transformer operates in 40°C average ambient, then the user must de-rate the nameplate kVA by 10% in order to meet the nameplate thermal-rating criteria. Conversely, operating in a 0°C average ambient environment allows the user to up-rate the transformer by approximately 30%.

Twenty-four-hour summer load cycles resemble a sine wave. The load peak occurs approximately when the temperature peak occurs at 3:00 to 5:00 in the evening. The load peak lasts less than 1 hr. The load valley occurs during predawn hours, when ambient temperature minimums occur at 4:00 to 6:00. The magnitude of the valley load is 50 to 60% of the peak load. The user can calculate the top-oil temperature at the valley in the same manner as at the peak. The value of making these calculations is to verify that a particular transformer's load response is correct. If the transformer responds properly, then it meets one of the criteria for recommending overload capability.

3.4.8 Science of Transformer Loading

The thermal rating of a power transformer differs from the thermal rating of other current-carrying elements in a substation. Examples of other elements are conductor, buswork, connectors, disconnects, circuit breakers, etc. The insulation system for these elements is air and solid support insulators. The cooling system is passive (ambient air). The thermal limits depend on the properties of the conductor itself. These elements are maximum-rated devices. In a power transformer, the cooling system is active. The thermal limits depend on the dielectric and mechanical properties of the cellulose and oil insulation system. As a maximum-rated device, the transformer capability may be 200% of the maximum nameplate rating (IEEE, 1995).

With overly conservative loading practices, cellulose insulation life due to thermal stress is practically limitless, theoretically more than 1000 years. In practice, deterioration of accessories and nonactive parts limits the practical life of the transformer. These elements include the tank, gauges, valves, fans, radiators, bushings, LTCs, etc. The average, practical life of a transformer is probably 30 to 50 years. Many fail beyond economical repair before 30 years. Therefore, it is reasonable and responsible to allow some thermal aging of the cellulose insulation system. The key is to identify the risk. Loss of insulation life is seldom the real risk. Most of the time, risks other than loss of insulation life are the limiting characteristics.

Paper insulation must have mechanical and dielectric strength. The mechanical strength allows it to withstand forces caused by through faults. When through faults occur, the winding conductors encounter forces to move. If the paper has sufficient mechanical strength, the coil assemblies will also have sufficient strength to withstand these forces. When aged paper loses its mechanical strength, through faults may cause winding movement sufficient to tear the paper. Physical damage due to excessive winding movement during through-fault conditions reduces the dielectric withstand of the paper insulation. At this point, the risk of dielectric failure is relatively high.

The Arrhenius reaction equation is the basic principle for thermal-aging calculations (Kelly et al., 1988a; Dakin, 1948):

$$L = Ae^{B/T} \tag{3.4.6}$$

L is the calculated insulation life in hours. A and B are constants that depend on the aging rate and end-of-life definition. They also depend on the condition of the insulation system. T is the absolute temperature in Kelvins (°C + 273) (Kelly et al., 1988a). There is a limited amount of functional-life test data available. Most available data is over 20 years old. Past calculations of life expectancy used extremely conservative numbers for the A and B constants (Kelly et al., 1988a).

Normal life of cellulose insulation is the time in years for a transformer operated with a constant 110°C hot-spot winding temperature to reach its defined end-of-life criteria. Degree of polymerization (DP) and tensile strength are properties that quantify aging of cellulose paper insulation (Kelly et al., 1988b). In the past, the industry accepted approximately $7^1/_2$ years for normal life. This is a misleadingly low value. Actual practice and more-recent studies show that normal life under these conditions is somewhere between 20 years and infinity (IEEE, 1995). It is not important for users to master these relationships and equations. However, the user should understand the following realities:

1. The relationship between life expectancy and temperature is logarithmic (McNutt, 1995). As the hottest-spot conductor temperature moves below 110°C, the life expectancy increases rapidly, and vice versa. An accepted rule of thumb is that the life doubles for every 8°C decrease in operating temperature. It halves for every 8°C increase in operating temperature (Kelly et al., 1988a).
2. Users seldom operate transformers such that the hot-spot winding temperature is above, at, or anywhere near 110°C. Even when peak loads cause 140°C hot-spot winding temperatures, the cumulative time that it operates above 110°C is relatively short, probably less than 200 to 400 hr per year.
3. Cellulose aging is a chemical reaction. As in all reactions, heat, water, and oxygen act as a catalyst. In the Arrhenius equation, the A and B constants are highly dependent on the presence of moisture and oxygen in the system (Kelly et al., 1988b). The expected life of the paper halves when the water content doubles (Bassetto et al., 1997; Kelly et al., 1988b). Also, tests indicate that the aging rate increases by a factor of 2.5 when oxygen content is high (Bassetto et al., 1997; Kelly et al., 1988b).

3.4.9 Water in Transformers under Load

The action of water in the insulation system of power transformers poses one of the major risks in loading transformers past their nameplate rating criteria. The insulation system inside a transformer consists of cellulose paper and oil working together. Water always exists in this system. Indeed, the paper must have some moisture content in order to maintain its tensile strength (Kelly et al., 1988b). However, the distribution of the water in this system is uneven. The paper attracts much more water than the oil (Kelly et al., 1988c). As the transformer cycles thermally throughout its life, the water redistributes itself. The water collects in the coldest part of the winding (bottom disks) and in the area of highest electrical stress (Kelly et al., 1988c). This redistribution of moisture is very unpredictable.

When a transformer is hot due to a heavy load, water moves from the paper to the oil. The heat in the conductor pushes the water out of the paper insulation, while the solubility of water in oil increases with temperature (Kelly et al., 1988b). When the transformer cools, water moves back to the paper. However, the water goes back into the paper much more slowly than it is driven out (Kelly et al., 1988b). As the oil cools, the water in the oil can approach saturation.

As long as changes take place gradually and temperatures do not reach extreme levels, the insulation system can tolerate the existence of a significant amount of water (Kelly et al., 1988b). However, loading transformers to a higher level causes higher temperatures and greater changes during the thermal cycling. Emergency loading causes greater temperature swings than normal loading. The key is to understand the point at which elevated temperatures introduce the risk of failure. That temperature level depends on the moisture content in the insulation system. There are two basic risks associated with the relationship between loading and water in the insulation system:

1. Reduction in dielectric strength due to saturation of moisture in oil
2. Bubble evolution

3.4.9.1 Dielectric Effects of Moisture in Oil

The dielectric strength of oil is a function of the average oil temperature and percent saturation of water in the oil (Moore, 1997). As long as the oil is hot, the water solubility is high (Kelly et al., 1988b). Therefore, a given amount of water in the oil produces a lower percent saturation at high temperature than at low temperature. As an example, consider a transformer with 1.5% water in the paper. A 100% load on a hot summer day produces a 70°C average oil temperature. The water content of the oil reaches 20 ppm as the moisture is driven from the paper into the oil (Kelly et al., 1988c). The water saturation at this point is 220 ppm, resulting in less than 10% saturation (Kelly et al., 1988b). The dielectric breakdown of the oil as measured by ASTM D1816 method is quite high, approximately 50 kV (Moore, 1997). In the evening, the load and temperature decrease, producing an average oil temperature of 50°C. The saturation level of water in the oil reduces to approximately 120 ppm (Kelly et al., 1988b). Since the oil goes back into the paper slowly, the water in oil is still 20 ppm, resulting in almost 20% saturation, which is still quite low. The dielectric breakdown is somewhat lower, approximately 45 kV (Moore, 1997). This is still quite high and poses no real problem.

Now consider the same example where the peak load increases to 120%. The resultant peak average oil temperature is 90°C, causing the moisture content of the oil to reach 60 ppm (Kelly et al., 1988c). The percent saturation at peak temperature is still less than 10% (Kelly et al., 1988b). However, during the evening the average oil temperature is 55°C. The water saturation level is 140 ppm, greater than 40% saturation (Kelly et al., 1988b). The dielectric breakdown reduces significantly to approximately 35 kV (Moore, 1997).

The conditions in this example should not cause problems, especially if the transformer in question is relatively new. However, the insulation systems in service-aged transformers have properties that magnify the risk illustrated in this example (Oommen et al., 1995). Severe emergency loading can produce higher risk levels, especially if they occur in winter. At 0°C, oil saturates at approximately 20 ppm (Kelly et al., 1988b). There is a point where the probability of failure increases to an unacceptable level.

3.4.9.2 Bubble Evolution

Bubble evolution is a function of conductor temperature, water content by dry weight of the paper insulation, and gas content of the oil (Oommen et al., 1995). Gas bubbles in a transformer insulation system are of concern because the dielectric strength of the gases is significantly lower than that of the cellulose insulation and oil. When bubbles evolve, they replace the liquid insulation. At this point, the dielectric strength decreases, and the risk of dielectric failure of the major and minor insulation systems increases.

Bubbles result from a sudden thermal change in the insulation at the hottest conductor. In a paper insulation system, there is an equilibrium state between the partial pressures of the dissolved gases in the paper. As temperature increases, the vapor pressure increases exponentially. When the equilibrium balance

TABLE 3.4.1 Loading Recommendations

Type of Load	ONAF [a] Maximum Top Oil Temperature, °C	OFAF [a] Maximum Top Oil Temperature, °C	Maximum Winding Temperature, °C	Maximum Load, %
Normal summer load	105	95	135	130
Normal winter load	80	70	115	140
Emergency summer load	115	105	150	140
Emergency winter load	90	80	130	150
Noncyclical load	95	85	115	110
Alarm Settings	ONAF [a] 65°C Rise	OFAF [a] 65°C Rise		
Top oil	105°C	95°C		
Hot spot	135°C	135°C		
Load amps	130%	130%		

[a] See Table 2.1.2 for a list of four-letter codes used in cooling-class designations.

tips, the water vapor pressure causes the cellulose insulation to suddenly release the water vapor as bubbles (Oommen et al., 1995).

According to past studies, moisture content is the most important factor influencing bubble evolution. The temperature at which bubbles evolve decreases exponentially as the moisture content in the cellulose insulation increases. Increasing content of other gases also significantly influences bubble evolution when high moisture content exists. Increasing content of other gases does not significantly influence bubble evolution at low moisture content (McNutt, 1995). Data show that in a dry transformer (less than 0.5% moisture by dry weight) bubble evolution from overload may not occur below 200°C. A service-aged transformer with 2.0% moisture may evolve bubbles at 140°C or less (Oommen et al., 1995).

3.4.10 Loading Recommendations

The IEEE loading guide and many published papers give guidelines and equations for calculating transformer loading. In theory, operators can load modern transformers temporarily up to 200% of nameplate rating or 180°C hottest-conductor temperature. These guidelines provide a tool to help a user understand the relationship between loading and the design limitations of a power transformer. However, understanding the relationship between loading and the general condition of and the moisture in the insulation system, along with voltage regulation, is also important. Unless conditions are ideal, these factors will limit a transformer's loading capability before quantified loss-of-life considerations.

The loading recommendations in Table 3.4.1 assume that the load is summer and winter peaking, that the fall and spring peak is approximately 60% of summer peak, that the average ambient temperature on a normal summer day is 30°C, that the daily load profile is similar to the cyclical load profile described earlier, that the winter peak is less than 120% of the summer peak, and that noncyclical loading is constant except for downtime.

1. The normal summer loading accounts for periods when temperatures are abnormally high. These might occur every three to five years. For every degree that the normal average ambient temperature during the hottest month of the year exceeds 30°C, derate the transformer 1% (i.e., 129% loading for 31°C average ambient).
2. The percent load is given on the basis of the current rating. For MVA loading, multiply by the per unit output voltage. If the output voltage is 0.92 per unit, the recommended normal summer MVA loading is 120%.
3. Exercise caution if the load power factor is less than 0.95 lagging. If the power factor is less than 0.92 lagging, then lower the recommended loading by 10% (i.e., 130% to 120%).
4. Verify that cooling fans and pumps are in good working order and oil levels are correct.

5. Verify that the oil condition is good: moisture is less than 1.5% (1.0% preferred) by dry weight, oxygen is less than 2.0%, acidity is less than 0.5, and CO gas increases after heavy load seasons are not excessive.

6. Verify that the gauges are reading correctly when transformer loads are heavy. If correct field measurements differ from manufacturer's test report data, then investigate further before loading past nameplate criteria.

7. During heavy load periods, verify with infrared camera or RTD that the LTC top-oil temperature relative to the main tank top-oil temperature is correct. For normal LTC operation, the LTC top oil is cooler than the main tank top oil. A significant and prolonged deviation from this indicates LTC abnormalities.

8. If the load current exceeds the bushing rating, do not exceed 110°C top-oil temperature (IEEE, 1995). If bushing size is not known, perform an infrared scan of the bushing terminal during heavy load periods. Investigate further if the temperature of the top terminal cap is excessive.

9. Use winding power-factor tests as a measure to confirm the integrity of a transformer's insulation system. This gives an indication of moisture and other contaminants in the system. High-BIL (basic insulation level) transformers require low winding power factors (<0.5%), while low-BIL transformers can tolerate higher winding power factors (<1.5%).

10. If the transformer is extremely dry (less than 0.5% by dry weight) and the load power factor is extremely good (0.99 lag to 0.99 lead), then add 10% to the above recommendations.

References

Bassetto, A., Mak, J., Batista, R.P., and de Faria, T.A., Economic Assessment of Power Transformer Loading, Doble Conference Proceedings, 64PAIC97, Boston, 1997, p. 8-10.1.

Chew, O., Operation of Transformers at Loads in Excess of Nameplate Ratings, presented at IEEE Power Engineering Society Transformer Seminar, New Orleans, LA, 1978.

Dakin, T.W., Electrical insulation deterioration treated as a chemical reaction rate phenomenon, *AIEE Trans.*, 67, 113–122, 1948.

IEEE, Standards Collection Distribution, Power and Regulating Transformers, IEEE Std. C57, Institute of Electrical and Electronics Engineers, Piscataway, NJ, 1994.

IEEE, Guide for Loading Mineral-Oil-Immersed Transformers, IEEE Std. C57.91-1995, Institute of Electrical and Electronics Engineers, Piscataway, NJ, 1995.

Kelly, J.J., Myers, S.D., and Parrish, R.H., *A Guide to Transformer Maintenance*, S.D. Myers, Akron, OH, 1988a, pp. 200–209.

Kelly, J.J., Myers, S.D., and Parrish, R.H., *A Guide to Transformer Maintenance*, S.D. Myers, Akron, OH, 1988b, pp. 211–258.

Kelly, J.J., Myers, S.D., and Parrish, R.H., *A Guide to Transformer Maintenance*, S.D. Myers, Akron, OH, 1988c, pp. 297–322.

McNutt, W.J., Discussion of the T.V. Oommen, E.M. Petrie, and S.R. Lindgren paper, Bubble Generation in Transformer Windings under Overload Conditions, Doble Conference Proceedings, 62PAIC95, Boston, 1995, p. 8-5A.1.

Moore, H.R., Water in Transformers, presentation to Southern Company, Atlanta, GA, 1997.

Oommen, T.V., Petrie, E.M., and Lindgren, S.R., Bubble Generation in Transformer Windings under Overload Conditions, Doble Conference Proceedings, 62PAIC95, Boston, 1995, p. 8-5.1.

Pierce, L.W., An investigation of the thermal performance of an oil-filled transformer winding, *IEEE Trans. Power Delivery*, 7, 1347–1358, 1992.

Pierce, L.W., Current Developments for Predicting Transformer Loading Capability, 1993 Minnesota Power Systems Conference, 1993.

Tillman, R.F., Relationships between Power Factor, Voltage Regulation, and Power Transformer Loading, Doble Conference Proceedings, 65PAIC98, Boston, 1998, p. 8-9.1.

3.5 Transformer Connections

Dan D. Perco

3.5.1 Introduction

In deciding the transformer connections required in a particular application, there are so many considerations to be taken into account that the final solution must necessarily be a compromise. It is therefore necessary to study in detail the various features of the transformer connections together with the local requirements under which the transformer will be operated. The advantages and disadvantages of each type of connection should be understood and taken into consideration.

This section describes the common connections for distribution, power, HVDC (high-voltage dc) converter, rectifier, and phase-shifting transformers. Space does not permit a detailed discussion of other uncommon transformer connections. The information presented in this section is primarily directed to transformer users. Additional information can be obtained from the IEEE transformer standards. In particular, reference is made to IEEE Std. C57.12.70, Terminal Markings and Connections for Distribution and Power Transformers; C57.105, Application of Transformer Connections in Three-Phase Distribution Systems; C57.129, General Requirements and Test Code for Oil-Immersed HVDC Converter Transformers; C57.18.10, Practices and Requirements for Semiconductor Power and Rectifier Transformers; C57.12.20, Overhead-Type Distribution Transformers; and C57.135, IEEE Guide for the Application, Specification, and Testing of Phase-Shifting Transformers.

3.5.2 Polarity of Single-Phase Transformers

The term *polarity* as applied to transformers is used to indicate the phase relationship between the primary and secondary windings of a given transformer or to indicate the instantaneous relative direction of voltage phasors in the windings of different transformers. This facilitates rapid and accurate connections of transformers in service. Transformer manufacturers have agreed to standardize the marking of terminals to indicate their polarity. For a single-phase, two-winding transformer, the high-voltage terminals are labeled H1 and H2, while the low-voltage terminals are labeled X1 and X2. When transformers are to be operated in parallel, like-marked terminals are to be joined together.

Transformers can be either subtractive or additive polarity. When like-numbered terminals such as H1 and X1 are joined together, the voltage between the other open terminals will be the difference of the individual impressed winding voltages for a transformer with subtractive polarity. For additive-polarity transformers, the voltage between the open terminals will be the sum of the individual winding voltages. The standards specify subtractive polarity for all transformers except for single-phase transformers 200 kVA and smaller and having high-voltage windings 8660 volts and below. In either case, the polarity of the transformer is identified by the terminal markings as shown in Figure 3.5.1. Subtractive polarity has correspondingly marked terminals for the primary and secondary windings opposite each other. For additive polarity, like-numbered winding terminal markings are diagonally disposed.

Transformers with subtractive polarity normally have the primary and secondary windings wound around the core in the same direction. However, the transformer can have subtractive-polarity terminal markings with the primary and secondary coils wound in the opposite directions if the internal winding leads are reversed.

3.5.3 Angular Displacement of Three-Phase Transformers

Angular displacement is defined as the phase angle in degrees between the line-to-neutral voltage of the reference-identified high-voltage terminal and the line-to-neutral voltage of the corresponding identified low-voltage terminal. The angle is positive when the low-voltage terminal lags the high-voltage terminal. The convention for the direction of rotation of the voltage phasors is taken as counterclockwise.

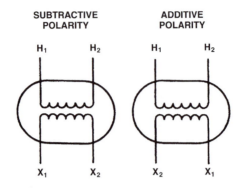

FIGURE 3.5.1 Single-phase transformer-terminal markings.

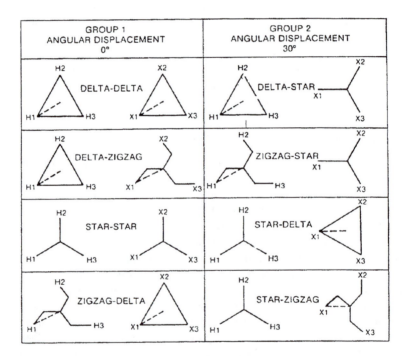

FIGURE 3.5.2 Standard angular displacement for three-phase transformers.

Since the bulk of the electric power generated and transmitted is three-phase, the grouping of trans-formers for three-phase transformations is of the greatest interest. Connection of three-phase transform-ers or three single-phase transformers in a three-phase bank can create angular displacement between the primary and secondary terminals. The standard angular displacement for two-winding transformers is shown in Figure 3.5.2. The references for the angular displacement are shown as dashed lines. The angular displacement is the angle between the lines drawn from the neutral to H1 and from the neutral to X1 in a clockwise direction from H1 to X1. The angular displacement between the primary and secondary terminals can be changed from 0° to 330° in 30° steps simply by altering the three-phase connections of the transformer. Therefore, selecting the appropriate three-phase transformer connections will permit connection of systems with different angular displacements. Figure 3.5.2 shows angular displacement for common double-wound three-phase transformers. Multicircuit and autotransformers are similarly connected.

3.5.4   Three-Phase Transformer Connections

Three-phase connections can be made either by using three single-phase transformers or by using a three-phase transformer. Advantages of the three-phase transformer is that it costs less, the weight is less, it requires less floor space, and has lower losses than three single-phase transformers. Circuit computations involving three-phase transformer banks under balanced conditions can be made by dealing with only one of the transformers or phases. It is more convenient to use line-to-neutral values, since transformer impedances can then be added directly to transmission-line impedances. All impedances must be converted to the same side of the transformer bank by multiplying them by the square of the voltage ratio.

There are two basic types of three-phase transformers, core type and shell type. The magnetic circuit of the shell type is very similar to three single-phase transformers. This type of transformer has a return circuit for each phase of the magnetic flux. Consequently, the zero-sequence impedance is equal to the positive-sequence impedance. The conditions with respect to magnetizing currents and zero-sequence impedances are essentially the same as for single-phase transformers connected in the same way. The center-phase coil is usually wound in a direction opposite to that of the two outer phases in order to decrease the core yoke required between phases. This reversal of polarity is corrected when the leads are terminated.

The magnetic circuit of each phase of a three-limb core-type transformer is mutually connected in that the flux of one phase must return through the other two phases. In this type of transformer, the total instantaneous magnetic flux in each core section due to the fundamental excitation current is zero. However, in the wye–wye-connected transformer, there are third-harmonic voltages in the phases caused by the third-harmonic current. These voltages and the resulting magnetic flux are all induced in the same direction. Since there is no return path for this flux in the core, the flux must return through the relatively low-reluctance path outside the core. The core-type transformer is occasionally manufactured with a five-limb core. In this case, the magnetic circuit and performance characteristics are similar to that of shell-type or single-phase transformers.

Unbalanced system faults and loads can cause significant zero-sequence magnetic flux to occur for some three-phase connections. Unless a magnetic return path for this flux is provided, the flux returns outside the core and can cause eddy-current heating in other transformer conductive components, such as the tank. The existence of zero-sequence flux either within or outside the core depends on both core configuration and winding connections. Three-phase transformer-core assemblies do not usually provide full-capacity return legs for zero-sequence flux. Thus, if sufficient zero-sequence flux occurs, it will be forced to return outside the core.

Three-phase transformer connections can be compared with each other with respect to:

- Ratio of kVA output to the kVA internal rating of the bank
- Degree of voltage symmetry with unbalanced phase loads
- Voltage and current harmonics
- Transformer ground availability
- System fault-current level
- Switching and system fault and transient voltages

In some cases, there may also be other operating characteristics to determine the most suitable connection for each application.

3.5.4.1   Double-Wound Transformers

The majority of three-phase transformer connections are made by connecting the individual phases either between the power-system lines, thus forming a delta connection, or by connecting one end of each phase together and the other ends to the lines, thus forming a wye (also referred to as star) connection. For these connections, the total rating of the internal windings is equal to the through-load rating. This accounts for the popularity of these connections. For all other double-wound transformer connections, the ratio is less than unity. For example, in the interconnected star or zigzag connection, the transformer is capable of delivering a load equal to only 86.6% of the internal winding rating. Since the cost of a transformer varies approximately with the three-quarter power of the internal kVA, the cost of an interconnected star or zigzag transformer is approximately 5% higher than for a similar double-wound transformer. All of these types of three-phase connections are shown in Figure 3.5.2.

3.5.4.1.1 Wye–Wye Connections of Transformers

Joining together the terminals of similar windings with the same polarity derives the neutral of the wye connection. This neutral point is available and can be brought out for any desired purpose, such as grounding or zero-sequence current measurements and protection. For high-voltage transmission systems, the use of the wye-connected transformer is more economical because the voltage across the phase of each winding is a factor of 1.73 less than the voltage between the lines. If the neutral point is grounded, it is not necessary to insulate it for the line voltage.

If the neutral is not grounded, the fault current during a system line to ground fault is insignificant because of the absence of a zero-sequence current path. If the neutral is grounded in the wye–wye transformer and the transformer is made with a three-limb core, the zero-sequence impedance is still high. As a result, the fault currents during a system line-to-ground fault are relatively low. For wye–wye transformers made of three single-phase units or with a shell-type or five-limb core-type, the zero-sequence impedance is approximately equal to the positive-sequence impedance. The fault current during a system ground fault for this case is usually the limiting factor in the design of the transformer. In all types of wye-wye transformer connections, only the transformer positive-sequence impedance limits the fault current during a system three-phase system fault.

With the wye connection, the voltages are symmetrical as far as the lines are concerned, but they introduce third-harmonic (or multiples of the third harmonic) voltage and current dissymmetry between lines and neutral. The third-harmonic voltage is a zero-sequence phenomenon and thereby is exhibited in the same direction on all phases. If the transformer and generator neutrals are grounded, third-harmonic currents will flow that can create interference in telephone circuits. If the transformer neutral is not grounded, the third-harmonic voltage at the neutral point will be additive for all three phases, and the neutral voltage will oscillate around the zero point at three times the system frequency. Third-harmonic voltages are also created on the lines, which can subject the power system to dangerous overvoltages due to resonance with the line capacitance. This is particularly true for shell-type three-phase transformers, five-limb core-type transformers, and three-phase banks of single-phase transformers. For any three-phase connection of three-limb core-form transformers, the impedance to third-harmonic flux is relatively high on account of the magnetic coupling between the three phases, resulting in a more stabilized neutral voltage. A delta tertiary winding can be added on wye–wye transformers to provide a path for third-harmonic and zero-sequence currents and to stabilize the neutral voltage. The tertiary in this application will be required to carry all of the zero-sequence fault current during a system line-to-ground fault.

The most common way to supply unbalanced loads is to use a four-wire wye-connected circuit. However, the primary windings of the transformer bank cannot be wye-connected unless the primary neutral is joined to the neutral of the generator. In this case, a third-harmonic voltage exists from each secondary line-to-neutral voltage because the generator supplies a sinusoidal excitation current. The third-harmonic currents created by the third-harmonic voltages can be a source of telephone interference. If the primary neutral is not connected to the generator, single-phase or unbalanced three-phase loads in the secondary cannot be supplied, since the primary current has to flow through the high impedance of the other primary windings.

3.5.4.1.2 Delta–Delta Connection

The delta–delta connection has an economic advantage over the wye–wye connection for low-voltage, high-current requirements because the winding current is reduced by a factor of 1.73 to 58% of that in the wye–wye connection.

Voltage and current symmetry with respect to the three lines is obtained only in the delta and zigzag connections. Delta-connected transformers do not introduce third harmonics or their multiples into the power lines. The third-harmonic-induced voltage components are 360° apart. They are therefore all in phase and cause a third-harmonic current to flow within the delta winding. This third-harmonic current acts as exciting current and causes a third-harmonic voltage to be induced in each winding that is in opposition to the third-harmonic component of voltage that was originally induced by the sinusoidal exciting current from the lines. As a result, the third harmonic is eliminated from the secondary voltage.

Another advantage of the delta–delta connection, if composed of three single-phase transformers, is that one transformer can be removed and the remaining two phases operated at 86.6% of their rating in the open delta connection.

The principle disadvantage of the delta–delta connection is that the neutral is not available. As a result, the phases cannot be grounded except at the corners. The insulation design is more costly because this type of three-phase transformer connection has higher ground voltages during system fault or transient voltages. Supplying an artificial neutral to the system with a grounding transformer can help to control these voltages. The delta-connection insulation costs increase with increasing voltage. Consequently, this type of connection is commonly limited to a maximum system voltage of 345 kV.

Differences in the voltage ratio of the individual phases causes a circulating current in both the primary and secondary deltas that is limited only by the impedance of the units. Differences in the impedances of the individual phases also causes unequal load division among the phases. When a current is drawn from the terminals of one phase of the secondary, it flows in the windings of all three phases. The current among the phases divides inversely with the impedance of the parallel paths between the terminals.

3.5.4.1.3 Delta–Wye and Wye–Delta Connections

The delta–wye or wye–delta connections have fewer objectionable features than any other connections. In general, these combine most of the advantages of the wye–wye and delta–delta connections. Complete voltage and current symmetry is maintained by the presence of the delta. The exciting third-harmonic current circulates within the delta winding, and no third-harmonic voltage appears in the phase voltages on the wye side. The high-voltage windings can be connected wye, and the neutral can be brought out for grounding to minimize the cost of the insulation.

Differences in magnetizing current, voltage ratio, or impedance between the single-phase units are adjusted by a small circulating current in the delta. All of these factors result in unbalanced phase voltages on the delta, which causes a current to circulate within the delta.

If the primary windings of a four-wire, wye-connected secondary that is supplying unbalanced loads are connected in delta, the unbalanced loading can be readily accommodated. There will be unbalanced secondary voltages caused by the difference in the regulation in each phase, but this is usually insignificant.

Although the delta–wye connection has most of the advantages of the wye–wye and delta–delta, it still has several disadvantages. This connection introduces a 30° phase shift between the primary and secondary windings that has to be matched for paralleling. A delta–wye bank cannot be operated with only two phases in an emergency. If the delta is on the primary side and should accidentally open, the unexcited leg on the wye side can resonate with the line capacitance.

3.5.4.2 Multiwinding Transformers

Transformers having more than two windings coupled to the same core are frequently used in power and distribution systems to interconnect three or more circuits with different voltages or to electrically isolate two or more secondary circuits. For these purposes, a multiwinding transformer is less costly and more efficient than an equivalent number of two winding transformers. The arrangement of windings can be varied to change the leakage reactance between winding pairs. In this way, the voltage regulation and the short-circuit requirements are optimized. The application of multiwinding transformers permits:

- Interconnection of several power systems operating at different voltages
- Use of a delta-connected stabilizing winding, which can also be used to supply external loads
- Control of voltage regulation and reactive power

- Electrical isolation of secondary circuits
- Duplication of supply to a critical load
- Connection for harmonic-filtering equipment
- A source for auxiliary power at a substation

Some of the problems with the use of multiwinding transformers are associated with the effect leakage impedance has on voltage regulation, short-circuit currents, and the division of load among the different circuits. All the windings are magnetically coupled to the leakage flux and are affected by the loading of the other windings. It is therefore essential to understand the leakage impedance behavior of this type of transformer to be able to calculate the voltage regulation of each winding and load sharing among the windings. For three-winding transformers, the leakage reactance between each pair of windings must be converted into a star-equivalent circuit. The impedance of each branch of the star circuit is calculated as follows:

$$Z_a = 0.5(Z_{ab} + Z_{ca} - Z_{bc}) \qquad (3.5.1)$$

$$Z_b = 0.5(Z_{bc} + Z_{ab} - Z_{ca}) \qquad (3.5.2)$$

and

$$Z_c = 0.5(Z_{ca} + Z_{bc} - Z_{ab}) \qquad (3.5.3)$$

where Z_a, Z_b, and Z_c are the star-equivalent impedances in each branch, and Z_{ab}, Z_{bc}, and Z_{ca} are the impedances as seen from the terminals between each pair of windings with the remaining winding left open circuit. The equivalent star-circuit reactances and resistances are determined in the same manner.

The four-winding transformer coupled to the same core is not commonly used because of the interdependence of the voltage regulation of each winding to the loading on the other windings. The equivalent circuit for a four-winding transformer is much more complicated, involving a complex circuit of six different impedances.

After the loading of each winding is determined, the voltage regulation and load sharing can be calculated for each impedance branch and between terminals of different windings. The currents in each winding during a system fault can also be calculated in a similar fashion.

3.5.4.3 Autotransformers Connections

It makes no difference whether the secondary voltage is obtained from another coil or from the primary turns. The same transformation ratio is obtained in either case. When the primary and secondary voltages are obtained from the same coil, schematically, the transformer is called an autotransformer. The performance of autotransformers is governed by the same fundamental considerations as for transformers with separate windings. The autotransformer not only requires less turns than the two-winding transformer; it also requires less conductor cross section in the common winding because it has to carry only the differential current between the primary and secondary. As a result, autotransformers deliver more external load than the internal-winding kVA ratings, depending on the voltage ratios of the primary and secondary voltages, as shown in Figure 3.5.3 and the following formula:

$$\text{Output/internal rating} = V_1/(V_1 - V_2) \qquad (3.5.4)$$

where

V_1 = voltage of the higher-voltage winding
V_2 = voltage of the lower-voltage winding

The internal rating, size, cost, excitation current, and efficiency of autotransformers are higher than in double-wound transformers. The greatest benefit of the autotransformer is achieved when the primary and secondary voltages are close to each other.

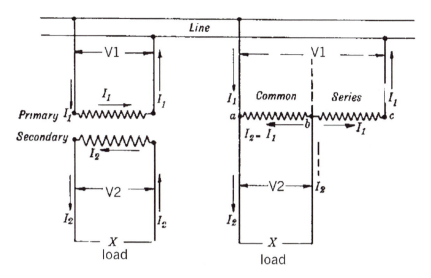

FIGURE 3.5.3 Current flow in double-wound transformer and autotransformer.

A disadvantage of the autotransformer is that the short-circuit current and forces are increased because of the reduced leakage reactance. In addition, most three-phase autotransformers are wye–wye connected. This form of connection has the same limitations as for the wye–wye double-wound transformers. Furthermore, there is no electrical isolation between the primary and secondary circuits with an autotransformer connection.

An autotransformer often has a delta-connected tertiary winding to reduce third-harmonic voltages, to permit the transformation of unbalanced three-phase loads, and to enable the use of supply-station auxiliary load or power-factor improvement equipment. The tertiary winding must be designed to accept all of these external loads as well as the severe short-circuit currents and forces associated with three-phase faults on its own terminals or single line-to-ground faults on either the primary or secondary terminals. If no external loading is required, the tertiary winding terminals should not be brought out except for one terminal to ground one corner of the delta during service operation. This eliminates the possibility of a three-phase external fault on the winding.

The problem of transformer insulation stresses and system transient protection is more complicated for autotransformers, particularly when tapping windings are also required. Transient voltages can also be more easily transferred between the power systems with the autotransformer connection.

3.5.4.4 Interconnected-Wye and Grounding Transformers

The interconnected-wye-wye connections have the advantages of the star–delta connections with the additional advantage of the neutral. The interconnected-wye or zigzag connection allows unbalanced-phase load currents without creating severe neutral voltages. This connection also provides a path for third-harmonic currents created by the nonlinearity of the magnetic core material. As a result, interconnected-wye neutral voltages are essentially eliminated. However, the zero-sequence impedance of interconnected-wye windings is often so low that high third-harmonic and zero-sequence currents will result when the neutral is directly grounded. These currents can be limited to an acceptable level by connecting a reactor between the neutral and ground. The interconnected-wye-wye connection has the disadvantage that it requires 8% additional internal kVA capacity. This and the additional complexity of the leads make this type of transformer connection more costly than the other common types discussed above.

The stable neutral inherent in the interconnected-wye or zigzag connection has made its use possible as a grounding transformer for systems that would be isolated otherwise. This is shown in Figure 3.5.4. The connections to the second set of windings can be reversed to produce the winding angular displacements shown in Figure 3.5.2.

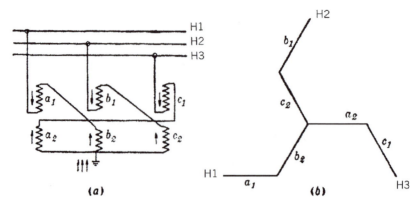

FIGURE 3.5.4 Interconnected star-grounding transformer: (a) current distribution in the coils for a line-to-ground fault and (b) normal operating voltages in the coils.

For a line-to-neutral load or a line-to-ground fault on the system, the current is limited by the leakage reactance between the two coils on each phase of the grounding transformer.

3.5.4.5 Phase-Shifting Transformers

The development of large, high-voltage power grids has increased the reliability and efficiency of electric power systems. However, the difference of voltages, impedance, loads, and phase angles between paralleled power lines causes unbalanced line-loading. The phase-shifting transformer is used to provide a phase shift between two systems to control the power flow. A phase-shifting transformer (PST) is a transformer that advances or retards the voltage phase-angle relationship of one circuit with respect to another circuit. In some cases, phase-shifting transformers can also control the reactive power flow by varying the voltages between the two circuits.

There are numerous different circuits and transformer designs used for this application. The two main type of PSTs used are shown in Figure 3.5.5a and Figure 3.5.5b. The single-core design shown in Figure 3.5.5a is most commonly used. With this design, it is generally accepted practice to provide two sets of three single-pole tap changers: one set on the source terminals and one set on the load terminals. This permits symmetrical voltage conditions while varying the phase angle from maximum advance to maximum retard tap positions. If only one set of three single-pole tap changers is used, the load voltage varies as the tap-changer phase-shift position is varied. The single-core design is less complicated, has less internal kVA, and is less costly than the other designs used. However, it has the following disadvantages:

- The LTC and tap windings are at the line ends of the power systems and are directly exposed to system transient voltages.
- The impedance of the PST varies directly with the square of the number of tap positions away from the mid-tap position. The impedance of this type of PST at the mid-tap or zero-phase-shift position is negligible. As a result, the short-circuit current at or near the mid-tap position is limited only by the system impedance.
- The maximum capacity of this type of PST is generally limited by the maximum voltage or current limitation of the tap changers. As a result, the maximum switching capacity of the tap changer cannot be fully utilized. The space required by these tap changers cause shipping restrictions in large-capacity PSTs.
- The transient voltage on the tap-changer reversing switch when switching through the mid-point position is very high. Usually, additional components are required in the PST or tap changer to limit these transfer voltages to an acceptable level.
- The cost of single-pole tap changers is substantially higher than three-pole tap changers used with some of the other PST designs.

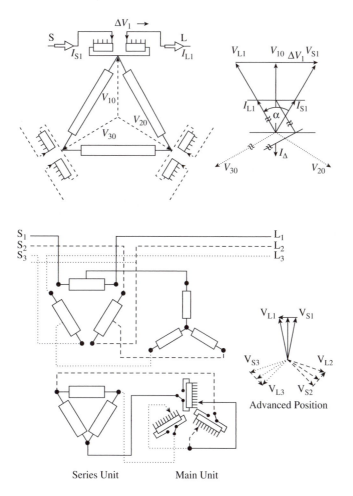

FIGURE 3.5.5 Two main type of PSTs: top) single core; bottom) dual core.

The other common PST circuit used is shown in Figure 3.5.5b. This PST design requires two separate cores, one for the series unit and one for the main or excitation unit. For large power, this PST design requires two separate tanks with oil-filled throat connection between them. This type of design does not have the technical limitations of the single-core design. Furthermore, another tap winding connected in quadrature to the phase-shifting tap windings can be readily added to provide voltage regulation as well as phase-shift control. This enables essentially independent control of the real and reactive power flow between the systems. However, the cost of this type of PST design is substantially higher because of the additional core, windings, and internal kVA capacity required.

3.5.5 Three-Phase to Six-Phase Connections

The three-phase connections discussed above are commonly used for six-pulse rectifier systems. However, for 12-pulse rectifier systems, three-phase to six-phase transformations are required. For low-voltage dc applications, there are numerous practical connection arrangements possible to achieve this. However, for high-voltage dc (HVDC) applications, there are few practical arrangements. The most commonly used connections are either a delta or wye primary with two secondaries: one wye- and one delta-connected.

3.5.6 Paralleling of Transformers

Transformers having terminals marked in the manner shown in Section 3.5.2, Polarity of Single-Phase Transformers, can be operated in parallel by connecting similarly marked terminals provided that their ratios, voltages, angular displacement, resistances, reactances, and ground connections are such as to permit parallel operation.

The difference in the no-load terminal voltages of the transformers causes a circulating current to flow between the transformers when paralleled. This current flows at any load. The impedance of the circuit, which is usually the sum of the impedances of the transformers that are operating in parallel, limits the circulating current. The inductive circulating current adds, considering proper phasor relationships, to the load current to establish the total current in the transformer. As a result, the capacity of the transformer to carry load current is reduced by the circulating current when the transformers are paralleled. For voltage ratios with a deviation of less than the 0.5%, as required by the IEEE standards, the circulating current between paralleled transformers is usually insignificant.

The load currents in the paralleled transformers divide inversely with the impedances of the paralleled transformers. Generally, the difference in resistance has an insignificant effect on the circulating current because the leakage reactance of the transformers involved is much larger than the resistance. Transformers with different impedance values can be made to divide their load in proportion to their load ratings by placing a reactor in series with one transformer so that the resultant impedance of the two branches creates the required load sharing.

When delta–delta connected transformer banks are paralleled, the voltages are completely determined by the external circuit, but the division of current among the phases depends on the internal characteristics of the transformers. Considerable care must be taken in the selection of transformers, particularly single-phase transformers in three-phase banks, if the full capacity of the banks is to be used when the ratios of transformation on all phases are not alike. In the delta–wye connection, the division of current is indifferent to the differences in the characteristics of individual transformers.

3.6   Transformer Testing

Shirish P. Mehta and William R. Henning

3.6.1 Introduction

Reliable delivery of electric power is, in great part, dependent on the reliable operation of power transformers in the electric power system. Power transformer reliability is enhanced considerably by a well-written test plan, which should include specifications for transformer tests. Developing a test plan with effective test specifications is a joint effort between manufacturers and users of power transformers. The written test plan and specifications should consider the anticipated operating environment of the transformer, including factors such as atmospheric conditions, types of grounding, and exposure to lightning and switching transients. In addition to nominal rating information, special ratings for impedance, sound level, or other requirements should be considered in the test plan and included in the specifications. Selection of appropriate tests and the specification of correct test levels, which ensure transformer reliability in service, are important parts of this joint effort.

Transformers can be subjected to a wide variety of tests for a number of reasons, including:

- Compliance with user specifications
- Assessment of quality and reliability
- Verification of design calculations
- Compliance with applicable industry standards

3.6.1.1 Standards

ANSI standards for power transformers are given in the C57 series of standards. Requirements that apply generally to all power and distribution transformers are given in the following two standards. These standards are particularly relevant and useful for those needing information on transformer testing.

IEEE C57.12.00, IEEE Standard General Requirements for Liquid-Immersed Distribution, Power, and Regulating Transformers [1].

IEEE C57.12.90, IEEE Standard Test Code for Liquid-Immersed Distribution, Power, and Regulating Transformers; and Guide for Short-Circuit Testing of Distribution and Power Transformers (ANSI) [2].

These two ANSI and IEEE standards, and others, will be cited frequently in this article. There are many other standards in the ANSI C57 series for transformers, covering the requirements for many specific products, product ranges, and special applications. They also include guides, tutorials, and recommended practices. These documents will be of interest to readers wanting to find out detailed, authoritative information on testing power and distribution transformers.

The standards cited above are very important documents because they facilitate precise communication and understanding between manufacturers and users. They identify critical features, provide minimum requirements for safe and reliable operation, and serve as valuable references of technical information.

3.6.1.2 Classification of Tests

According to ANSI and IEEE standards [1], all tests on power transformers fall into one of three categories: (1) routine tests, (2) design tests, and (3) other tests. The manufacturer may perform additional testing to ensure the quality of the transformer at various stages of the production cycle. For this discussion, tests on power transformers are categorized as shown in Table 3.6.1.

Some of these tests are performed before the transformer core and coil assembly is placed in the tank, while other tests are performed after the transformer is completely assembled and ready for "final testing." However, what are sometimes called "final tests" are not really final. Additional tests are performed just before transformer shipment, and still others are carried out at the customer site during installation and commissioning. All of these tests, the test levels, and the accept/reject criteria represent an important aspect of the joint test plan development effort made between the manufacturer and the purchaser of the transformer.

3.6.1.3 Sequence of Tests

The sequence in which the various tests are performed is also specified. An example of test sequence is as follows:

Tests before tanking
- Preliminary ratio, polarity, and connection of the transformer windings
- Core insulation tests
- Ratio and polarity tests of bushing-current transformers

Tests after tanking (final tests)
- Final ratio, polarity, and phase rotation
- Insulation capacitance and dissipation factor
- Insulation resistance
- Control-wiring tests
- Lightning-voltage impulse tests
- Applied-voltage tests
- Induced-voltage tests and partial-discharge measurements
- No-load-loss and excitation-current measurements
- Winding-resistance measurements
- Load-loss and impedance-voltage measurements

TABLE 3.6.1 Transformer Tests by Category

Dielectric Tests		Tests of Performance Characteristics	Thermal Tests	Other Tests
Transients	Low (Power) Frequency			
Lightning impulse - Full wave - Chopped wave - Steep wave	Applied voltage	No-load loss	Winding resistance	Insulation capacitance and dissipation factor [a]
Switching impulse	Single-phase induced	Excitation current, %	Heat-run test - Oil rise - Winding rise - Hottest-spot rise	Sound-level tests
	Three-phase induced Partial discharge	Load loss Impedance, % Zero-sequence impedance Ratio test Short-circuit test	Overload heat run Gas in oil Thermal scan	10-kV excitation current Megger Core ground Electrical center [a] Recurrent surge Dew point Core loss before impulse Control-circuit test [a] Test on series transformer [a] LTC tests [a] Preliminary ratio tests [a] Test on bushing CT [a] Oil preservation system [a]

[a] Quality control tests

- Temperature-rise tests (heat runs)
- Tests on gauges, accessories, LTCs, etc.
- Sound-level tests
- Other tests as required

Tests before shipment
- Dew point of gas
- Core-ground megger test
- Excitation-frequency-response test

Commissioning tests
- Ratio, polarity, and phase rotation
- Capacitance, insulation dissipation factor, and megger tests
- LTC control settings check
- Test on transformer oil
- Excitation-frequency-response test
- Space above the oil in the transformer tank.

3.6.1.4 Scope of This Chapter Section

This chapter section (Section 3.6, Transformer Testing) will cover testing to verify or measure the following:

- Voltage ratio and proper connections
- Insulation condition

- Control devices and control wiring
- Dielectric withstand
- Performance characteristics
- Other tests

3.6.2 Voltage Ratio and Proper Connections

3.6.2.1 The Purpose of Ratio, Polarity, and Phase-Relation Tests

Tests for checking the winding ratios, and for checking the polarity and phase relationships of winding connections, are carried out on all transformers during factory tests. The purpose of these tests is to ensure that all windings have the correct number of turns according to the design, that they are assembled in the correct physical orientation, and that they are connected properly to provide the desired phase relationship for the case of polyphase transformers. If a transformer is equipped with either a de-energized tap changer (DETC) or a load tap changer (LTC), or both, then ratio tests are also carried out at the various positions of the tap changer(s). The objective of ratio tests at different tap positions is to ensure that all winding taps are made at the correct turns and that the tap connections are properly made to the tap-changing devices.

The standards [2] define three separate tests:

1. Ratio test
2. Polarity test
3. Phase-relation test

However, test sets are available that, once all the connections are made, determine ratio and polarity simultaneously and facilitate testing of three-phase transformers to determine phase relationship by switch selection of the required leads. In this sense, the three tests can be combined.

3.6.2.2 Ratio Test

ANSI and IEEE standards [1] require that the measured voltage ratio between any two windings be within ±0.5% of the value indicated on the nameplate. To verify this requirement, ratio tests are performed in which the actual voltage ratio is determined through measurements. Ratio tests can be made by energizing the transformer with a low ac test voltage and measuring the voltage induced in other windings at various tap settings, etc. In each case, the voltage ratio is calculated and compared with the voltage ratio indicated on the transformer nameplate. More commonly, transformer-turns-ratio test sets (TTRs) are used for making the tests. In this method, the transformer to be tested is energized with a low ac voltage at power frequency in parallel with the high-turn winding of a standard reference transformer in the test set. The induced voltage in the low-voltage (LV) winding of the transformer under test is compared against the induced voltage of the variable low-turn reference winding, in both magnitude and phase, to verify voltage ratio and polarity.

3.6.2.3 Polarity Test

Polarity is usually checked at the same time as turns ratio, using the test set. At balance, a transformer-turns-ratio test set displays the voltage ratio between two windings and also indicates winding polarity.

There are three other ways to check polarity:

1. Inductive kick
2. Alternating voltage
3. Comparison with transformer of known polarity and same ratio

Information on these methods can be found in the IEEE test code [2].

3.6.2.4 Phase Relationship Test

The transformer test sets used to determine voltage ratio and polarity are designed for single-phase operation, but supplemental switching arrangements are employed to facilitate testing of three-phase transformers. Based on the phasor diagram of the three-phase transformer, appropriate high and low voltages (HV and LV) are selected by switches for determining the phase relationships.

The IEEE test code [2] discusses additional methods that can be used to determine or verify phase relationship.

3.6.3 Insulation Condition

3.6.3.1 The Purpose of Insulation Condition Tests

Full-scale dielectric testing of the transformer insulation system is discussed in a later section of this article. To be discussed here are insulation tests, performed at voltages of about 10 kV, to verify the condition of the insulation system. Initial measurements at the factory can be recorded and compared with later measurements in the field to assess changes in the condition of the transformer insulation.

The quality of the transformer insulation and the efficacy of the insulation processing for moisture removal are evaluated through the results of insulation power-factor tests and insulation resistance tests.

3.6.3.2 Insulation Power Factor

Insulation power-factor tests are performed in one of two ways. In the first general method, a bridge circuit is used to measure the capacitance and dissipation factor (tan δ) of the insulation system. In the second general method, a specially designed test set measures voltage, current, and power when tests are made at about 10 kV. From these measurements, the insulation power factor (cos ϕ) is determined. For a well-processed insulation system, the dielectric loss will be low, with a correspondingly small loss angle. When the loss angle is small, then cos ϕ is nearly equal to tan δ, and the "insulation power factor" and "tan δ" methods are therefore equivalent.

3.6.3.2.1 Insulation Capacitance and Dissipation Factor (tan δ) Test

The measurement of insulation capacitance and dissipation factor (tan δ) is carried out using a capacitance bridge. A transformer ratio arm bridge or a Schearing bridge can be used for this purpose. In a two-winding transformer, there are three measurements of capacitance: (1) HV to ground, (2) LV to ground, and (3) HV to LV. These values of capacitance and their respective values of insulation dissipation factor (tan δ) are to be measured.

For performing these tests, the following connections are made. All HV line terminals are connected together and labeled (H); all LV line terminals are connected together and labeled (L); and a connection is made to a ground terminal, usually a connection to the transformer tank, which is labeled (G). Leads from the measuring instrument or bridge are connected to one or both terminals and ground. Either grounded specimen measurements or guarded measurements are possible, so that all capacitance values and dissipation-factor values can be determined. Figure 3.6.1 shows the measurement of capacitance and dissipation factor of the HV-to-LV capacitance, using guarded measurements. Figure 3.6.2 is for the low-to-ground capacitance. These measurements are usually made at voltages of 10 kV or less, at or near power frequency. In substations and factory test floors, interference control circuits may be required to achieve the desired sensitivity at balance.

3.6.3.2.2 Volt-Ampere-Watt Test

Test sets, specifically designed for this purpose, determine the required parameters by measurements of voltage, current, and power when tests are made with about 10-kV excitation. Insulation power factor is computed from the measured quantities, based on the definition of power factor, W/VA.

3.6.3.2.3 Interpretation of Results

Insulation power factor values of less than 0.5% are considered an indication of well-processed, dried-out transformer insulation. Although there is general recognition and agreement that power-factor and

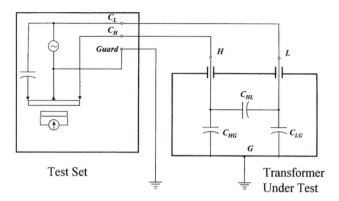

FIGURE 3.6.1 Measurement of capacitance and dissipation factor — high to low.

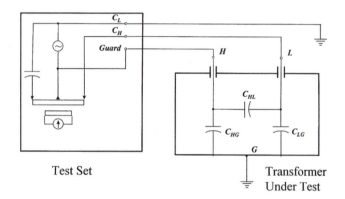

FIGURE 3.6.2 Measurement of capacitance and dissipation factor — low to ground.

dissipation-factor values are dependent on temperature, the exact relationship is not uniformly agreed upon. When insulation power factor is measured, temperature should also be recorded.

3.6.3.3 Insulation Resistance

Insulation resistance of a two-winding transformer insulation system — HV to ground, LV to ground, and HV to LV — is determined with a "megger" type of instrument, so named because the expected value for resistance is in the millions of ohms. Historically, insulation-resistance measurements are also made to assess the amount of moisture in transformer insulation. However, the measurement of insulation dissipation factor has shown to be a better indicator of the overall condition of insulation in a power transformer.

3.6.4 Control Devices and Control Wiring

Modern power transformers often incorporate equipment-condition-monitoring devices and substation-automation-related devices as part of their control systems. Testing of the devices and the functional verification of the control system is a very critical aspect of the controls test plan. The controls test should include performance verification of control devices with various equipment protection, monitoring, and supervisory functions. Such devices may include: liquid-temperature gauges, winding-temperature simulator gauges, liquid-level gauges, pressure-vacuum gauges, conservator Buchholz relays, sudden-pressure relays, pressure-relief devices, etc. Many power transformers are equipped with-load tap changers, and tests to verify operation of the LTC at rated current during factory tests are now included in the latest

ANSI standards. LTC control settings for voltage level, bandwidth, time delay, first-customer protection, backup protection, etc., should also be verified during factory tests and at commissioning.

Bushing current transformer leads are usually routed to the control cabinet for connection by the customer. That these leads are connected to the proper terminals is usually checked during the load-loss test on power transformers by monitoring the secondary current in various tap positions of the bushing CTs.

Applied voltage tests, commonly called "hipot" tests, are performed to check the dielectric integrity of the control wiring. A 2.5-kV ac test is recommended for current transformer leads, and a 1.5-kV ac test is required for other control circuit wiring.

3.6.5   Dielectric Withstand

In actual operation on a power system, a transformer is subjected to both normal and abnormal dielectric stresses. For example, a power transformer is required to operate continuously at 105% of rated voltage when delivering full-load current and at 110% of rated voltage under no-load for an indefinite duration [1]. These are examples of conditions defined as *normal operating conditions* [1]. The voltage stresses associated with normal conditions as defined above, although higher than stresses at rated values, are nonetheless considered normal stresses.

A transformer may be subjected to *abnormal* dielectric stresses, arising out of various power system events or conditions. Sustained power-frequency overvoltage can result from Ferranti rise, load rejection, and ferroresonance. These effects can produce abnormal turn-to-turn and phase-to-phase stresses. On the other hand, line-to-ground faults can result in unbalance and very high terminal-to-ground voltages, depending upon system grounding. Abnormal transient overvoltages of short duration arise out of lightning-related phenomena, and longer duration transient overvoltages can result from line-switching operations.

Even though these dielectric stresses are described as *abnormal*, the events causing them are expected to occur, and the transformer insulation system must be designed to withstand them. To verify the transformer capability to withstand these kinds of abnormal but expected transient and low-frequency dielectric stresses, transient and low-frequency dielectric tests are routinely performed on all transformers. The general IEEE transformer standard [1] identifies the specific tests required. It also defines test levels for each test. The IEEE test code [2] describes exactly how the tests are to be made; it defines pass-fail criteria; and it provides valid methods of corrections to the results.

3.6.5.1   Transient Dielectric Tests

3.6.5.1.1   *Purpose of Transient Dielectric Tests*
Transient dielectric tests consist of lightning-impulse tests and switching-impulse tests. They demonstrate the strength of the transformer insulation system to withstand transient voltages impinged upon the transformer terminals during surge-arrester discharges, line-shielding flashovers, and line-switching operations.

Power transformers are designed to have certain transient dielectric strength characteristics based on basic impulse insulation levels (BIL). The general IEEE transformer standard [1] provides a table listing various system voltages, BIL, and test levels for selected insulation classes. The transient dielectric tests demonstrate that the power transformer insulation system has the necessary dielectric strength to withstand the voltages indicated in the tables.

3.6.5.1.2   *Lightning-Impulse Test*
Impulse tests are performed on all power transformers. In addition to verification of dielectric strength of the insulation system, impulse tests are excellent indicators of the quality of insulation, workmanship, and processing of the paper and insulating-oil system. The sequence of tests, test connections, and applicable standards is described below.

Lightning-impulse voltage tests simulate traveling waves due to lightning strikes and line flashovers. The full-wave lightning-impulse voltage waveshape is one where the voltage reaches crest magnitude in

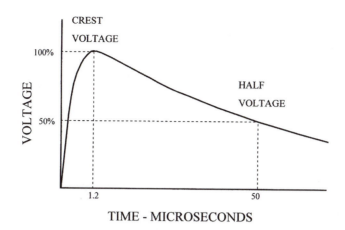

FIGURE 3.6.3 Standard full-wave lightning impulse.

1.2 μs, then decays to 50% of crest magnitude in 50 μs. This is shown in Figure 3.6.3. Such a wave is said to have a *waveshape* of 1.2 × 50.0 μs. The term *waveshape* will be used in this article to refer to the test wave in a general way. The term *waveform* will be used when referring to detailed features of the test voltage or current records, such as oscillations, "mismatches," and chops. The difference in meaning of these two terms can be found in the IEEE dictionary [3].

In addition to the standard-impulse full wave, a second type of lightning-impulse wave, known as the chopped wave, or sometimes called the tail-chopped wave, is used in transformer work. The chopped wave employs the same waveshape as a full-wave lightning impulse, except that its crest value is 10% greater than that of the full wave, and the wave is chopped at about 3 μs. The chop in the voltage wave is accomplished by the flashover of a rod gap, or by using some other chopping device, connected in parallel with the transformer terminal being tested. This wave is shown in Figure 3.6.4. The chopped-wave test simulates the sudden external flashover (in air) of the line insulation to ground. When the voltage applied to a transformer terminal suddenly collapses, the step change in voltage causes internal oscillations that can produce high dielectric stresses in specific regions of the transformer winding. The chopped-wave test demonstrates ability to withstand the sudden collapse of instantaneous voltage.

In addition to the full-wave test and the chopped-wave test, a third type of test known as front-of-wave test is sometimes made. (The test is sometimes called the steepwave test or front-chopped test.) The front-of-wave test simulates a direct lightning strike on the transformer terminals. Although direct strokes to transformer terminals in substations of modern design have very low probabilities of occurrence, front-of-wave tests are often specified. The voltage wave for this test is chopped on the front of the wave before the prospective crest value is reached. The rate of rise of voltage of the wave is set to about 1000 kV/μs. Chopping is set to occur at a chop time corresponding to an assigned instantaneous crest value. Front-of-wave tests, when required, must be specified.

Lightning-impulse tests, including full-wave impulse and chopped-wave impulse test waves, are made on each line terminal of power transformers. The recommended sequence is:

1. One reduced-voltage, full-wave impulse, with crest value of 50 to 70% of the required full-wave crest magnitude (BIL) to establish reference pattern waveforms (impulse voltage and current) for failure detection.
2. Two chopped-wave impulses, meeting the requirements of crest voltage value and time to chop, followed by:
3. One full-wave impulse with crest value corresponding to the BIL of the winding line terminal

When front-of-wave tests are specified, impulse tests are carried out in the following sequence: one reduced full-wave impulse, followed by two front-of-wave impulses, two chopped-wave impulses, and one full-wave impulse.

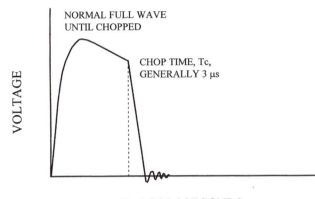

FIGURE 3.6.4   Standard chopped-wave lightning impulse.

Generally, impulse tests are made on line terminals of windings, one terminal at a time. Terminals not being tested are usually solidly grounded or grounded through resistors with values of resistance in the range of 300 to 450 Ω. The voltage on terminals not being tested should be limited to 80% of the terminal BIL. Details about connections, tolerances on waveshapes, voltage levels, and correction factors are given in the IEEE test code [2] and the IEEE impulse guide [4].

3.6.5.1.2.1 Lightning-Impulse Test Equipment — The generation, measurement, and control of impulse voltage waves is a very specialized subject. In this section, only a very brief general introduction to the subject is provided. Most impulse-generator designs are based on the Marx circuit. Figure 3.6.5 shows a schematic diagram of a typical Marx-circuit impulse generator with four stages. In principle, voltage multiplication is obtained by charging a set of parallel-connected capacitors in many stages of the impulse generator to a predetermined dc voltage, then momentarily reconnecting the capacitor stages in series to make the individual capacitor voltages add. The reconnection from parallel to series is accomplished through the controlled firing of a series of adjustable sphere gaps, adjusted to be near breakdown at the dc charging voltage. After the capacitors are charged to the proper dc voltage level, a sphere gap in the first stage is made to flash over by some means. This initiates a cascade flashover of all the sphere gaps in the impulse generator. The gaps function as switches, reconnecting the capacitor stages from parallel to series, producing a generator output voltage that is approximately equal to the voltage per stage times the number of stages.

The desired time to crest value on the front of the wave and the time to half-crest value on the tail of the wave are controlled by wave-shaping circuit elements. These elements are indicated as R_s, R_p, and $C_{Loading}$ in Figure 3.6.5. Generally, control of the time to crest on the front of the wave is realized by changing the values of series resistance, the impulse-generator capacitance, and the load capacitance. Control of the time to 50% magnitude on the tail of the wave is realized by changing the values of parallel resistors and the load capacitance. Control of the voltage crest magnitude is provided by adjustment of the dc charging voltage and by changing the load on the impulse generator. The time of flashover for chopped waves is controlled by adjustment of gap spacings of the chopping gaps or the rod gaps.

The capacitor-charging current path for the impulse generator is shown in Figure 3.6.6. At steady state, each of the capacitors is charged to a voltage equal to the dc supply voltage. After the cascade firing of the sphere gaps, the main discharging current path becomes, in simplified form, that of Figure 3.6.7. The RC time constants of the dc charging resistors, R_c as defined in Figure 3.6.5, have values typically expressed in seconds, while the waveshape control elements, R_p and R_s as defined in Figure 3.6.5, have RC time constants typically expressed in microseconds. Hence, for the time period of the impulse-generator discharge, the relatively high resistance values of the charging resistors represent open circuits for the relatively short time period of the generator discharge. This is indicated by dotted lines in Figure 3.6.7.

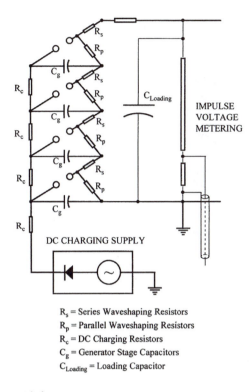

R$_s$ = Series Waveshaping Resistors
R$_p$ = Parallel Waveshaping Resistors
R$_c$ = DC Charging Resistors
C$_g$ = Generator Stage Capacitors
C$_{Loading}$ = Loading Capacitor

FIGURE 3.6.5 Marx generator with four stages.

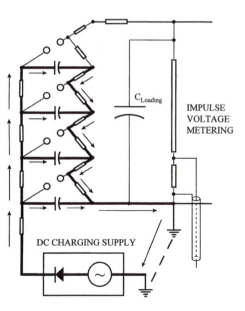

FIGURE 3.6.6 Charging the capacitors of an impulse generator.

The discharge path shown in the figure is somewhat simplified for clarity: Significant currents do flow in the shunt wave-shaping resistors, R$_p$, and significant current also flows in the loading capacitor, C$_{Loading}$. These currents, which are significant in controlling the waveshape, are ignored in Figure 3.6.7.

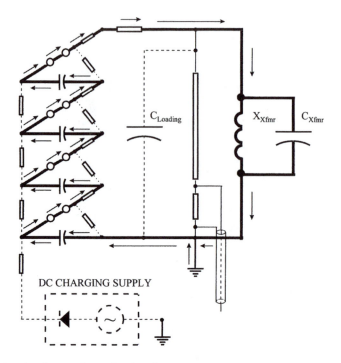

FIGURE 3.6.7 Discharging the capacitors of an impulse generator.

The measurement of impulse voltage in the range of a million volts in magnitude requires the use of voltage dividers. Depending upon requirements, either resistive, capacitive, or optimally damped (RC) types of dividers, having stable ratios and fast response times, are utilized to scale the high-voltage impulses to provide a suitable input for instruments. Most impulse-test facilities utilize specially designed impulse oscilloscopes or, more recently, specially designed transient digitizers, for accurate measurement of impulse voltages.

Measurement of the transient currents associated with impulse voltages is carried out with the aid of special noninductive shunts or wideband current transformers included in the path of current flow. Usually, voltages proportional to impulse currents are measured with the impulse oscilloscopes or transient digitizers described earlier.

3.6.5.1.2.2 Impulse Test Setup — For consistent results it is important that the test setup be carefully made, especially with respect to grounding, external clearances, and induced voltages produced by impulse currents. Otherwise, impulse-failure detection analysis could be flawed. One example of proper impulse-test setup is shown in Figure 3.6.8. This figure illustrates proper physical arrangement of the impulse generator, main circuit, chopping circuit, chopping gap, test object, current shunt, voltage measuring circuit, and voltage divider. High voltages and currents at high frequencies in the main circuit and the chopping circuit can produce rapidly changing electromagnetic fields, capable of inducing unwanted noise and error voltages in the low-voltage signal circuits connected to the impulse-recorder inputs. The purpose of this physical arrangement is to minimize these effects.

3.6.5.1.2.3 Impulse-Test Failure Detection — To accomplish failure detection or to verify the absence of a dielectric failure in the transformer insulation system, the impressed impulse-voltage waveforms and the resulting current waveforms of the full-wave test are compared with the reduced full-wave test reference waveforms. The main idea behind failure detection in transformers is that if no dielectric breakdowns or partial discharges occur, then the final full-wave test voltage and current waves will exhibit waveforms identical to the initial reduced full-wave reference tests when appropriately scaled. The occurrence of a dielectric breakdown would produce a sudden change in the inductance-capacitance network

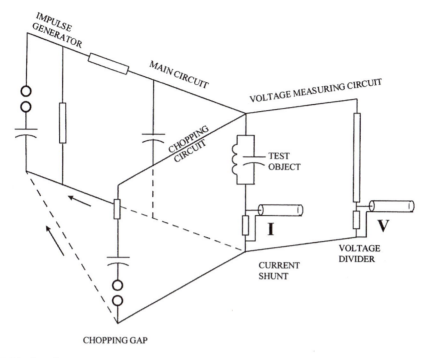

FIGURE 3.6.8 Impulse-test setup.

seen at the tested terminals of the transformer, causing a deviation in the test waveforms compared with the reference waveforms. The act of comparing the reduced full-wave records and the full-wave records is sometimes called "matching" the waves. If they are identical, the waves are said to be *matched.* Any differences in the waves, judged to be significant, are said to be *mismatches.* If there are mismatches, something is not correct, either in the test setup or in the dielectric system of the transformer. Various waveforms of voltages and currents associated with different types of defects are presented in great detail in the IEEE impulse guide [4]. When digital recorders are employed, methods of waveform analysis using the frequency dependence of the transformer impedance, transformer transfer function, and other digital waveform-analysis tools are now being developed and used to aid failure detection. Measurements of the voltages and currents in various parts of the transformer under test can aid in location of dielectric defects. These schemes are summarized in Figure 3.6.9.

3.6.5.1.3 Switching-Impulse Test
Man-made transients, as opposed to nature-made transients, are often the result of switching operations in power systems. Switching surges are relatively slow impulses. They are characterized by a wave that:

1. Rises to peak value in no less than 100 μs
2. Falls to zero voltage in no less than 1000 μs
3. Remains above 90% of peak value, before and after time of crest, for no less than 200 μs

This is shown in Figure 3.6.10. Generally, the crest value of the switching-impulse voltage is approximately 83% of the BIL.

Voltages of significant magnitude are induced in all windings due to core-flux buildup that results from the relatively long duration of the impressed voltage during the switching-impulse test. The induced voltages are approximately proportional to the turns ratios between windings. Depending upon the transformer construction, shell-form versus core-form, three-leg versus five-leg construction, etc., many connections for tests are possible. Test voltages at the required levels can be applied directly to the winding under test, or they can be induced in the winding under test by application of switching impulse voltage of suitable magnitude across another winding, taking into consideration the turns ratio between the two

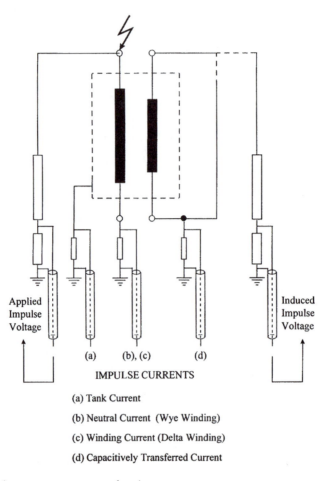

(a)   (b), (c)   (d)

IMPULSE CURRENTS

(a) Tank Current

(b) Neutral Current (Wye Winding)

(c) Winding Current (Delta Winding)

(d) Capacitively Transferred Current

FIGURE 3.6.9 Impulse-current measurement locations.

windings. The magnitudes of voltages between windings and between different phases depend on the connections. This is discussed in great detail in the IEEE impulse guide [4].

Because of its long duration and high peak-voltage magnitude, application of switching impulses on windings can result in saturation of the transformer core. When saturation of the core occurs, the resulting waves exhibit faster-falling, shorter-duration tails. By reversing polarity of the applied voltages between successive shots, the effects of core saturation can be reduced. Failures during switching-impulse tests are readily visible on voltage wave oscillograms and are often accompanied by loud noises and external flashover.

Switching-impulse tests are generally carried out with impulse generators having adequate energy capacity and appropriate wave-shaping resistors and loading capacitors.

3.6.5.2 Low-Frequency Dielectric Tests

3.6.5.2.1 Purpose of Low-Frequency Dielectric Tests

When high-frequency impulse voltages are applied to transformer terminals, the stress distributions within the windings are not linear but depend on the inductances and capacitances of the windings. Also, the effects of oscillations penetrating the windings produce complex and changing voltage distributions. Low-frequency stresses that result from power-frequency overvoltage, on the other hand, result in stresses with a linear distribution along the winding. Because the insulation system is stressed differently at low frequency, a second set of tests is required to demonstrate dielectric withstand under power-frequency conditions. The low-frequency dielectric tests demonstrate that the power transformer insulation system

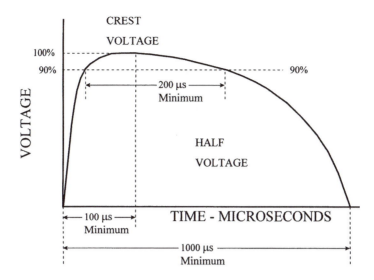

FIGURE 3.6.10 Standard switching-impulse wave.

has the necessary dielectric strength to withstand the voltages indicated in the tables of the standards for low-frequency tests.

3.6.5.2.2 Applied-Voltage Test

The applied-voltage test is often called the hipot test. The purpose of this test is to verify the major insulation in transformers. More specifically, the purpose is to ensure that the insulation between windings and the insulation of windings to ground can withstand the required power-frequency voltages, applied for 1-min duration. For fully insulated windings, the test voltage levels are related to the BIL of the windings. For windings with graded insulation, test levels correspond to the winding terminal with the lowest BIL rating.

For two-winding transformers, there are two applied tests. Test 1 is the applied-voltage test of the HV winding, which is carried out by connecting all HV terminals together and connecting them to a high-voltage test set, as shown in Figure 3.6.11. The LV terminals are connected together and connected to ground. Power frequency voltage of the correct level with respect to ground is applied to the HV terminals with LV terminals grounded. In this test, the insulation system between HV to ground and between HV to LV is stressed. A second test, Test 2, is made at the correct level of test voltage for the LV winding, with the LV winding terminals connected together and connected to the high-voltage test set. In Test 2, the HV winding terminals are connected together and to ground. In the second test, the insulation system LV to ground and LV to HV are stressed. This is also shown in Figure 3.6.11.

A dielectric failure during this test is the result of internal or external flashover or tracking. Internal failures may be accompanied by a loud noise heard on the outside and "smoke and bubbles" seen in the oil.

3.6.5.2.3 Induced-Voltage Test

The induced-voltage test, with monitoring of partial discharges during the test, is one of the most significant tests to demonstrate the integrity of the transformer insulation system. During this test, the turn-to-turn, disc-to-disc, and phase-to-phase insulation systems are stressed simultaneously at levels that are considerably higher than during normal operation. Weaknesses in dielectric design, processing, or manufacture may cause partial discharge activity during this test. Partial discharges are generally monitored on all line terminals rated 115 kV or higher during the induced-voltage test. This test produces the required voltages in the windings by magnetic induction. Because the test voltages are significantly higher than rated voltages, the core would ordinarily saturate if 60-Hz voltage were employed. To avoid core saturation, the test frequency is increased to a value normally in the range of 100 to 400 Hz.

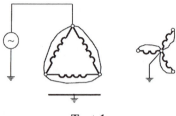

Test 1

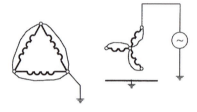

Test 2

FIGURE 3.6.11 Applied-voltage tests on two-winding transformers.

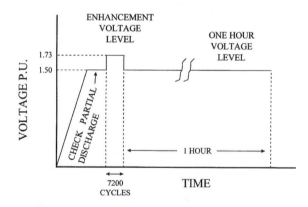

FIGURE 3.6.12 Induced-voltage test for class II power transformers.

Tests for Class II power transformers require an extended duration induced test. Class II is defined in ANSI and IEEE standards [1]. The test is usually made by raising the voltage on the LV windings to a value that induces 1.5 times maximum operating voltage in the HV windings. If no partial discharge activity is detected, the voltage is raised to the enhancement level (usually 1.73 per unit [p.u.]) for a time period of 7200 cycles at the test frequency. Voltage is then reduced to the 1.5-p.u level and held for a period of one hour. This test is described graphically in Figure 3.6.12. In the U.S., PD (partial discharge) level measurements are carried out with wideband PD detectors or narrowband RIV (radio influence voltage) meters. The wideband PD detectors read apparent charge in picocoulombs (pC). The narrowband RIV meters read the radio influence voltage in microvolts (μV). Often both instruments are used, taking wideband pC and narrowband μV readings simultaneously.

During the 1-h period, partial discharge readings are taken every 5 min. The criteria for a successful test are:

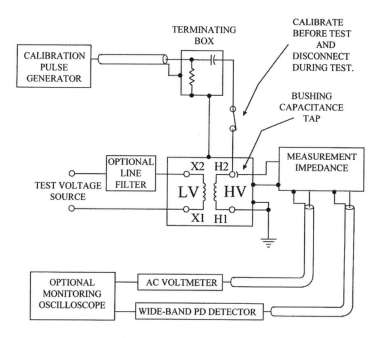

FIGURE 3.6.13 Test circuit for partial-discharge measurement during the induced-voltage test.

- The magnitude of partial discharge activity does not exceed 100 μV (500 pC).
- After the PD level has stabilized, any increase in partial discharge activity does not exceed 30 μV (150 pC).
- The partial discharge levels during the 1-h period do not show an increasing trend.
- There is no sustained increase in partial discharge level during the last 20 min of the test.

The test circuit for PD measurement during the induced-voltage test is shown in Figure 3.6.13. The bushing capacitance tap is connected to a coupling impedance unit, which provides PD signals to the PD detection unit. Methods for measurement and calibration are described in detail in the IEEE test code [2] and in the IEEE partial discharge measurement guide [5].

If there is significant level of PD activity during the test, it is desirable to identify the location of the PD source inside the transformer so that the problem can be corrected. This will ensure PD-free operation. Methods for location of PD sources inside the transformer often are based on triangulation. Ultrasonic acoustic waves arriving at transformer tank surfaces due to PD activity are monitored at various locations on the tank walls. Knowing wave propagation velocities and travel paths, it is possible to locate sources of PD activity with reasonable accuracy.

3.6.6 Performance Characteristics

3.6.6.1 No-Load Loss and Excitation-Current Measurements

3.6.6.1.1 Purpose of No-Load Loss Measurements

A transformer dissipates a constant no-load loss as long as it is energized at constant voltage, 24 hours a day, for all conditions of loading. This power loss represents a cost to the user during the lifetime of the transformer. Maximum values of the no-load loss of transformers are specified and often guaranteed by the manufacturer. No-load-loss measurements are made to verify that the no-load loss does not exceed the specified or guaranteed value.

3.6.6.1.2 Nature of the Quantity Being Measured

Transformer no-load loss, often called core loss or iron loss, is the power loss in a transformer excited at rated voltage and frequency but not supplying load. The no-load loss comprises three components:

1. Core loss in the core material
2. Dielectric loss in the insulation system
3. I^2R loss due to excitation current in the energized winding

The no-load loss of a transformer is primarily caused by losses in the core steel (item 1, above). The remaining two sources are sometimes ignored. As a result, the terms *no-load loss, core loss,* and *iron loss* are often used interchangeably. *Core loss* and *iron loss,* strictly speaking, refer only to the power loss that appears within the core material. The following discussion on no-load loss, or core loss, will explain why the average-voltage voltmeter method, to be described later, is recommended. The magnitude of no-load loss is a function of the magnitude, frequency, and waveform of the impressed voltage. These variables affect the magnitude and shape of the core magnetic flux waveform and hence affect the value of the core loss. It has been verified through measurements on power and distribution transformers that core loss also depends, to some extent, upon the temperature of the core. According to the IEEE test code [2], the approximate rate of change of no-load loss with core temperature is 0.00065 per-unit core loss increase for each °C reduction in core temperature. The two main components of the core loss are hysteresis loss and eddy-current loss. The change in eddy-current loss, due to a change in the resistivity of the core steel as temperature changes, appears to be one factor that contributes to the observed core-loss temperature effect. The hysteresis loss magnitude is a function of the peak flux density in the core-flux waveform. When the impressed voltage waveform is distorted (not a pure sine wave), the resulting peak flux density in the flux waveform depends on the average absolute value of the impressed voltage wave. Eddy-current loss is a function of the frequency of the power source and the thickness of the core-steel laminations. Eddy loss is strongly influenced by harmonics in the impressed voltage. The IEEE transformer test code [2] recommends the average-voltage voltmeter method, to be described below, for measuring no-load loss.

3.6.6.1.3 How No-Load Loss Is Measured

The measurement of no-load loss, according to the average-voltage voltmeter method, is illustrated in Figure 3.6.14. Voltage and current transformers are required to scale the inputs for voltmeters, ammeters, and wattmeters. Three-phase no-load-loss measurements are carried out the same way, except that three sets of instruments and instrument transformers are utilized. The test involves raising voltage on one winding, usually the low-voltage winding, to its rated voltage while the other windings are in open circuit. Two voltmeters connected in parallel are employed. The voltmeter labeled V_a in Figure 3.6.14 represents an average-responding, rms-calibrated voltmeter. The voltmeter labeled V_r represents a true rms-responding voltmeter. Harmonics in the impressed voltage will cause the rms value of the waveform to be different from the average-absolute (rms-scaled) value, and the two voltmeter readings will differ. When the voltage reading, as measured by the average-responding voltmeter, reaches a value corresponding to the rated

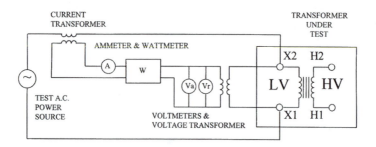

FIGURE 3.6.14 Test circuit for no-load-loss measurement.

voltage of the excited winding, readings are taken of the rms current, the rms voltage, and the no-load power. The ratio of the measured rms current to the rated load current of the excited winding, expressed in percent, is commonly referred to as the percent excitation current. The measured no-load loss is corrected to a sine-wave basis by a formula given in the IEEE test code [2], using the readings of the two voltmeters. The correction is shown below. The corrected value is reported as the no-load loss of the transformer.

$$P_c = \frac{P_m}{P1 + \left(\dfrac{V_r}{V_a}\right)^2 P2} \tag{3.6.1}$$

where

P_c is the corrected (reported) value of no-load loss
P_m is the measured value of no-load loss
V_a is the reading of the average-responding, rms-calibrated voltmeter
V_r is the reading of the true-rms-responding voltmeter
$P1$ and $P2$ are the per-unit hysteresis and per-unit eddy-current losses, respectively

According to the IEEE test code [2], if the actual values of $P1$ and $P2$ are not available, it is suggested that the two loss components be assumed equal in value, assigning each a value of 0.5 p.u.

3.6.6.2 Load Loss and Impedance Measurements

3.6.6.2.1 *Purpose of Load Loss Measurements*
A transformer dissipates a load loss that depends upon the transformer load current. Load loss is a cost to the user during the lifetime of the transformer. Maximum values of the load loss of transformers at rated current are specified and often guaranteed by the manufacturer. Load-loss measurements are made to verify that the load loss does not exceed the specified or guaranteed value.

3.6.6.2.2 *Nature of the Quantity Being Measured*
The magnitude of the load loss is a function of the transformer load current. Its magnitude is zero when there is no load on the transformer. Load loss is always given for a specified transformer load, usually at rated load. Transformer load loss, often called copper loss or winding loss, includes I²R losses due to load current in the winding conductors and stray losses in various metallic transformer parts due to eddy currents induced by leakage fields. Stray losses are produced in the winding conductors, in core clamps, in metallic structural parts, in magnetic shields, and in tank walls due to the presence of leakage fields. Stray losses also include power loss due to circulating currents in parallel windings and in parallel conductors within windings.

Because winding resistance varies with conductor temperature, and because the resistivities of the structural parts producing stray losses vary with temperature, the transformer load losses are a function of temperature. For this reason, a standard reference temperature (usually 85°C) for reporting the load loss is established in ANSI and IEEE standards [1]. To correct the load-loss measurements from the temperature at which they are measured to the standard reference temperature, a correction formula is provided in the IEEE test code [2]. This correction involves the calculation of winding I²R losses, where I is the rated current of the winding in amperes, and R is the measured dc resistance of the winding. The I²R losses and the stray losses are separately corrected and combined in the formula given in the standard. The measurement of the dc winding resistance, R, is covered in Section 3.6.6.3. Stray losses are determined by subtracting the I²R losses from the measured total load losses. All of this is covered in detail in the IEEE test code [2]. The formula for this conversion is stated in general in Equation 3.6.2:

$$W_{LL} = \frac{(I_p^2 R_p + I_s^2 R_s)(T_k + T_r)}{(T_k + T_m)} + \frac{(W_m - I_p^2 R_p - I_s^2 R_s)(T_k + T_m)}{(T_k + T_r)} \tag{3.6.2}$$

where

T_r is the reference temperature (°C)

T_m is the temperature at which the load loss is measured (°C)

W_m is the load loss (W) as measured at temperature T_m (°C)

W_{LL} is the load loss (W) corrected to a reference temperature T_r (°C)

T_k is a winding material constant: 234.5 for copper, 225 for aluminum

R_p and R_s are, respectively, the primary and secondary resistances (Ω)

I_p and I_s are the primary and secondary rated currents (A)

It can be seen from Equation 3.6.2 that the measurement of load losses and correction to the reference temperature involves the measurement of five separate quantities:

1. Electric power (the load loss as measured)
2. Temperature (the temperature at time of test)
3. Resistance of the primary winding (at a known temperature)
4. Resistance of the secondary winding (at a known temperature)
5. Electric current (needed to adjust the current to the required values)

Another characteristic of transformer load loss is its low power factor. Most often, when considering electric-power measurement applications, the power factor of the load being measured is relatively high, usually exceeding 80%. For large modern power transformers, the power factor can be very low, in the range from 1 to 5%. The measurement of electric power at very low power factor requires special consideration, as discussed in the next section.

3.6.6.2.3 How Load Loss Is Measured

Load losses are normally measured by connecting one winding, usually the low-voltage winding, to a short circuit with adequately sized shorting bars and connecting the other winding, usually the high-voltage winding, to a power-frequency voltage source. The source voltage is adjusted until the impressed voltage causes rated current to flow in both windings. Input rms voltage, rms current, and electric power are then measured. Figure 3.6.15 shows a circuit commonly used for measurement of load losses of a single-phase transformer. Three-phase measurement is carried out in the same way but with three sets of instruments and instrument transformers. Precision scaling devices are usually required because of the high magnitudes of current, voltage, and power involved.

The applied test voltage when the transformer is connected as in Figure 3.6.15, with rated currents in the windings, is equal to the impedance voltage of the transformer. Hence impedance, or impedance voltage, is also measured during the load-loss test. The ratio of the measured voltage to the rated voltage of the winding, multiplied by 100, is the percent impedance voltage of the transformer. This quantity is commonly called "percent impedance" or, simply, "impedance."

3.6.6.2.4 Discussion of the Measurement Process

The equivalent circuit of the transformer being tested in the load-loss test and the phasor diagram of test voltage and current during the test are shown in Figure 3.6.16. The load-loss power factor for the load-loss test is $\cos \theta = E_R / E_Z$. Because the transformer leakage impedance, consisting of R and X in Figure 3.6.16, is mainly reactive, and more so the larger the transformer, the power factor during the load-loss test is very low. In addition, there is a trend in modern transformers to create designs with lower losses due to increased demands for improved efficiency and transformer load-loss evaluations for optimal life-cycle costs. Transformer designs for low values of load loss lead to reductions in the equivalent resistance shown in Figure 3.6.16, and hence to low values for the quantity, E_R, which translates to lower values for the load-loss power factor in modern transformers. In fact the load-loss power factors for large modern transformers are often very low, in the range from 0.01 to 0.05 per unit. Under circuit conditions with very low power factor, the accurate measurement of electric power requires special scaling devices having very low phase-angle error and power measuring instruments having high accuracy at low power factor. In addition to the IEEE test code [2], a Transformer Loss Measurement Guide, C57.123, has

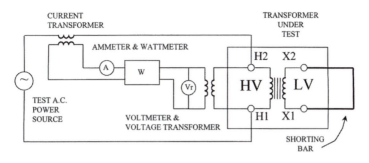

FIGURE 3.6.15 Test circuit for load-loss measurement.

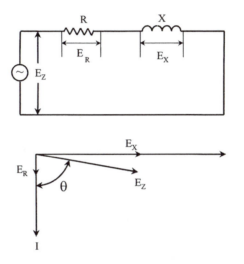

FIGURE 3.6.16 Equivalent circuit and phasor diagram for load-loss test.

recently been developed by a working group of the IEEE PES Transformers Committee. When generally available, it will provide a complete background and the basis for carrying out measurements and calibrations to ensure accurate values of reported losses as required by ANSI and IEEE standards [1].

Because the magnitudes of load losses and impedance depend upon the tap positions of the de-energized tap changer (DETC) and the load tap changer (LTC), if present, load-loss and impedance-voltage measurements are usually carried out in the rated voltage connection and at the tap extremes. If the transformer under test has multiple MVA ratings that depend on the type of cooling, the tests are normally carried out at all ratings.

3.6.6.3 Winding Resistance Measurements

3.6.6.3.1 *Purpose of Winding Resistance Measurements*
Measurements of dc winding resistance are of fundamental importance because they form the basis for determining the following:

Resistance measurements, taken at known temperatures, are used in the calculation of winding conductor I^2R losses. The I^2R losses at known temperatures are used to correct the measured load losses to a standard reference temperature. Correction of load losses is discussed in section 3.6.6.2.

Resistance measurements, taken at known temperatures, provide the basis to determine the temperature of the same winding at a later time by measuring the resistance again. From the change in resistance, the change in temperature can be deduced. This measurement is employed to determine

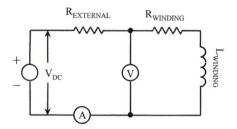

FIGURE 3.6.17 Circuit for measuring winding resistance.

average winding temperatures at the end of heat run tests. Taking resistance measurements after a heat run test is discussed in section 3.6.6.4.

Resistance measurements across the transformer terminals provide an assessment of the quality of internal connections made to the transformer windings. Loose or defective connections are indicated by unusually high or unstable resistance readings.

3.6.6.3.2 *Nature of the Quantity Being Measured*

The dc winding resistance differs from the value of resistance indicated for the resistor shown in Figure 3.6.16 or the resistors that appear in textbook illustrations of the *PI* or *T* equivalent circuits of transformers to represent the resistance of the windings. The resistors in the equivalent circuits include the effects of winding I^2R loss, eddy loss in the windings, stray losses in structural parts, and circulating currents in parallel conductors — namely, they represent the resistive components of the load loss. The resistors shown in the equivalent circuits can be thought of as representing an equivalent ac resistance of the windings. The dc resistance of the windings is a different quantity, one that is relevant for calculating I^2R, for determining average winding temperature, and for evaluating electrical connections.

3.6.6.3.3 *How Winding-Resistance Measurements Are Made*

The measurement of power transformer winding resistance is normally done using the voltmeter-ammeter method or using a ratiometric method to display the voltage–current ratio directly. A circuit for the measurement of winding resistance is shown in Figure 3.6.17. A dc source is used to establish the flow of steady direct current in the transformer winding to be measured. After the R-L transient has subsided, simultaneous readings are taken of the voltage across the winding and the current through the winding. The resistance of the winding is determined from these readings based on Ohm's law.

3.6.6.3.4 *Discussion of the Measurement Process*

If a dc voltage is applied as a step to a series R-L circuit, the current will rise exponentially with a time constant of *L/R*. This is familiar for the case where both resistance and inductance remain constant during the transient period. For a transformer winding, however, it is possible for the true resistance, the apparent resistance, and the inductance of the winding to change with time. The true resistance may change if the direct current is of high-enough magnitude and is applied long enough to heat the winding substantially, thereby changing its resistance during the measurement. The inductance changes with time because of the nonlinear B-H curve of the core steel and varies in accordance with the slope of the core-steel saturation curve. In addition, there is an apparent resistance, *Ra*, during the transient period.

$$Ra = \frac{V}{I} = R + \frac{L}{I}\frac{dI}{dt} \qquad (3.6.3)$$

Note that the apparent resistance, *Ra*, is higher than the true resistance, *R*, during the transient period and that the apparent resistance derived from the voltmeter and ammeter readings equals the true resistance only after the transient has subsided.

Resistance measurement error due to heating of the winding conductor is usually not a problem in testing transformers, but the possibility of this effect should be taken into consideration, especially for

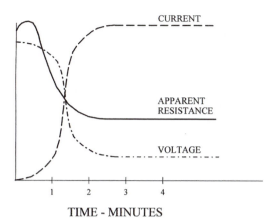

TIME - MINUTES

FIGURE 3.6.18 Current, voltage, and apparent resistance with time.

some low-current distribution transformer windings where the dc current can be significant compared with the rated current. It is more likely that errors will occur because of meter readings taken before core saturation is achieved. The process involved in core saturation is described below.

Compared with the exponential current-vs.-time relationship for the R-L circuit with constant R and constant L, the current in a transformer winding, when a dc voltage is first applied, rises slowly. The slow rate of rise comes about because of the high initial impedance of the winding. The initially high impedance results from the large effective inductance of the winding with its iron core. As the current slowly increases, the flux density in the core slowly rises until the core begins to saturate. At this point, the winding no longer behaves like an iron-core coil and instead behaves like an air-core coil, with relatively low inductance. The rate of rise of the current increases for a period as the core saturates; then the current levels off at a steady-state value. Typical shapes for the voltage, current, and apparent resistance are shown in Figure 3.6.18. The magnitude of the dc voltage affects the rate at which flux builds up in the core, since $V = N(d\Phi/dt)$. The higher the magnitude of the dc voltage, the shorter is the time to saturation because of a higher value for $d\Phi/dt$. At the same time, though, the coil must be able to provide the required magnetomotive force in ampere turns, $N \times I$, needed to force the core into saturation, which leads to a minimum value for the dc current. Of course, there is an upper limit to the value for dc current, namely the point at which conductor heating would disturb the resistance measurement.

Note the time scale of the graph in Figure 3.6.18. It is very important that the steady-state dc current be attained before meter readings are taken. If this is not done, errors in excess of 20% are easily realized.

3.6.6.3.5 Winding Resistance and Average Winding Temperature

Two of the three purposes listed above for measuring the dc resistance of a transformer winding inherently involve a concomitant measurement of temperature. When measuring resistance for the purpose of calculating I²R at a given temperature, the I²R value obtained will be used to determine the load-loss value at a different temperature. When the winding resistance is measured before and during a heat run, the determination of average winding temperature at the end of a heat run test requires knowledge of winding resistance at two temperatures.

The winding dc resistance at two temperatures, $T1$ and $T2$, will have values of $R1$ and $R2$, respectively, at the two temperatures. The functional relationship between winding resistance and average temperature is shown in Equation 3.6.4:

$$\frac{R1}{R2} = \frac{T1 + T_k}{T2 + T_k} \tag{3.6.4}$$

where

> $R1$ is the value of winding resistance, corresponding to average winding temperature of $T1$
> $R2$ is the value of winding resistance, corresponding to average winding temperature of $T2$
> T_k is 234.5°C for copper, 225°C for aluminum

Correction of load loss for temperature is covered in section 3.6.6.2. Determination of average winding temperature in a heat run test is covered in section 3.6.6.4.

3.6.6.4   Heat Run Tests

3.6.6.4.1   *Purpose of Heat Run Tests*

The maximum allowable average and hottest-spot temperature rises of the windings over ambient temperature and the maximum allowable temperature rise of the top oil of the transformer are specified by ANSI and IEEE standards and are guaranteed by the manufacturer. The purpose of temperature-rise tests is to demonstrate that the transformer will deliver rated load without exceeding the guaranteed values of the temperature rises of the windings and oil. According to the ANSI and IEEE standards, these tests are performed at the minimum and maximum load ratings of a transformer.

3.6.6.4.2   *Test Methods*

For factory testing, it is not practical to connect the transformer to a load impedance with full rated secondary voltage applied to the simulated load. Although this would most directly simulate service conditions, most of the total test input power would dissipate in the load impedance. The load power, which equals the load rating of the transformer, is much larger than the sum of no-load and load losses that dissipate in the transformer. The electrical heating of the load would not contribute to transformer heating. Electric power consumption for the test would be excessive, and the test would not be practical for routine testing.

In factory tests according to methods specified in the ANSI and IEEE standards, several artificial loading schemes can be used to simulate heat dissipation caused by the load and no-load losses of the transformer. The back-to-back loading method, described in the IEEE test code [2], requires two identical transformers and is often used for heat runs of distribution transformers. For power transformers, it is most common to use the short-circuit loading method, as specified in the IEEE test code [2]. The test setup is similar to that used for measurement of load loss and impedance voltage. One winding is connected to a short circuit, and sufficient voltage is applied to the other winding to result in currents in both windings that generate the required power loss to heat the oil and windings.

3.6.6.4.3   *Determination of the Top, Bottom, and Average Oil Rises*

This discussion applies to the short-circuit loading method. Initially, the test current is adjusted to provide an input power loss equal to the no-load loss plus the load loss. This can be called the total loss. The total loss is corrected to the guaranteed temperature rise plus ambient temperature. During this portion of the heat run, the windings provide a heat source for the oil and the oil cooling system. The winding temperatures will be higher than expected because higher-than-rated currents are applied to the windings, but here only the oil temperature rises are being determined. For an oil-filled power transformer, the total power loss for a given rating is maintained until the top-oil temperature rise is stabilized. Stabilization is defined as no more than 1°C change in three consecutive 1-h periods.

After stabilization is achieved, the top- and bottom-oil readings are used to determine the top, bottom, and average oil temperature rises over ambient at the specified load.

3.6.6.4.4   *Determination of the Average Winding-Temperature Rise*

After the top, bottom, and average oil temperature rises are determined, the currents in the windings are reduced to rated value for a period of 1 h. Immediately following this 1-h period, the ac power leads are disconnected, and resistance measurements are carried out on both windings. The total time from disconnection of power and the first resistance readings should be as short as possible, typically less than 2 min — certainly not more than 4 min. Repeated measurements of winding resistance are carried out for a period of 10 to 15 min as the windings cool down toward the surrounding average oil temperature. The average oil temperature is itself falling, but at a much slower rate (with a longer time constant). Data for the changing resistance vs. time is then plotted and extrapolated back to the instant of shutdown. With computers, the extrapolation can be done using a regression-based curve-fitting approach. The extrapolated value of winding resistance at the instant of shutdown is used to calculate winding temperature, using the method discussed below, with some correction for the drop in top-oil rise during the 1-h loading at rated current. The winding temperature, thus determined, minus ambient temperature is equal to the winding temperature rise at a given loading.

Similar tests are repeated for all ratings for which temperature rise tests are required.

3.6.6.4.5 *Determining the Average Temperature by Resistance*
The average winding temperature, as determined by the following method, is sometimes called the "average winding temperature by resistance." The word *measurement* is implied at the end of this phrase. The conversion of measured winding resistance to average winding temperature is accomplished as follows. Initial resistance measurements are made at some time before commencement of the heat run when the transformer is in thermal equilibrium. When in equilibrium, the assumption can be made that the temperature of the conductors is uniform and is equal to that of the transformer oil surrounding the coils. Initial resistance measurements are made and recorded, along with the oil temperature. This measurement is sometimes called the cold-resistance test, so the winding resistance measured during the test will be called the cold resistance R_c, and the temperature will be called the cold temperature T_c. At the end of the heat run, R_h, the hot resistance is determined from the time series of measured resistance values by extrapolation to the moment of shutdown. The formula given below is used to determine T_h, the hot temperature, knowing the hot resistance, cold resistance, and the cold temperature. This calculated temperature is the average winding temperature by resistance.

$$T_h = \frac{R_h}{R_c}[T_c + T_k] - T_k \tag{3.6.5}$$

where

      T_h is the "hot" temperature
      T_c is the "cold" temperature
      R_h is the "hot" resistance
      R_c is the "cold" resistance
      T_k is a material constant: 234.5°C for copper, 225°C for aluminum

Accurate measurements of R_c and T_c during the cold-resistance test, as well as accurate measurements of hot winding resistance, R_h, at the end of the heat run are extremely critical for accurate determination of average winding temperature rises by resistance. The reason for this will be evident by analyzing the above formula. The following discussion illustrates how measurement errors in the three measured quantities, R_h, R_c, and T_c, affect the computed quantity, T_h, via the functional relationship by which T_h is computed.

Shown in Table 3.6.2 are computed values for T_h and the resulting error, e_{Th}, in the computation of T_h for sample sets of the measured quantities R_h, R_c, and T_c, measured in error by the amounts e_{Rh}, e_{Rc}, and e_{Tc}, respectively. Let us examine this table row by row. The way that measurement error propagates in the calculation is shown in the formula below for copper conductors.

$$(T_h + e_{Th}) = \frac{(R_h + e_{Rh})}{(R_c + e_{Rc})}[(T_c + e_{Tc}) + 234.5] - 234.5 \tag{3.6.6}$$

TABLE 3.6.2 Effect of Measurement Error in Average Winding Temperature by Resistance

Row	$T_h + e_{Th}$ (°C)	$R_h + e_{Rh}$ (Ω)	$R_c + e_{Rc}$ (Ω)	$T_c + e_{Th}$ (°C)
1	87.375 + 0	0.030 + 0	0.024 + 0	23.0 + 0
2	87.375 − 3.187 = 84.188 (−3.6%)	0.030 + 0	0.024 + 0.00024 = 0.02424 (+1.0%)	23.0 + 0
3	87.375 + 3.218 = 90.594 (+3.7%)	0.030 + 0.0003 = 0.0303 (+1.0%)	0.024 + 0	23.0 + 0
4	87.375 + 1.25 = 88.625 (+1.43%)	0.030 + 0	0.024 + 0	23.0 + 1.0 = 24.0 (4.35 %)

In row 1 of Table 3.6.2, there is no measurement error. The sample shows a set of typical measured values. The value for T_h in this row can be considered the "correct answer." In row 2 the cold-resistance measurement was 1% too high, causing the calculated value of T_h to be 3.6% too low. This amplification of the relative measurement error is due to the functional relationship employed to perform the calculated result. This example illustrates the importance of measuring the resistance very carefully and accurately. Similarly, in row 3 a hot-resistance reading 1% too high results in a calculated hot temperature that is 3.7% too high. Row 4 shows the result if the cold-resistance reading is 1°C too high. The result is that the determined hot resistance is 1.25°C too high. In this case, while there is a reduction in the error expressed as percent, the absolute error in °C is in fact greater than the original temperature error in the cold-temperature reading. These examples show that all three measured quantities — R_h, R_c, and T_c — must be measured accurately to obtain an accurate determination of the average winding temperature.

Other methods for correction to the instant of shutdown based on W/kg, or W/kg and time, are given in the IEEE test code [2]. The cooling-curve method, however, is preferred.

3.6.7 Other Tests

3.6.7.1 Short-Circuit-Withstand Tests

3.6.7.1.1 Purpose of Short-Circuit Tests

Short-circuit currents during through-fault events expose the transformer to mechanical stresses caused by magnetic forces, with typical magnitudes expressed in thousands of kilograms. Heating of the conductors and adjacent insulation due to I^2R losses also occurs during a short-circuit fault. The maximum mechanical stress is primarily determined by the square of the peak instantaneous value of current. Hence, the short-circuit magnitude and degree of transient offset are specified in the test requirements. Fault duration and frequency of occurrence also affect mechanical performance. Therefore, the number of faults, sometimes called "shots," during a test and the duration of each fault are specified. Conductor and insulation heating is for the most part determined by the rms value of the fault current and the fault duration.

Short-circuit-withstand tests are intended primarily to demonstrate the mechanical-withstand capability of the transformer. Thermal capability is demonstrated by calculation using formulas provided by IEEE C57.12.00 [1].

3.6.7.1.2 Transformer Short-Circuit Categories

The test requirements and the pass-fail evaluation criteria for short-circuit tests depend upon transformer size and construction. For this purpose, transformers are separated into four categories as shown in Table 3.6.3. The IEEE standards [1] and [2] refer to these categories while discussing the test requirements and test results evaluation.

3.6.7.1.3 Configurations

A short circuit is applied using low-impedance connections across either the primary or the secondary terminals. A secondary fault is preferred, since it more directly represents the system conditions. The fault can be initiated in one of two ways:

TABLE 3.6.3 Transformer Short-Circuit Test Categories

Category	Single-Phase, kVA	Three-Phase, kVA
I	5 to 500	15 to 500
II	501 to 1,667	501 to 5,000
III	1,668 to 10,000	5,001 to 30,000
IV	above 10,000	above 30,000

1. Starting with an open circuit at the terminal to be faulted, energize the transformer and then apply the short circuit by closing a breaker across the terminal.
2. Starting with a short circuit across the terminal to be faulted, close a breaker at the source terminal to energize the prefaulted transformer.

The first fault-initiation method is the preferred method. Given in the order of preference as listed in C57.12.90 [2], the following fault types are permissible:

- Three-phase source, three-phase short circuit.
- Single-phase source, single phase-to-ground short circuit.
- Single-phase source, simulated three-phase short circuit.
- Single-phase source, single-phase short circuit, applied one phase at a time

The simulated three-phase short circuit, using single-phase source, is conducted as follows, depending on whether the connection is delta or wye. For wye-connected windings, the fault or source is applied between one terminal and the other two terminals connected together. For delta-connected windings, the fault or source is connected between two line terminals, with no connection to the other terminal, which must be repeated for each of the three phases.

3.6.7.1.4 *Short-Circuit Current Duration, Asymmetry, and Number of Tests*
The required short-circuit current magnitude and degree of asymmetry of the first cycle of current is given in IEEE C57.1200 [1]. Each phase of the transformer is subjected to a total of six tests that satisfy the symmetrical current requirements. At least two of the six must also satisfy the asymmetrical current tests. Five of the six tests must have a fault duration of 0.25 sec. One of the tests, which is sometimes called the "long duration test," must be of longer duration for Category I, II, and III transformers. The longer duration must be determined as follows:

For Category I transformers,

$$t = 1250/I^2$$

where

t = duration, sec
I = fault current, per unit of load current

For Category II transformers,
$t = 1.0$ sec

For Category III transformers,

$t = 0.5$ sec

Clause 7 of IEEE C57.12.00-1993 [1] and Clause 12 of IEEE C57.12.90-1993 [2] provide a detailed and excellent description of the requirements and procedures for short-circuit testing.

3.6.7.1.5 *Evaluation of Short-Circuit Test Results*
Clause 12.5 of IEEE C57.12.90-1993 [2], Proof of Satisfactory Performance, spells out the requirements for passing or failing the tests. These include:

- Visual inspection
- Dielectric tests following short-circuit tests
- Observations of terminal voltage and current waveforms — no abrupt changes
- Changes in leakage impedance are limited to values specified in the standards
- Low-voltage impulse tests may be performed
- Increases in excitation current are limited by the standards

Clause 12.5 of the test code discusses these criteria in detail.

3.6.7.2 Special Tests

Many additional tests are available to obtain certain information about the transformer, usually to address a specific application issue. These are listed below.

- Overload heat run
- Gas-in-oil sampling and analysis (in conjunction with other tests)
- Extended-duration no-load-loss tests
- Zero-sequence impedance measurements
- Tests of the load tap changer
- Short-circuit-withstand tests
- Fault-current capability of enclosures (overhead distribution transformers)
- Telephone line voice frequency electrical noise (overhead distribution transformers)
- Tests on controls

These tests and others, of a specific nature, are beyond the scope of this general article. Interested readers will find more information about these and other tests in the national and international standards dealing with transformers.

References

1. IEEE, Standard General Requirements for Liquid-Immersed Distribution, Power, and Regulating Transformers, IEEE Std. C57.12.00-1993, Institute of Electrical and Electronics Engineers, Piscataway, NJ, 1993.
2. IEEE, Standard Test Code for Liquid-Immersed Distribution, Power, and Regulating Transformers; and Guide for Short-Circuit Testing of Distribution and Power Transformers (ANSI), IEEE/ANSI Std. C57.12.90-1992, Institute of Electrical and Electronics Engineers, Piscataway, NJ, 1992.
3. IEEE, The New IEEE Standard Dictionary of Electrical and Electronics Terms, Std. 100-1992, Institute of Electrical and Electronics Engineers, Piscataway, NJ, 1992.
4. IEEE, Guide for Transformer Impulse Tests, IEEE Std. C57.98-1993, Institute of Electrical and Electronics Engineers, Piscataway, NJ, 1993.
5. IEEE, Guide for Partial Discharge Measurement in Liquid-Filled Power Transformers and Shunt Reactors, IEEE Std. C57.113-1991, Institute of Electrical and Electronics Engineers, Piscataway, NJ, 1991.

3.7 Load-Tap-Change Control and Transformer Paralleling

James H. Harlow

3.7.1 Introduction

Tap changing under load (TCUL), be it with load-tap-changing (LTC) power transformers or step-voltage regulators, is the primary means of dynamically regulating the voltage on utility power systems. Switched

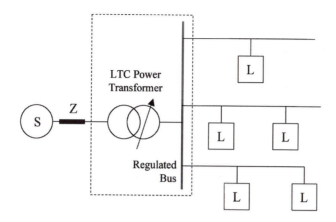

FIGURE 3.7.1A Three-phase bus regulation with three-phase LTC transformer.

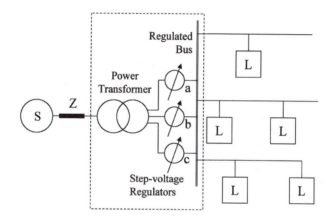

FIGURE 3.7.1B Independent-phase bus regulation with three single-phase step-voltage regulators.

shunt capacitors can also be used for this purpose, but in the context of this discussion, shunt capacitors are presumed to be applied with the objective of improving the system power factor.

The control of a tap changer is much more involved than simply responding to a voltage excursion at the transformer secondary. Modern digital versions of LTC control include so many ancillary functions and calculated parameters that it is often used with its communications capability to serve as the means for system-condition monitoring.

The control of the tap changer in a transformer or step-voltage regulator is essentially the same. Unless stated otherwise, the use of the term "transformer" in this section applies equally to step-voltage regulators. It should be recognized that either type of product can be constructed as single-phase or three-phase apparatus, but that transformers are more often three phase, while step-voltage regulators are more commonly single phase.

3.7.2   System Perspective, Single Transformer

This discussion is patterned to a typical utility distribution system, the substation and the feeder, although much of the material is also applicable to transmission applications. This first discussion of the control considers that the control operates only one LTC in isolation, i.e., there is not the opportunity for routine operation of transformers in parallel.

The system can be configured in any of several ways according to the preference of the user. Figure 3.7.1 depicts two common implementations. In the illustrations of Figure 3.7.1, the dashed-line box depicts a substation enclosing a transformer or step-voltage regulators. The implementations illustrated accomplish bus-voltage regulation on a three-phase (Figure 3.7.1a) or single-phase (Figure 3.7.1b) basis. Another common application is to use voltage regulators on the distribution feeders. A principal argument for the use of single-phase regulators is that the voltages of the three phases are controlled independently, whereas a three-phase transformer or regulator controls the voltage of all phases based on knowledge of the voltage of only one phase. For Figure 3.7.1:

> S = source, the utility network at transmission or subtransmission voltage, usually 69 kV or greater
> Z = impedance of the source as "seen" looking back from the substation
> L = loads, distributed on the feeders, most often at 15 kV to 34.5 kV

The dominance of load-tap-changing apparatus involves either 33 voltage steps of 5/8% voltage change per step, or 17 voltage steps of $1^1/_4$% voltage change per step. In either case, the range of voltage regulation is ±10%.

3.7.3   Control Inputs

3.7.3.1   Voltage Input

The voltage of the primary system is unimportant to the LTC control. The system will always include a voltage transformer (VT) or other means to drop the system voltage to a nominal 120 V for use by the control. Because of this, the control is calibrated in terms of 120 V, and it is common to speak of the voltage being, say, 118 V, or 124 V, it being understood that the true system voltage is the value stated times the VT ratio. Presuming a single tap step change represents 5/8% voltage, it is easily seen that a single tap step change will result in 0.00625 per unit (p.u.) × 120 V = 0.75-V change at the control.

The control receives only a single 120-V signal from the voltage transformer, tracking the line-ground voltage of one phase or a line-line voltage of two phases. For the case of three-phase apparatus, the user must take great care, as later described, in selecting the phase(s) for connecting the VT.

3.7.3.2   Current Input

The current transformers (CT) are provided by the transformer manufacturer to deliver control current of "not less than 0.15 A and not more than 0.20 A … when the transformer is operating at the maximum continuous current for which it is designed." (This per ANSI Std. C57.12.10, where the nominal current is 0.2 A. Other systems may be based on a different nominal current, such as 5.0 A.) As with the voltage, the control receives only one current signal, but it can be that of one phase or the cross connection (the paralleling of two CTs) of two phases.

3.7.3.3   Phasing of Voltage and Current Inputs

In order for the control to perform all of its functions properly, it is essential that the voltage and current input signals be in phase for a unity power factor load or, if not, that appropriate recognition and corrective action be made for the expected phasing error.

Figure 3.7.2 depicts the three possible CT and VT orientations for three-phase apparatus. Note that for each of the schemes, the instrument transformers could be consistently shifted to different phases from that illustrated without changing the objective. The first scheme, involving only a line-to-ground-rated VT and a single CT, is clearly the simplest and least expensive. However, it causes all control action to be taken solely on the basis of knowledge of conditions of one phase. Some may prefer the second scheme, as it gives reference to all three phases, one for voltage and the two not used for voltage to current. The third scheme is often found with a delta-connected transformer secondary. Note that the current signal derived in this case is √3 times the magnitude of the individual CT secondary currents. This must be scaled before the signal is delivered to the control.

The user must assure the proper placement of the VT(s) to be consistent with the CT(s) provided inside the transformer. This is not a concern with step-voltage regulators, where all instrument transformers are provided (per ANSI/IEEE C57.15) internal to the product.

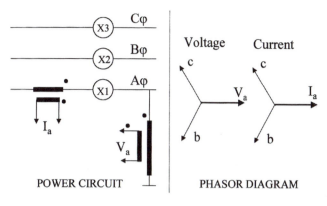

VT (L-G), 1 CT on same phase

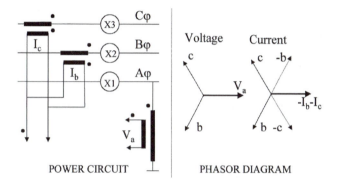

VT (L-G), 2 CTs on other phases

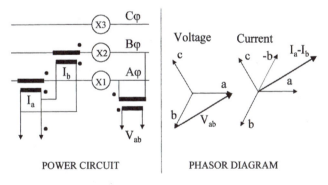

VT (L-L), 2 CTs on same phases

FIGURE 3.7.2 Phasing of voltage and current inputs.

3.7.4 The Need for Voltage Regulation

Referring to the circuits of Figure 3.7.1, note that system conditions will change over time, with the result that the voltage at the substation bus, and as delivered to the load, will change. From Figure 3.7.1, the source voltage, the source impedance, and the load conditions will be expected to change with time. The most notable of these, the load, must be recognized to consist of two factors — the magnitude and the power factor, or what is the same point, the real (watt) and reactive (var) components.

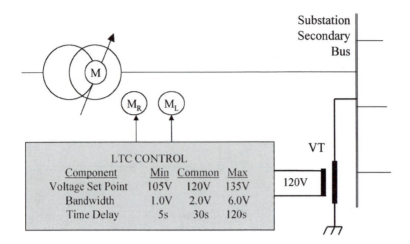

FIGURE 3.7.3 Control for voltage regulation of the bus.

3.7.4.1 Regulation of the Voltage at the Bus

Many times, the object of the LTC is simply to hold the substation bus voltage at the desired level. If this is the sole objective, it is sufficient to bring only a VT signal that is representative of the bus voltage into the control. The secondary of the VT is usually 120 V at the nominal bus voltage, but other VT secondary voltages, especially 125 V, 115 V and 110 V, are used. Figure 3.7.3 shows the circuit. The figure shows a motor (M) on the LTC, which is driven in either the raise or lower direction by the appropriate output (M_R or M_L) of the control. This first control, Figure 3.7.3, is provided with only three settings, these being those required for the objective of regulating the substation secondary-bus voltage:

3.7.4.1.1 Voltage Set Point

The voltage set point on the control voltage base, e.g., 120 V, is the voltage desired to be held *at the load*. (The load location for this first case is the substation secondary bus because line-drop compensation is not yet considered and is therefore zero.) This characteristic is also commonly referred to as "voltage band center," this being illustrative of the point that there is a band of acceptable voltage and that this is the midpoint of that band. If line-drop compensation (LDC, as detailed later) is not used, the set point will often be somewhat higher than 120 V, perhaps 123 V to 125 V; with use of LDC the setting will be lower, perhaps 118 V.

3.7.4.1.2 Bandwidth

The bandwidth describes the voltage range, or band, that is considered acceptable, i.e., in which there is not a need for any LTC corrective action. The bandwidth is defined in the ANSI/IEEE C57.15 standard as a voltage with one-half of the value above and one-half below the voltage set point. Some other controls adjust the bandwidth as a percentage of the voltage set point, and the value represents the band on each side of the band center. The bandwidth voltage selected is basically determined by the LTC voltage change per step. Consider a transformer where the voltage change per step is nominally 0.75 V (5/8% of 120 V). Often this is only the average; the actual voltage change per step can differ appreciably at different steps. Clearly, the bandwidth must be somewhat greater than the maximum step-change voltage, because if the bandwidth were less than the voltage change per step, the voltage could pass fully across the band with a single step, causing a severe hunting condition. The minimum suggested bandwidth setting is twice the nominal step-change voltage, plus 0.5 V, for a 2.0-V minimum setting for the most common 5/8% systems. Many users choose somewhat higher bandwidths when the voltage is not critical and there is a desire to reduce the number of daily tap changes.

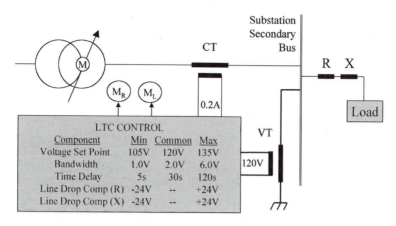

FIGURE 3.7.4 Control for voltage regulation at the load.

3.7.4.1.3 Time Delay

All LTC controls incorporate an intentional time delay from the time the voltage is "out of band," until a command is given for tap-changer action. Were it not for the control delay, the LTC would respond to short-lived system voltage sags and swells, causing many unwanted and unnecessary tap changes. Most applications use a linear time-delay characteristic, usually set in the range of 30 to 60 sec, although controls with inverse time-delay characteristics are also available, where the delay is related inversely to the voltage digression from the set point

3.7.4.2 Regulation of the Voltage at the Load

It is recognized that if it were easy to do so, the preferred objective would be to regulate the voltage at the load rather than at the substation bus. The difficulty is that the voltage at the load is not commonly measured and communicated to the control; therefore it must be calculated in the control using system parameters calculated by the user. Basically, the calculation involves determining the line impedance $(R + jX_L)$ between the substation and the load, the location of which is itself usually nebulous.

The procedure used is that of line-drop compensation (LDC), i.e., the boosting of the voltage at the substation to compensate for the voltage drop on the line. The validity of the method is subject to much debate because of (1) the uncertainty of where to consider the load to be when it is in fact distributed, and (2) the inaccuracies encountered in determining the feeder line resistance and reactance.

The principle upon which LDC is based is that there is one concentrated load located a sufficient distance from the LTC transformer for the voltage drop in the line to be meaningful. Consider Figure 3.7.4, which is similar to Figure 3.7.3 with the addition of a load located remote from the substation and a current signal input to the control. The distance from the substation to the load must be defined in terms of the electrical distance, the resistance (R), and inductive reactance (X) of the feeder. The means of determining the line R and X values is available from numerous sources (Beckwith Electric Co., 1995; Harlow Engineering Associates, 2001).

The LDC resistance and reactance settings are expressed as a value of volts on the 120-V base. These voltage values are the voltage drop on the line (R = in-phase component; X = quadrature component) when the line-current magnitude is the CT primary rated current.

The manner is which the control accounts for the line-voltage drop is illustrated with phasor diagrams. Figure 3.7.5 consists of three illustrations showing the applicable phasor diagram as it changes by virtue of the load magnitude and power factor. In the illustrations, the voltage desired at the load, V_{Load} is the reference phasor; its magnitude does not change. All of the other phasors change when the load current changes in magnitude or phase angle. For all of the diagrams:

IR = voltage drop on the line due to line resistance; in-phase with the current
IX = voltage drop on the line due to line inductive reactance; leads the current by 90°
IZ = total line-voltage drop, the phasor sum of IR and IX

In the first illustration (Figure 3.7.5a), the power-factor angle, φ, is about 45°, lagging, for an illustration of an exaggerated low power factor of about 0.7. It is seen that the voltage at the bus needs to be boosted to the value V_{Bus} to overcome the IR and IX voltage drops on the line. The second illustration (Figure 3.7.5b) shows simply that if the line current doubles with no change of phase angle, the IR and IX phasors also double, and a commensurately greater boost of V_{Bus} is required to hold the V_{Load} constant. The third illustration (Figure 3.7.5c) considers that the line-current magnitude is the same as the first case, but the angle is now leading. The IZ phasor simply pivots to reflect the new phase angle. It is interesting to note that in this case, the V_{Load} magnitude exceeds V_{Bus}. This is modeling the real system. Too much shunt capacitance on the feeder (excessive leading power factor) results in a voltage rise along the feeder. The message for the user is that LDC accurately models the line drop, both in magnitude and phase.

No "typical" set-point values for LDC can be given, unless it is zero, as the values are so specific to the application. Perhaps due to the difficulty in calculating reasonable values, line-drop compensation is not used in many applications. An alternative to line-drop compensation, Z-compensation, is sometimes preferred for its simplicity and essential duplication of LDC. To use Z-comp, the control is programmed to simply raise the output voltage, as a linear function of the load current, to some maximum voltage boost. This method is not concerned with the location of the load, but it also does not compensate for changes in the power factor of the load.

3.7.5 LTC Control with Power-Factor-Correction Capacitors

Many utility distribution systems include shunt capacitors to improve the load power factor as seen from the substation and to reduce the overall losses by minimizing the need for volt-amperes reactive (vars) from the utility source. This practice is often implemented both with capacitors that are fixed and others that are switched in response to some user-selected criterion. Further, the capacitors can be located at the substation, at the load, or at any intermediate point.

The capacitors located at the load and source are illustrated in Figure 3.7.6. The position of the capacitors is important to the LTC control when LDC is used.

With the capacitors located at the load, it is presumed that the var requirement of the load is matched exactly by the capacitor, Figure 3.7.6a, and that the transformer CT current is exactly the current of the line. Here LDC is correctly calculated in the control because the control LDC circuit accurately represents the current causing the line voltage drop.

In the second illustration, Figure 3.7.6b, the capacitor supplying the var requirement of the load is located at the substation bus. There is an additional voltage drop in the line due to the reactive current in the line. This reactive current is not measured by the LDC CT. In this case, the line drop voltage is not correctly calculated by the control. To be accurate for this case, it is necessary to determine the voltage drop on the feeder due to the capacitor-produced portion of the load current and then add that voltage to the control set-point voltage to account for the drop not recognized by the LDC circuit.

The matter is further confused when the capacitor bank is switched. With the capacitor bank in the substation, it is possible devise a control change based on the presence of the bank. No realistic procedure is recognized for the case where the bank is not in close physical proximity to the control, keeping in mind that the need is lessened to zero as the capacitor location approaches the load location.

3.7.6 Extended Control of LTC Transformers and Step-Voltage Regulators

A LTC control often includes much more functionality than is afforded by the five basic set points. Much of the additional functionality can be provided for analog controls with supplemental hardware packages, or it can be provided as standard equipment with the newer digital controls. Some functionality, most notably serial communications, is available only with digital controls.

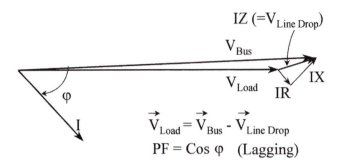

FIGURE 3.7.5A   Load phasor diagram, normal load, lagging power factor.

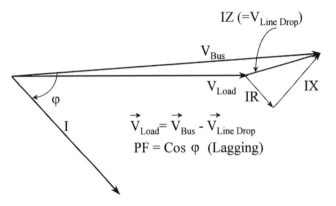

FIGURE 3.7.5B   Load phasor diagram, heavy load, lagging power factor.

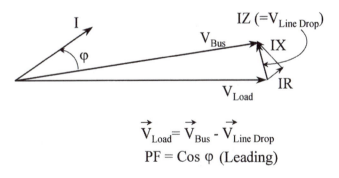

FIGURE 3.7.5C   Load phasor diagram, normal load, leading power factor.

3.7.6.1   Voltage Limit Control

Perhaps the most requested supplemental LTC control function is voltage limit control, also commonly referred to as "first house (or first customer) protection." This feature can be important with the use of line-drop compensation where excessive line current could result in the voltage at the substation bus becoming excessive. System realities are such that this means that the voltage at the "first house," or the load immediately outside the substation, is also exposed to this high-voltage condition.

Review again Figure 3.7.4, where LDC is used and it is desired to hold 118 V at the load. The voltage at the source (the secondary substation bus) is boosted as the load increases in order to hold the load at 118 V. If the load continues to grow, the voltage at the source rises accordingly. At some point, the load could increase to the point where the first customer is receiving power at excessive voltage. Recognize

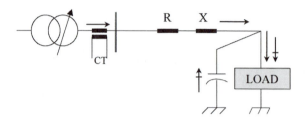

FIGURE 3.7.6A Feeder with power-factor-correction capacitors, capacitors at the load.

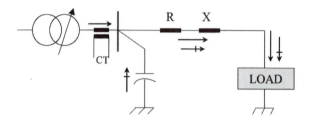

FIGURE 3.7.6B Feeder with power-factor-correction capacitors, capacitors at the source bus.

that the control and the load tap changer are performing properly; the bus voltage is too high because of the unanticipated excessive load on the system.

The voltage limit control functions only in response to the actual local bus voltage, opening the "raise" circuit to the drive motor at a user-selected voltage, thereby defeating the basic control "raise" signal when the bus voltage becomes excessive. In this way, the LTC action will not be responsible for first-customer overvoltage.

These controls provide a degree of protection for the case where the bus voltage further exceeds the selected voltage set-point cutoff voltage, as might occur with a sudden loss of load or other system condition unrelated to the LTC. If the voltage exceeds a second value, usually about 2 V higher than the selected voltage set-point cutoff voltage, the voltage limit control will, of itself, command a "lower" function to the LTC.

Most of these controls provide a third capability, that of undervoltage LTC blocking so that the LTC will not run the "lower" function if the voltage is already below a set-point value. This function mirrors the "raise" block function described above.

The voltage-limit-control functionality described is built into all of the digital controls. This is good in that it is conveniently available to the user, but caution needs to be made regarding its expected benefit. The function can not be called a backup control unless it is provided as a physically separate control. Consider that the bus voltage is rising, not correctly because of LDC action, but because of a failure of the LTC control. The failure mode of the control may cause the LTC to run high and the bus voltage to rise accordingly. This high voltage will not be stopped if the voltage-limit-control function used is that which is integral to the defective digital control. Only a supplemental device will stop this and thereby qualify as a backup control.

3.7.6.2 Voltage Reduction Control

Numerous studies have reported that, for the short term, a load reduction essentially proportional to a voltage reduction can be a useful tool to reduce load and conserve generation during critical periods of supply shortage. It is very logical that this can be implemented using the LTC control. Many utilities prepare for this with the voltage-reduction capability of the control.

With analog controls, voltage reduction is usually implemented using a "fooler" transformer at the sensing-voltage input of the control. This transformer could be switched into the circuit using SCADA (supervisory control and data acquisition) to boost the voltage at the control input by a given percentage,

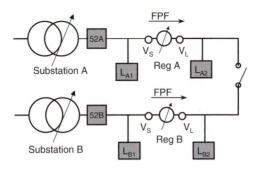

FIGURE 3.7.7A Step-voltage regulator in routine application.

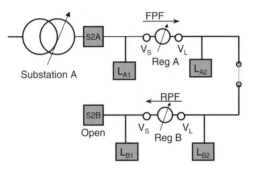

FIGURE 3.7.7B Step-voltage regulator in reverse power-flow application.

or often up to three different percentages, using different taps on the fooler transformer. Having the sensed voltage boosted by, say, 5% without changing the voltage set point of the control will cause the control to run the tap position down by 5% voltage, thus accomplishing the desired voltage reduction. The percentage of three reduction steps, most commonly 2.5, 5.0, and 7.5%, is preestablished by the design of the fooler transformer.

Digital controls do the same function much more conveniently, more accurately, and more quickly. The voltage reduction applicable to steps 1, 2, and 3 are individually programmed and, upon implementation, effectively lower the voltage set point. Control based on the new set point is implemented without intentional time delay, reverting to panel time delay after the voltage reduction has been implemented. Most often, controls provide for three steps of voltage reduction, each step individually programmed up to 10.0%.

3.7.6.3 Reverse Power Flow

Voltage regulators as used on distribution feeders are sometimes subjected to reverse power flow due to system switching. A common practice is to install the distribution system per the illustration of Figure 3.7.7. Normally, step-voltage regulators A and B both usually "see" forward power flow, as in Figure 3.7.7a. Line switching, such as might be necessary due to service required in Substation B, causes Feeder B to be served from Substation A and the Regulator B to operate in reverse power flow mode, as in Figure 3.7.7b. Were special precautions not taken, Regulator B would operate incorrectly, inevitably running to a "raise" or "lower" tap limit. This would have occurred because the regulator would continue to attempt to operate based on knowledge of the voltage at its load terminal, but now that terminal is, in fact, the source. Tap-changer operation does nothing to correct that voltage, but it severely disturbs the voltage to loads L_{B1} between Regulator B and Substation B.

The remedy is to operate Regulator B based on the voltage at its usual source (now the load) terminal. Analog controls have been provided to automatically switch between the load and source terminals upon recognition of reverse power flow, but the expense of doing so, involving in most cases the addition of a new VT, seldom justified the avoidance of simply manually switching off automatic operation during the planned power-reversal event. Digital controls accomplish this feature automatically with no additional hardware. These controls calculate the voltage at the normal source terminal using knowledge of the measured load terminal voltage, the recognized tap position plus, in some cases, an approximation of the internal regulation of the regulator. Accomplishing voltage regulation with Regulator B during reverse power-flow operation in this manner adds only about 0.5% to the error of the control vis a vis having a supplemental VT for the application.

The apparatus and procedures defined above for step-voltage regulators are not correct for most transformer applications where reverse power flow can occur. The basic difference is that the feeder-regulator application remains a radial system after the line switching is complete. Reverse power in transformers is more likely to occur on a system where the reverse power is due to a remote generator that is operating continuously in parallel with the utility. The proper operation of the LTC in this case must be evaluated for the system. Perhaps the preferred operation would be to control the LTC so as to minimize the var exchange between the systems. Some systems are simply operated with the LTC control turned off of automatic operation during reverse power flow so that the LTC says fixed on position until forward power flow resumes.

3.7.7 Introduction to Control for Parallel Operation of LTC Transformers and Step-Voltage Regulators

For a variety of reasons, it may be desirable to operate LTC transformers or regulators in parallel with each other. This may be done simply to add additional load-handling capability to an existing overloaded transformer, or it may be by initial design to afford additional system reliability, anticipating that there may be a failure of one transformer.

Most common paralleling schemes have the end objective of having the load tap changers operate on the same, or on nearly the same, tap position at all times. For more-complex schemes, this may not be the objective. A knowledge of the system is required to assess the merits of the various techniques.

3.7.7.1 The Need for Special Control Considerations

To understand why special control consideration needs to be given to paralleling, consider two LTC transformers operating in parallel, i.e., the primary and secondary of the transformers are bused together as in Figure 3.7.8. If the transformers are identical, they will evenly divide the load between them at all times while they are operating on the same tap position. Consider that the voltage at the secondary bus goes out of band. Even if the controls are set the same, they are never truly identical, and one will command tap-change operation before the other. Later, when the voltage again changes so as to require a second voltage correction, the same control that operated more quickly the first time would be expected to do so again. This can continue indefinitely, with one LTC doing all of the operation. As the tap positions of the LTC transformers digress, the current that circulates in the substation increases. This current simply circulates around the loop formed by the buses and the two transformers doing no useful work, but it causes an increase in losses, perhaps causing one or both of the transformers to overheat. To put the matter in perspective, consider the case of a distribution substation where there are two 5/8%-step, 15-MVA LTC transformers of 8.7% impedance in parallel. The secondary voltage is 12.47 kV. For a tap difference of only one step, a circulating current of 25 A flows in the substation LV bus. This current magnitude increases linearly with tap-position difference.

In the illustration above for LTC transformers, the circulating current was limited by the impedances of the transformers. It is very important to recognize that the same procedure cannot be done with step-voltage regulators, as the impedance of a regulator is very low at even the extreme tap positions, and it may be essentially zero at the neutral tap position. In this case, if one regulator is on neutral and the other moves to position 1R, the circulating current will be expected to be sufficiently large to cause catastrophic failure of the regulators. The user is cautioned: Step-voltage regulators can only be operated in parallel when there is adequate supplemental impedance included in the current loop. This is most

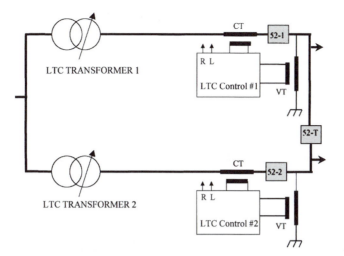

FIGURE 3.7.8 LTC transformers in parallel with no interconnections.

often the impedance of the transformers that are in series with each regulator bank, or it could be a series current-limiting reactor.

3.7.7.2 Instrument Transformer Considerations

Refer again to Figure 3.7.2. Most techniques for paralleling use the voltage and current signals derived for line-drop compensation. It is essential that the paralleled transformers deliver voltage and current signals that are in phase with each other when the system has no circulating current. This means that the instrument transformers must deliver signals from equivalent phases. Further, the ratios of the instrument transformers must be the same, except under very special conditions that dictate otherwise.

3.7.8 Defined Paralleling Procedures

Numerous procedures have been identified over the years to accomplish LTC transformer paralleling using electronic control. These are listed with some limited description along with alternative names sometimes used:

1. Negative reactance (reverse reactance): Seldom used today except in some network applications, this is one of the oldest procedures accomplished by other than mechanical means. This means of paralleling is the reason LTC controls are required by the standard to provide negative X capability on the line-drop compensation.

   Advantages: Simplicity of installation. The system requires no apparatus other than the basic control, with LDC X set as a negative value. There is no control interconnection wiring, so transformers can be distant from each other.

   Disadvantages: Operation is with a usually high −X LDC set point, meaning that the bus voltage will be *lowered* as the load increases. This can be sufficiently compensated in particular cases with the additional use of +R LDC settings.

2. Cross-connected current transformers: Unknown in practice today, the system operates on precepts similar to the negative reactance method. The LDC circuits of two controls are fed from the line CT of the opposite transformer.

   Advantages: System requires no apparatus other than the basic control, but it does require CT circuits to pass between the transformers.

   Disadvantages: System may need to be operated with a value of +X LDC much higher than that desired for LDC purposes, thereby boosting the voltage too much. The system can be used on two transformers, only.

3. Circulating current (current balance): The most common method in use in the U.S. today; about 90% of new installations use this procedure. It has been implemented with technical variations by several sources.

   Advantages: Generally reliable operation for any reasonable number of paralleled transformers. Uses the same CT as that provided for LDC, but it operates independently of the line-drop compensation.

   Disadvantages: Control circuits can be confusing, and they must be accurate as to instrument transformer polarities, etc. Proper operation is predicated on the system being such that any significant difference in CT currents must be due only to circulating current. Matched transformers will at times operate unbalanced, i.e., at differing tap steps under normal conditions.

4. Master/follower (master/slave, electrical interlock, lock-in-step): Used by most of the 10% of new installations in the U.S. that do not use circulating current. It is used much more commonly elsewhere in the world than in the U.S.

   Advantages: Matched transformers will always be balanced, resulting in minimal system losses.

   Disadvantages: As usually implemented, involves numerous auxiliary relays that may fail, locking out the system.

5. Reactive current balance (delta var): Generally used only when special system circumstances require it. Operates so as to balance the reactive current in the transformers.

   Advantages: Can be made to parallel transformers in many more-complex systems where other methods do not work.

   Disadvantages: May be more expensive than more common means.

The two most common paralleling procedures are master/follower and circulating-current minimization. The negative-reactance method also deserves summary discussion.

3.7.8.1 Master/Follower

The master/follower method operates on the simple premise: Designate one control as the master; all other units are followers. Only the master needs to know the voltage and need for a tap change. Upon recognition of such a need, the master so commands a tap change of the LTC on the first transformer. Tap-changer action of #1 activates contacts and relays that make the circuit for LTC following action in all of the followers, temporarily locking out further action by the master. The LTC action of the followers in turn activates additional contacts and relays, thus freeing the master to make a subsequent tap change when required.

3.7.8.2 Circulating-Current Method

All of the common analog implementations of the circulating-current method provide an electronic means (a "paralleling balancer") of extracting the load current and the circulating current from the total transformer current. These currents are then used for their own purposes, the load current serving as the basis for line-drop compensation and the circulating current serving to bias the control to favor the next LTC action that will tend to keep the circulating current to a minimum.

Consider Figure 3.7.9. The balancers receive the scaled transformer current and divide it into the components of load current and circulating current. The load-current portion of the transformer current is that required for line-drop compensation. The circulating-current portion is essentially totally reactive and is the same in magnitude in the two controls, but of opposite polarity. Presuming a lagging power-factor load, the control monitoring the transformer on the higher tap position will "see" a more lagging current, while that on the lower tap position sees a less lagging, or perhaps leading, current. This circulating current is injected into the controls. The polarity difference serves to bias them differently. The next LTC to operate is the one that will correct the voltage while tending to reduce the circulating current, i.e., the one that brings the tap positions into closer relation to each other.

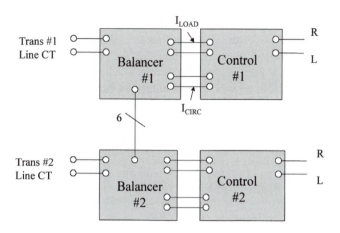

FIGURE 3.7.9 Control block diagram, paralleling by usual circulating-current method.

3.7.8.3 Negative Reactance Method

The negative reactance method (Harlow, 2002) has a very distinct advantage over all others: it involves no supplemental apparatus beyond the basic LTC control. It also involves no wiring between controls, and it is very forgiving of mismatched transformers, including the built-in instrument transformers. It fell from favor about 50 years ago because of an inherent inaccuracy in its operation. The error, in the all-important voltage-regulation function, is a function of the power factor of the load. If the load power factor changes from about 1.0 to 0.7 lag, there will be a maximum error of perhaps ±1.5 V (3 V total swing) in the regulated voltage. This, coupled with the overall inaccuracy of controls of 50 years past (the standard of the day allowed up to 5% error!), apparently made for an intolerable condition.

Today's controls are far superior in accuracy, and the power system typically operates at a much improved and consistent power factor. These two points combine to afford an order-of-magnitude improvement in the operating accuracy of the technique. The technique is little known in the industry, perhaps to the detriment of many who would find applications to benefit from the procedure.

3.7.9 Characteristics Important for LTC Transformer Paralleling

There are many transformer characteristics that must be known and evaluated when it is planned to parallel LTC transformers. Some of the more notable follow:

1. Impedance and MVA. "Impedance" as a criterion for paralleling is more correctly stated as the percent impedance referred to a common MVA base. Two transformers of 10% impedance —one of 10 MVA and the other of 15 MVA — are not the same. Two other transformers, one 10% impedance and 10 MVA and another of 15% impedance and 15 MVA, are suitably matched per this criterion. There is no definitive difference in the impedances that will be the limit of acceptability, but a difference of no greater than 7.5% is realistic.
2. Voltage rating and turns ratio. It may not be essential that the voltage ratings and turns ratios be identical. If one transformer is 69–13.8 kV and the other 69–12.47 kV, the difference may be tolerated by recognizing and accepting a fixed-step tap discrepancy, or it may be that the ratios can be made more nearly the same using the de-energized tap changers.
3. Winding configuration. The winding configuration, as delta–wye, wye–wye, etc., is critical, yet transformers of different configurations can be paralleled if care is taken to ensure that the phase shift through the transformers is the same.
4. Instrument transformers. The transformers must have VTs and CTs that produce in-phase signals of the correct ratio to the control, and they must be measuring the same phase in the different transformers.

3.7.10 Paralleling Transformers with Mismatched Impedance

Very often it is desired to use two existing transformers in parallel, even when it is recognized that the impedance mismatch is greater than that recommended for proper operation. This can usually be accomplished, although some capacity of one transformer will be sacrificed.

If the impedances of the transformers in parallel are not equal, the current divides inversely with the impedances so that the same voltage appears across both impedances. The impedances — effectively the transformer impedance as read from the nameplate — can be taken to be wholly reactive. The problem when dealing with mismatched transformers in parallel is that while the current divides per the impedances, the control — if operating using the conventional circulating-current method — is attempting to match the currents.

Realize that the LTC control and the associated paralleling equipment really have no knowledge of the actual line current. They know only a current on its scaled base, which could represent anything from 100 to 3000 A. The objective of the special considerations to permit the mismatched-impedance transformers to be paralleled is to supply equal current signals to the controls when the transformers are carrying load current in inverse proportion to their impedances.

To illustrate, consider the paralleling of two 20-MVA transformers of 9% and 11% impedance, which is much more than the 7.5% difference criterion stated earlier. We establish that $Z_1 = k \times Z_2$, so:

$$k = Z_1/Z_2 = 9/11 = 0.818$$

And since $I_1 = I_2/k$:

$$I_1 = I_2/k = 1.222\ I_2$$

Transformer T1 will carry 22.2% more load than transformer T2, even though they are of the same MVA rating. A resolution is to fool the controls to act as though the current is balanced when, in fact, it is mismatched by 22%. A solution is found by placing a special ratio auxiliary CT in the control-current path of T2 that boosts its current by 22%. This will cause both controls to see the same current when, in fact, T1 is carrying 22% more current. In this way, the controls are fooled into thinking that the load is balanced when it is actually unbalanced due to the impedance mismatch. Transformer T2 has effectively been derated by 22% in order to have the percent impedances match.

The circulating-current paralleling described here is commonly used in the U.S. Another basis for implementing circulating-current paralleling is now available. The new scheme does not require the breakout of the circulating current and the load current from the total transformer current. Rather, the procedure is to recognize the apparent power factors as seen by the transformers and then act so as to make the power factors be equal. The control configuration used is that of Figure 3.7.10, where the phasor diagram, Figure 3.7.11, shows typical loading for mismatched transformers. The fundamental difference in this manner of circulating-current paralleling is that the principle involves the equalization of the apparent power factors as seen by the transformers, i.e., the control acts to make φ1 ≈ φ2. The benefit of this subtle difference is that it is more amenable to use where the transformers exhibit mismatched impedance.

References

American National Standard for Transformers 230 kV and Below 833/958 through 8333/10417 kVA, Single-Phase; and 750/862 through 60000/80000/100000 kVA, Three-Phase without Load Tap Changing; and 3750/4687 through 60000/80000/100000 kVA with Load Tap Changing, ANSI Standard C57.12.10, National Electrical Manufacturers Association, Rosslyn, VA, 1998.

Beckwith Electric Co., Basic Considerations for the Application of LTC Transformers and Associated Controls, Application Note 17, Beckwith Electric, Largo, FL, 1995.

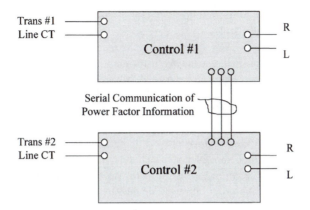

FIGURE 3.7.10 Control block diagram, paralleling by equal power factor, circulating-current method.

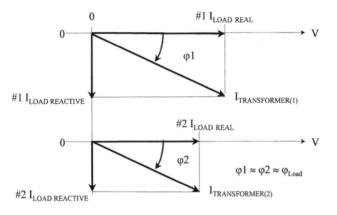

FIGURE 3.7.11 Phasor diagram — mismatched transformers in parallel by equal-power-factor method.

Harlow Engineering Associates, LTC Control and Transformer Paralleling, Class Notes, Chap. 1: Basic Control of LTC Transformers and Step-Voltage Regulators, Harlow Engineering, Mentone, AL, 2001, pp. 17–21.

Harlow, J.H., Let's Rethink Negative Reactance Transformer Paralleling, in Proceedings of IEEE T&D Latin America Conference, Sao Paulo, Brazil, March 2002.

IEEE, Standard Requirements, Terminology, and Test Code for Step-Voltage Regulators, IEEE C57.15-1999, Institute of Electrical and Electronics Engineers, Piscataway, NJ, 1999.

3.8 Power Transformer Protection

Armando Guzmán, Hector J. Altuve, and Gabriel Benmouyal

3.8.1 Introduction

Three characteristics generally provide means for detecting transformer internal faults [1]. These characteristics include an increase in phase currents, an increase in the differential current, and gas formation caused by the fault arc [2,3]. When transformer internal faults occur, immediate disconnection of the faulted transformer is necessary to avoid extensive damage and/or preserve power system stability and power quality. Three types of protection are normally used to detect these faults: overcurrent protection for phase currents, differential protection for differential currents, and gas accumulator or rate-of-pressure-rise protection for arcing faults.

Overcurrent protection with fuses or relays provided the first type of transformer fault protection [4], and it continues to be applied in small-capacity transformers. The differential principle for transformer protection was introduced by connecting an inverse-time overcurrent relay in the paralleled secondaries of the current transformers (CT) [4]. The percentage-differential principle [5], which was immediately applied to transformer protection [4,6,7], provided excellent results in improving the security of differential protection for external faults with CT saturation.

The analysis presented here focuses primarily on differential protection. Differential relays are prone to misoperation in the presence of transformer inrush currents, which result from transients in transformer magnetic flux. The first solution to this problem was to introduce an intentional time delay in the differential relay [4,6]. Another proposal was to desensitize the relay for a given time to override the inrush condition [6,7]. Others suggested adding a voltage signal to restrain [4] or to supervise the differential relay [8].

Researchers quickly recognized that the harmonic content of the differential current provided information that helped differentiate faults from inrush conditions. Kennedy and Hayward proposed a differential relay with only harmonic restraint for bus protection [9]. Hayward [10] and Mathews [11] further developed this method by adding percentage-differential restraint for transformer protection. These early relays used all of the harmonics for restraint. With a relay that used only the second harmonic to block, Sharp and Glassburn introduced the idea of harmonic blocking instead of restraining [12].

Many modern transformer differential relays use either harmonic restraint or blocking methods. These methods ensure relay security for a very high percentage of inrush and overexcitation cases. However, these methods do not work in cases with very low harmonic content in the operating current. Common harmonic restraint or blocking, introduced by Einval and Linders [13], increases relay security for inrush, but it could delay operation for internal faults combined with inrush in the nonfaulted phases.

Transformer overexcitation is another possible cause of differential-relay misoperation. Einval and Linders proposed the use of an additional fifth-harmonic restraint to prevent such misoperations [13]. Others have proposed several methods based on waveshape recognition to distinguish faults from inrush, and these methods have been applied in transformer relays [14–17]. However, these techniques do not identify transformer overexcitation conditions.

This chapter describes an improved approach for transformer differential protection using current-only inputs. The approach ensures security for external faults, inrush, and overexcitation conditions, and it provides dependability for internal faults. It combines harmonic restraint and blocking methods with a waveshape recognition technique. The improved method uses even harmonics for restraint and the dc component and the fifth harmonic to block operation.

3.8.2 Transformer Differential Protection

Percentage-restraint differential protective relays [5,6] have been in service for many years. Figure 3.8.1 shows a typical differential-relay connection diagram. Differential elements compare an operating current with a restraining current. The operating current (also called differential current), I_{OP}, can be obtained as the phasor sum of the currents entering the protected element:

$$I_{OP} = \left| \vec{I}_{W1} + \vec{I}_{W2} \right| \tag{3.8.1}$$

I_{OP} is proportional to the fault current for internal faults and approaches zero for any other operating (ideal) conditions.

There are different alternatives for obtaining the restraining current, I_{RT}. The most common ones include the following:

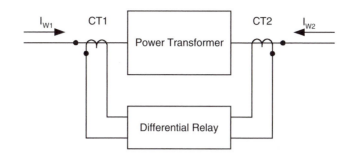

FIGURE 3.8.1 Typical differential-relay connection diagram.

$$I_{RT} = k \left| \vec{I}_{W1} - \vec{I}_{W2} \right| \qquad (3.8.2)$$

$$I_{RT} = k \left(\left| \vec{I}_{W1} \right| + \left| \vec{I}_{W2} \right| \right) \qquad (3.8.3)$$

$$I_{RT} = Max \left(\left| \vec{I}_{W1} \right|, \left| \vec{I}_{W2} \right| \right) \qquad (3.8.4)$$

where k is a compensation factor, usually taken as 1 or 0.5.

Equation 3.8.3 and Equation 3.8.4 offer the advantage of being applicable to differential relays with more than two restraint elements.

The differential relay generates a tripping signal if the operating current, I_{OP} is greater than a percentage of the restraining current, I_{RT}:

$$I_{OP} > SLP \cdot I_{RT} \qquad (3.8.5)$$

Figure 3.8.2 shows a typical differential-relay operating characteristic. This characteristic consists of a straight line having a slope (SLP) and a horizontal straight line defining the relay minimum pickup current, I_{PU}. The relay operating region is located above the slope characteristic (Equation 3.8.5), and the restraining region is below the slope characteristic.

Differential relays perform well for external faults as long as the CTs reproduce the primary currents correctly. When one of the CTs saturates, or if both CTs saturate at different levels, false operating current appears in the differential relay and could cause relay misoperation. Some differential relays use the harmonics caused by CT saturation for added restraint and to avoid misoperations [9]. In addition, the slope characteristic of the percentage-differential relay provides further security for external faults with CT saturation. A variable-percentage or dual-slope characteristic, originally proposed by Sharp and Glassburn [12], further increases relay security for heavy CT saturation. Figure 3.8.2 shows this characteristic as a dotted line.

CT saturation is only one of the causes of false operating current in differential relays. In the case of power transformer applications, other possible sources of error include the following:

- Mismatch between the CT ratios and the power transformer ratio
- Variable ratio of the power transformer caused by a tap changer
- Phase shift between the power transformer primary and secondary currents for delta–wye connections
- Magnetizing inrush currents created by transformer transients because of energization, voltage recovery after the clearance of an external fault, or energization of a parallel transformer
- High exciting currents caused by transformer overexcitation

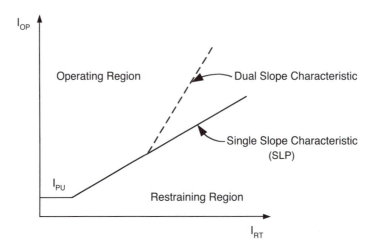

FIGURE 3.8.2 Differential relay with dual-slope characteristic.

The relay percentage-restraint characteristic typically solves the first two problems. A proper connection of the CTs or emulation of such a connection in a digital relay (auxiliary CTs historically provided this function) addresses the phase-shift problem. A very complex problem is that of discriminating internal fault currents from the false differential currents caused by magnetizing inrush and transformer overexcitation.

3.8.3 Magnetizing Inrush, Overexcitation, and CT Saturation

Inrush or overexcitation conditions in a power transformer produce false differential currents that could cause relay misoperation. Both conditions produce distorted currents because they are related to transformer core saturation. The distorted waveforms provide information that helps to discriminate inrush and over-excitation conditions from internal faults. However, this discrimination can be complicated by other sources of distortion such as CT saturation, nonlinear fault resistance, or system resonant conditions.

3.8.3.1 Inrush Currents

The study of transformer excitation inrush phenomena has spanned more than 50 years [18–26]. Magnetizing inrush occurs in a transformer whenever the polarity and magnitude of the residual flux do not agree with the polarity and magnitude of the ideal instantaneous value of steady-state flux [22]. Transformer energization is a typical cause of inrush currents, but any transient in the transformer circuit can generate these currents. Other causes include voltage recovery after the clearance of an external fault or after the energization of a transformer in parallel with a transformer that is already in service. The magnitudes and waveforms of inrush currents depend on a multitude of factors and are almost impossible to predict [23]. The following list summarizes the main characteristics of inrush currents:

- Generally contain dc offset, odd harmonics, and even harmonics [22,23]
- Typically composed of unipolar or bipolar pulses, separated by intervals of very low current values [22,23]
- Peak values of unipolar inrush current pulses decrease very slowly. Time constant is typically much greater than that of the exponentially decaying dc offset of fault currents.
- Second-harmonic content starts with a low value and increases as the inrush current decreases.
- Relay currents are delta currents (a delta winding is encountered in either the power- or current-transformer connections, or is simulated in the relay), which means that currents of adjacent windings are subtracted, and the following occur:
    - DC components are subtracted.
    - Fundamental components are added at 60°.

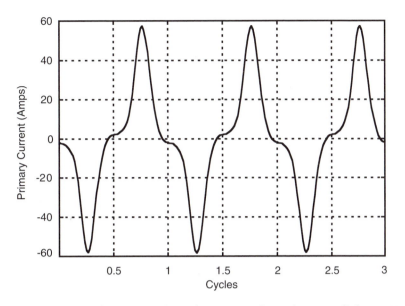

FIGURE 3.8.3 Exciting current of an overexcited transformer; overvoltage of 150% applied to a 5-KVA, 230/115-V single-phase transformer.

- Second harmonics are added at 120°.
- Third harmonics are added at 180° (they cancel out), etc.

Sonnemann et al. initially claimed that the second-harmonic content of the inrush current was never less than 16 to 17% of the fundamental [22]. However, transformer energization with reduced voltages can generate inrush currents with second-harmonic content less than 10%, as will be discussed later in this section.

3.8.3.2 Transformer Overexcitation

The magnetic flux inside the transformer core is directly proportional to the applied voltage and inversely proportional to the system frequency. Overvoltage and/or underfrequency conditions can produce flux levels that saturate the transformer core. These abnormal operating conditions can exist in any part of the power system, so any transformer can be exposed to overexcitation.

Transformer overexcitation causes transformer heating and increases exciting current, noise, and vibration. A severely overexcited transformer should be disconnected to avoid transformer damage. Because it is difficult, with differential protection, to control the amount of overexcitation that a transformer can tolerate, transformer differential protection tripping for an overexcitation condition is not desirable. Instead, use separate transformer overexcitation protection, and the differential element should not trip for these conditions. One alternative is a V/Hz relay that responds to the voltage/frequency ratio.

Overexcitation of a power transformer is a typical case of ac saturation of the core that produces odd harmonics in the exciting current. Figure 3.8.3 shows the exciting current recorded during a real test of a 5-kVA, 230/115-V, single-phase laboratory transformer [24]. The current corresponds to an overvoltage condition of 150% at nominal frequency. For comparison purposes, the peak value of the transformer nominal current is 61.5 A, and the peak value of the exciting current is 57.3 A.

Table 3.8.1 shows the most significant harmonics of the current signal depicted in Figure 3.8.3. Harmonics are expressed as a percentage of the fundamental component. The third harmonic is the most suitable for detecting overexcitation conditions, but either the delta connection of the CTs or the delta-connection compensation of the differential relay filters out this harmonic. The fifth harmonic, however, is still a reliable quantity for detecting overexcitation conditions.

TABLE 3.8.1 Harmonic Content of the Current Signal Shown in Figure 3.8.3

Frequency Component	Magnitude (primary amps)	Percentage of Fundamental
Fundamental	22.5	100.0
Third	11.1	49.2
Fifth	4.9	21.7
Seventh	1.8	8.1

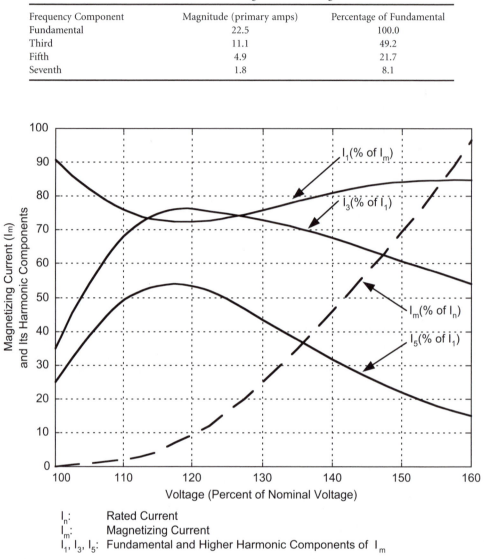

I_n: Rated Current
I_m: Magnetizing Current
I_1, I_3, I_5: Fundamental and Higher Harmonic Components of I_m

FIGURE 3.8.4 Harmonic content of transformer exciting current as a function of the applied voltage. (From Cooper Power Systems, *Electric Power System Harmonics: Design Guide*, Cooper Power Systems, Bulletin 87011, Franksville, WI, 1990. With permission.)

Einval and Linders [13] were the first to propose using the fifth harmonic to restrain the transformer differential relay. They recommended setting this restraint function at 35% of fifth harmonic with respect to the fundamental. Figure 3.8.4 [25] shows the harmonic content of the excitation current of a power transformer as a function of the applied voltage. As the voltage increases, saturation and the exciting current increase. The odd harmonics, expressed as a percentage of the fundamental, first increase and then begin to decrease at overvoltages on the order of 115 to 120%. Setting the differential-relay fifth-harmonic restraint to 35% ensures security for overvoltage conditions less than 140%. For greater overvoltages, which could quickly destroy the transformer in a few seconds, it is desirable to have the differential-relay fast tripping added to that of the transformer-overexcitation relay.

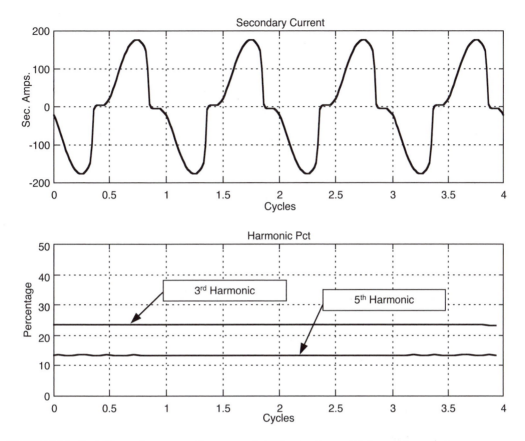

FIGURE 3.8.5 Typical secondary current for symmetrical CT saturation and its harmonic content.

3.8.3.3 CT Saturation

CT saturation during faults and its effect on protective relays have received considerable attention [26–31]. In the case of transformer-differential protection, the effect of CT saturation is double-edged. For external faults, the resulting false differential current can produce relay misoperation. In some cases, the percentage restraint in the relay addresses this false differential current. For internal faults, the harmonics resulting from CT saturation could delay the operation of differential relays having harmonic restraint or blocking.

The main characteristics of CT saturation are the following:

- CTs reproduce faithfully the primary current for a given time after fault inception [30]. The time to CT saturation depends on several factors, but it is typically one cycle or longer.
- The worst CT saturation is produced by the dc component of the primary current. During this dc saturation period, the secondary current can contain dc offset and odd and even harmonics [10,28].
- When the dc offset dies out, the CT has only ac saturation, characterized by the presence of odd harmonics in the secondary current [9,10,26].

Figure 3.8.5 shows a typical secondary current waveform for computer-simulated ac symmetrical CT saturation. This figure also depicts the harmonic content of this current. The figure confirms the presence of odd harmonics and the absence of even harmonics in the secondary current.

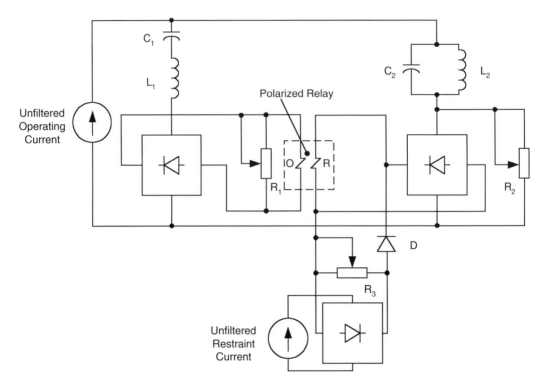

FIGURE 3.8.6 Transformer differential relay with harmonic restraint.

3.8.4 Methods for Discriminating Internal Faults from Inrush and Overexcitation Conditions

Early transformer differential-relay designs used time delay [4,6] or a temporary desensitization of the relay [6,7] to override the inrush current. An additional voltage signal to restrain [4] or to supervise (block) [8] the differential relay has also been proposed. These proposals increased operating speed at the cost of higher complexity. Recent methods use voltage information to provide transformer protection [32–35]. It is recognized, however, that while an integrated digital substation-protection system provides voltage information, this is not the case for a stand-alone differential relay. Adding voltage signals to such a relay requires voltage transformers that are normally not available in the installation.

The current-based methods for discriminating internal faults from inrush and overexcitation conditions fall into two groups: those using harmonics to restrain or block [9–13] and those based on wave-shape identification [14–17].

3.8.4.1 Harmonic-Based Methods

The harmonic content of the differential current can be used to restrain or to block the relay, providing a means to differentiate between internal faults and inrush or overexcitation conditions. The technical literature on this topic has not clearly identified the differences between restraint and blocking.

The first harmonic-restrained differential relays used all harmonics to provide the restraint function [9–11]. The resulting high level of harmonic restraint provided security for inrush conditions at the expense of operating speed for internal faults with CT saturation.

Figure 3.8.6 shows a simplified schematic diagram of a well-known transformer differential relay with harmonic restraint [36]. Auxiliary current transformers (not shown) form the operating and restraint currents of the relay. The operating current is the phasor sum of the currents entering the protected transformer (Equation 3.8.1). For two-winding transformers, the restraint current is formed in a through-current transformer having two primary circuits, one of which is connected in each of the main-current transformer circuits. The resulting restraint current is the phasor difference of the currents entering the transformer (Equation 3.8.2). For three-winding transformers, three independent through-current transformers are provided, one for each of the current transformer windings. The secondary currents of the through-current transformers are independently rectified and summed to form a restraint current with the following form:

$$I_{RT} = k\left(\left|\vec{I}_{W1}\right| + \left|\vec{I}_{W2}\right| + \left|\vec{I}_{W3}\right|\right)$$
(3.8.6)

The main operating unit of the transformer differential relay is a polarized relay (see Figure 3.8.6). This unit performs an amplitude comparison of the rectified currents applied to the operating (O) and restraint (R) coils. The unit trips when the current in the operating coil is greater than the current in the restraint coil.

The circuit supplying the operating coil of the polarized relay includes a band-pass series filter (L_1, C_1) tuned to 60 Hz. The circuit supplying the restraint coil of the polarized relay contains a notch-type parallel filter (L_2, C_2) tuned to 60 Hz. Restraint current is also supplied to the restraint coil to provide a percentage-differential characteristic. As a result, the polarized unit compares the fundamental component of the operating current with a restraint signal consisting of the harmonics of the operating current plus the unfiltered restraint current.

The differential relay operating condition can be expressed as

$$I_{OP} > SLP \cdot I_{RT} + K_2 I_2 + K_3 I_3 + ...$$
(3.8.7)

where I_{OP} represents the fundamental component of the operating current; I_2, I_3, ... are the higher harmonics; I_{RT} is the unfiltered restraint current; and K_2, K_3, ... are constant coefficients.

Resistor R_1 (see Figure 3.8.6) provides adjustment of the minimum pickup current, I_{PU}, of the differential relay (see Figure 3.8.2). Resistor R_2 controls the level of harmonic restraint in the relay. Resistor R_3 provides the slope percentage adjustment; it has three taps corresponding to the slope values of 15, 25, and 40%.

During transformer-magnetizing inrush conditions, the unfiltered operating current may, in addition to harmonics, contain a dc-offset component. The band-pass filter (L_1, C_1) blocks the dc component, but the notch filter (L_2, C_2) passes the dc component, thus providing an additional temporary restraint that increases relay security for inrush.

Diode D (see Figure 3.8.6) did not appear in the initial version of this transformer differential relay [11]. For high values of the restraint current, diode D conducts, and Equation 3.8.7 is valid. On the other hand, the diode cuts off for low-restraint currents, and only the operating-current harmonics are applied to the restraint coil of the polarized relay. In this case, the differential-relay operation condition is as follows:

$$I_{OP} > K_2 I_2 + K_3 I_3 + ...$$
(3.8.8)

The transformer differential relay also contains an instantaneous overcurrent element (not shown) that provides instantaneous tripping for heavy internal faults, even if the current transformers saturate.

Einval and Linders [13] designed a three-phase differential relay with second- and fifth-harmonic restraint. This design complemented the idea of using only the second harmonic to identify inrush currents (originally proposed by Sharp and Glassburn [12]), by using the fifth harmonic to avoid misoperations for transformer overexcitation conditions.

The relay [13] includes air-gap auxiliary current transformers that produce voltage secondary signals and filter out the dc components of the input currents. A maximum-voltage detector produces the percentage-differential-restraint voltage, so the restraint quantity is of the form shown in Equation 3.8.4.

The relay forms an additional restraint voltage by summing the second- and fifth-harmonic components of a voltage proportional to the operating current. The basic operation equation for one phase can be expressed according to the following:

$$I_{OP} > SLP \cdot I_{RT} + K_2 I_2 + K_5 I_5 \qquad (3.8.9)$$

Einval and Linders [13] first introduced the concept of common harmonic restraint in this relay. The harmonic-restraint quantity is proportional to the sum of the second- and fifth-harmonic components of the three relay elements. The relay-operation equation is of the following form:

$$I_{OP} > SLP \cdot I_{RT} + \sum_{n=1}^{3} \left(K_2 I_{2n} + K_5 I_{5n} \right) \qquad (3.8.10)$$

Sharp and Glassburn [12] were first to propose harmonic blocking. Figure 3.8.7 depicts a simplified schematic diagram of the transformer differential relay with second-harmonic blocking [12]. The relay consists of a differential unit, DU, and a harmonic blocking unit, HBU. Differential-relay tripping requires operation of both DU and HBU units.

In the differential unit (Figure 3.8.7a), an auxiliary current transformer (not shown) forms the operating current according to Equation 3.8.1. This current is rectified and applied to the operating coil of a polarized relay unit. Auxiliary air-gap current transformers (not shown) produce secondary voltages that are proportional to the transformer winding currents. These voltages are rectified by parallel-connected rectifier bridges, which behave as a maximum voltage detector. The resulting restraint current, applied to the restraint coil of the polarized relay unit, has the form of Equation 3.8.4. The polarized relay unit performs an amplitude comparison of the operating and the restraint currents and generates the relay percentage-differential characteristic (Equation 3.8.5). Resistor R_1 (see Figure 3.8.7a) provides the slope percentage adjustment for the differential relay. An auxiliary saturating transformer (not shown) connected in the operating circuit provides a variable slope characteristic.

In the harmonic blocking unit (Figure 3.8.7b), an auxiliary air-gap current transformer (not shown) generates a version of the operating current (Equation 3.8.1) without the dc-offset component, which is blocked by the air-gap transformer. The fundamental component and higher harmonics of the operating current are passed to two parallel circuits, rectified, and applied to the operating and restraint coils of the polarized relay unit. The circuit supplying the operating coil of the polarized relay unit includes a notch-type parallel filter (L_1, C_1) tuned to 120 Hz. The circuit supplying the restraint coil of the polarized relay contains a low-pass filter (L_3) combined with a notch filter (L_2, C_2) tuned to 60 Hz. The series combination of both filters passes the second harmonic and rejects the fundamental component and remaining harmonics of the operating current. As a result, the polarized relay compares an operating signal formed by the fundamental component plus the third- and higher-order harmonics of the operating current, with a restraint signal that is proportional to the second harmonic of the operating current. The operating condition of the harmonic blocking unit, HBU, can be expressed as follows:

$$I_{OP} + K_3 I_3 + K_4 I_4 + \ldots > K_2 I_2 \qquad (3.8.11)$$

Figure 3.8.7c shows a simplified diagram of the relay contact logic. Transient response of the filters for inrush currents with low second-harmonic content can cause differential-relay misoperation. A time-delay auxiliary relay, T, shown in Figure 3.8.7c, prevents this misoperation. The relay also includes an instantaneous overcurrent unit (not shown) to provide fast tripping for heavy internal faults.

Typically, digital transformer differential relays use second- and fifth-harmonic blocking logic. Figure 3.8.8a shows a logic diagram of a differential element having second- and fifth-harmonic blocking. A tripping signal requires fulfillment of the operation depicted in Equation 3.8.5 without fulfillment of the following blocking conditions (Equation 3.8.12 and Equation 3.8.13):

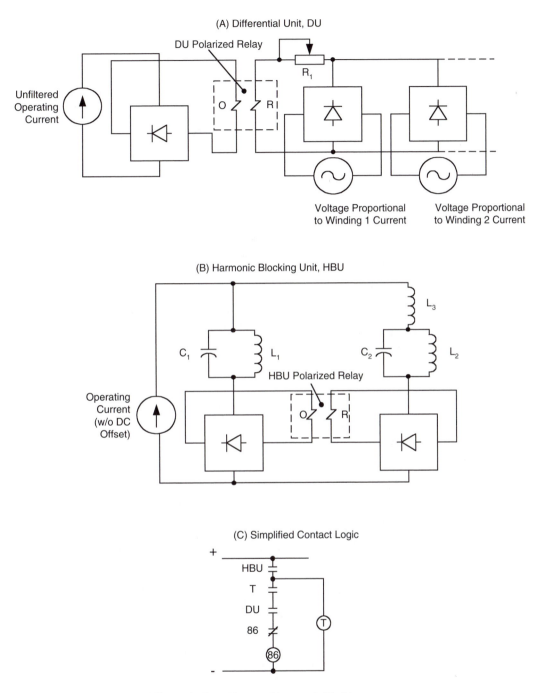

FIGURE 3.8.7 Transformer differential relay with second-harmonic blocking.

$$I_{OP} < K_2 I_2 \tag{3.8.12}$$

$$I_{OP} < K_5 I_5 \tag{3.8.13}$$

Figure 3.8.8b depicts the logic diagram of a differential element using second- and fifth-harmonic restraint. Figure 3.8.9 shows the three-phase versions of the transformer differential relay with independent

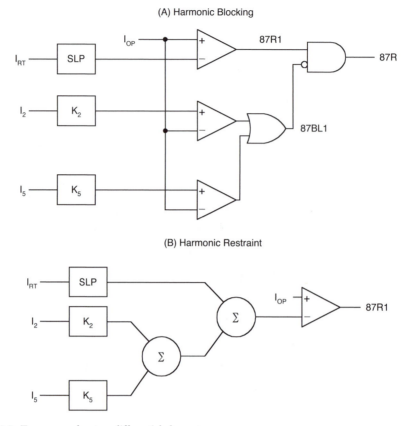

FIGURE 3.8.8 Two approaches to a differential element.

harmonic blocking or restraint. The relay is composed of three differential elements of the types shown in Figure 3.8.8. In both cases, a tripping signal results when any one of the relay elements asserts.

Note that in the harmonic-restraint element (see Figure 3.8.8b), the operating current, I_{OP}, should overcome the combined effects of the restraining current, I_{RT}, and the harmonics of the operating current. On the other hand, in the harmonic-blocking element, the operating current is independently compared with the restraint current and the harmonics. Table 3.8.2 summarizes the results of a qualitative comparison of the harmonic restraint (using all harmonics) and blocking methods for transformer differential protection.

The comparison results given in Table 3.8.2 suggest use of the blocking method, if security for inrush can be guaranteed. However, it is not always possible to guarantee security for inrush, as is explained later in this chapter. Therefore, harmonic restraint is an alternative method for providing relay security for inrush currents having low harmonic content.

Another alternative is to use common harmonic restraint or blocking. This method is simple to implement in a blocking scheme and is the preferred alternative in present-day digital relays. Figure 3.8.10 shows a logic diagram of the common harmonic blocking method.

A method that provides a compromise in reliability between the independent- and common-harmonic blocking methods, described earlier, forms a composite signal that contains information on the harmonics of the operating currents of all relay elements. Comparison of this composite signal with the operating current determines relay operation.

The composite signal, I_{CH}, may be of the following form:

$$I_{CH} = \sum_{n=1}^{3} K_2 I_{2n} + K_3 I_{3n} + \dots \tag{3.8.14}$$

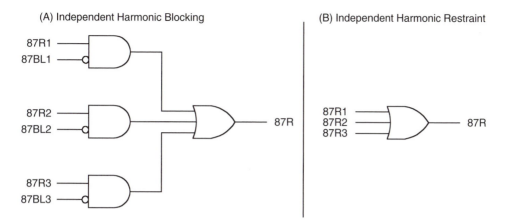

FIGURE 3.8.9 Three-phase differential relay with: (a) independent harmonic blocking and (b) independent harmonic restraint.

TABLE 3.8.2 Comparison of Harmonic Restraint and Blocking Methods

	All-Harmonic Restraint (HR)	Harmonic Blocking (HB)	Remarks
Security for external faults	higher	lower	HR always uses harmonics from CT saturation for additional restraint; HB only blocks if the harmonic content is high
Security for inrush	higher	lower	HR adds the effects of percentage and harmonic restraint; HB evaluates the harmonics independently
Security for overexcitation	higher	lower	Same as above; however, a fifth-harmonic blocking scheme is the best solution, as will be shown in a later section
Dependability	lower	higher	Harmonics from CT saturation reduce the sensitivity of HR for internal faults; the use of only even harmonics solves this problem
Speed	lower	higher	Percentage differential and harmonic blocking run in parallel in HB
Slope characteristic	harmonic dependent	well defined	HB slope characteristic is independent from harmonics
Testing	results depend upon harmonics	straightforward	Same as above

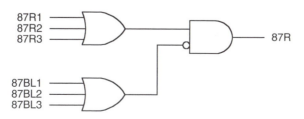

FIGURE 3.8.10 Common harmonic blocking method.

I_{CH} may contain all or only part of the harmonics of the operating current. Another possibility is to calculate the root-mean-square (rms) value of the harmonics for each relay element, I_{Hn}:

$$I_{Hn} = \sqrt{I_{2n}^2 + I_{3n}^2 + ...}$$ (3.8.15)

The composite signal, I_{CH}, can then be calculated as some type of an average value using Equation 3.8.16 or Equation 3.8.17.

$$I_{CH} = \frac{1}{3} \sum_{n=1}^{3} I_{Hn}$$ (3.8.16)

$$I_{CH} = \frac{1}{3} \sqrt{\sum_{n=1}^{3} I_{Hn}^2}$$ (3.8.17)

The relay blocking condition is the following:

$$I_{OP} < K_{CH} I_{CH}$$ (3.8.18)

Common-harmonic-blocking logic provides high security but sacrifices some dependability. Energization of a faulted transformer could result in harmonics from the inrush currents of the nonfaulted phases, and these harmonics could delay relay operation.

3.8.4.2   Wave-Shape Recognition Methods

Other methods for discriminating internal faults from inrush conditions are based on direct recognition of the waveshape distortion of the differential current.

Identification of the separation of differential current peaks represents a major group of waveshape recognition methods. Bertula [37] designed an early percentage-differential relay in which the contacts vibrated for inrush current (because of the low current intervals) and remained firmly closed for symmetrical currents corresponding to internal faults. Rockefeller [16] proposed blocking relay operation if successive peaks of the differential current fail to occur at about 7.5 to 10 msec.

A well-known principle [14,38] recognizes the length of the time intervals during which the differential current is near zero. Figure 3.8.11 depicts the basic concept behind this low-current-detection method. The differential current is compared with positive and negative thresholds having equal magnitudes. This comparison helps to determine the duration of the intervals during which the absolute value of the current is less than the absolute value of the threshold. The time intervals are then electronically compared with a threshold value equal to one-quarter cycle. For inrush currents (Figure 3.8.11a), the low-current intervals, t_A, are greater than one-quarter cycle, and the relay is blocked. For internal faults (Figure 3.8.11b), the low-current intervals, t_B, are less than one-quarter cycle, and the relay operates.

Using the components of the rectified differential current provides an indirect way to identify the presence of low-current intervals. Hegazy [39] proposed comparing the second harmonic of the rectified differential current with a given threshold to generate a tripping signal. Dmitrenko [40] proposed issuing a tripping signal if the polarity of a summing signal remains unchanged. This signal is the sum of the dc and amplified fundamental components of the rectified differential current.

Another group of methods makes use of the recognition of dc offset or asymmetry in the differential current. Some early relays [15,41,42] used the saturation of an intermediate transformer by the dc offset of the differential current as a blocking method. A transient additional restraint based on the dc component was an enhancement to a well-known harmonic-restraint transformer differential relay [11]. Michelson [43] proposed comparing the amplitudes of the positive and negative semicycles of

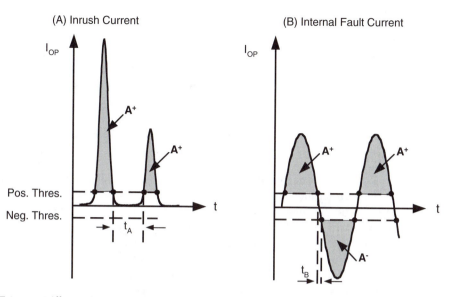

FIGURE 3.8.11 Differential-relay blocking based on recognizing the duration time of low-current intervals.

the differential current with given thresholds in two different polarized elements. Both elements must pick up to produce a trip. Rockefeller [16] suggested extending this idea to a digital relay.

Another alternative [44] is to use the difference of the absolute values of the positive and negative semicycles of the differential current for restraint. It has also been proposed [44] to use the amplitude of the negative semicycles of the differential current as the relay operating quantity. The negative semicycle is that having the opposite polarity with respect to the dc component. More recently, Wilkinson [17] proposed making separate percentage-differential comparisons on both semicycles of the differential current. Tripping occurs if an operation condition similar to Equation 3.8.7 is true for both semicycles.

3.8.5 An Improved Approach for Transformer Protection

The evaluation in the previous section of existing harmonic-restraint/blocking methods makes clear that independent restraint/blocking methods may fail to ensure security for some real-life inrush conditions. Common harmonic restraint/blocking could provide solutions, but the behavior of these methods for internal faults combined with inrush currents requires further study.

Combining restraint and blocking into an independent restraint/blocking method provides an improved approach to transformer differential protection. Even harmonics of the differential current provide restraint, while both the fifth harmonic and the dc component block relay operation.

3.8.5.1 Even-Harmonic Restraint

In contrast to the odd harmonics ac CT saturation generates, even harmonics are a clear indicator of magnetizing inrush. Even harmonics resulting from dc CT saturation are transient in nature. It is important to use even harmonics (and not only the second harmonic) to obtain better discrimination between inrush and internal fault currents.

Tests suggest use of even harmonics (second and fourth) in a restraint scheme that ensures security for inrush currents having very low second-harmonic current. The operation equation is:

$$I_{OP} > SLP \cdot I_{RT} + K_2 I_2 + K_4 I_4 \qquad (3.8.19)$$

3.8.5.2 Fifth-Harmonic Blocking

It is a common practice to use the fifth harmonic of the operation current to avoid differential-relay operation for transformer overexcitation conditions [13]. A preferred solution may be a harmonic blocking scheme in which the fifth harmonic is independently compared with the operation current. In this scheme, a given relay setting, in terms of fifth-harmonic percentage, always represents the same overexcitation condition. In a fifth-harmonic restraint scheme, a given setting may represent different overexcitation conditions, depending on the other harmonics that may be present.

Relay tripping in this case requires fulfillment of Equation 3.8.19 and not Equation 3.8.13.

3.8.5.3 DC Blocking

The improved method of even-harmonic restraint and fifth-harmonic blocking provides very high relay security for inrush and overexcitation conditions. There are, however, some inrush cases in which the differential current is practically a pure sine wave. One of the real cases analyzed later exhibits such a behavior. Any harmonic-based method could cause relay misoperation in such extreme inrush cases.

The dc component of inrush current typically has a greater time constant than that for internal faults. The presence of dc offset in the inrush current is an additional indicator that can be used to guarantee relay security for inrush. This waveshape recognition method is relatively easy to apply in a digital relay, because extraction of the dc component is a low-pass filtering process.

The improved method splits the differential current into its positive and negative semicycles and calculates one-cycle sums for both semicycles. Then, the method uses the ratio of these sums to block relay operation. The one-cycle sum of the positive semicycle is proportional to the area A^+ (see Figure 3.8.11); the one-cycle sum of the negative semicycle is proportional to the area A^-.

The sum, S^+, of the positive-current samples is given by the following equations:

$$S^+ = \left| \sum_{k=1}^{N} i_k \right| \rightarrow \left(i_k > 0 \right) \tag{3.8.20}$$

$$S^+ = 0 \rightarrow \left(i_k \leq 0 \right) \tag{3.8.21}$$

where i_k represents the current samples, and N is the number of samples per cycle.

The sum S^- of the negative-current samples is given by:

$$S^- = \left| \sum_{k=1}^{N} i_k \right| \rightarrow \left(i_k < 0 \right) \tag{3.8.22}$$

$$S^- = 0 \rightarrow \left(i_k \geq 0 \right) \tag{3.8.23}$$

Calculate the dc ratio, DCR, according to Equation 3.8.24, to account for both positive and negative dc offsets:

$$DCR = \frac{Min\left(S^+, S^-\right)}{Max\left(S^+, S^-\right)} \tag{3.8.24}$$

Equation 3.8.24 gives a DCR value that is normalized (the value of DCR is always between 0 and 1) and also avoids division by zero. By comparing DCR with a 0.1 threshold, the relay dc-blocking method is implemented:

$$DCR < 0.1 \tag{3.8.25}$$

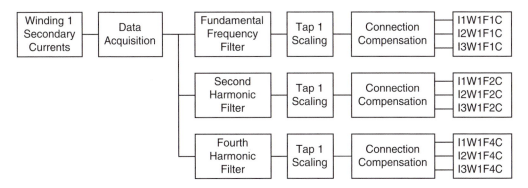

FIGURE 3.8.12 Data acquisition, filtering, scaling, and compensation for Winding 1 currents.

Relay tripping requires the fulfillment of Equation 3.8.19 but not Equation 3.8.13 or Equation 3.8.25.

Selecting a value for the threshold in Equation 3.8.25 means deciding on a compromise between security and speed. A high value (near 1) affords high security but is detrimental to speed. From tests, a value of 0.1 is selected as a good solution. The delay is practically negligible for system X/R ratios as great as 40.

The response of this dc blocking method depends on the dc signal information apart from the harmonic content of the differential current. For example, the method ensures dependability for internal faults with CT saturation and maintains its security during inrush conditions with low even-harmonic content.

3.8.6   Current Differential Relay

The relay consists of three differential elements. Each differential element provides percentage-differential protection with independent even-harmonic restraint and fifth-harmonic and dc blocking. The user may select even-harmonic blocking instead of even-harmonic restraint. In such a case, two blocking modes are available: (1) independent harmonic and dc blocking and (2) common harmonic and dc blocking.

3.8.6.1   Data Acquisition, Filtering, Scaling, and Compensation

Figure 3.8.12 shows the block diagram of the data acquisition, filtering, scaling, and compensation sections for Winding 1 currents. The input currents are the CT secondary currents from Winding 1 of the transformer. The data-acquisition system includes analog low-pass filters and analog-to-digital converters. The digitalized current samples are the inputs to four digital band-pass filters. These filters extract the samples corresponding to the fundamental component and to the second, fourth, and fifth (not shown) harmonics of the input currents. A dc filter (not shown) also receives the current samples as inputs and forms the one-cycle sums of the positive and negative values of these samples. The outputs of the digital filters are then processed mathematically to provide the scaling and connection compensation required by the power and current transformers.

3.8.6.2   Restraint-Differential Element

Figure 3.8.13 shows a schematic diagram of one of the percentage-differential elements with even-harmonic restraint (Element 1). Inputs to the differential element are the filtered, scaled, and compensated sets of samples corresponding to the fundamental component and second and fourth harmonics of the currents from each of the transformer windings. The magnitude of the sum of the fundamental-component currents forms the operating current, IOP1, according to Equation 3.8.1. The scaled sum of the magnitudes of the fundamental-component currents forms the restraint current, IRT1, according to Equation 3.8.3, with k = 0.5. The magnitudes of the sums of the second-and fourth-harmonic currents represent the second- (I1F2) and fourth- (I1F4) harmonic restraint currents.

Restraint current, IRT1, is scaled to form the restraint quantity IRT1·f (SLP). Comparator 1 and switch S1 select the slope value as a function of the restraint current to provide a dual-slope percentage

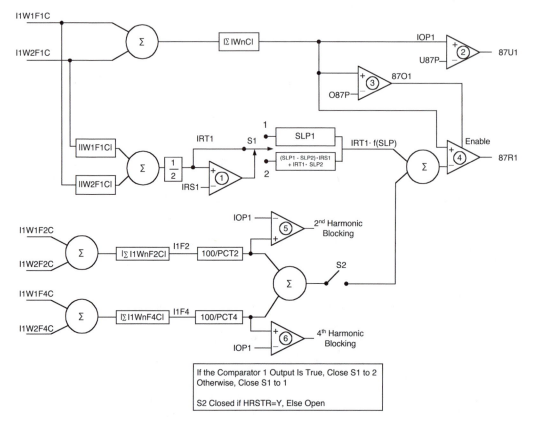

FIGURE 3.8.13 Even-harmonic restraint, 87R1, and unrestraint, 87U1, differential elements.

characteristic. Harmonic-restraint currents are scaled to form the second- and fourth-harmonic restraint quantities. The scaling factors 100/PCT2 and 100/PCT4 correspond to K_2 and K_4, respectively (Equation 3.8.19).

Comparator 4 compares the operating current to the sum of the fundamental and harmonic restraint quantities. The comparator asserts for fulfillment of Equation 3.8.19. Comparator 3 enables Comparator 4 if the operating current, IOP1, is greater than a threshold value, O87P. Assertion of Comparator 3 provides the relay minimum pickup current, I_{PU}. Switch S2 permits enabling or disabling of even-harmonic restraint in the differential element.

Comparators 5 and 6 compare the operating current to the second- and fourth-harmonic restraint quantities, respectively, to generate the second- and fourth-harmonic blocking signals. Comparison of the operating current with the fifth-harmonic restraint quantity (not shown) permits generation of the fifth-harmonic blocking signal (5HB1).

The differential element includes an unrestrained, instantaneous differential overcurrent function. Comparator 2, which compares the operating current, IOP1, with a threshold value, U87P, provides the unrestrained differential overcurrent function.

Figure 3.8.14 depicts the operating characteristic of the restraint-differential element. The characteristic can be set as either a single-slope, percentage-differential characteristic or as a dual-slope, variable percentage-differential characteristic. Figure 3.8.14 shows recommended setting values.

3.8.6.3 DC Filtering and Blocking Logic

Figure 3.8.15 shows a schematic diagram of the dc blocking logic for Element 1. The positive, S^+, and negative, S^-, one-cycle sums of the differential current are formed. Then determine the minimum and

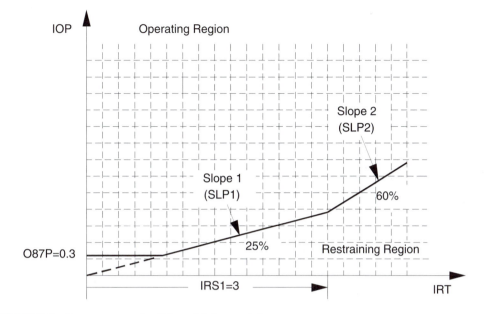

FIGURE 3.8.14 Percentage restraint differential characteristic.

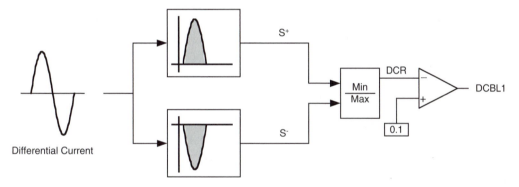

FIGURE 3.8.15 DC blocking logic.

the maximum of the absolute values of the two one-cycle sums and calculate the dc ratio, DCR, by dividing the minimum one-cycle sum value by the maximum one-cycle sum value. When DCR is less than a threshold value of 0.1, the relay issues a blocking signal, DCBL1. Then, the relay blocking condition is according to Equation 3.8.25.

By defining DCR as the ratio of the minimum to the maximum values of the one-cycle sums, an accounting is made for differential currents having positive or negative dc-offset components. In addition, the resulting DCR value is normalized.

Relay tripping requires the fulfillment of Equation 3.8.19 but not Equation 3.8.13 or Equation 3.8.25.

3.8.6.4 Relay Blocking Logic

Figure 3.8.16 depicts the blocking logic of the differential elements. If the even-harmonic restraint is not in use, switch S1 closes to add even-harmonic blocking to the fifth-harmonic and dc blocking functions. In this case, the differential elements operate in a blocking-only mode. Switches S2, S3, S4, and S5 permit enabling or disabling each of the blocking functions. The output (87BL1) of the differential-element blocking logic asserts when any one of the enabled logic inputs asserts.

Figure 3.8.17 shows the blocking logic of the differential relay. The relay can be set to an independent blocking mode (IHBL=Y) or a common blocking mode (IHBL=N).

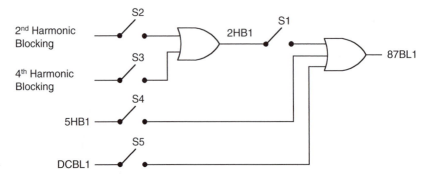

FIGURE 3.8.16 Differential-element blocking logic.

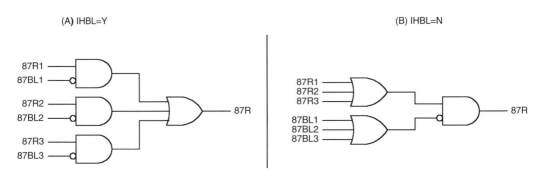

FIGURE 3.8.17 Differential-relay blocking logic.

3.8.7 Differential-Element Performance during Inrush Conditions

Following is a study of the performance of the differential elements for three field cases of transformer energization. These cases are special because they cause some of the traditional differential elements to misoperate.

3.8.7.1 Case 1

Figure 3.8.18 shows a transformer energization case while A-phase is faulted and the transformer is not loaded. The transformer is a three-phase, delta–wye-connected distribution transformer; the CT connections are wye at both sides of the transformer.

Figure 3.8.19 shows the differential element 1 inrush current; this element uses I_{AB} current. This signal looks like a typical inrush current. The current signal has low second-harmonic content and high dc content compared with the fundamental. Another interesting fact is that this signal also has high third-harmonic content. Figure 3.8.20 shows the second, third, and fourth harmonic as percentages of fundamental. Notice that the second harmonic drops below 5%. Figure 3.8.21 shows the dc content as a percentage of fundamental of the inrush current. The dc content is high during the event; this is useful information for adding security to the differential relay.

The differential elements operate as follows:

3.8.7.1.1 Second- and Fourth-Harmonic Blocking

The low second- and fourth-harmonic content produces misoperation of the differential element that uses independent harmonic blocking.

3.8.7.1.2 All-Harmonic Restraint

The harmonic-restraint relay that uses all harmonics maintains its security because of the high third-harmonic content of the inrush current.

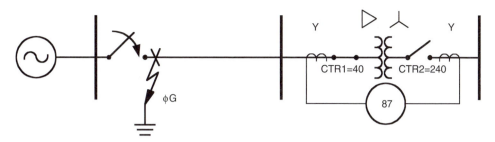

FIGURE 3.8.18 Transformer energization while A-phase is faulted.

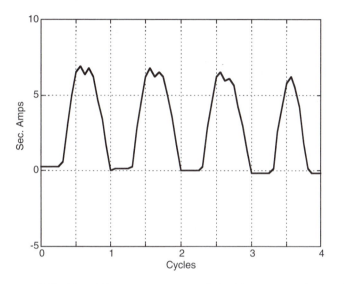

FIGURE 3.8.19 Element 1 high-side winding current, I_{AB}, recorded while energizing the transformer with an A-phase external fault.

3.8.7.1.3 Low-Current Detection

The waveform has a low-differential-current section that lasts one-quarter of a cycle each cycle, the minimum time that the element requires for blocking; this element marginally maintains its security.

3.8.7.1.4 Second- and Fourth-Harmonic Restraint

The low second- and fourth-harmonic content produces misoperation of the differential element that uses independent harmonic restraint.

3.8.7.1.5 DC-Ratio Blocking

The ratio of the positive to the negative dc value is zero, so this element properly blocks the differential element.

3.8.7.2 Case 2

This case is similar to Case 1, but it differs in that the transformer is loaded while being energized with reduced A-phase voltage. Figure 3.8.22 shows the delta–wye distribution transformer; the CT connections are wye and delta to compensate for transformer phase shift. In this application, the differential relay does not need to make internal phase-shift compensation.

Figure 3.8.23 shows the differential Element 1 inrush current from the high- and low-side transformer windings after relay scaling. The two signals are 180° out of phase, but they have different instantaneous values. These values create the differential current shown in Figure 3.8.24.

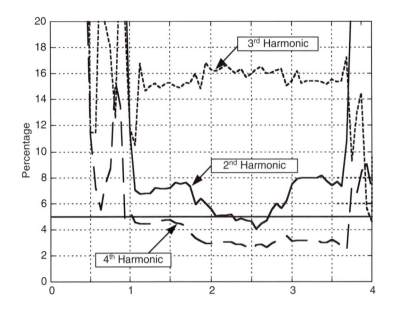

FIGURE 3.8.20 Second, third, and fourth harmonic as percentages of fundamental of the inrush current where the third-harmonic content is greater than the even-harmonic content.

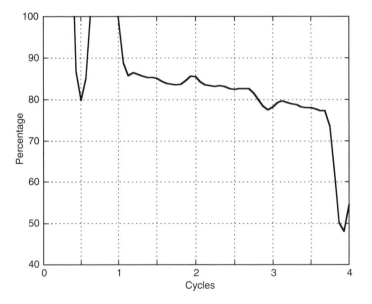

FIGURE 3.8.21 DC component as percentage of the rms value of fundamental during inrush conditions.

Figure 3.8.25 and Figure 3.8.26 show the harmonic and dc content, respectively, of the differential current as a percentage of fundamental. This signal has a second-harmonic content that drops to 7%, while the fourth harmonic drops to approximately 10%. In this case, the even harmonics, especially the fourth, provide information to properly restrain or block the differential element. The dc content also provides information that adds security to the differential element.

The differential elements operate as follows:

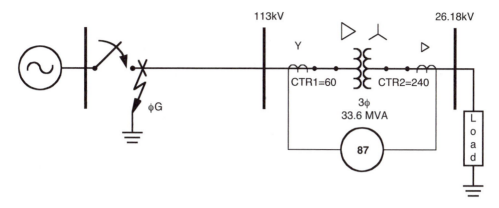

FIGURE 3.8.22 Transformer energization while A-phase is faulted and the transformer is loaded.

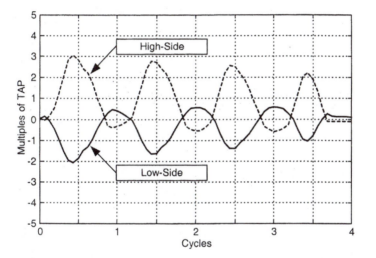

FIGURE 3.8.23 Element 1 inrush currents from the high- and low-side transformer windings after relay scaling.

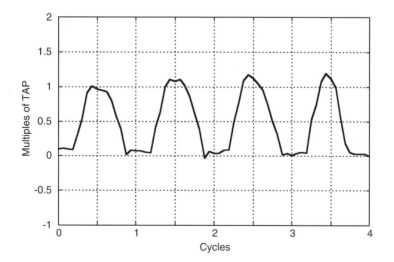

FIGURE 3.8.24 Differential current during transformer energization with the power transformer loaded.

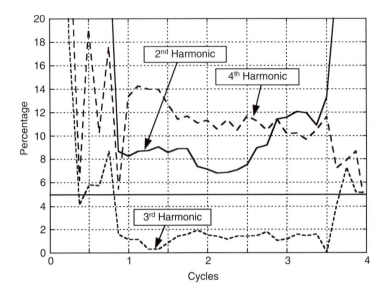

FIGURE 3.8.25 When the loaded transformer is energized with reduced voltage, the fourth harmonic provides information to restrain or block the differential element.

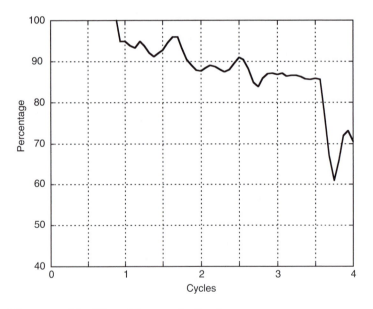

FIGURE 3.8.26 DC content of the differential current for Case 2.

3.8.7.2.1 Second- and Fourth-Harmonic Blocking

The second and fourth harmonics properly block the differential element. Notice that the second-harmonic percentage must be set to 6% for independent harmonic blocking applications.

3.8.7.2.2 All-Harmonic Restraint

The harmonic-restraint relay that uses all harmonics maintains its security because of the even-harmonic content of the signal.

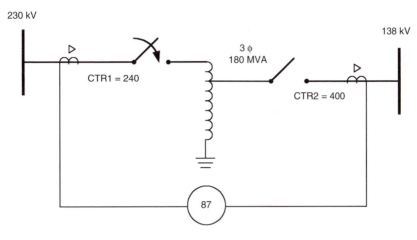

FIGURE 3.8.27 Transformer energization during commissioning.

3.8.7.2.3   Low-Current Detection
The waveform has a low-differential-current section that lasts longer than one-quarter cycle, so this logic properly blocks the differential element.

3.8.7.2.4   Second- and Fourth-Harmonic Restraint
The even-harmonic content of the signal restrains the differential relay from tripping.

3.8.7.2.5   DC-Ratio Blocking
The ratio of the positive to the negative dc value is close to zero at the beginning of the event and increases to values greater than 0.1, so this element only blocks at the beginning of the event.

3.8.7.3   Case 3

Figure 3.8.27 shows a field case of the energization during commissioning of a three-phase, 180-MVA, 230/138-kV autotransformer. The autotransformer connection is wye–wye; the CTs are connected in delta at both sides of the autotransformer.

Figure 3.8.28 shows the relay secondary currents from the autotransformer high side. These currents result from autotransformer energization with the low-side breaker open. The currents are typical inrush waves with a relatively small magnitude. Notice that the signal low-current intervals last less than one-quarter cycle.

Figure 3.8.29 shows the harmonic content of the inrush current. We can see that the inrush current has a relatively small second-harmonic percentage, which drops to approximately 9%.

As in previous cases, Figure 3.8.30 shows that the dc content of the inrush current is high during the event.

All differential elements except the low-current detector operate correctly for this case. The low-current zone in this case lasts less than the one-quarter cycle required to determine blocking conditions.

Table 3.8.3 summarizes the performance of the different inrush-detection methods discussed earlier.

The all-harmonic restraint method performs correctly for all three cases, as seen in Table 3.8.3. This method sacrifices relay dependability during symmetrical CT saturation conditions. Combining the even-harmonic restraint method and the dc-ratio blocking method provides a good compromise of speed and reliability.

3.8.8   Conclusions

1. Most transformer differential relays use the harmonics of the operating current to distinguish internal faults from magnetizing inrush or overexcitation conditions. The harmonics can be used to restrain or to block relay operation. Harmonic-restraint and -blocking methods ensure relay

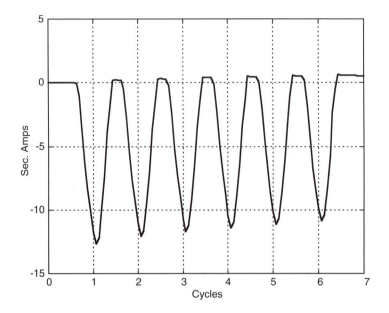

FIGURE 3.8.28 Inrush current with low-current intervals lasting less than one-quarter cycle.

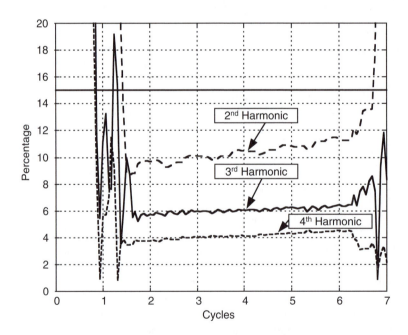

FIGURE 3.8.29 Second-harmonic percentage drops to approximately 9%.

security for a very high percentage of inrush and overexcitation cases. However, these methods fail for cases with very low harmonic content in the operating current.

2. Common harmonic restraint or blocking increases differential-relay security, but it could delay relay operation for internal faults combined with inrush currents in the nonfaulted phases.

3. Waveshape-recognition techniques represent another alternative for discriminating internal faults from inrush conditions. However, these techniques fail to identify transformer overexcitation conditions.

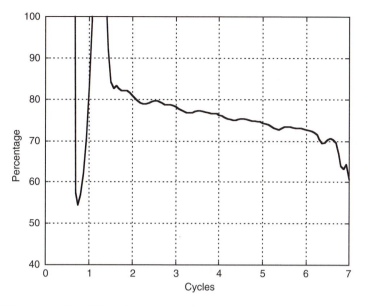

FIGURE 3.8.30 DC content of the differential current for Case 3.

TABLE 3.8.3 Inrush Detection Methods Performance during Inrush Conditions

Method	Case 1	Case 2	Case 3
Second- and fourth-harmonic blocking	☒ Low even-harmonic content	☑ Second-harmonic setting 6%	☑ Second-harmonic setting 8%
All-harmonic restraint	☑ High third-harmonic content	☑ Even-harmonic content	☑ Harmonic content
Low-current detection	☑ Low-current interval = $^1/_4$ cycle	☑ Low-current interval > $^1/_4$ cycle	☒ Low-current interval < $^1/_4$ cycle
Even-harmonic restraint	☒ Low even-harmonic content	☑ Even-harmonic content	☑ Even-harmonic content
DC-ratio blocking	☒ DC ratio = zero	☒ DC ratio > 0.1 after 1 cycle	☑ DC ratio = zero

Note: ☑ = Correct operation; ☒ = Misoperation.

4. An improved approach for transformer protection that combines harmonic-restraint and -blocking methods with a waveshape-recognition technique provides added security to the independent harmonic-restraint element without sacrificing dependability. This new method uses even harmonics for restraint, plus dc component and fifth harmonic for blocking.

5. Using even-harmonic restraint ensures security for inrush currents with low second-harmonic content and maintains dependability for internal faults with CT saturation. The use of fifth-harmonic blocking guarantees an invariant relay response to overexcitation. Using dc-offset blocking ensures security for inrush conditions with very low total harmonic distortion.

References

1. Klingshirn, E.A., Moore, H.R., and Wentz, E.C., Detection of faults in power transformers, *AIEE Trans.*, 76, 87–95, 1957.

2. Madill, J.T., Typical transformer faults and gas detector relay protection, *AIEE Trans.*, 66, 1052–1060, 1947.
3. Bean, R.L. and Cole, H.L., A sudden gas pressure relay for transformer protection, *AIEE Trans.*, 72, 480–483, 1953.
4. Monseth, I.T. and Robinson, P.H., *Relay Systems: Theory and Applications*, McGraw Hill, New York, 1935.
5. Sleeper, H.P., Ratio differential relay protection, *Electrical World*, 827–831, October 1927.
6. Cordray, R.E., Percentage differential transformer protection, *Electrical Eng.*, 50, 361–363, 1931.
7. Cordray, R.E., Preventing false operation of differential relays, *Electrical World*, 160–161, July 25, 1931.
8. Harder, E.L. and Marter, W.E., Principles and practices of relaying in the United States, *AIEE Trans.*, 67, 1005–1023, 1948.
9. Kennedy, L.F., and Hayward, C.D., Harmonic-current-restrained relays for differential protection, *AIEE Trans.*, 57, 262–266, 1938.
10. Hayward, C.D., Harmonic-current-restrained relays for transformer differential protection, *AIEE Trans.*, 60, 377–382, 1941.
11. Mathews, C.A., An improved transformer differential relay, *AIEE Trans.*, 73, 645–650, 1954.
12. Sharp, R.L. and Glassburn, W.E., A transformer differential relay with second-harmonic restraint, *AIEE Trans.*, 77, 913–918, 1958.
13. Einval, C.H. and Linders, J.R., A three-phase differential relay for transformer protection, *IEEE Trans.*, PAS-94, 1971–1980, 1975.
14. Dmitrenko, A.M., Semiconductor pulse-duration differential restraint relay, *Izv. Vysshikh Ucheb-nykh Zavedenii, Elektromekhanika*, No. 3, 335–339, March 1970 (in Russian).
15. Atabekov, G.I., *The Relay Protection of High Voltage Networks*, Pergamon Press, London, 1960.
16. Rockefeller, G.D., Fault protection with a digital computer, *IEEE Trans.*, PAS-98, 438–464, 1969.
17. Wilkinson, S.B., Transformer Differential Relay, U.S. Patent 5,627,712, May 6, 1997.
18. Hayward, C.D., Prolonged inrush currents with parallel transformers affect differential relaying, *AIEE Trans.*, 60, 1096–1101, 1941.
19. Blume, L.F., Camilli, G., Farnham, S.B., and Peterson, H.A., Transformer magnetizing inrush currents and influence on system operation, *AIEE Trans.*, 63, 366–375, 1944.
20. Specht, T.R., Transformer magnetizing inrush current, *AIEE Trans.*, 70, 323–328, 1951.
21. AIEE Committee Report, Report on transformer magnetizing current and its effect on relaying and air switch operation, *AIEE Trans.*, 70, 1733–1740, 1951.
22. Sonnemann, W.K., Wagner, C.L., and Rockefeller, G.D., Magnetizing inrush phenomena in transformer banks, *AIEE Trans.*, 77, 884–892, 1958.
23. Berdy, J., Kaufman, W., and Winick, K., A Dissertation on Power Transformer Excitation and Inrush Characteristics, presented at Symposium on Transformer Excitation and Inrush Characteristics and Their Relationship to Transformer Protective Relaying, Houston, TX, 1976.
24. Zocholl, S.E., Guzmán, A., and Hou, D., Transformer Modeling as Applied to Differential Protection, presented at 22nd Annual Western Protective Relay Conference, Spokane, WA, 1995.
25. Cooper Power Systems, *Electric Power System Harmonics: Design Guide*, Cooper Power Systems, Bulletin 87011, Franksville, WI, 1990.
26. Waldron, J.E. and Zocholl, S.E., Design Considerations for a New Solid-State Transformer Differential Relay with Harmonic Restraint, presented at Fifth Annual Western Protective Relay Conference, Sacramento, CA, 1978.
27. Marshall, D.E. and Langguth, P.O., Current transformer excitation under transient conditions, *AIEE Trans.*, 48, 1464–1474, 1929.
28. Wentz, E.C. and Sonnemann, W.K., Current transformers and relays for high-speed differential protection with particular reference to offset transient currents, *AIEE Trans.*, 59, 481–488, 1940.

29. Concordia, C. and Rothe, F.S., Transient characteristics of current transformers during faults, *AIEE Trans.*, 66, 731–734, 1947.

30. IEEE Power Engineering Society, Transient Response of Current Transformers, Special Publication 76 CH 1130-4 PWR, Institute of Electrical and Electronics Engineers, Piscataway, NJ, 1976.

31. IEEE, Guide for the Application of Current Transformers Used for Protective Relaying Purposes, IEEE C37.110-1996, Institute of Electrical and Electronics Engineers, Piscataway, NJ, 1996.

32. Sykes, J.A., A New Technique for High Speed Transformer Fault Protection Suitable for Digital Computer Implementation, IEEE Paper No. C72 429-9, presented at Power Engineering Society Summer Meeting, San Francisco, 1972.

33. Thorp, J.S. and Phadke, A.G., A Microprocessor-Based, Voltage-Restrained, Three-Phase Transformer Differential Relay, presented at Proceedings of the Southeastern Symposium on System Theory, Blacksburg, VA, April 1982, pp. 312–316.

34. Thorp, J.S. and Phadke, A.G., A new computer based, flux restrained, current differential relay for power transformer protection, *IEEE Trans.*, PAS-102, 3624–3629, 1983.

35. Inagaki, K., Higaki, M., Matsui, Y., Kurita, K., Suzuki, M., Yoshida, K., and Maeda, T., Digital protection method for power transformers based on an equivalent circuit composed of inverse inductance, *IEEE Trans. Power Delivery*, 3, 1501–1510, 1998.

36. General Electric Co., Transformer Differential Relay with Percentage and Harmonic Restraint Types BDD 15B, BDD16B, Document GEH-2057F, GE Protection and Control, Malvern, PA.

37. Bertula, G., Enhanced transformer protection through inrush-proof ratio differential relays, *Brown Boveri Review*, 32, 129–133, 1945.

38. Giuliante, A. and Clough, G., Advances in the Design of Differential Protection for Power Transformers, presented at 1991 Georgia Tech. Protective Relaying Conference, Atlanta, 1991, pp. 1–12.

39. Hegazy, M., New principle for using full-wave rectifiers in differential protection of transformers, *IEE Proc.*, 116, 425–428, 1969.

40. Dmitrenko, A.M., The use of currentless pauses for detuning differential protection from transient imbalance currents, *Elektrichestvo*, No. 1, 55–58, January 1979 (in Russian).

41. Robertson, D., Ed., *Power System Protection Reference Manual — Reyrolle Protection*, Oriel Press, London, 1982.

42. Edgeley, R.K. and Hamilton, F.L., The application of transductors as relays in protective gear, *IEE Proc.*, 99, 297, 1952.

43. Michelson, E.L., Rectifier relay for transformer protection, *AIEE Trans.*, 64, 252–254, 1945.

44. Podgornyi, E.V. and Ulianitskii, E.M., A comparison of principles for detuning differential relays from transformer inrush currents, *Elektrichestvo*, No. 10, 26–32, October 1969 (in Russian).

3.9   Causes and Effects of Transformer Sound Levels

Jeewan Puri

In many cities today there are local ordinances specifying maximum allowable sound levels at commercial and residential property lines. Consequently, the sound energy radiated from transformers has become a factor of increasing importance to the neighboring residential areas. It is therefore appropriate that a good understanding of sound power radiation and its measurement principles be developed for appropriately specifying sound levels in transformers. An understanding of these principles can be helpful in minimizing community complaints regarding the present and future installations of transformers.

3.9.1   Transformer Sound Levels

An understanding of the following basic principles is necessary to evaluate a sound source to quantify the sound energy.

3.9.1.1 Sound Pressure Level

The main quantity used to describe a sound is the size or amplitude of the pressure fluctuations at a human ear. The weakest sound a healthy human ear can detect has an amplitude of 20 millionths of a pascal (20 µPa). A pressure change of 20 µPa is so small that it causes the eardrum to deflect a distance less than the diameter of a single hydrogen molecule. Amazingly, the ear can tolerate sound pressures more than a million times higher. Thus, if we measured sound in Pa, we would end up with some quite large, unmanageable numbers. To avoid this, another scale is used — the decibel or dB scale.

The decibel is not an absolute unit of measurement. It is a ratio between a measured quantity and an agreed reference level. The dB scale is logarithmic and uses the hearing threshold of 20 µPa as the reference level. This is defined as 0 dB. A sound pressure level L_p can therefore be defined as:

$$L_p = 10 \ \log \ (P/P_0)^2 \tag{3.9.1}$$

where

L_p = sound pressure level, dB
P_0 = reference level = 20 µPa

One useful aspect of the decibel scale is that it gives a much better approximation of the human perception of relative loudness than the pascal scale.

3.9.1.2 Perceived Loudness

We have already defined sound as a pressure variation that can be heard by a human ear. A healthy human ear of a young person can hear frequencies ranging from 20 Hz to 20 kHz. In terms of sound pressure level, audible sounds range from the threshold of hearing at 0 dB to the threshold of pain, which can be over 130 dB.

Although an increase of 6 dB represents a doubling of the sound pressure, in actuality, an increase of about 10 dB is required before the sound subjectively appears to be twice as loud. The smallest change in sound level we can perceive is about 3 dB.

The subjective or perceived loudness of a sound is determined by several complex factors. One such factor is that the human ear is not equally sensitive at all frequencies. It is most sensitive to sounds between 2 kHz and 5 kHz and less sensitive at higher and lower frequencies.

3.9.1.3 Sound Power

A source of sound radiates energy, and this results in a sound pressure. Sound energy is the cause. Sound pressure is the effect. Sound power is the rate at which energy is radiated (energy per unit time). The sound pressure that we hear (or measure with a microphone) is dependent on the distance from the source and the acoustic environment (or sound field) in which sound waves are present. This in turn depends on the size of the room and the sound absorption characteristics of its wall surfaces. Therefore the measuring of sound pressure does not necessarily quantify how much noise a machine makes. It is necessary to find the sound power because this quantity is more or less independent of the environment and is the unique descriptor of the noisiness of a sound source.

3.9.1.4 Sound Intensity Level

Sound intensity describes the rate of energy flow through a unit area. The units for sound intensity are watts per square meter.

Sound intensity also gives a measure of direction, as there will be energy flow in some directions but not in others. Therefore, sound intensity is a vector quantity, as it has both magnitude and direction. On the other hand, pressure is a scalar quantity, as it has magnitude only. Usually we measure the intensity in a direction normal (at 90°) to a specified unit area through which the sound energy is flowing.

Sound intensity is measured as the time-averaged rate of energy flow per unit area. At some points of measurements, energy may be traveling back and forth. If there is no net energy flow in the direction of measurement, there will be no net recorded intensity.

Like sound pressure, sound intensity level L_I is also quantified using a dB scale, where the measured intensity I in W/m² is expressed as a ratio to a reference intensity level I_0 as follows:

$$L_I = 10 \log (I/I_0) \tag{3.9.2}$$

where

L_I = sound intensity level, dB
I_0 = reference level = 10^{-12} W/m²

3.9.1.5 Relationship between Sound Intensity and Sound Pressure Level

For any free progressive wave, there is a unique relation between the mean-square sound pressure and the intensity. This relation at a particular point and in the direction of the wave propagation is described as follows:

$$I = P^2_{rms}/\rho c \tag{3.9.3}$$

where

I = sound intensity, W/m²
P^2_{rms} = mean-square sound pressure, (N/m²)², measured at that particular point where I is desired in the free progressive wave
ρc = characteristic resistance, mks rayls

Note that for air, at T = 20°C and atmospheric pressure = 0.751 m of Hg, ρc = 406 mks rayls. This value does not change significantly over generally encountered ambient temperature and atmospheric pressure conditions.

As described in Equation 3.9.2, sound intensity level, in decibels is:

$$L_I = 10 \log (I/I_0)$$

where I = sound intensity (power passing in a specified direction through a unit area), W/m²

Combining the above equations, the sound intensity level can be expressed as:

$$L_I = 10 \log [(P^2_{rms}/\rho c)/I_0] = 10 \log (P_{rms}/P_0)^2 + 10 \log [(P^2_0/\rho c)/I_0] \tag{3.9.4}$$

From this expression, L_I can be defined as follows:

$$L_I = L_p - 10 \log K \tag{3.9.5}$$

where K = constant = $I_0 \times \rho c/P^2_0$, which is dependent upon ambient pressure and temperature
By definition,

$$P^2_0/I_0 = (20 \times 10^{-6})^2/10^{-12} = 400 \text{ mks rayls}$$

Note that the quantity 10 log K will equal zero when K = 1 or when ρc equals 400. As described earlier, under commonly encountered temperature and atmospheric conditions, $\rho c \approx 400$. Therefore, in free field measurements, $L_p \approx L_I$. That is, noise pressure and noise intensity measurement, in free space, yield the same numerical value.

3.9.2 Sound-Energy Measurement Techniques

Sound level of a source can be measured by directly measuring sound pressure or sound intensity at a known distance. Both of these measurement techniques are quite equivalent and acceptable. In most of the industry worldwide, sound-pressure measurements have been used for quantifying sound levels in transformers. As a result of the recent work completed by CIGRE, sound-intensity measurements are now being incorporated as an alternative in IEC Std. 60076-10.

3.9.2.1 Sound-Pressure-Level Measurement

A sound-level meter is an instrument designed to respond to sound in approximately the same way as the human ear and to give objective, reproducible measurements of sound pressure level. There are many different sound measuring systems available. Although different in detail, each system consists of a microphone, a processing section and a readout unit.

The microphone converts the sound signal to an equivalent electrical signal. The most suitable type of microphone for sound-level meters is the condenser microphone, which combines precision with stability and reliability. The electrical signal produced by the microphone is quite small. It is therefore amplified by a preamplifier before being processed.

Several different types of processing may be performed on the signal. The signal may pass through a weighting network of filters. It is relatively simple to build an electronic circuit whose sensitivity varies with frequency in the same way as the human ear, thus simulating the equal-loudness contours. This has resulted in three different internationally standardized characteristics termed the A, B, and C weightings. The A-weighting network is the most widely used, since the B and C weightings do not correlate well with subjective tests.

3.9.2.2 Sound-Intensity Measurements

Until recently, only sound pressure that was dependent on the sound field could be measured. Sound power can be related to sound pressure only under carefully controlled conditions where special assumptions are made about the sound field. Therefore, a noise source had to be placed in specially constructed rooms, such as anechoic or reverberant chambers, to measure sound power levels with the desired accuracy.

Sound intensity, however, can be measured in any sound field. This property allows all the measurements to be done directly in situations where a plurality of sound sources are present. Measurements on any sound source can be made even when all the others are radiating noise simultaneously. Sound intensity measurements are directional in nature and only measure sound energy radiated form a sound source. Therefore, steady background noise makes no contribution to the sound power of the source determined with sound-intensity measurements.

Sound intensity is the time-averaged product of the pressure and particle velocity. A single microphone can measure pressure. However, measuring particle velocity is not as simple. With Euler's linearized equation, the particle velocity can be related to the pressure gradient (i.e., the rate at which the instantaneous pressure changes with distance).

Euler's equation is essentially Newton's second law applied to a fluid. Newton's second law relates the acceleration given to a mass to the force acting on it. If the force and the mass are known, the acceleration can be found. This can then be integrated with respect to time to find the velocity.

With Euler's equation, it is the pressure gradient that accelerates a fluid of density ρ. With the knowledge of pressure gradient and density of the fluid, the particle acceleration (or deceleration) can be calculated as follows.

$$a = -1/\rho \ \partial P/\partial r \qquad (3.9.6)$$

where

a = particle deceleration due to a pressure change ∂P in a fluid of density ρ across a distance ∂r

Integrating Equation 3.9.6 gives the particle velocity, u, as follows:

$$u = \int \circ 1/\rho \ \partial P/\partial r \ dt \qquad (3.9.7)$$

It is possible to measure the pressure gradient with two closely spaced microphones facing each other and relate it to the particle velocity using the above equation. With two closely spaced microphones A

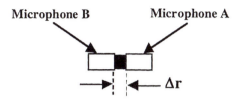

FIGURE 3.9.1 Measurement of pressure gradient using two closely spaced microphones.

and B separated by a distance Δr (Figure 3.9.1), it is possible to obtain a linear approximation of the pressure gradient by taking the difference in their measured pressures, P_A and P_B, and dividing it by the distance Δr between them. This is called a finite-difference approximation.

The pressure-gradient signal must now be integrated to give the particle velocity u as follows:

$$u = -1/\rho \int [(P_A - P_B)/\Delta r]dt \qquad (3.9.8)$$

Since intensity I is the time-averaged product of pressure P and particle velocity u,

$$I = -P/\rho \int [(P_A - P_B)/\Delta r]dt \qquad (3.9.9)$$

where $P = (P_A + P_B)/2$.

This is the basic principle of signal processing in sound-intensity-measuring equipment.

3.9.3 Sources of Sound in Transformers

Unlike cooling-fan or pump noise, the sound radiated from a transformer is tonal in nature, consisting of even harmonics of the power frequency. It is generally recognized that the predominant source of transformer noise is the core. The low frequency, tonal nature of this noise makes it harder to mitigate than the broadband higher frequency noise that comes from the other sources. This is because low frequencies propagate farther with less attenuation. Also, tonal noise can be perceived more acutely than broadband levels, even with high background noise levels. This combination of low attenuation and high perception makes tonal noise the dominant problem in the neighboring communities around transformers. To address this problem, most noise ordinances impose penalties or stricter requirements for tonal noise.

Even though the core is the principal noise source in transformers, the load noise, which is primarily caused by the electromagnetic forces in the windings, can also be a significant influence in low-sound-level transformers. The cooling equipment (fans and pumps) noise typically dominates the very low- and very high-frequency ends of the sound spectrum, whereas the core noise dominates in the intermediate range of frequencies between 100 and 600 Hz. These sound-producing mechanisms can be further characterized as follows.

3.9.3.1 Core Noise

When a strip of iron is magnetized, it undergoes a very small change in its dimensions (usually only a few parts in a million). This phenomenon is called magnetostriction. The change in dimension is independent of the direction of magnetic flux; therefore, it occurs at twice the line frequency. Because the magnetostriction curve is nonlinear, higher harmonics of even order also appear in the resulting core vibration at higher induction levels (above 1.4 T).

Flux density, core material, core geometry, and the wave form of the excitation voltage are the factors that influence the magnitude and frequency components of the transformer core sound levels. The mechanical resonance in transformer mounting structure as well as in core and tank walls can also have a significant influence on the magnitude of transformer vibrations and, consequently, on the acoustic noise generated.

3.9.3.2 Load Noise

Load noise is caused by vibrations in tank walls, magnetic shields, and transformer windings due to the electromagnetic forces resulting from leakage fields produced by load currents. These electromagnetic forces are proportional to the square of the load currents.

The load noise is predominantly produced by axial and radial vibration of transformer windings. However, marginally designed magnetic shielding can also be a significant source of sound in transformers. A rigid design for laminated magnetic shields with firm anchoring to the tank walls can greatly reduce their influence on the overall load sound levels. The frequency of load noise is usually twice the power frequency. An appropriate mechanical design for laminated magnetic shields can be helpful in avoiding resonance in the tank walls. The design of the magnetic shields should take into account the effects of overloads to avoid saturation, which would cause higher sound levels during such operating conditions.

Studies have shown that except in very large coils, radial vibrations do not make any significant contribution to the winding noise. The compressive electromagnetic forces produce axial vibrations and thus can be a major source of sound in poorly supported windings. In some cases, the natural mechanical frequency of winding clamping systems may tend to resonate with electromagnetic forces, thereby severely intensifying the load noise. In such cases, damping of the winding system may be required to minimize this effect.

The presence of harmonics in load current and voltage, most especially in rectifier transformers, can produce vibrations at twice the harmonic frequencies and thus a sizeable increase in the overall sound level of a transformer.

Through several decades, the contribution of the load noise to the total transformer noise has remained moderate. However, in transformers designed with low induction levels and improved core designs for complying with low sound-level specifications, the load-dependent winding noise of electromagnetic origin can become a significant contributor to the overall sound level of the transformer. In many such cases, the sound power of the winding noise is only a few dB below that of the core noise.

3.9.3.3 Fan and Pump Sound

Power transformers generate considerable heat because of the losses in the core, coils, and other metallic structural components of the transformer. This heat is removed by fans that blow air over radiators or coolers. Noise produced by the cooling fans is usually broadband in nature. Cooling fans usually contribute more to the total noise for transformers of smaller ratings and for transformers that are operated at lower levels of core induction. Factors that affect the total fan noise output include tip speed, blade design, number of fans, and the arrangement of the radiators.

3.9.4 Sound Level and Measurement Standards for Transformers

In NEMA Publication TR-1, Tables 02 thorough 04 list standard sound levels for liquid-filled power, liquid-filled distribution, and dry-type transformers. These sound-level requirements must be met unless special lower sound levels are specified by the customer.

The present sound-level measurement procedures as described in IEEE standards C57.12.90 and C57.12.91 specify that the sound-level measurements on a transformer shall be made under no-load conditions. Sound-pressure measurements shall be made to quantify the total sound energy radiated by a transformer. Sound-intensity measurements have already been incorporated into IEC Sound Level Measurement Standard 60076-10 as an acceptable alternative. It is anticipated that this method will be adopted in the IEEE standards also in the near future. The following is a brief description of the procedures used for this determination.

3.9.4.1 Transformer Connections during Test

This test is performed by exciting one of the transformer windings at rated voltage of sinusoidal wave shape at rated frequency while all the other windings are open circuited. The tap changer (if any) is at the rated-voltage tap position. In some cases (e.g., transformers equipped with reactor-type on-load tap changers), a tap position other than the rated may be used if the transformer produces maximum sound levels at this position.

3.9.4.2 Principal Radiating Surface for Measurements

The principal radiating surface is that from which the sound energy is emanating toward the receiver locations. The location of the radiating surface is determined based on the proximity of the cooling equipment to the transformer.

For transformers with no cooling equipment (or with cooling equipment mounted less than 3 m from the transformer tank) or dry-type transformers with enclosures provided with cooling equipment (if any) inside the enclosure, the principal radiating surface is obtained by taking the vertical projection of a string contour surrounding the transformer and its cooling equipment (if any), as shown in Figure 3.9.2. The vertical projection begins at the tank cover and terminates at the base of the transformer.

Separate radiating surfaces for the transformer and its cooling equipment are determined if the cooling equipment is mounted more than 3 m from the transformer tank. The principal radiating surface for the cooling equipment is determined by taking the vertical projection of the string perimeter surrounding the cooling equipment, as shown in Figure 3.9.3. The vertical projection begins at the top of the cooling structure and terminates at its base.

3.9.4.3 Prescribed Contour Location for Measurements

All sound-level measurements are made on a prescribed contour located 0.3 m away from the radiating surface. The location of this contour depends on the radiating surface as determined by the proximity of the cooling equipment to the transformer, as shown in Figure 3.9.2 and Figure 3.9.3. The location of the prescribed contours above the base of the transformer shall be at half the tank height for transformer tanks <2.5 m high or at one-third and two-thirds the tank height for transformer tanks >2.5 m high.

3.9.4.4 Measuring Positions on Prescribed Contour

The first microphone position is located on the prescribed contour opposite the main tank drain valve. Proceeding in a clockwise direction (as viewed from the top of the transformer) additional measuring positions on the prescribed contour are located no more than 1 m apart.

The minimum number of measurements as stipulated in IEEE C57.12.90 or IEEE C57.12.91 for North American practice are taken on each prescribed contour. These standards specify that sound-level measurements shall be made with and without the cooling equipment in operation. IEC 60076-10 standard should be consulted for European practices, which are slightly different.

3.9.4.5 Sound-Pressure-Level Measurements

A-weighted sound-pressure-level measurements are the most commonly used method for determining sound levels in transformers. Sound pressure measurements are quite sensitive to the ambient sound levels on the test floor. Therefore, appropriate corrections for the ambient sound level and reflected sound from the surrounding surfaces must also be quantified to determine the true sound level of the transformer.

It is recommended that acceptable ambient sound level conditions should be met for obtaining reliable measurements on transformers. For this reason, industry standards specify that A-weighted ambient sound pressure levels must be measured immediately before and after the measurements on the transformer. The ambient noise level readings are taken at each microphone position on the prescribed contours with the transformer and cooling equipment (if any) de-energized. These measurements are used to correct the measurements made on the transformer. The magnitude of this correction depends on the difference between the ambient and the transformer sound levels. This difference should not be less than 5 dB for a valid measurement. No correction is necessary if the ambient sound level is more than 10 dB lower than the transformer sound level.

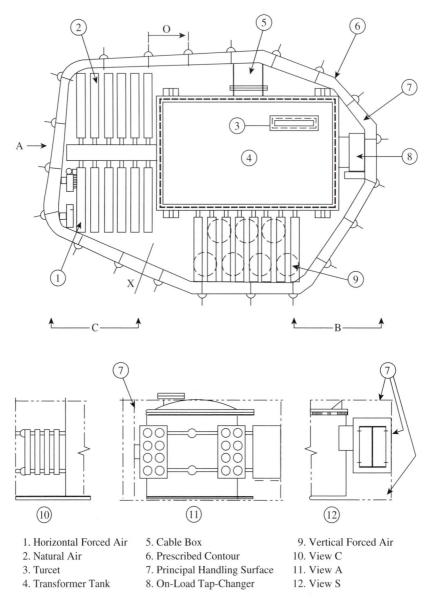

1. Horizontal Forced Air   5. Cable Box   9. Vertical Forced Air
2. Natural Air   6. Prescribed Contour   10. View C
3. Turcet   7. Principal Handling Surface 11. View A
4. Transformer Tank   8. On-Load Tap-Changer   12. View S

FIGURE 3.9.2 Typical microphone positions for sound measurement on transformers having cooling auxiliaries mounted either directly on the tank or on a separate structure spaced <3 m from the principal radiating surface of the main tank. (From IEC 60076-10. With permission.)

From the measured sound-pressure levels, L_{pAi}, at each microphone position on the prescribed contour, an A-weighted average sound-pressure level, L_{pA}, can be calculated using Equation 3.9.10:

$$L_{pA} = 10 \log \left[1/N \sum_{i=1}^{N} 100^{.1 \, L_{pAi}} \right] - K \; dB \tag{3.9.10}$$

where

L_{pA} = A-weighted average sound-pressure level, dB (reference = 20μPa)

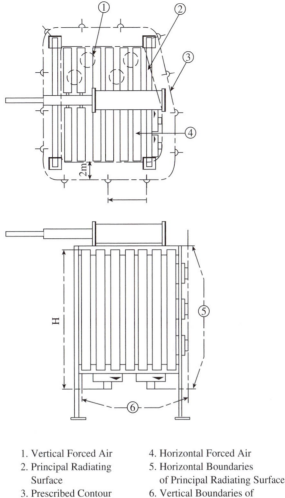

1. Vertical Forced Air
2. Principal Radiating Surface
3. Prescribed Contour
4. Horizontal Forced Air
5. Horizontal Boundaries of Principal Radiating Surface
6. Vertical Boundaries of Principal Radiating Surface

FIGURE 3.9.3 Typical microphone positions for sound measurement on cooling auxiliaries mounted on a separate structure spaced ≥3 m from the principal radiating surface of the transformer. (From IEC 60076-10. With permission.)

L_{pAi} = A-weighted sound-pressure level measured at the ith position and corrected for the ambient noise level per Table 1, dB (reference = 20μPa)

N = total number of measuring positions on the prescribed contour

K = environmental correction for the influence of reflected sound and ambient sound level, dB (IEEE and IEC standards should be consulted for details)

3.9.4.6 Sound-Intensity Measurements

The equipment for these measurements has only recently emerged in the industry. By definition, the A-weighted sound-intensity measurements provide a measure of sound power radiated in watts through a unit area per unit time. This type of measurement yields a vector quantity that represents the sound energy radiated in a direction normal to the principal radiating surface of the transformer.

The noise-intensity measuring probes use two matched microphones that respond to sound pressure so that the readings taken by them does not differ by more than 1.5 dB at any location.

From the measured sound intensity levels, L_{IAi}, at each position of the prescribed contour, an A-weighted average sound intensity level in decibels, L_{IA}, can be calculated using Equation 3.9.11:

$$L_{IA} = 10 \log \left[1/N \sum_{i=1}^{N} 100^{.1 \, Lia} \right] \text{dB} \qquad (3.9.11)$$

where

L_{IA} = A-weighted average sound-intensity level, dB (reference = 10^{-12} W/m²)
L_{ia} = A-weighted sound-intensity level measured at the *i*th position for the ambient noise level, (reference = 10^{-12} W/m²)
N = total number of measuring positions on the prescribed contour

Unlike sound pressure, the noise-intensity measurements are not influenced by the ambient noise level, provided that the ambient noise remains constant as the measurements are taken around the prescribed contour. Under such conditions, noise-intensity measurements can be made in ambient sound level even higher than the sound level of the transformer. For this reason, this type of measurement is especially suitable for transformers designed for very low noise levels.

It should be recognized that, at this time, the transformer industry's experience in sound intensity measurement is rather limited. Actual measurements published in CIGRE publications have demonstrated that the reliability of the sound-intensity measurements depends on the difference ΔL between the average sound intensity and pressure measurements made by the same probes. This work suggests that for optimum results, ΔL should not be more than 8dB.

3.9.4.7   Calculation of Sound Power Level

It is necessary to calculate sound pressure levels at property lines located away from the transformer in order to demonstrate compliance with the local ordinances.

Sound power level provides a measure of the total sound energy radiated by a transformer. With this quantity, the sound pressure levels at any desired distance (outside the prescribed contour) from the transformer can be calculated.

Since sound-pressure or sound-intensity measurements yield the same numerical result, the following equations can be used to calculate sound power levels in decibels from the measured sound-pressure or sound-intensity levels.

$$L_{WA} = L_{IA} + 10 \log (S) \qquad (3.9.12)$$

or

$$L_{WA} = L_{pA} + 10 \log (S) \qquad (3.9.13)$$

where

S = area of the radiating surface, m²
L_{WA} = 10 log (W/W$_0$) = A-weighted sound power level, dB (W$_0$ = 10^{-12} W)
L_{pA} = A-weighted average sound pressure level, dB (reference = 20 μPa)
L_{IA} = A-weighted average sound intensity level, dB (reference = 10^{-12} W/m²)

The calculation of effective radiating surface area S is also a function of the distance at which the sound measurements were made. IEEE or IEC standards can be consulted for details.

3.9.4.8   Sound-Pressure-Level Calculations at Far Field Receiver Locations

Sound-level requirements at a specific receiver location in the far field play a major role in specifying sound levels for transformers.

The following basic method can be considered for estimating transformer sound pressure levels for simple installations. These calculations provide an accurate estimate of transformer noise emissions at distances roughly greater than twice the largest dimension of the transformer.

Given the sound-pressure-level requirement of L_{pD} dB at a distance D from the transformer, the maximum allowable sound power L_W (in watts) for the transformer can be estimated using Equation 3.9.14:

$$L_W = L_{pD} + 10 \log (2 \Pi D^2) \tag{3.9.14}$$

Therefore, the maximum allowable sound-pressure level L_{pT} at the measurement surface near the transformer will be:

$$L_{pT} = L_W - 10 \log (\text{measurement surface area}) \tag{3.9.15}$$

where
measurement surface area, $m^2 = 1.25 \times$ transformer tank height $\times$ measurement contour length

3.9.5 Factors Affecting Sound Levels in Field Installations

To ensure repeatability, the factory measurements are made under controlled conditions specified by sound-measurement standards. Every effort is made to maintain core excitation voltage constant and sinusoidal to ensure the presence of a known core induction while making sound-level evaluations in transformers.

The effects of ambient sound levels and reflections must be subtracted from the overall sound-level measurements in order to determine the true sound power of a transformer. Therefore, it is essential that the ambient sound levels are constant and that they are measured accurately. Many times it becomes necessary to make these measurements in a sound chamber, where ambient sound level is very low and the effects of reflecting surfaces can also be eliminated.

The sound-level measurements at the transformer site can be drastically different, depending on its operating conditions. The effects of the following factors should therefore be considered while defining the sound-level requirements for transformers.

3.9.5.1 Load Power Factor

In the factory, core and winding sound levels are measured separately at rated voltage and full-load current, respectively. These sound levels are produced by core and winding vibrations of twice the power frequency and its even harmonics. It is assumed that these vibrations are in phase with each other, and therefore their power levels are added to predict the overall sound level of the transformer. However, this assumption applies only when the transformer is carrying purely resistive load.

Under actual operating conditions, depending on the power factor of the load, the phase angle between voltage and load current may induce a change in the factory-predicted transformer sound level.

3.9.5.2 Internal Regulation

The magnitude and the phase angle of the load currents also change the internal voltage drop in the transformer windings. The transformer loading conditions therefore can change the core induction level and significantly influence the core sound levels.

3.9.5.3 Load Current and Voltage Harmonics

During factory tests, only sinusoidal load current is simulated for measuring winding noise. This noise is produced by magnetic forces that are proportional to the square of the load current. However, harmonic content in the load current has a larger impact on the sound level than might be expected from the amplitude of the harmonic currents alone, since they interact with the power-frequency load current. In such cases, the magnetic force is proportional to the cross product between the power-frequency current and the harmonic current in addition to the force that is proportional to the square of load current and the square of the harmonic current. Thus, the highest contribution to the sound level due to the harmonic current occurs when the product of the load current and the harmonic current reaches the maximum.

The resulting audible tones are made up of frequencies of the harmonic current ± the fundamental power frequency.

Current harmonics are a major source of increase in sound levels in HVDC and rectifier transformers. Nonlinear loads cause harmonics in the excitation voltage, resulting in an increase in core sound levels. This influence must be considered when specifying sound level for a transformer.

3.9.5.4 DC Magnetization

Even a moderate dc magnetization of a transformer core will result in a significant increase in the transformer audible sound level. In addition to increasing the power level of the normal harmonics in the transformer vibrations (i.e., even harmonics of the power frequencies), dc magnetization will add odd harmonic tones to the overall sound level of the transformer.

Modern cores have high remnant flux density. Upon energization, the core sound levels may be as much as 20 dB higher than the factory test value. It is therefore recommended that a transformer should be energized for approximately six hours before evaluating its sound levels.

Traditionally, circuits like dc feeders to the transportation systems have been a source of dc fields in transformers. However, with the increased used of power electronic equipment in power transmission systems and in the industry, the number of possible sources for dc magnetization is increasing. Geomagnetic storms can also cause severe dc magnetization in transformers connected to long transmission lines.

3.9.5.5 Acoustical Resonance

Dry-type transformers are most frequently applied inside buildings. In a room with walls of low sound-absorption coefficient, the sound from the transformer will reflect back and forth between walls, resulting in a buildup of sound level in the room.

The number of dBs by which the sound level at the transformer will increase can be approximated using Equation 3.9.16:

$$\text{dB buildup} = 10 \log \left[1 + 4(1 - a) \, A_T/(a \, A_U) \right] \qquad (3.9.16)$$

where

A_T = surface area of the transformer
A_U = area of the reflecting surface
a = average absorption coefficient of the surfaces

In a room with concrete walls (with an absorption coefficient of 0.01) and with sound-reflecting surface area four times that of the transformer ($A_U/A_T = 4$), the increase in sound level at the transformer can be 20 dB. However, covering the reflecting surfaces of this room with sound-absorbing material with absorption coefficient of 0.3 will reduce this buildup to 5.5 dB.

Sound propagation is affected by many factors, such as atmospheric absorption, interceding barriers, and reflective surfaces. An explanation of these factors is beyond the scope of this text; however, they are mentioned to make the reader aware of their potential influence. If the existing site conditions will influence the sound propagation, the reader is advised to reference textbooks dealing with the subject of acoustic propagation or to consult an expert in conducting more accurate sound-propagation calculations.

References

Anon., *Measuring Sound*, Brüel & Kjaer, Nerum, Denmark, 1984.

Anon., *Sound Intensity*, Brüel & Kjaer, Nerum, Denmark, 1993.

Beranek, L., Ed., *Noise and Vibration Control*, Institute of Noise Control Engineering, Washington, D.C.

Fanton, J.P. et al., Transformer noise: determination of sound power level using the sound intensity measurement method, IEC SC 12-WG 12 (April 1990), *Electra*, 144, 1992.

Gade, S., *Sound Intensity Theory*, Technical Review No. 3, Brüel & Kjaer, Nerum, Denmark, 1982.

Gade, S., *Sound Intensity Instrumentation and Applications*, Technical Review No. 4, Brüel & Kjaer, Nerum, Denmark, 1982.

IEEE Audible Sound and Vibration Subcommittee Working Group, Guide for Sound Level Abatement and Determination for Liquid Immersed Power Transformers and Shunt Reactors Rated to be 500kVA, IEEE C57.156 (draft 9), Institute of Electrical and Electronics Engineers, Piscataway, NJ.

Replinger, E., Study of Noise Emitted by Power Transformers Based on Today's Viewpoint, CIGRE Paper 12-08, Siemens AG, Transformatorenwerk, Nürnberg, Germany, 1988.

Specht, T.R., Noise Levels of Indoor Transformers, Westinghouse Corp. Transformer Division, 1955.

Westinghouse Electric Corp., Bolt Beranek, and Newman Inc., Power Transformer Noise Abatement, ESERC Report EP 9-14, Empire State Electric Energy Research Corp., New York, 1981.

3.10   Transient-Voltage Response

Robert C. Degeneff

3.10.1   Transient-Voltage Concerns

3.10.1.1   Normal System Operation

Transformers are normally used in systems to change power from one voltage (or current) to another. This is often driven by a desire to optimize the overall system characteristics, e.g., economics, reliability, or performance. To achieve these system goals, a purchaser must specify — and a designer must configure — the transformer to meet a desired impedance, voltage rating, power rating, thermal characteristic, short-circuit strength, sound level, physical size, and voltage-withstand capability. Obviously, many of these goals will produce requirements that are in conflict, and prudent compromise will be required. Failure to achieve an acceptable characteristic for any of these goals will make the overall transformer design unacceptable. Transformer characteristics and the concomitant design process are outlined in the literature [1–4].

Normally, a transformer operates under steady-state voltage excitation. Occasionally, a transformer (in fact all electrical equipment) experiences a dynamic or transient overvoltage. Often, it is these infrequent transient voltages that establish design constraints for the insulation system of the transformer. These constraints can have a far-reaching effect on the overall equipment design. The transformer must be configured to withstand any abnormal voltages covered in the design specification and realistically expected in service. Often, these constraints have great impact on other design issues and, as such, have significant effect on the overall transformer cost, performance, and configuration. In recent years, engineers have explored the adverse effect of transient voltages on the reliability of transformers [5–7] and found them to be a major cause of transformer failure.

3.10.1.2   Sources and Types of Transient-Voltage Excitation

The voltages to which a transformer's terminals are subjected can be broadly classed as steady state and transient. The transient voltages a transformer experiences are commonly referred to as dynamic, transient, and very fast transients.

The majority of the voltages a transformer experiences during its operational life are steady state, e.g., the voltage is within ±10% of nominal, and the frequency is within 1% of rated. As power-quality issues grow, the effect of harmonic voltages and currents on performance is becoming more of an issue. These harmonics are effectively reduced-magnitude steady-state voltages and currents at harmonic frequencies (say 2nd to the 50th). These are addressed in great detail in IEEE Std. 519, Recommended Practices and Requirements for Harmonic Control in Electrical Power Systems [8]. Strictly speaking, all other voltage excitations are transients, e.g., dynamic, transient, and very fast transient voltages.

Dynamic voltages refer to relatively low-frequency (60 to 1500 Hz), damped oscillatory voltage. Magnitudes routinely observed are from one to three times the system's peak nominal voltage. Transient voltage refers to the class of excitation caused by events like lightning surges, switching events, and line faults causing voltages of the chopped waveform [9]. Normally, these are aperiodic waves. Occasionally, the current chopping of a vacuum breaker will produce transient periodic excitation in the 10- to 200-kHz range [10]. The term very fast transient encompasses voltage excitation with rise times in the range of 50 to 100 ηsec and frequencies from 0.5 to 30 MHz. These types of voltages are encountered in gas-insulated stations. The voltages produced within the transformer winding structure by the system is a part of the problem that must be addressed and understood if a successful insulation design is to be achieved [11]. Since transient voltages affect system reliability, in turn affecting system safety and economics, a full understanding of the transient characteristic of a transformer is warranted.

3.10.1.3 Addressing Transient-Voltage Performance

Addressing the issue of transient-voltage performance can be divided into three activities: recognition, prediction, and mitigation. By 1950 over 1000 papers had been written to address these issues [12–14]. The first concern is to appreciate that transient-voltage excitation can produce equipment responses different than one would anticipate at first glance. For example, the addition of more insulation around a conductor may, in fact, make the transient-voltage distribution worse and the insulation integrity of the design weaker. Another example is the internal voltage amplification a transformer experiences when excited near its resonant frequency. The transient-voltage distribution is a function of the applied voltage excitation and the shape and material content of the winding being excited. The capability of the winding to withstand the transient voltage is a function of the specific winding shape, the material's voltage-vs.-time characteristic, the past history of the structure, and the statistical nature of the voltage-withstand characteristic of the structure.

The second activity is to assess or predict the transient voltage within the coil or winding. Today, this generally is accomplished using a lumped-parameter model of the winding structure and some form of computer solution method that facilitates calculation of the internal transient response of the winding. Once this voltage distribution is known, its effect on the insulation structure can be computed with a two- or three-dimensional finite element method (FEM). The resultant voltages, stresses, and creeps are examined in light of the known material and geometrical capability of the system, with consideration for desired performance margins.

The third activity is to establish a transformer structure or configuration that — in light of the anticipated transient-voltage excitation and material capability, variability, and statistical nature — will provided acceptable performance margins. Occasionally, nonlinear resistors are used as part of the insulation system to achieve a cost-effective, reliable insulation structure. Additionally, means of limiting the transient excitation include the use of nonlinear resistors, capacitors, and snubbers.

3.10.1.4 Complex Issue to Predict

The accurate prediction of the transient-voltage response of coils and winding has been of interest for almost 100 years. The problem is complex for several reasons. The form of excitation varies greatly. Most large power transformer are unique designs, and as such each transformer's transient-response characteristic is unique. Each has its own impedance-vs.-frequency characteristic. As such, the transient-response characteristic of each transformer is different. Generally, the problem is addressed by building a large lumped-parameter model of inductances, capacitances, and resistances. Constructing the lumped-parameter model is challenging. The resultant mathematical model is ill-conditioned, e.g., the resultant differential equation is difficult to solve. The following sections outline how these challenges are currently addressed.

It should be emphasized that the voltage distribution within the winding is only the first component of the insulation design process. The spatial distribution of the voltages within the winding must be determined, and finally the ability of the winding configuration in view of its voltage-vs.-time characteristic must be assessed.

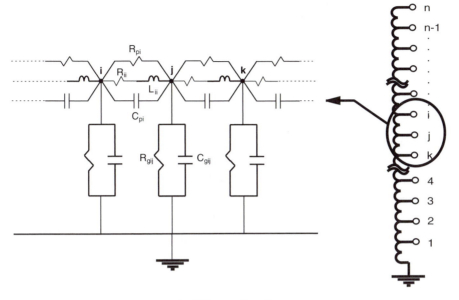

Mutual Inductances should be considered,

i.e. M_{ji} is between segments j and i.

FIGURE 3.10.1 Sample of section used to model example coil.

3.10.2 Surges in Windings

3.10.2.1 Response of a Simple Coil

Transformer windings are complex structures of wire and insulation. This is the result of many contradictory requirements levied during the design process. In an effort to introduce the basic concepts of transient response, a very simple disk coil was modeled and the internal transient response computed. The coil consisted of 100 identical continuous disk sections of 24 turns each. The inside radius of the coil is 318.88 mm; the space between each disk coil is 5.59 mm; and the coil was assumed to be in air with no iron core. Each turn was made of copper 7.75 mm in height, 4.81 mm in the radial direction, with 0.52 mm of insulation between the turns. For this example, the coil was subjected to a full wave with a 1.0 per-unit voltage. Figure 3.10.1 provides a sketch of the coil and the node numbers associated with the calculation. For this example, the coil has been subdivided into 50 equal subdivisions, with each subdivision a section pair. Figure 3.10.2 contains the response of the winding as a function of time for the first 200 μsec. It should be clear that the response is complex and a function of both the applied excitation voltage and the characteristics of the coil itself.

3.10.2.2 Initial Voltage Distribution

If the voltage distribution along the helical coil shown in Figure 3.10.2 is examined at times very close to time zero, it is observed that the voltage distribution is highly nonuniform. For the first few tenths of a μsecond, the distribution is dominated exclusively by the capacitive structure of the coil. This distribution is often referred to as the initial (or short time) distribution, and it is generally highly nonuniform. This initial distribution is shown in Figure 3.10.3. For example, examining the voltage gradient over the first 10% of the winding, one sees that the voltage is 82% rather that the anticipated 10%, or one sees a rather large enhancement or gradient in some portions of the winding.

The initial distribution shown in Figure 3.10.3 is based on the assumption that the coil knows how it is connected, i.e., it requires some current to flow in the winding, and this requires some few tens of a nanosecond. The initial distribution can be determined by evaluating the voltage distribution for the

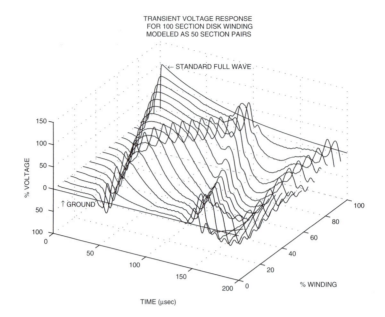

FIGURE 3.10.2 Voltage versus time for helical winding.

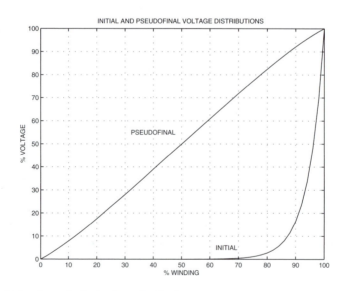

FIGURE 3.10.3 Initial and pseudo-final voltage distribution.

capacitive network of the winding and ignoring both the inductive and resistive components of the transformer. This discussion is applicable for times greater than approximately 0.25 μsec. This is the start of the transient response for the winding. For times smaller than 0.25 μsec, the distribution is still dictated by capacitance, but the transformer's capacitive network is unaware that it is connected. This is addressed in the chapter on very fast transients and in the work of Narang et al. [11].

3.10.2.3 Steady-State Voltage Distribution

The steady-state voltage distribution depends primarily on the inductance and losses of the windings structure. This distribution, referred to by Abetti [15] as the pseudo-final, is dominated primarily by the self-inductance and mutual inductance of the windings and the manner in which the winding is connected. This steady-state voltage distribution is very nearly (but not identical) to the turns ratio. For the simple winding shown in Figure 3.10.1, the distribution is known by inspection and shown in Figure 3.10.3, but for more complex windings it can be determined by finding the voltage distribution of the inductive network and ignoring the effect of winding capacitances.

3.10.2.4 Transient-Voltage Distribution

Figure 3.10.2 shows the transient response for this simple coil. The transient response is the voltage the coil experiences when the voltage within the coil is in transition between the initial voltage distribution and the steady-state distribution. It is the very same idea as pulling a rubber-band away from its stretched (but steady-state position) and then letting it go and monitoring its movement in space and time. What is of considerable importance is the magnitude and duration of these transient voltages and the ability of the transformer's insulation structure to consistently survive these voltages.

The coil shown in Figure 3.10.1 is a very simple structure. The challenge facing transformer design engineers since the early 1900s has been to determine what this voltage distribution would be for a complex winding structure of a commercial transformer design.

3.10.3 Determining Transient Response

3.10.3.1 History

Considerable research has been devoted to determining the transformer's internal transient-voltage distribution. These attempts started at the beginning of this century and have continued at a steady pace for almost 100 years [12–14]. The earliest attempts at using a lumped-parameter network to model the transient response was in 1915. Until the early 1960s, these efforts were of limited success due to computational limitations encountered when solving large numbers of coupled stiff differential equations. During this period the problem was attacked in the time domain using either a standing-wave approach or the traveling-wave method. The problem was also explored in the frequency domain and the resultant individual results combined to form the needed response in the time domain. Abetti introduced the idea of a scale model, or analog, for each new design. Then, in the 1960s, with the introduction of the high-speed digital computer, major improvements in computational algorithms, detail, accuracy, and speed were obtained.

In 1956 Waldvogel and Rouxel [16] used an analog computer to calculate the internal voltages in the transformer by solving a system of linear differential equations with constant coefficients resulting from a uniform, lossless, and linear lumped-parameter model of the winding, where mutual and self-inductances were calculated assuming an air core. One year later, McWhirter et al. [17], using a digital and analog computer, developed a method of determining impulse voltage stresses within the transformer windings; their model was applicable, to some extent, for nonuniform windings. But it was not until 1958 that Dent et al. [18] recognized the limitations of the analog models and developed a digital computer model in which any degree of nonuniformity in the windings could be introduced and any form of applied input voltage applied. During the mid 1970s, efforts at General Electric [19–21] focused on building a program to compute transients for core-form winding of completely general design. By the end of the 1970s, an adequate linear lossless model of the transformer was available to the industry. However, adequate representation of the effects of the nonlinear core and losses was not available. Additionally, the transformer models used in insulation design studies had little relationship with lumped-parameter transformer models used for system studies.

Wilcox improved the transformer model by including core losses in a linear-frequency-domain model where mutual and self-impedances between winding sections were calculated considering a grain-oriented conductive core with permeability μ_r [22]. In his work, Wilcox modeled the skin effect, the losses associated with the magnetizing impedance, and a loss mechanism associated with the effect of the flux radial component on the transformer core during short-circuit conditions. Wilcox applied his modified modal theory to model the internal voltage response on practical transformers. Vakilian [23] modified White's inductance model [19] to include the saturable characteristics of the core and established a system of ordinary differential equations for the linearized lumped-parameter induction resistive and capacitive (LRC) network. These researchers then used Gear's method to solve for the internal voltage response in time. During the same year, Gutierrez and Degeneff [24,25] presented a transformer-model reduction technique as an effort to reduce the computational time required by linear and nonlinear detailed transformer models and to make these models small enough to fit into the electromagnetic transients program (EMTP). In 1994 de Leon and Semlyen [26] presented a nonlinear three-phase-transformer model including core and winding losses. This was the first attempt to combine frequency dependency and nonlinearity in a detailed transformer model. The authors used the principle of duality to extend their model to three-phase transformers.

3.10.3.2 Lumped-Parameter Model

Transient response is a result of the flow of energy between the distributed electrostatic and electromagnetic characteristics of the device. For all practical transformer winding structures, this interaction is quite complex and can only be realistically investigated by constructing a detailed lumped-parameter model of the winding structure and then carrying out a numerical solution for the transient-voltage response. The most common approach is to subdivide the winding into a number of segments (or groups of turns). The method of subdividing the winding can be complex and, if not addressed carefully, can affect the accuracy of the resultant model. The resultant lumped-parameter model is composed of inductances, capacitances, and losses. Starting with these inductances, capacitances, and resistive elements, equations reflecting the transformer's transient response can be written in numerous forms. Two of the most common are the basic admittance formulation of the differential equation and the state-variable formulation. The admittance formulation [27] is given by:

$$[I(s)] = \left[\frac{1}{s}[\Gamma_n] + [G] + s[C]\right][E(s)] \tag{3.10.1}$$

The general state-variable formulation is given by Vakilian [23] describing the transformer's lumped-parameter network at time t:

$$\begin{bmatrix} [L] & 0 & 0 \\ 0 & [C] & 0 \\ 0 & 0 & [U] \end{bmatrix} \begin{bmatrix} [di_e/dt] \\ [de_n/dt] \\ [df_e/dt] \end{bmatrix} = \begin{bmatrix} -[r] & [T]^t \\ -[T] & -[G] \\ -[r] & [T]^t \end{bmatrix} \begin{bmatrix} [i_e] \\ [e_n] \end{bmatrix} - \begin{bmatrix} 0 \\ [I_s] \\ 0 \end{bmatrix} \tag{3.10.2}$$

where the variables in Equation 3.10.1 and Equation 3.10.2 are:

$[i_e]$ = vector of currents in the winding
$[e_n]$ = model's nodal voltages vector
$[f_e]$ = windings' flux-linkages vector
$[r]$ = diagonal matrix of windings series resistance
$[T]$ = windings connection matrix
$[T]^t$ = transpose of $[T]$
$[C]$ = nodal capacitance matrix
$[U]$ = unity matrix
$[I(s)]$ = Laplace transform of current sources
$[E(s)]$ = Laplace transform of nodal voltages
$[\Gamma_n]$ = inverse nodal inductance matrix = $[T][L]^{-1}[T]^t$
$[L]$ = matrix of self and mutual inductances
$[G]$ = conductance matrix for resistors connected between nodes
$[I_s]$ = vector of current sources

In a linear representation of an iron-core transformer, the permeability of the core is assumed constant regardless of the magnitude of the core flux. This assumption allows the inductance model to remain constant for the entire computation. Equation 3.10.1 and Equation 3.10.2 are based on this assumption that the transformer core is linear and that the various elements in the model are not frequency dependent. Work in the last decade has addressed both the nonlinear characteristics of the core and the frequency-dependent properties of the materials. Much progress has been made, but their inclusion adds considerably to the model's complexity and increases the computational difficulty. If the core is nonlinear, the permeability changes as a function of the material properties, past magnetization history, and instantaneous flux magnitude. Therefore, the associated inductance model is time dependent. The basic strategy for solving the transient response of the nonlinear model in Equation 3.10.1 or Equation 3.10.2 is to linearize the transformer's nonlinear magnetic characteristics at an instant of time based on the flux in the core at that instant.

Two other model formulations should be mentioned. De Leon addressed the transient problem using a model based on the idea of duality [26]. The finite element method has found wide acceptance in solving for electrostatic and electromagnetic field distributions. In some instances, it is very useful in solving for the transient distribution in coils and windings of complex shape.

3.10.3.3   Frequency-Domain Solution

A set of linear differential equations representing the transient response of the transformer can be solved either in the time domain or in the frequency domain. If the model is linear, the resultant solution will be the same for either method. The frequency-domain solution requires that the components of the input waveform at each frequency be determined; then these individual sinusoidal waves are applied individually to the transformer, and the resultant voltage response throughout the winding determined; and finally the total response in the time domain is determined by summing the component responses at each frequency by applying superposition. An advantage of this method is that it allows the recognition of frequency-dependent losses to be addressed easily. Disadvantages of this method are that it does not allow the modeling of time-dependent switches, nonlinear resistors like ZnO, or the recognition of nonlinear magnetic-core characteristics.

3.10.3.4   Solution in the Time Domain

The following briefly discusses the solution of Equation 3.10.1 and Equation 3.10.2. There are numerous methods to solve Equation 3.10.1, but it has been found that when solving the stiff differential equation model of a transformer, a generalization of the Dommel method [28,29] works very well. A lossless lumped-parameter transformer model containing n nodes has approximately $n(n + 1)/2$ inductors and $3n$ capacitors. Since the total number of inductors far exceeds the number of capacitors in the network, this methodology reduces storage and computational time by representing each capacitor as an inductor in parallel with a current source. The following system of equations results:

$$\left[\hat{Y}\right]\left[F(t)\right]=\left[I(t)\right]-\left[H(t)\right]$$

(3.10.3)

where

$[F(t)]$ = nodal integral of the voltage vector
$[I(t)]$ = nodal injected current vector
$[H(t)]$ = past history current

and

$$\left[\hat{Y}\right]=\frac{4}{\Delta t^2}[C]+\frac{2}{\Delta t}[G]+\left[\Gamma_n\right]$$

(3.10.4)

The lumped-parameter model is composed of capacitances, inductances, and losses computed from the winding geometry, permittivity of the insulation, iron core permeability, and the total number of sections into which the winding is divided. Then the matrix $[\hat{Y}]$ is computed using the integration step size, Δt. At every time step, the above system of equations is solved for the unknowns in the integral of the voltage vector. The unknown nodal voltages, $[E(t)]$, are calculated by taking the derivative of $[F(t)]$. Δt is selected based on the detail of the model and the highest resonant frequency of interest. Normally, Δt is smaller than one-tenth the period of this frequency.

The state-variable formulation shown as Equation 3.10.2 can be solved using differential equation routines available in IMSL or others based on the work of Gear or Adams [30]. The advantage of these routines is that they are specifically written with the solution of stiff systems of equations in mind. The disadvantage of these routines is that they consume considerable time during the solution.

3.10.3.5 Accuracy Versus Complexity

Every model of a physical system is an approximation. Even the simplest transformer has a complex winding and core structure and, as such, possesses an infinite number of resonant frequencies. A lumped-parameter model, or for that matter any model, is at best an approximation of the actual device of interest. A lumped-parameter model containing a structure of inductances, capacitances, and resistances will produce a resonant-frequency characteristic that contains the same number of resonant frequencies as nodes in the model. The transient behavior of a linear circuit (the lumped-parameter model) is determined by the location of the poles and zeros of its terminal-impedance characteristic. It follows then that a detailed transformer model must posses two independent characteristics to faithfully reproduce the transient behavior of the actual equipment. First, accurate values of R, L, and C, reflecting the transformer geometry. This fact is well appreciated and documented. Second, the transformer must be modeled with sufficient detail to address the bandwidth of the applied waveshape. In a valid model, the highest frequency of interest would have a period at least ten times larger than the travel time of the largest winding segment in the model. If this second characteristic is overlooked, a model can produce results that appear valid, but may have little physical basis. A final issue is the manner in which the transformer structure is subdivided. If care is not taken, the manner in which the model is constructed will itself introduce significant errors, with a mathematically robust computation but an inaccurate approximation of the physically reality.

3.10.4 Resonant Frequency Characteristic

3.10.4.1 Definitions

The steady-state and transient behavior of any circuit, for any applied voltage, is established by the location of the poles and zeros of the impedance function of the lumped-parameter model in the complex plane. The zeros of the terminal-impedance function coincide with the natural frequencies of the model, by definition. McNutt [31] defines terminal resonance as the terminal current maximum and a terminal impedance minimum. In a physical system, there are an infinite number of resonances. In a lumped-parameter model of a system, these are as many resonances as nodes in the model (or the order of the system). Terminal resonance is also referred to as series resonance [32,33]. Terminal antiresonance is defined as a terminal current minimum and a terminal impedance maximum [31]. This is also referred to as parallel resonance [32,33]. McNutt defines internal resonance as an internal voltage maximum and internal antiresonance as an internal voltage minimum.

3.10.4.2 Impedance versus Frequency

The terminal resonances for a system can be determined by taking the square root of the eigenvalues of the system matrix, $[A]$, shown in the state-variable representation for the system shown in Equation 3.10.5.

$$[\dot{q}] = [A][q] + [B][u]$$

$$[y] = [C][q] + [D][u] \tag{3.10.5}$$

where

  [A] = state matrix
  [B] = input matrix
  [C] = output matrix
  [D] = direct transmission matrix
  q = vector of state variables for system
  $\dot{q}$ = first derivative of [q]
  u = vector of input variables
  y = vector of output variables

The impedance-vs.-frequency characteristic requires a little more effort. In light of the previous definitions, terminal resonance can be defined as occurring when the reactive component of the terminal impedance is zero. Equivalently, terminal resonance occurs when the imaginary component of the quotient of the terminal voltage divided the injected terminal current is zero. Recalling that, in the Laplace domain, s is equivalent to $j\omega$ with a system containing n nodes with the excited terminal node j, one can rewrite Equation 3.10.1 to obtain:

$$\begin{bmatrix} e_i(s) \\ e_2(s) \\ - \\ e_j(s) \\ - \\ e_n(s) \end{bmatrix} = \begin{bmatrix} Z_{1j}(s) \\ Z_{2j}(s) \\ - \\ Z_{jj}(s) \\ - \\ Z_{nj}(s) \end{bmatrix} \begin{bmatrix} i_j(s) \end{bmatrix} \tag{3.10.6}$$

The voltage at the primary (node j) is in operational form. Rearranging the terminal impedance is given by:

$$Z_t(\omega) = Z_{jj}(j\omega) = Z_{jj}(s) = \frac{e_j(s)}{i_j(s)} \tag{3.10.7}$$

In these equations, the unknown quantities are the voltage vector and the frequency. It is a simple matter to assume a frequency and solve for the corresponding voltage vector. Solving Equation 3.10.7 over a range of frequencies results in the well-known impedance-vs.-frequency plot. Figure 3.10.4 shows the impedance versus frequency for the example used in Figure 3.10.1 and Figure 3.10.2.

3.10.4.3 Amplification Factor

The amplification factor or gain function is defined as:

$$[N_{lm,j}] = \frac{Voltage\ between\ points\ l\ and\ m\ at\ frequency\ \omega}{Voltage\ applied\ at\ input\ node\ j\ at\ frequency\ \omega} \tag{3.10.8}$$

As shown by Degeneff [27], this results in:

$$N_{lm,j} = \frac{[Z_{lj}(j\omega) - Z_{mj}(jw)]}{Z_{jj}(j\omega)} \tag{3.10.9}$$

It is a simple matter to assume a frequency and solve for the corresponding voltage-vs.-frequency vector. If one is interested in the voltage distribution within a coil at one of the resonant frequencies, this can be found from the eigenvector of the coil at the frequency of interest. If one is interested in the distribution at any other frequency, Equation 3.10.9 can be utilized. This is shown in Figure 3.10.5.

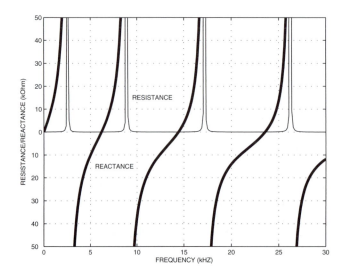

FIGURE 3.10.4 Terminal impedance for a helical winding.

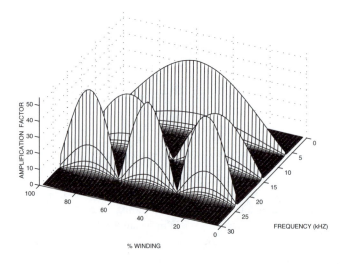

FIGURE 3.10.5 Amplification factor at 5 from 0 to 30 kHz.

3.10.5 Inductance Model

3.10.5.1 Definition of Inductance

Inductance is defined as:

$$L = \frac{d\lambda}{dI} \tag{3.10.10}$$

and if the system is linear:

$$L = \frac{\lambda}{I} \tag{3.10.11}$$

where

> L = inductance, H
> λ = flux linkage caused by current I, weber turns
> I = current producing flux linkages, A
> $d\lambda$ = first derivative of λ
> dI = first derivative of I

L is referred to as self-inductance if the current producing the flux is the same current being linked. L is referred to as mutual inductance if the current producing the flux is other than the current being linked [3]. Grover [34] published an extensive work providing expressions to compute the mutual and self-inductance in air for a large number of practical conductor and winding shapes.

One of the most difficult phenomena to model is the magnetic flux interaction involving the different winding sections and the iron core. Historically, this phenomenon has been modeled by dividing the flux into two components: the common and leakage flux. The common flux dominates when the transformer behavior is studied under open-circuit conditions, and the leakage flux dominates the transient response when the winding is shorted or loaded heavily. Developing a transformer model capable of representing the magnetic-flux behavior for all conditions the transformer will see in factory test and in service requires the accurate calculation of the mutual and self-inductances.

3.10.5.2 Transformer Inductance Model

Until the introduction of the computer, there was a lack a practical analytical formulas to compute the mutual and self-inductances of coils with an iron core. Rabins [35] developed an expression to calculate mutual and self-inductances for a coil on an iron core based on the assumption of a round core leg and infinite core yokes, both of infinite permeability. Fergestad and Heriksen [36] improved Rabins's inductance model in 1974 by assuming an infinite permeable core except for the core leg. In their approach, a set of state-variable equations was derived from the classic lumped-parameter model of the winding.

White [19,37] derived an expression to calculate the mutual and self-inductances in the presence of an iron core with finite permeability under the assumption of an infinitely long iron core. White's inductance model had the advantage that the open-circuit inductance matrix could be inverted [37]. White derived an expression for the mutual and self-inductance between sections of a transformer winding by solving a two-dimensional problem in cylindrical coordinates for the magnetic vector potential assuming a nonconductive and infinitely long open core. He assumed that the leakage inductance of an open-core configuration is the same as the closed core [19,37].

Starting from the definition of the magnetic vector potential $\vec{B} = \nabla \times \vec{A}$ and $\nabla \times \vec{A} = 0$ and using Ampere's law in differential form, $\vec{J}_f = \nabla \times \vec{H}$, White solved the following equation:

$$\nabla^2 \vec{A}(r,z) = \mu_o \vec{J}(r,z) \tag{3.10.12}$$

The solution broke into two parts: the air-core solution and the change in the solution due to the insertion of the iron core, as shown in the following equations:

$$\vec{A}(r,z) = \begin{cases} \vec{A}_o(r,z) + \vec{A}_1(r,z) & ,0 \le r < R_c \\ \vec{A}_o(r,z) + \vec{A}_2(r,z) & ,r > R_c \end{cases}$$

$\vec{A}_0(r,z)$ is the solution when the core is present, and $\vec{A}_1(r,z)$ and $\vec{A}_2(r,z)$ are the solutions when the iron core is added. Applying Fourier series to Equation 3.10.12, the solution for $\vec{A}_0(r,z)$ was found first and then $\vec{A}_2(r,z)$.

Knowing the magnetic vector potential allows the flux linking a filamentary turn at (r,z) to be determined by recalling $\phi(r,z) = \oint \vec{A}(r,z) \cdot d\vec{l}$. The flux for the filamentary turn is given by:

$$\phi(r,z) = 2\pi r\left\{\vec{A}_o(r,z) + \vec{A}_2(r,z)\right\} = \phi_o(r,z) + 2\pi r\vec{A}_2(r,z) \tag{3.10.13}$$

The flux in air, $\phi_0(r,z)$, can be obtained from known formulas for filaments in air [49], therefore it is only necessary to obtain the change in the flux linking the filamentary turn due to the iron core. If the mutual inductance L_{ij} between two coil sections is going to be calculated, then the average flux linking section i need to be calculated. This average flux is given by

$$\phi_{ave} = \frac{\int_{R_i}^{\bar{R}_i}\int_{Z_i}^{\bar{Z}_i}\phi(r,z)dzdr}{H_i\left(R_i - \bar{R}_i\right)} \tag{3.10.14}$$

Knowing the average flux, the mutual inductance can be calculated using the following expression

$$L_{ij} = \frac{N_i N_j \phi_{ave}}{I_j} \tag{3.10.15}$$

White's final expression for the mutual inductance between two coil segments is:

$$L_{ij} = L_{ijo} + 2N_i N_j\left(1 - v_r\right)\mu_o R_c\int_0^\infty \frac{I_0\left(\omega R_c\right)I_1\left(\omega R_c\right)F(\omega)}{v_r + \left(1 - v_r\right)\omega R_c I_1\left(\omega R_c\right)K_0\left(\omega R_c\right)}d\omega \tag{3.10.16}$$

where

$$F(\omega) = \frac{1}{\omega}\left[\frac{1}{\omega\left(\bar{R}_i - R_i\right)}\int_{\omega R_i}^{\omega \bar{R}_i} xK_1(x)dx\right]\frac{2}{\omega H_i}\sin\left(\frac{\omega H_i}{2}\right)$$

$$\left[\frac{1}{\omega\left(\bar{R}_i - R_i\right)}\int_{\omega R_i}^{\omega \bar{R}_i} xK_1(x)dx\right]\frac{2}{\omega H_j}\sin\left(\frac{\omega H_j}{2}\right)\cos\left(\omega d_{ij}\right) \tag{3.10.17}$$

and where L_{ijo} is the air-core inductance; $v_r = 1/\mu_r$ is the relative reluctivity; and $I_0(\omega R_c)$, $I_1(\omega R_c)$, $K_0(\omega R_c)$, and $K_1(\omega R_c)$ are modified Bessel functions of first and second kind.

3.10.5.3 Inductance Model Validity

The ability of the inductance model to accurately represent the magnetic characteristic of the transformer can be assessed by the accuracy with which it reproduces the transformer electrical characteristics, e.g., the short-circuit and open-circuit inductance and the pseudo-final (turns ratio) voltage distribution. The short-circuit and open-circuit inductance of a transformer can be determined by several methods, but the simplest is to obtain the inverse of the sum of all the elements in the inverse nodal inductance matrix, Γ_n. This has been verified in other works [38,39]. The pseudo-final voltage distribution is defined in a work by Abetti [15]. It is very nearly the turns-ratio distribution and must match whatever voltage distribution the winding arrangement and number of turns dictates. An example of this is available in the literature [39].

3.10.6 Capacitance Model

3.10.6.1 Definition of Capacitance

Capacitance is defined as:

$$C = \frac{Q}{V} \tag{3.10.18}$$

where

C = capacitance between the two plates, F
Q = charge on one of the capacitor plates, C
V = voltage between the capacitor plates, V

Snow [40] published an extensive work on computing the capacitance for unusual shapes of conductors. Practically, however, most lumped-parameter models of windings are created by subdividing the winding into segments with small radial and axial dimensions and large radiuses, thus enabling the use of a simple parallel plate formula [20] to compute both the series and shunt capacitance for a segment. For example:

$$C = \varepsilon_o \varepsilon_r \frac{RadCir}{nD} \qquad (3.10.19)$$

where

ε_0 = permittivity of freespace, 8.9×10^{-12}, C·m
ε_r = relative permittivity between turns
Rad = radial build of the segments turns, m
Cir = circumference of the mean turn within segment, m
n = number of turns with the segment
D = separation between turns, m

In computing these capacitances, the relative permittivity, ε_r, of the materials must be recognized. This is a function of the material, moisture content, temperature, and effective age of the material. A large database of this type of information is available [41,42]. Since most lumped-parameter models assume the topology has circular symmetry, if the geometry is unusually complex, it may be appropriate to model the system with a three-dimensional FEM. It should be emphasized that all of the above is based on the assumption that the capacitive structure of the transformer is frequency invariant. If the transient model is required to be valid over a very large bandwidth, then the frequency characteristic of dielectric structure must be taken into account.

3.10.6.2 Series and Shunt Capacitance

In order to construct a lumped-parameter model, the transformer is subdivided into segments (or groups of turns). Each of these segments contains a beginning node and an exit node. Between these two nodes there will generally be associated a capacitance, traditionally called the series capacitance. These are the intrasection capacitances. In most cases, it is computed using the simple parallel plate capacitance given by Fink and Beaty [43]. An exception to this is the series capacitance of disk winding segments, and expressions for their series capacitance is given in the next section.

Additionally, each segment will have associated with it capacitances between adjacent sections of turns or to a shield or earth. These are the intersection capacitances. These capacitances are generally referred to as shunt capacitances and are normally divided in half and connected to each end of the appropriate segments. This is an approximation, but if the winding is subdivided into relatively small segments, the approximation is acceptable and the error introduced by the model is small.

3.10.6.3 Equivalent Capacitance for Disk Windings

This section presents simplified expressions to compute the series capacitance for disk winding section pairs. Since most lumped-parameter models are not turn-to-turn models, an electrostatic equivalent of the disk section is used for the series capacitance. It is well known that as the series capacitance of disk winding sections becomes larger with respect to the capacitances to ground, the initial distribution becomes more linear (straight line) and the transient response in general more benign. Therefore, since it is possible to arrange the turns within a disk section in many ways without affecting the section's

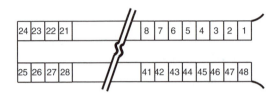

A. Continuous Disk Winding

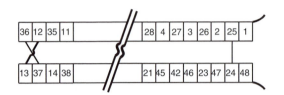

B. Interleaved Disk Winding

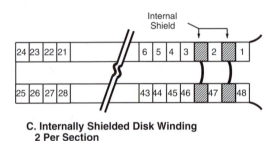

**C. Internally Shielded Disk Winding
2 Per Section**

FIGURE 3.10.6 Common disk winding section pairs: A, continuous; B, interleaved; and C, internally shielded.

inductance characteristics or the space or material it requires, the industry has offered many arrangements in an effort to increase this effective series capacitance.

The effective series capacitance of a disk winding is a capacitance that, when connected between the input and output of the disk winding section pair, would store the same electrostatic energy the disk section pair would store (between all turns) if the voltage were distributed linearly within the section. A detailed discussion of this modeling strategy is available in the literature [44,45]. Figure 3.10.6 illustrated the cross section of three common disk winding configurations. The series capacitance of the continuous disk is given by:

$$C_{continuous} = \frac{2}{3}Cs + [\frac{n-2}{n^2}]C_t \qquad (3.10.20)$$

The series capacitance of the interleaved disk section pair is given by:

$$C_{interleaved} = 1.128C_s + [\frac{n-4}{4}]C_t \qquad (3.10.21)$$

The interleaved disk provides a greater series capacitance than the continuous disk, but it is more difficult to produce. A winding that has a larger series capacitance than the continuous disk but that is simpler to manufacture than the interleaved is the internally shielded winding. Its series capacitance is given by:

$$C_{internalshield} = \frac{2}{3}C_s + [\frac{n-2-2n_s}{n^2}]C_t + 4C_{ts}\sum_{i=1}^{n_s}[\frac{n_i}{n}]^2 \qquad (3.10.22)$$

where

> C_s = capacitance between sections
> C_t = capacitance between turns
> C_{ts} = capacitance between turn and internal shield
> n = turns in section pair
> n_i = location of shield within section
> n_s = internal shields within section pair

Selecting the disc winding section is often a compromise of electrical performance, economics, and manufacturing preference for a given firm.

3.10.6.4 Initial Voltage Distribution

The initial voltage distribution can be determined experimentally by applying a voltage wave with a fairly fast rise time (e.g., 0.5 µsec) and measuring the normalized distribution within the winding structure at an intermediate time (e.g., 0.3 µsec). The initial distribution can be computed analytically by injecting a current into the excited node and determining the normalized voltage throughout the transformer winding structure. This computational method is outlined in detail in another work [39]. If one is considering a single coil, it is common practice to determine the gradient of the transient voltage near the excited terminal (which is the most severe). This gradient is referred to as α and is found by an equation from Greenwood [46]:

$$\alpha = \sqrt{\frac{C_g}{C_s}}$$ (3.10.23)

where

> α = winding gradient
> C_g = capacitance to ground, F
> C_s = effective series capacitance, F

For the coil with the initial distribution shown in Figure 3.10.3, the α is on the order of 12.

3.10.7 Loss Model

At steady state, losses are a costly and unwanted characteristic of physical systems. At high frequency, losses produce a beneficial effect in that they reduce the transient-voltage response of the transformer by reducing the transient-voltage oscillations. In general, the oscillations are underdamped. The effect of damping on the resonant frequency is to reduce the natural frequencies slightly. Losses within the transformer are a result of a number of sources, each source with a different characteristic.

3.10.7.1 Copper Losses

The losses caused by the current flowing in the winding conductors are referred to as series losses. Series losses are composed of three components: dc losses, skin effect, and proximity effect.

3.10.7.1.1 DC Resistance

The conductor's dc resistance is given by:

$$R_{dc} = \rho \frac{l}{A}$$ (3.10.24)

where

> ρ = conductor resistivity, Ω·m
> l = length of conductor, m
> A = conductor area, m²

The variable ρ is a function of the conductor material and its temperature.

3.10.7.1.2 Skin Effect

Lammeraner and Stafl [47] give an expression for the skin effect in a rectangular conductor. The impedance per unit length of the conductor (Z, Ω/m) is given by:

$$Z = \frac{k}{4h\sigma} \coth kb \quad \Omega/m \tag{3.10.25}$$

where

$$k = \frac{1+j}{a} \tag{3.10.26}$$

and where

$$a = \sqrt{\frac{2}{\omega\sigma\mu}} \tag{3.10.27}$$

h = half the conductor height, m
b = half the conductor thickness, m
σ = conductivity of the conductor, S/m
μ = permeability of the material, H/m
ω = frequency, rad/sec

Defining ξ as follows,

$$\xi = b\sqrt{j\omega\sigma\mu} \tag{3.10.28}$$

then Equation 3.10.25 can be expressed as

$$Z_{\text{skin}} = R_{\text{dc}}\xi \coth \xi \tag{3.10.29}$$

where R_{dc} is the dc resistance per unit length of the conductor. Equation 3.10.29 is used to calculate the impedance due to the skin effect as a function of frequency.

3.10.7.1.3 Proximity Effect

Proximity effect is the increase in losses in one conductor due to currents in other conductors produced by a redistribution of the current in the conductor of interest by the currents in the other conductors. A method of finding the proximity-effect losses in the transformer winding consists in finding a mathematical expression for the impedance in terms of the flux cutting the conductors of an open winding section due to an external magnetic field. Since windings in large power transformers are mainly built using rectangular conductors the problem reduces to the study of eddy-current losses in a packet of laminations. Lammeraner and Stafl [47] derived an expression for the flux as a function of frequency in a packet of laminations. It is given in the following equation:

$$\Phi = \frac{2al\mu}{1+j} H_o \tanh(1+j)\frac{b}{a} \tag{3.10.30}$$

where l is the conductor length, H_0 is the rms value of the magnetic flux intensity, and the remaining variables are the same as defined in Equation 3.10.27.

Assuming H_0 in Equation 3.10.30 represents the average value of the magnetic field intensity inside the conductive region represented by the winding section I, and defining L_{ijo} as

$$L_{ijo} = N_i N_j \phi_{ijo} \tag{3.10.31}$$

where ϕ_{ijo} is the average flux cutting each conductor in section *i* due to the current I_j, and where N is the number of turns in each section, then the inductance (L_{ij}, H) as a function of frequency is:

$$L_{ij} = \frac{L_{ijo}}{(1+j)\dfrac{b}{a}} \tanh(1+j)\frac{b}{a} \quad H \tag{3.10.32}$$

The impedance ($Z_{prox,ij}$, Ω) is obtained by multiplying the inductance by the complex variable *s*. Using the same notation as in Equation 3.10.29, the impedance of the conductor due to the proximity effect is given as

$$Z_{prox_{ij}} = s\frac{L_{ij_o}}{\xi} \tanh\xi \quad \Omega \tag{3.10.33}$$

3.10.7.2   Core Losses

The effect of eddy currents in the core have been represented in various works [26,48,49] by the well-tested formula:

$$Z = \frac{4N^2 A}{ld^2\sigma} x \tanh x \tag{3.10.34}$$

where

$$x = \frac{d\sqrt{j\omega\mu\sigma}}{2} \tag{3.10.35}$$

and where

  l = length of the core limb (axial direction), m
  d = thickness of the lamination, m
  μ = permeability of the material, H/m
  N = number of turns in the coil
  A = total cross-sectional area of all laminations
  ω = frequency, rad/sec

This formula represents the equivalent impedance of a coil wound around a laminated iron core limb. The expression was derived [49] by solving Maxwell's equations assuming the electromagnetic field distribution is identical in all laminations and an axial component of the magnetic flux.

The total hysteresis loss in core volume, V, in which the flux density is everywhere uniform and varying cyclically at a frequency of ω, can be expressed as:

$$P_h = 2\pi\omega\eta \ V \ \beta^n_{max} \tag{3.10.36}$$

where

  P_h = total hysteresis loss in core
  η = constant, a function of material
  V = core volume
  β = flux density
  n = exponent, dependent upon material, 1.6 to 2.0
  ω = frequency, rad/sec

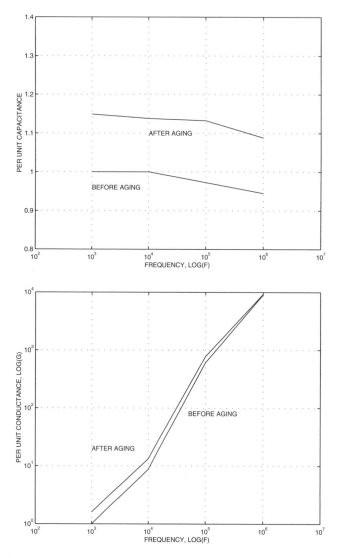

FIGURE 3.10.7 Top) Oil-soaked paper capacitance as a function of frequency; bottom) Oil-soaked paper conductance as a function of frequency.

3.10.7.3 Dielectric Losses

The capacitive structure of a transformer has associated with it parallel losses. At low frequency, the effect of capacitance on the internal voltage distribution can be ignored. As such, the effect of the losses in the dielectric structure can be ignored. However, at higher frequencies the losses in the dielectric system can have a significant effect on the transient response. Batruni et al. [50] explore the effect of dielectric losses on the impedance-vs.-frequency characteristic of the materials in power transformers. These losses are frequency dependent and are shown in Figure 3.10.7.

3.10.8 Winding Construction Strategies

3.10.8.1 Design

The successful design of a commercial transformer requires the selection of a simple structure so that the core and coils are easy to manufacture. At the same time, the structure should be as compact as possible to reduce required materials, shipping concerns, and footprint. The form of construction should allow convenient removal of heat, sufficient mechanical strength to withstand forces generated during system faults, acceptable noise characteristics, and an electrical insulation system that meets the system steady-state and transient requirements. There are two common transformer structures in use today. When the magnetic circuit is encircled by two or more windings of the primary and secondary, the transformer is referred to as a core-type transformer. When the primary and secondary windings are encircled by the magnetic material, the transformer is referred to as a shell-type transformer.

3.10.8.2 Core Form

Characteristics of the core-form transformer are a long magnetic path and a shorter mean length of turn. Commonly used core-form magnetic circuits are single-phase transformers with two-legged magnetic path, with turns wound around each leg; three-legged magnetic path, with the center leg wound with conductor; and a legged magnetic path, with the two interior legs wound with conductors [2,3]. Three-phase core-form designs are generally three-legged magnetic cores, with all three legs possessing windings, and a five-legged core arrangement, with the three center legs possessing windings. The simplest winding arrangement has the low-voltage winding nearest the core and the high-voltage winding wound on top of the low. Normally, in the core-form construction the winding system is constructed from helical, layer, or disk-type windings. Often the design requirements call for a winding arrangement that is a more complex arrangement, e.g., interleaving high- and low-voltage windings, interwound taps, or bifurcated windings having entry and exit points other than the top or bottom of the coil. Each of these variations has, to one degree or another, an effect on the transformer's transient-voltage response. To ensure an adequate insulation structure, during the design stage each possible variation must be explored to evaluate its effect on the transient overvoltages.

3.10.8.3 Shell Form

Shell-form transformer construction features a short magnetic path and a longer mean length of electrical turn. Fink and Beaty [43] point out that this results in the shell-form transformer having a larger core area and a smaller number of winding turns than the core-form of the same output and performance. Additionally, the shell form generally has a larger ratio of steel to copper than an equivalently rated core-form transformer. The most common winding structure for shell-form windings is the primary-secondary-primary (P-S-P), but it is not uncommon to encounter shell-form winding of P-S-P-S-P. The winding structure for both the primary and secondary windings is normally of the pancake-type winding structure [2].

3.10.8.4 Proof of Design Concept

The desire of the purchaser is to obtain a transformer at a reasonable price that will achieve the required performance for an extended period of time. The desire of the manufacturer is to construct and sell a product, at a profit, that meets the customer's goals. The specification and purchase contract are the document that combines both purchaser's requirements and manufacturer's commitment in a legal format. The specification will typically address the transformer's service condition, rating, general construction, control and protection, design and performance review, testing requirements, and transportation and handling. Since it is impossible to address all issues in a specification, the industry uses standards that are acceptable to purchaser and supplier. In the case of power transformers, the applicable standards would include those found in IEEE C57, IEC 76, and NEMA TR-1.

3.10.8.5 Standard Winding Tests

ANSI/IEEE C57.12.00 [51] defines routine and optional test and testing procedures for power transformers. The following are listed as routine tests for transformers larger than 501 kVA:

Winding resistance
Winding-turns ratio
Phase-relationship tests: polarity, angular displacements, phase sequence
No-load loss and exciting current

Load loss and impedance voltage
Low-frequency dielectric tests (applied voltage and induced voltage)
Leak test on transformer tank

The following are listed as tests to be performed on only one of a number of units of similar design for transformers 501 kVA and larger:

Temperature-rise tests
Lightning-impulse tests (full and chopped wave)
Audible sound test
Mechanical test from lifting and moving of transformer
Pressure tests on tank

Other tests are listed [51] that include, for example, short-circuit forces and switching-surge-impulse tests. Additionally, specific tests may be required by purchasers base on their application or field experience.

The variety of transient voltages a transformer may see in its normal useful life are virtually unlimited [9]. It is impractical to proof test each transformer for every conceivable combination of transient voltages. However, the electrical industry has found that it is possible, in most instances, to assess the integrity of the transformer's insulation systems to withstand transient voltages with the application of a few specific, aperiodic voltage waveforms. Figure 3.10.8 illustrates the forms of the full, chopped, and switching surge waves. IEEE C57-1990 [51] contains the specific wave characteristics, relationships and acceptable methods and connections required for these standard tests. Each of these tests is designed to test the insulation structure for a different transient condition. The purpose of applying this variety of tests is to substantiate adequate performance of the total insulation system for all the various transient voltages it may see in service.

The insulation integrity for steady-state and dynamic voltages is also assessed by factory tests called out in ANSI/IEEE C57.12.00 [51]. One should not be lulled to sleep and think that a transformer that has passed all factory voltage tests (both impulse and low frequency) can withstand all transient voltages to which it may be subjected by the system. One should always assess the environment the transformer is applied in and determine if there may be unusual transient-voltage excitation present in an application that is not covered in the standards (see [9,10,31]).

3.10.8.6 Design Margin

The actual level of insulation requested, e.g., basic impulse insulation level (BIL) and switching impulse level (BSL), is determined by recognizing the system within which the transformer will operate and the arrester's level of protection. Normally, a minimum protective margin of 15 to 20% between the arrester peak voltage and the transformer capability at three sigma (3σ) is established. This is illustrated in Figure 3.10.9 for a 230-kV transformer at 750-kV BIL protected with a 180-kV-rated arrester [52]. The curve designated A in Figure 3.10.9 is used to represent the transformer's insulation-coordination characteristic (insulation capability) when subjected to aperiodic and oscillatory wave forms. The curve to be used to represent the transformer volt–time insulation-coordination characteristic when subjected to aperiodic wave forms with a time to failure between 0.1 and 2000 μsec is to be based on five points [53]. The five points are:

1. Front-of-wave voltage plotted at its time of crest (about 0.5 μsec). If the front-of-wave voltage is not available, a value of 1.3 times the BIL should be plotted at 0.5 μsec.
2. Chopped-wave voltage at its time of crest (about 3.0 μsec)
3. Full-wave voltage (or BIL) plotted at about 8 μsec
4. Switching-surge voltage (or BSL) plotted at about 300 μsec
5. A point at 2000 μsec, where its magnitude is established with the following expression:

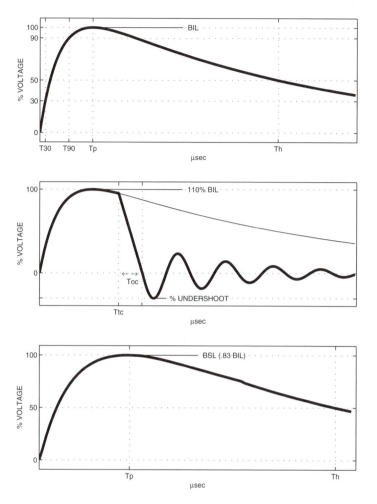

FIGURE 3.10.8 Standard voltage waveforms for impulse tests: full, chopped, switching surge.

$$log\,V_{2000} = \frac{log\,T_{BSL} - log\,T_{2000}}{m} + log\,V_{BSL} = \frac{log\,\dfrac{300}{2000}}{m} + log\,V_{BSL}$$

where V_{2000} is the voltage at 2000 μsec, T_{2000}; V_{BSL} is the BSL voltage (or 0.83 the BIL); and T_{BSL} is equal to 300 μsec. The value of m is established as the inverse of the slope of a straight line drawn on log-log paper from the BSL point to a point established by the peak of the 1-h induced test voltage plotted at a time the induced voltage exceeds 90% of its peak value (i.e., 28.7% of 3600 sec or 1033.2 sec).

The connection between all points is made with a smooth continuous curve. The first four points in the curve establish an approximate level of insulation voltage capability for which one would anticipate only one insulation failure out of 1000 applications of that voltage level, e.g., at 3σ the probability of failure is $1.0 - 0.99865$ or 0.001. Experience has shown that the standard deviation for transformer insulation structures is on the order of 10 to 15%. Figure 3.10.9 assumes that σ is 10%. Curve B, or the 50% failure-rate curve, is established by increasing the voltage in Curve A by 30%. Therefore, for Curve B, on average the unit would be expected to fail one out of two times if it were subjected to this level of voltage. Curves C and D establish 1- and 2-σ curves, or 16% and 2.3% failure-rate curves, respectively. The inserted normal distribution on the right of Figure 3.10.9 illustrate this concept. All of this discussion is based on the assumption that the transformer is new.

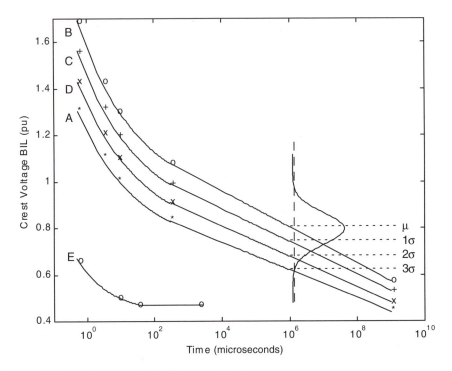

FIGURE 3.10.9 Voltage–time curve for insulation coordination.

3.10.8.7 Insulation Coordination

In a field installation, an arrester is normally placed directly in front of the transformer to afford it protection from transient voltages produced on the system. Curve E in Figure 3.10.9 is a metal-oxide-arrester protective curve established in a manner similar to that described in IEEE Std. C62.2. The curve is specified by three points:

1. The front-of-wave voltage held by the arrester plotted at 0.5 μsec
2. The 8 × 20-μsec voltage plotted at 8.0 μsec
3. The switching-surge voltage plotted in straight line from 30 to 2000 μsec

The protective ratio is established by dividing the transformer insulation capability by the arrester protective level for the waveshape of interest. For example, in Figure 3.10.9 the protective level for a switching surge is on the order of 177% or (0.83/0.47) × 100.

3.10.8.8 Additional System Considerations

The current standards reflect the growing and learning within the industry, and each year they expand in breadth to address issues that are of concern to the industry. However, at present the standards are silent in regard to the effects of system voltage on transient response, multiphase surges, aging or mechanical movement of insulation structures, oscillatory voltage excitation, temperature variations, movement of oil, and loading history. A prudent user will seek the advice of users of similar products and explore their experience base.

3.10.9 Models for System Studies

3.10.9.1 Model Requirements

The behavior of large power transformers under transient conditions is of interest to both transformer designers and power engineers. The transformer designer employs detailed electrical models to establish a reliable and cost-effective transformer insulation structure. The power engineer models not only the transformer, but also the system in order to investigate the effects of power-system transients.

Considerable effort has been devoted to computing the transformer's internal transient response. Models of this type may contain several hundred nodes for each phase. This detail is necessary in order to compute the internal response in enough detail to establish an adequate transformer insulation design. The utility engineer usually is not interested in the internal response, but is concerned only with the transformer's terminal response. Even if the transformer's detailed model were available, its use would create system models too large to be effectively used in system studies. What has been the normal practice is to create a reduced-order model of the transformer that represents the terminal response of the transformer. Experience has shown that great care must be taken to obtain a terminal model that provides a reasonable representation of the transformer over the frequency range of interest [24,27].

The challenge in creating a high-fidelity reduced model lies in the fact that as the size of the model is reduced, the number of valid eigenvalues must also decrease. In effect, any static reduction technique will produce a model that is intrinsically less accurate than the more detailed lumped-parameter model [54,55].

3.10.9.2 Reduced-Order Model

McNutt et al. [31] suggested a method of obtaining a reduced-order transformer model by starting with the detailed model and appropriately combining series and shunt capacitances. This suggestion was extended by de Leon [55]. This method is limited to linear models and can not be used to eliminate large proportions of the detailed models without effecting the resulting model accuracy. Degeneff [38] proposed a terminal model developed from information from the transformer's name plate and capacitance as measured among the terminals. This model is useful below the first resonant frequency, but it lacks the necessary accuracy at higher frequencies for system studies. Dommel et al. [56] proposed a reduced model for EMTP described by branch impedance or admittance matrix calculated from open- and short-circuit tests. TRELEG and BCTRAN matrix models for EMTP can be applied only for very-low-frequency studies. Morched et al. [54] proposed a terminal transformer model, composed of a synthesized LRC network, where the nodal admittance matrix approximates the nodal admittance matrix of the actual transformer over the frequency range of interest. This method is appropriate only for linear models.

Other references [24,25] present a method for reducing both a detailed linear and nonlinear lumped-parameter model to a terminal model with no loss of accuracy. The work starts with equation 1, progresses to equation 4, and then applies Kron reduction to obtain a terminal model of the transformer that retains all the frequency fidelity of the initial transformer lumped-parameter model. All of the appropriate equations to apply this technique within EMTP are available in the literature [24,25].

References

1. Blume, L.F., Boyajian, A., Camilli, G., Lennox, T.C., Minneci, S., and Montsinger, V.M., *Transformer Engineering*, 2nd ed., John Wiley and Sons, New York, 1954, pp. 416–423.
2. Bean, R.L., Crackan, N., Moore, H.R., and Wentz, E., *Transformers for the Electric Power Industry*, McGraw-Hill, New York, 1959. Reprint, Westinghouse Electric Corp., New York, 1959.
3. Massachusetts Institute of Technology, Department of Electrical Engineering: *Magnetic Circuits and Transformers*, John Wiley and Sons, New York, 1943.
4. Franklin, A.C., *The J & P Transformer Book*, 11th ed., Butterworth, London, 1983, chap. 15, pp. 351–367.
5. Kogan, V.I., Fleeman, J.A., Provanzana, J.H., and Shih, C.H., Failure analysis of EHV transformers, *IEEE Trans. Power Delivery*, PAS-107, 672–683, 1988.
6. An international survey on failures in large power transformers in service, final report of Working Group 05 of Study Committee 12 (Transformers), *Electra*, 88, 49–90, 1983.

7. Kogan, V.I., Fleeman, J.A., Provanzana, J.H., Yanucci, D.A., and Kennedy, W.N., Rationale and Implementation of a New 765-kV Generator Step-Up Transformer Specification, CIGRE Paper 12-202, Cigre, Paris, August 1990.

8. IEEE, Recommended Practices and Requirements for Harmonic Control in Electrical Power Systems, IEEE Std. 519, Institute of Electrical and Electronics Engineers, Piscataway, NJ, 1999.

9. Degeneff, R.C., Neugebaur, W., Panek, J., McCallum, M.E., and Honey, C.C., Transformer response to system switching voltages, *IEEE Trans. Power Appar. Syst.*, 101, 1457–1465, 1982.

10. Greenwood, A., *Vacuum Switchgear*, Institute of Electrical Engineers, Short Run Press, Exeter, U.K., 1994.

11. Narang, A., Wisenden, D., and Boland, M., Characteristics of Stress on Transformer Insulation Subjected to Very Fast Transient Voltages, CEA No. 253 T 784, Canadian Electricity Association, Hull, Quebec, Canada, July 1998.

12. Abetti, P.A., Survey and classification of published data on the surge performance of transformers and rotating machines, *AIEE Trans.*, 78, 1403–1414, 1959.

13. Abetti, P.A., Survey and classification of published data on the surge performance of transformers and rotating machines, 1st supplement, *AIEE Trans.*, 81, 213, 1962.

14. Abetti, P.A., Survey and classification of published data on the surge performance of transformers and rotating machines, 2nd supplement, *AIEE Trans.*, 83, 855, 1964.

15. Abetti, P.A., Pseudo-final voltage distribution in impulsed coils and windings, *AIEE Trans.*, 87–91, 1960.

16. Waldvogel, P. and Rouxel, R., A new method of calculating the electric stresses in a winding subjected to a surge voltage, *Brown Boveri Rev.*, 43, 206–213, 1956.

17. McWhirter, J.H., Fahrnkopf, C.D., and Steele, J.H., Determination of impulse stresses within transformer windings by computers, *AIEE Trans.*, 75, 1267–1273, 1956.

18. Dent, B.M., Hartill, E.R., and Miles, J.G., A method of analysis of transformer impulse voltage distribution using a digital computer, *IEEE Proc.*, 105, 445–459, 1958.

19. White, W.N., An Examination of Core Steel Eddy Current Reaction Effect on Transformer Transient Oscillatory Phenomena, GE Technical Information Series, No. 77PTD012, General Electric Corp., Pittsfield, MA, April 1977.

20. Degeneff, R.C., Blalock, T.J., and Weissbrod, C.C., Transient Voltage Calculations in Transformer Winding, GE Technical Information Series, No. 80PTD006, General Electric Corp., Pittsfield, MA, 1980.

21. White, W.N., Numerical Transient Voltage Analysis of Transformers and LRC Networks Containing Nonlinear Resistive Elements, presented at 1977 PICA Conference, Toronto, pp. 288–294.

22. Wilcox, D.J., Hurley, W.G., McHale, T.P., and Conton, M., Application of modified modal theory in the modeling of practical transformers, *IEEE Proc.*, 139, 472–481, 1992.

23. Vakilian, M., A Nonlinear Lumped Parameter Model for Transient Studies of Single Phase Core Form Transformers, Ph.D. thesis, Rensselaer Polytechnic Institute, Troy, New York, August 1993.

24. Gutierrez, M., Degeneff, R.C., McKenny, P.J., and Schneider, J.M., Linear, Lumped Parameter Transformer Model Reduction Technique, IEEE Paper No. 93 SM 394-7 PWRD, Institute of Electrical and Electronics Engineers, Piscataway, NJ, 1993.

25. Degeneff, R.C., Gutierrez, M., and Vakilian, M., Nonlinear, Lumped Parameter Transformer Model Reduction Technique, IEEE Paper No. 94 SM 409-3 PWRD, Institute of Electrical and Electronics Engineers, Piscataway, NJ, 1994.

26. de Leon, F. and Semlyen, A., Complete transformer model for electromagnetic transients, *IEEE Trans. Power Delivery*, 9, 231–239, 1994.

27. Degeneff, R.C., A general method for determining resonances in transformer windings, *IEEE Trans. Power Appar. Syst.*, 96, 423–430, 1977.

28. Dommel, H.W., Digital computer solution of electromagnetic transients in single and multiphase networks, *IEEE Trans. Power Appar. Syst.*, PAS-88, 388–399, 1969.

29. Degeneff, R.C., Reducing Storage and Saving Computational Time with a Generalization of the Dommel (BPA) Solution Method, presented at IEEE PICA Conference, Toronto, 1977, pp. 307–313.

30. FORTRAN Subroutines for Mathematical Applications, Version 2.0, MALB-USM-PERFCT-EN9109-2.0, IMSL, Houston, TX, September 1991.

31. McNutt, W.J., Blalock, T.J., and Hinton, R.A., Response of transformer windings to system transient voltages, *IEEE Trans. Power Appar. Syst.*, 93, 457–467, 1974.

32. Abetti, P.A., Correlation of forced and free oscillations of coils and windings, *IEEE Trans. Power Appar. Syst.*, 78, 986–996, 1959.

33. Abetti, P.A. and Maginniss, F.J., Fundamental oscillations of coils and windings, *IEEE Trans. Power Appar. Syst.*, 73, 1–10, 1954.

34. Grover, F.W., *Inductance Calculations — Working Formulas and Tables*, Dover Publications, New York, 1962.

35. Rabins, L., Transformer reactance calculations with digital computers, *AIEE Trans.*, 75, 261–267, 1956.

36. Fergestad, P.I. and Henriksen, T., Transient oscillations in multiwinding transformers, *IEEE Trans. Power Appar. Syst.*, 93, 500–507, 1974.

37. White, W.N., Inductance Models of Power Transformers, GE Technical Information Series, No. 78PTD003, General Electric Corp., Pittsfield, MA, April 1978.

38. Degeneff, R.C., A Method for Constructing Terminal Models for Single-Phase n-Winding Transformers, IEEE Paper A78 539-9, Summer Power Meeting, Los Angeles, 1978.

39. Degeneff, R.C. and Kennedy, W.N., Calculation of Initial, Pseudo-Final, and Final Voltage Distributions in Coils Using Matrix Techniques, Paper A75-416-8, presented at Summer Power Meeting, San Francisco, 1975.

40. Snow, C., Formulas for Computing Capacitance and Inductance, NBS Circular 544, National Institutes of Standards and Technology, Gaithersburg, MD, 1954.

41. Clark, F.M., *Insulating Materials for Design and Engineering Practice*, Wiley, New York, 1962.

42. von Hippel, A., *Dielectric Materials and Applications*, Massachusetts Institute of Technology, Cambridge, 1954.

43. Fink, D.G. and Beaty, H.W., *Standard Handbook for Electrical Engineers*, 12th ed., McGraw-Hill, New York, 1987.

44. Scheich, A., Behavior of Partially Interleave Transformer Windings Subject to Impulse Voltages, Bulletin Oerlikon, No. 389/390, Oerlikon Engineering Co., Zurich, 1960, pp. 41–52.

45. Degeneff, R.C., Simplified Formulas to Calculate Equivalent Series Capacitances for Groups of Disk Winding Sections, GE TIS 75PTD017, General Electric Corp., Pittsfield, MA, 1976.

46. Greenwood, A., *Electrical Transients in Power Systems*, John Wiley & Sons, New York, 1991.

47. Lammeraner, J. and Stafl, M., *Eddy Currents*, Chemical Rubber Co. Press, Cleveland, 1966.

48. Tarasiewicz, E.J., Morched, A.S., Narang, A., and Dick, E.P., Frequency dependent eddy current models for the nonlinear iron cores, *IEEE Trans. Power Appar. Syst.*, 8, 588–597, 1993.

49. Avila-Rosales, J. and Alvarado, F.L., Nonlinear frequency dependent transformer model for electromagnetic transient studies in power systems, *IEEE Trans. Power Appar. Syst.*, 101, 4281–4289, 1982.

50. Batruni, R., Degeneff, R., and Lebow, M., Determining the effect of thermal loading on the remaining useful life of a power transformer from its impedance versus frequency characteristic, *IEEE Trans. Power Delivery*, 11, 1385–1390, 1996.

51. IEEE, Guide and Standards for Distribution, Power, and Regulating Transformers, C57-1990, Institute of Electrical and Electronics Engineers, Piscataway, NJ, 1990.

52. Balma, P.M., Degeneff, R.C., Moore, H.R., and Wagenaar, L.B., The Effects of Long-Term Operation and System Conditions on the Dielectric Capability and Insulation Coordination of Large Power Transformers, Paper No. 96 SM 406-9 PWRD, presented at the summer meetings of IEEE/PES, Denver, CO, 2000.

53. IEEE, Guide for the Application of Metal-Oxide Surge Arrester for Alternating-Current Systems, C62.22-1991, Institute of Electrical and Electronics Engineers, Piscataway, NJ, 1991.

54. Morched, A., Marti, L., and Ottevangers, J., A High Frequency Transformer Model for EMTP, No. 925M 359-0, presented at IEEE 1992 Summer Meeting, Seattle, WA.

55. de Leon, F. and Semlyen, A., Reduced order model for transformer transients, *IEEE Trans. Power Delivery,* 7, 361–369, 1992.

56. Dommel, H.W., Dommel, I.I., and Brandwajn, V., Matrix representation of three-phase n-winding transformers for steady state and transient studies, *IEEE Trans. Power Appar. Syst.*, PAS-101, 1982.

3.11   Transformer Installation and Maintenance

Alan Oswalt

3.11.1   Transformer Installation

The first priority is to hire a reliable contractor to move and assemble the transformer. There are many stories where the contractors, lacking experience or proper equipment, drop the transformer or do not assemble the components correctly. Accepting a low bid could cost your company more than securing a competent contractor.

Do not assume that all manufacturers have the same methods of installation. Your understanding of the manufacturer's transformer installation book and reviewing the complete set of drawings in advance will help you to understand "their" procedures. Some manufacturers have a toll-free number which allows you to clarify drawings and/or the assembly methods. Others have put together a series of videos and/ or CDs that will assist you to understand the complete assembly. Then you should review all of the information with your assembly contractor.

3.11.1.1   Receiving Inspection

Prior to unloading a transformer and the accessories, a complete inspection is necessary. If any damage or problems are found, contact the transformer manufacturer *before* unloading. Freight damage should be resolved, as it may be required to return the damaged transformer or the damaged accessories. Photographs of the damage should be sent to the manufacturer. Good receiving records and photographs are important, should there be any legal problems.

Three important inspections checks are (1) loss of pressure on the transformer, (2) above zone 3 on the impact recorder, and (3) signs of movement by the transformer or its accessories. If any of the three inspection checks indicate a problem, an internal inspection is recommended.

A shorted core reading could also mean a bad transit ride. With a railroad shipment, if any of the checks indicate problems and an internal inspection does not reveal the problem, get an *exception report* filled out by a railroad representative. This report will assist you later if hidden damages are found.

Low core meggar readings (200 Megohms) could be an indication of moisture in the unit and require extra costs to remove.[2] The moisture could have entered the unit through a cracked weld caused by the bad transit ride. (See Figs. 3.11.1 and 3.11.2.)

Entering a unit requires good confined entry procedures and can be done after contacting the manufacture, as they may want to have a representative present to do the inspection. Units shipped full of oil require a storage tanker and the costs should be agreed upon before starting.

Assuming that we now have a good transformer and it is setting on its substation pad, there are some items that are essential for assembly. First ground the transformer before starting the assembly. Static

---

[2]A dew point test will determine the moisture in the transformer. A dew point reading should be used with the winding temperature value (insulation temperature) to determine the percentage of moisture. (See Figs. 3.11.1 and 3.11.2.)

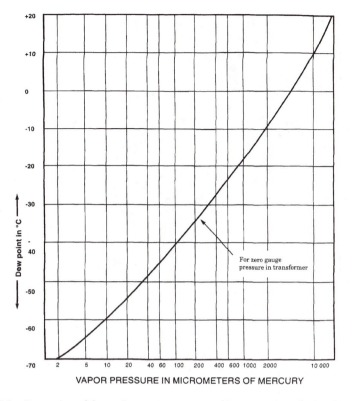

VAPOR PRESSURE IN MICROMETERS OF MERCURY

FIGURE 3.11.1 Conversion of dew point to vapor pressure. (Courtesy of Waukesha Electric Systems.)

electricity can build up in the transformer and cause a problem for the assembly crew. A static discharge could cause a crew member to jump or move and lose their balance while assembling parts.

Another item is to have all accessories to be assembled set close to the unit, as this eliminates a lot of lost time moving parts closer or from a storage yard. With the contractor setting the accessories close to the unit, you can usually save a day of assembly time. Keep in mind that some transformer manufactures "match-mark" each item. This means that each part has a specific location on the unit. Do not try to interchange the parts. Some manufacturers do not have this requirement, which allows bushings, radiators, and other parts to be assembled at the contractor or customer's discretion.

Weather is a major factor during the assembly of any transformer. Always have an ample supply of dry air flowing through the unit during the assembly. Be ready to seal the unit on positive pressure at the end of the day or if the weather turns bad. If the weather is questionable, keep the openings to a minimum and have everything ready to seal the unit.

There are many types of contaminants that can cause a transformer to fail. Foreign objects dropped into the windings, dirt brought into the unit on the assemblers' shoes, moisture left in the assembled parts, and misplaced or forgotten tools left inside are just a few items that could cause a failure. Take time to caution the assemblers about the preventions of contaminants and to follow good safety procedures. Again, an experienced contractor should have experienced assemblers and good assembly procedures in place.

If weather conditions are a problem, there are many other assembly operations that can be done during marginal weather. The following are a few:

1. Uncrate all items.
2. Locate and count all hardware.
3. Clean bushings with denatured alcohol and cover.
4. Count and replace gaskets — if possible.

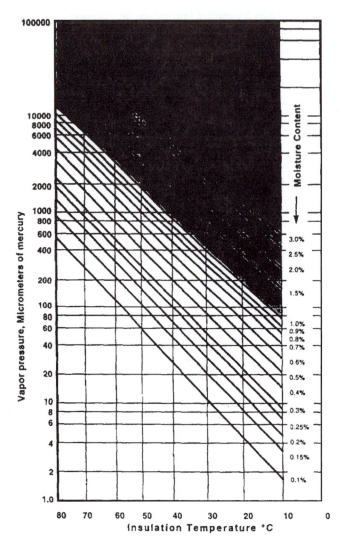

FIGURE 3.11.2   Moisture equilibrium chart (with moisture content in percent of dry weight of insulation). (Courtesy of Waukesha Electric Systems.)

Caution: Do not supply power to the control cabinet as it could back-feed into the inside current transformers, which could energize the primary and secondary bushings. A shock from the bushings could cause serious injury.

3.11.1.2   Bushings

You need to read the installation manual to understand the correct lifting and assembly methods. There are a variety of bushings so it will save time to have read this information. All bushing surfaces should be cleaned again with denatured alcohol. This also includes the inside tube (draw-through type bushing). During shipment, even though the bushings may be protected, contaminants, such as moisture, could be found inside the bushing tube. The draw-through bushings have a conductive cable, or a rod, that has to be pulled through the bushing while it is being installed. In some cases, the corona shield on HV bushings should be removed and cleaned. There are also bottom connected bushings that require copper bus, a terminal, and hardware to secure the connection to the bushing and winding.

All connections should be cleaned and free of oxidation or corrosion and then wiped down with denatured alcohol.[3]

After the installation of all bushings and all internal connections are made, another inspection should be made for the following:

1. Lead clearances. During the internal assembly work, some leads may have been moved. Check the manufacturer's installation book for the necessary clearances. The information should include the basic installation level (bil) rating along with the clearances needed.
2. Bolted connections, done by the assemblers, should be inspected for proper clearances.
3. Wipe down and vacuum clean the inside of the unit around the assembly area to remove any dirt or oil smudges.
4. Operate the de-energized tap changer (DETC) and check its mechanical operation.
5. Check for items, such as tools, that may have been left inside during the assembly.
6. Replace man-hole gaskets, if required.

Some units have conservators and require gas piping and oil piping connected to the transformer, after the man-hole covers are installed on the transformer and before pressure or vacuum cycles are started.

3.11.1.3 Oil Conservators

Conservators are usually mounted on one end of the transformer and well above the cover and bushings. Conservators normally have a rubber bladder inside. This bladder expands or retracts due to the temperature of the oil vs. the ambient temperature. The inside of the bladder is connected to external piping, and then to a silica gel breather. All exposure of the oil to the air is eliminated, yet the bladder can flex. (See Fig. 3.11.3.)

The oil supply piping, from the conservator to the transformer, should have at least one valve. The valve(s) must be closed during the vacuum cycle as the vacuum will try to pull the rubber bladder through the piping. The oil piping should have been cleaned prior to installation and the valves inspected. The conservator should have an inspection cover and the inside bladder inspected. While making this inspection, also check the operation of the oil float. (See Fig. 3.11.3.)

3.11.1.4 Gas Monitoring and Piping

The piping is used to bring any combustible gases to a monitor. The monitor is usually located on the cover where it is visible from ground level. All gas piping should be cleaned by blowing dry air through them, or cleaning with a rag and denatured alcohol. Gas pipes are usually not connected to the gas monitor until after the vacuum/oil filling. The gas monitor could have tubing running down the side of the unit to allow ground-level sampling or bleeding of the line. There are other types of oil/gas monitors than the one shown in Fig. 3.11.4.

3.11.1.5 Radiators

All radiators should be free of moisture and contaminants such as rust. If anything is found, the radiators should be cleaned and oil flushed with new transformer oil. The radiators may have to be replaced with new ones. Take time to inspect each radiator for bent fins or welding defects. If a problem is found, the repair should be made before installation. Touch-up painting, if needed, should be done, as it is difficult to reach all areas after the radiators are installed. During the radiator installation, all of the radiator valves need to be tested on at least 1 kg (2 lb) of pressure, or under oil, for a good seal.

Some gaskets for mounting the radiator/valve mounting flange may have to be replaced. Coating the outside of the gasket with petroleum jelly protects the surface of the gasket during the radiator assembly. The radiator surface will then slide without damaging the gasket.

---

[3]Some manufacturers will require all bushing gaskets to be replaced. Others furnish a Buta-N O-Ring that, in most cases, will not need replacing.

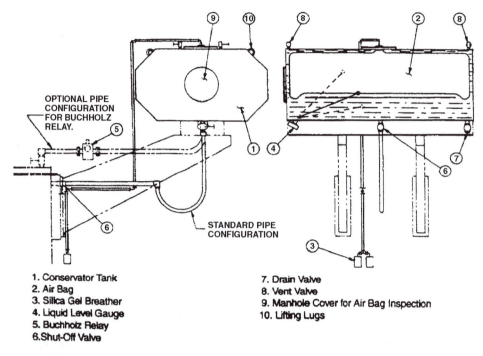

1. Conservator Tank
2. Air Bag
3. Silica Gel Breather
4. Liquid Level Gauge
5. Buchholz Relay
6. Shut-Off Valve

7. Drain Valve
8. Vent Valve
9. Manhole Cover for Air Bag Inspection
10. Lifting Lugs

FIGURE 3.11.3   Outline of typical relay installation on transformer cover. (Courtesy of Waukesha Electric Systems.)

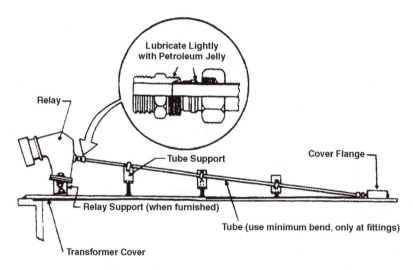

FIGURE 3.11.4   Conservator tank construction. (Courtesy of Waukesha Electric Systems.)

3.11.1.6   Coolers

Coolers are oil to air heat exchangers which require oil pumps and usually less space around the transformer. Forced oil cooling can be controlled by a top oil or a winding temperature gauge, or both. All pumps, piping, and coolers should be inspected for contaminants before assembly. The correct pump rotation is an important checkpoint.

3.11.1.7   Load Tap Changers (LTC)

Some LTCs mounted external to the main tank are shipped full of oil but you may want to make an internal inspection. After removing the oil you can inspect for problems. Check the manufacturer's installation book for information concerning vacuum oil filling of the unit. Some LTCs require a vacuum line to main tank for equalizing the pressure. *Do not* operate the LTC mechanism while the unit is on vacuum, as severe damage could occur to the mounting board.

If the LTC requires oil, *do not* add the oil while the main tank is on vacuum as the unequal pressure could damage the LTC. The process of adding oil to the external LTC tank will put pressure inside the tank. This added positive pressure along with the negative gage pressure of the main tank could cause the LTC barrier boards to rupture. No additional work should be attempted while the main transformer tank is under vacuum.

Look for loose hardware or any misalignment of the contacts. Operate the LTC through all positions and check each contact for alignment. Refer to the supplier's instruction manual for the allowable variance. Perfect "center line" alignment during the complete range of operation from 16R to 16L will be difficult to achieve.

3.11.1.8 Positive Pressure System

This system consists of a cabinet with regulating equipment and alarms with an attached nitrogen bottle. A positive supply of nitrogen is kept on the transformer. With positive pressure on the tank, the possibility of moisture entering is reduced. Loss of nitrogen pressure, without any oil spillage, is usually found in the transformer "gas space" or the nitrogen supply system.

3.11.1.9 Control Cabinet

All control equipment must be inspected for loose wiring or problems caused by the shipping. The fan, gauges, LTC controls, and monitoring equipment must be tested or calibrated. Information for the installation and/or the calibration should be supplied by the manufacturer.

3.11.1.10 Accessories

There are many items that could be required for your particular transformer. A few are listed below.

1. HV and LV arrestors
2. External current transformers
3. Discharge meters
4. Neutral grounding resistors
5. Bushing potential devices
6. Cooling fan (hi speed)
7. Gas monitors (various types)

Caution: All oil handling equipment, transformer bushings, and the transformer should be grounded before starting the vacuum oil cycle. Special requirements are needed for vacuum oil filling in cold weather. Check your manufacturer's manual.

3.11.1.11 Vacuum Cycle

Pulling vacuum on a transformer is usually done through the mechanical relief flange or a special vacuum valve located on the cover of the transformer. **A vacuum sensor, to send a signal to the vacuum recording gauge, should be at the highest location on the transformer's cover.** This position reduces the risk of the sensor being contaminated with oil, which would let the vacuum gauge give a false reading. All readings from this gauge should be recorded at least every hour. *Note:* All radiator and cooler valves should be open prior to starting the vacuum cycle.

3.11.1.12 Vacuum Filling System

Manufacturers differ on the duration of vacuum required and the method to add oil to the unit. It is important that the vacuum crew doing this process follow the correct procedure as stated by the manufacturer or the warranty can be invalid. Good record-keeping during this process is just as important for your information as it is for supplying the manufacturer with information that validates the warranty.

The length of time (pulling vacuum) will vary as to the exposure time to atmospheric air, the transformer rating, and the dew point/moisture calculations. Most of the needed information as to the vacuum cycle time, should be furnished in the installation book. (See Figs. 3.11.1 and 3.11.2.)

Always consider that some contractors have used their equipment on older and/or failed transformers. The equipment needs to be thoroughly cleaned with new transformer oil and a new filter medium added to the oil filtering equipment. The vacuum oil pump should have new vacuum oil installed and it should be able to "pull-down" against a closed valve to below 1 mm of pressure.

3.11.1.13 Transformer Oil

The oil supplied should be secured from an approved source and meet the IEEE C57 106-1991 guide for acceptance. When requested, an inhibitor can be added to the oil, to a level of 0.3% by weight.

3.11.1.14 Adding the Oil

All oil from tankers should be field tested for acceptable dielectric level prior to pumping oil through the oil handling equipment. The oil temperature should be at least 0°C while pumping directly from an oil tank car. A superior method that will assist in the removal of moisture involves heating the oil to 50 to 70°C and passing the oil through a filter. Oil filling a conservator transformer takes a lot more time as the piping and the conservator have to be slowly filled while air is "bled" out of the piping, bushings, CT turrets, and gas monitor. Methods vary for adding the oil to the conservator because of the risk to the oil bladder. Weeks later, the air should be "bled" again. If this is not done, you could receive a false signal that may take the transformer out of service.

3.11.1.15 Field Test

After the vacuum oil filling cycle, the transformer should be field tested and compared to some of the factory tests. The field test also will give you a baseline record for future reference.

These are a few of the tests that are routinely done:

1. Current transformer ratio
2. Turns ratio of the transformer, including tests at al LTC tap positions
3. Power factor bushings
4. Power factor transformer
5. Winding resistance
6. Core ground
7. Lab test of the oil
8. Test all gauges
9. Test all pressure switches

A field test report may be required by the manufacturer in order to validate the warranty. Any questionable test values should be brought to the attention of the transformer manufacturer.

3.11.2 Transformer Maintenance

The present maintenance trend is to reduce cost, which in some cases means lengthening the intervals of time to do maintenance or eliminating the maintenance completely. The utility, or company, realizes some savings on manpower and material by lengthening the maintenances cycle, but by doing this, the risk factor is increased. A few thousand dollars for a maintenance program could save your utility or company a half-million dollar transformer. Consider the following:

1. The length of time to have a transformer rebuilt or replaced
2. The extra load on your system
3. Rigging costs to move the transformer
4. Freight costs for the repair, or buying a new one
5. Disassembly and reassembly costs

6. Costs to set up and use a mobile transformer
7. Costs for oil handling of a failed unit
8. Vacuum oil filling of the rebuilt or new transformer
9. Customer's dissatisfaction with outage
10. Labor costs, which usually cover a lot of overtime or employees pulled away from their normal work schedule

You will have to answer questions such as: Why did this happen? Could it have been prevented? Time spent on a scheduled maintenance program is well worth the expense.

There are many systems available to monitor the transformer which can assist you in scheduling your maintenance program. Many transformer manufacturers can supply monitoring equipment that alerts the owner to potential problems. However, relying solely on monitoring equipment may not give your notice or alert you to mechanical problems. Some of these problems can be: fans that fail, pressure switches that malfunction, or oil pumps that cease to function. You could also have oil leaks that need to be repaired. Annual inspections can provide a chance for correcting a minor repair before it becomes a major repair.

3.11.2.1 Maintenance Tests

There are two important tests that could prevent a field failure. Using an infrared scan on a transformer could locate "hot spots". The high temperature areas could be caused by a radiator valve closed, low oil in a bushing, or an LTC problem. Early detection could allow time to repair the problem.

Another important test is dissolved gas analysis test of the oil by a lab. A dissolved gas analysis lab test will let you know if high levels of gases are found and they will inform you as to the recommended action. Following the lab report could let you plan your course of action. If there seems to be a problem, it would be worthwhile to take a second dissolved gas in oil sample and send it to a different lab and compare the results (IEEE C57 104-1991). Maintenance inspection and tests can be divided into two sections: (1) minor and, after a set period of years, (2) major inspection. Annual tests are usually done while the transformer is in service, and consist of the following:

1. Check the operation of the LTC mechanism for misalignment or excessive noise.
2. Take an oil sample from the LTC.
3. Change silica gel in breathers.
4. Inspect fan operation.
5. Take an oil sample from the main tank.
6. Check oil level in bushings.
7. Check tank and radiators for oil leaks.
8. Check for oil levels in main tank and the LTC.
9. Make sure all control heaters are operating.
10. Check all door gaskets.
11. Record the amount of LTC operations and operate through a couple of positions.
12. Most importantly, have your own check-off list and take time to do each check. This record (check-off list) can be used for future reference.

Major inspections require the transformer to be out of service. Both primary and secondary bushings should be grounded before doing the work. Besides the annual inspection checks that should be made, the following should also be done:

1. Power factor the bushings and compare to the values found during the installation tests.
2. Power factor the transformer and check these valves.
3. Make a complete inspection of the LTC and replace any questionable parts. If major repair is required during this inspection, a turns ratio test should be done.
4. Painting rusty areas may be necessary.

5. Test all pressure switches and alarms.
6. Check the tightness of all bolted connections.
7. Check and test the control cabinet components.

The lists for annual and major inspections may not be complete for your transformer and you may want to make a formal record for your company's reference. Some transformer installation/installation books furnished by the manufacturer may have a list of their inspections areas which you should utilize in making your own formal record.

References

Waukesha Electric Systems Instruction Book.

3.12   Problem and Failure Investigation

Wallace Binder and Harold Moore

3.12.1   Introduction

The investigation of transformer problems or failures is in many respects similar to medical procedures. The health of a transformer can be monitored using the many diagnostic tools available today. Ignoring a minor problem can lead to a more severe failure. Documenting and recording the results of operation and diagnostic testing is essential to a complete evaluation. Early detection and mitigation of developing problems can, in the long term, save the cost of major repairs. Many transformer operators with significant numbers of transformers perform field diagnostics and share this information with other operators to establish benchmark performance for transformers in different applications and with different designs.

Elements of transformer design, application, and operation are involved in failure investigation. The elements to be investigated depend on the nature and the severity of the problem. If a failure is involved, all of the elements are usually investigated. The analyses required can be quite complex and involved. It would be impossible to describe the many details involved in complex failure investigations within this one section; indeed, a complete book could be written on this subject. In the space allotted here, it will only be possible to describe the processes involved in such investigations.

Excellent references are ANSI/IEEE C57.117, IEEE Guide for Reporting Failure Data for Power Transformers and Shunt Reactors on Electrical Power Systems, and ANSI/IEEE C57.125, IEEE Guide for Failure Investigation, Documentation, and Analysis for Power Transformers and Shunt Reactors.

The following steps are involved in problem and failure investigations.

- Collect pertinent background data.
- Visit the site to obtain application and operational data.
- Interview all persons that may have relevant knowledge.
- Inspect the transformer and perform a partial or complete dismantling if a failure is involved.
- Analyze the available information and transformer history.
- Prepare a preliminary report and review with persons who are involved or who have a direct interest to generate additional inputs.
- Write a final report.

As a general statement, no details should be neglected in such investigations. Experience has indicated that what may appear to be minor details sometimes hold the essential clues for solutions. Collection of *all* relevant data is the most important aspect of problem and failure investigation.

3.12.2 Background Investigation

The background investigation usually involves exploring the following.

- Identification of failure. How did the transformer problem manifest itself or how did the failure occur? This starting point assumes that some event has occurred, such as a trip, malfunction, or abnormal diagnostic test results.
- The original transformer design
- A history of similar transformer designs
- The original manufacturing and testing of the transformer
- System application and operational data

3.12.2.1 Transformer Records

3.12.2.1.1 Transformer Application

Cooperation between parties involved will speed the investigation and lead more easily to a correct conclusion. The manufacturer should be contacted even when the transformer is many years old. Most equipment manufacturers track problems with their designs, and awareness of problems can help them improve designs and manufacturing processes. The manufacturer may also be able to provide information leading to a solution.

Application of the transformer in the power system should be determined. A party with expertise in transformer application should confirm that the transformer was properly rated for the purpose for which it was intended and that subsequent operation has been consistent with that intent.

Transformer problems can lead to service outages. Careful analysis of the situation before restoring service can avoid a repetition of the outage and avoid increasing damage to the transformer.

The performance of transformers made by the manufacturer should be studied. Industry records of transformers of similar rating and voltage class may be helpful in establishing base data for the investigation. For example, there have been transformers made by certain manufacturers that have a history of short-circuit failures in service. At one time, the failure record of extra high voltage (EHV) transformers designed with three steps of reduced basic impulse insulation level (BIL) had a higher failure rate than those with higher BILs.

3.12.2.1.2 Transformer Design

The specifications for the transformer, instruction books and literature, nameplate, and drawings such as the outline and internal assembly should be examined. If failures involving the core and windings have occurred, the investigative process is much like a design review in which the details of the insulation design, winding configuration, lead configurations and clearances, short-circuit capability, and the core construction and normal operating induction must be studied. If components are involved, bushings and bushing clearances, tap changers, heat exchanger equipment, and control equipment should be investigated.

3.12.2.1.2.1 Design Review Process — An explanation of the design review process would require a separate chapter and is beyond the scope of this subject. The process involves a detailed study of the winding and insulation designs, short-circuit capability, thermal design, magnetic-circuit characteristics, leakage flux losses and heating analysis, materials used, oil preservation systems, etc. In some cases, it may be desirable to conduct part or all of a design review as a part of a problem investigation.

3.12.2.1.2.2 Determine if Similar Designs Experienced Problems or Failures — This can be difficult, since there is no agency that collects and distributes data on all industry problems. It has been made even more difficult by the closing of several major transformer manufacturing plants. However, some information can usually be obtained by discussions with users having transformers made by the same manufacturer.

3.12.2.1.3 *Manufacturing and Testing of the Transformer*

The manufacturing and test records should be studied to determine if discrepancies occurred. Of particular importance are deviations from normal manufacturing specifications or practices. Such deviations could be involved in the problem or failure. All parts of the test data must be studied to determine if discrepancies or deviations existed. Partial-discharge and impulse test data that have any deviation from good practices should be recorded. Approvals of deviations made during testing, especially relating to dielectric and other test standards, are sometimes indications of difficulty during testing.

3.12.2.1.4 *Transformer Installation*

The records of installation and initial field tests should be used as a benchmark for any future test results. These initial field tests are more easily compared with subsequent field tests than with factory tests. Factory tests are performed at full load current or full voltage, and adjustments must be made to field test results to account for differences in losses at reduced load and differences in excitation at the operating voltage.

3.12.2.2 Transformer Protection

The protection of the transformer is as important a part of the application as the rating values on the transformer. Entire texts, as well as Chapter 3.8 in this text, are devoted to the subject of transformer protection. When investigating a failure, one should collect all the protection-scheme application and confirm that the operation of any tripping function was correct.

3.12.2.2.1 *Surge Arresters*

Surge arrester protective level must be coordinated with the BIL of the transformer. Their purpose, to state what may seem obvious, is to protect the transformer from impulse voltages and high-frequency transients. Surge arresters do not eliminate voltage transients. They clip the voltages to a level that the transformer insulation system is designed to tolerate. However, repeated impulse voltages can have a harmful effect on the transformer insulation.

3.12.2.2.2 *Overcurrent Protection*

Overcurrent devices must adequately protect the transformer from short circuits. Properly applied, the time–current characteristic of the device should coordinate with that of the transformer. These characteristics are described in IEEE C57.109-1993, Guide for Liquid-Immersed Transformer Through-Fault Duration. Overcurrent devices may be as simple as power fuses or more complex overcurrent relays. Modern overcurrent relays contain recording capability that may contain valuable information on the fault being investigated.

3.12.2.2.3 *Differential Protection*

Differential relays, if applied, should be coordinated with the short-circuit current available, the transformer turns ratio and connection, and the current transformers employed in the differential scheme. If differential relays have operated correctly, a fault occurred within the protected zone. One must determine if the protected zone includes only the transformer, or if other devices, such as buswork or circuit breakers, might have faulted.

3.12.2.3 Recording Devices

The occurrence of unusual trends or events that indicated a possible problem should be recorded. Operation of relays or protective equipment that indicated a failure should be studied. Copies of oscillographic or computer records of the events surrounding the problem or failure should be obtained.

Records of events immediately prior to the observation of a problem or immediately prior to a trip are often only the final events in a series that led to the failure. Investigations at the transformer location should concentrate on collecting such data as relay targets, event recordings, and on-line monitoring records as well as making observations of the condition of the transformer and associated equipment.

Records of relay operation must be analyzed to determine the sequence of events. A failure of a transformer due to a through fault, if uncorrected, can lead to the failure of a replacement unit. This can be a very expensive oversight.

Based on initial observations and recordings, a preliminary cause may be determined. Subsequent tests should focus on confirming or refuting this cause.

3.12.2.4 Operational History

3.12.2.4.1 Diagnostic Testing

The operation records of the transformer should be examined in detail. Records of field tests such as insulation resistance and power factor, gas-in-oil analyses, oil test data, and bushing tests should be studied. Any trends from normal such as increasing power factor, water in oil, or deterioration of oil properties should be noted. Any internal inspections that may have been performed, changes in oil, and any repairs or modifications are of interest. Maintenance records should be examined for evidence of either good or poor maintenance practices. Proper application of arresters and other protective schemes should be verified.

3.12.2.4.2 Severe Duty

Operational history such as switching events, numbers and type of system short circuits, known lightning strikes, overloads, etc., should be recorded.

3.12.2.4.2.1 Overloads — The effect of loading a transformer beyond nameplate is not apparent without disassembly. The history of transformer operation will be the first indication of damage from overloads. The transformer may have some inherent capability to handle loading in excess of nameplate rating. However, the result of loading beyond nameplate can result in accelerated loss of life. This loss of life is cumulative and cannot be restored. Loading recommendations are more fully described in IEEE C57.91-1995, Guide for Loading Mineral-Oil-Immersed Transformers, and in IEEE C57.96-1999, Guide for Loading Dry-Type Distribution and Power Transformers.

3.12.2.4.2.2 Overvoltages — Maximum continuous transformer operating voltage should not exceed the levels specified in ANSI C84.1-1995. Overexcitation will result in core heating and subsequent damage. The results of core heating may not be readily discerned until a transformer is untanked. Examination of the operating records may provide insight into this as a cause of damage.

3.12.2.4.2.3 Short Circuits — IEEE C57.12.00-2000, IEEE Standard General Requirements for Liquid-Immersed Distribution, Power, and Regulating Transformers, and IEEE C57.12.01, IEEE Standard General Requirements for Dry-Type Distribution and Power Transformers Including Those with Solid-Cast and/or Resin-Encapsulated Windings, state limits for short-circuit durations and magnitude. Not all short-circuits reach the magnitude of those limits. However, like overloads, the effect can be cumulative. This effect is further expanded in IEEE C57.109-1993, IEEE Guide for Liquid-Immersed Transformer Through-Fault-Current Duration. The effect is described as an "extremely inverse time-current characteristic" upon which the overcurrent protection of the transformer should be based.

3.12.3 Problem Analysis where No Failure Is Involved

3.12.3.1 Transformer Components

Component failures can lead to major failures if left uncorrected. All of these should be corrected as soon as identified to avoid letting them develop into problems that are more serious.

3.12.3.1.1 Transformer Tank

Transformer tank welds can crack after thermal cycling or even after withstanding the stresses of through-faults. Bushing and radiator gaskets can deteriorate over time, leading to leaks. Left unattended, these leaks can result in moisture entering the transformer insulation.

3.12.3.1.2 *Transformer Radiators and Coolers*

Radiators can leak, leading to ingress of moisture, or they can clog, reducing their cooling efficiency. Fan motors can fail, resulting in a loss of cooling. Today's modern low-loss transformer designs can have a dual rating ONAN/ONAF with as few as one or two cooling fans. Loss of a large percentage of the fans necessitates using the ONAN rating for transformer capacity. (See Section 2.1.2.3, Cooling Classes, for a discussion of cooling-class terminology.)

3.12.3.1.3 *Transformer Bushings*

Bushings occasionally develop problems such as leaks or high power factor and thermal problems. Problems detected during periodic diagnostic testing should be remedied as quickly as possible.

3.12.3.1.4 *Transformer Gauges*

Gauges and indicating devices should be repaired or replaced as soon as it is suspected that they are giving false indication. Confirm that the abnormal indication is false and replace the device. Letting a false indication continue may disguise the development of a more severe problem.

3.12.3.2 Severe Duty Investigations

It is advisable to make some investigations after severe operating events such as direct lightning strikes (if known), high-magnitude short-circuit faults, and inadvertent high-magnitude overloads. Problem investigations are sometimes more difficult than failure analyses because some of the internal parts cannot be seen without dismantling the transformer. The same data-collection process is recommended for problems as for failures. It is recommended that a plan be prepared when the investigation is initiated, including a checklist to guide the study and to ensure that no important steps are omitted. A recommended checklist is in ANSI/IEEE C57.125. Specific steps that are recommended are as follows:

3.12.3.2.1 *Communication with Persons Involved or Site Visit*

- Obtain the background of the events indicating a problem.
- If there is any external evidence such as loss of oil, overheated parts, or other external indications, a site visit may be advisable to get first-hand information and to discuss the events with persons who operated the transformer.
- Determine if there have been any unusual events such as short circuits, overvoltages, or overloads.

3.12.3.2.2 *Diagnostic Testing*

If the utility has a comprehensive field-testing program, obtain and study the following data. If the data are not available, arrange for tests to be made.

3.12.3.2.2.1 Gas-in-Oil Analyses — Obtain test results for several years prior to the event, if possible. Many good papers and texts have discussed the interpretation of dissolved gas-in-oil analysis (DGA) results. One must also remember that once an internal failure has occurred, the initiating cause may be masked by the resulting fault gases. IEEE C57.104-1991, IEEE Guide for Interpretation of Gases Generated in Oil-Immersed Transformers, can provide the latest interpretation of DGA results.

3.12.3.2.2.2 Oil Test Data — Dielectric strength in accordance with ASTM D 1816

- Water in oil
- Power factor at 25 and 100°C, if available
- General characteristics such as color, inter-facial tension (IFT), etc.

3.12.3.2.2.3 Turns Ratio — This test should match the factory results and be within the standard 0.5% of calculated value (except as noted in Clause 9.1 of IEEE C57.12.00-2000). Any deviation indicates a partially shorted turn.

3.12.3.2.2.4 Resistance — Such measurements must be made using suitable high-accuracy instruments; these measurements are not easy to make under service conditions.

3.12.3.2.2.5 Insulation Power Factor, Resistance, and Capacitance — A polarization index can be determined in the same manner. Polarization index results that are less than 2.0 indicate deterioration of the insulation.

3.12.3.2.2.6 Low-Voltage Exciting Current (if Previous Data Are Available) — This information is an excellent indication of winding movement, but the test must be performed under circumstances similar to the benchmark test to be of value.

3.12.3.2.2.7 Short-Circuit Impedance (if Faults Have Been Involved) — This test is an indicator of winding movement that may also give indication of shorted turns.

3.12.3.2.3 Internal Inspection

If the problem cannot he identified from the test data and behavior analyses, an internal inspection may be necessary. In general, internal inspections should be avoided because the probability of failure increases after persons have entered transformers. The following items should be checked when the transformer is inspected.

- Is there evidence of carbon tracks indicating flashovers or severe partial discharges?
- Check leads for evidence of overheating. The insulation will be tan, brown, or black in extreme cases.
- Check for evidence of partial-discharge trees or failure paths. Do odors indicate burned insulation or oil?
- Check bolted connections for proper tightness.
- Are leads and bushing lower shields in position? Is the lead insulation tight?
- Inspect the windings for evidence of distortion or movement.
- Check the end insulation on core-form designs for evidence of movement or looseness.
- Are the support members at the ends of the phases tight in shell-form designs?
- Are coil clamping devices tight and in position?
- Check tap changers for contact deterioration. Is there evidence of problems in the operating mechanisms?
- In core-form units, check visible parts of the core for evidence of heating.

After the inspection has been completed, look over everything again for evidence of "does anything look abnormal." Such final inspections frequently reveal something of value.

3.12.4 Failure Investigations

The same processes are required in failure investigations as for problem analyses. In fact, it is recommended that the approaches described for problem investigations be performed before dismantling of the transformer to determine possible causes for the failure. Performing this work in advance usually results in hypotheses for the failure and frequently indicates directions for the dismantling. Thus, the only additional steps to be performed following a failure are to dismantle the transformer and make a determination whether it should be repaired. There is never a 100% certainty that the cause of failure can be determined. All theories should be tested against the facts available, and when assumptions come into doubt, the information must be replaced with data that is confirmed by the other facts available.

As a general rule, all steps of the process should be documented with photographs. All too often it happens that something that was not considered relevant in the early stages of the investigation is later determined to be important. A photographic record will ensure that evidence that has been destroyed in the dismantling process is available for later evaluation.

3.12.4.1 Dismantling Process

Complete dismantling is usually performed when a failure has occurred. In a few instances, such operations are performed to determine the cause for a problem, such as excessive gassing that could not be explained by other investigations.

The following steps are recommended for this process.

3.12.4.1.1 *Transformer Expert*

The process should be directed by a person or persons having knowledge of transformer design. If it is done in the original manufacturer's plant, the manufacturer's experts will usually be available. However, it is recommended that the user have experts available if the failure mechanism is in doubt. It is good to have two experts available because they may look into the failure from different perspectives and will provide the opportunity to discuss various aspects of the investigation from differing viewpoints.

3.12.4.1.2 *Inspection before Untanking*

Inspect the tank for distortion resulting from high internal pressures that sometimes result from failures. Check the position of leads and connections. Determine if there has been movement of bushings.

3.12.4.1.3 *Inspection after Removal of Core and Windings*

- Make a detailed inspection of the top ends of windings and cores.
- Inspect the mechanical and electrical condition of the interphase insulation and the insulation between the windings and the tank.
- Check the core ground.
- Check lead entrances to windings for mechanical, thermal, and electrical condition.
- Check the general condition of leads, connections, and tap changers.
- Check the leakage flux shields on tank walls or on frames for heating or arcing.
- Examine the wedging between phases and from the windings to the core on shell-form units and the winding clamping structures on core-form units to determine if there is still pressure on the windings.

3.12.4.1.4 *Detailed Inspection of the Windings*

- Is there evidence of electrical failure paths from the windings over the major insulation to ground?
- Is there evidence of distortion of the coils or windings resulting from short-circuit failures?
- Has electrical failure occurred between turns, and has there been mechanical distortion of the turns?
- Have failures occurred between windings or between coils? Are there weaknesses in nonfailed portions that could affect the electrical strength? Such instances include damage to turn insulation, distorted disks or coils, or insulation pieces not properly assembled.
- Is there evidence of metallic contamination on the insulation and windings?
- Is there evidence of hot spots at the ends of windings or in cap leads inside the disks or coils?
- Is there any evidence of partial discharges or overheating in the leads or connections?
- Were the windings properly supported for short circuit?
  - Are spacers in alignment?
  - Is insulation between windings and between the inner winding and the core tight?

3.12.4.1.5 *Examination of the Magnetic Circuit*

- Have electrical arcs occurred to the core? If so, determine if the fault current flowed from the failure point to ground, resulting in damage to core laminations in this path. Note that any such damage will usually require scrapping of the laminations.
- Is there evidence of leakage flux heating in the outer laminations?

- Has heating occurred as the result of large gaps at the joints, excessive burrs at slit, or cut edges and joints?
- Does mechanical distortion exist in any parts of the core?
- Is there evidence of heating in the lock plates used for mechanical support of the frames?
- Is the core ground in good condition, with no evidence of heating or burning? Is there any evidence of a second (unintended) core ground having developed?

3.12.4.1.6 Mechanical Components

- Is there distortion in the mechanical supports?
- Is there evidence of leakage flux heating in the frames or frame shields?

3.12.5 Analysis of Information

Information on interpretation of the data could take volumes, and there is much information on this subject in the technical literature. Some simple guides are listed below for reference.

- The presence of high carbon monoxide and carbon dioxide are indications of thermal or oxidative damage to cellulose insulation. If there is high CO and there have been no overloads or previous indications of thermal problems, the problem may be excessive oxygen in the oil.
- High oxygen is usually an indication of inadequate oil processing, gasket leaks, or leak of air through the rubber bag in expansion tanks.
- Acetylene is an indication of arcing or very high temperatures.
- Deterioration of oil dielectric strength usually results from particulate contamination or excessive water.
- High power factor or low resistance between windings or from the windings to ground is usually the result of excessive water in the insulation.
- High water in oil may result from excessive water in the paper. Over 95% of the total water in the system is in the paper, so that high water in oil is a reflection of the water in the paper.
- Turns ratio different from previous measurements is an indication of shorts in a winding. The shorts can be between turns or between parts of windings, such as disk-to-disk.
- A measurable change in the leakage impedance is an indication of winding movement or distortion.
- Open circuits result from major burning in the windings or possibly a tap-changer malfunction.
- High hydrogen, with methane being about 20% of the hydrogen, is an indication of partial discharges. "Spitting" or "cracking" noises noted prior to the failure are sometimes indicators of intense partial discharges.

3.12.5.1 Interpretation and Analysis of Information

The most important part of the process is analysis of the information gathered. The objective is determining the cause of the problem or failure, and adequate analysis is obviously necessary if problems are to be solved and failures are to be prevented. There is no one process that is best for all situations. However, there are two helpful steps for reaching conclusions in such matters.

- Make a systematic analysis of the data.
- Compare data analyses to known problem and failure modes.

3.12.5.1.1 Systematic Analysis of Data

- Prepare a list of known facts. List also the unknowns. Attempt to find answers for the unknowns that appear to be of importance.
- Analyze known facts to determine if a pattern indicates the nature of the problem.

- Prepare a spreadsheet of test data and observations, including inspections. Note items that appear to indicate the cause of the problem or failure.
- Use problem-solving techniques.

3.12.5.1.2 *Comparison to Known Problem and Failure Modes*

There are many recognized possible failure modes; a few are listed here as examples and for guidance.

3.12.5.1.2.1 Dielectric or Insulation Failures — Surface or creepage over long distances. If the design is shown to be adequate, this phenomenon is usually caused by contamination. If the design is marginal, slight amounts of contamination may initiate the discharges or failure.

- Oil space breakdown. This can occur in any part of the insulation, since oil is the weak link in the insulation system. If the design is marginal, discharges can be initiated by particulate contamination or water in the oil. This type of breakdown usually occurs at interfaces with paper, such as at the edge of a radial spacer in a disk-to-disk space or at the edge of a spacer in a high-voltage winding to low-voltage winding space.
- Oil breakdown over long distances, as from a bushing shield to tank wall or from a lead to ground. This problem type is usually caused by overstress in the large oil gap. It can occur in marginal situations if particles or gas bubbles are present in the gap. The dielectric strength of oil is lower at low temperatures if there is an appreciable amount of water in the oil. If such breakdowns occur in very low temperature conditions, investigate the oil strength at the low temperature as a function of the water in the oil. Consider also that the oil level may have been low by virtue of the very low temperature, causing parts normally under the oil to be exposed.
- Turn-to-turn failures. If the design is adequate, such failures can result from mechanical weakness in the paper or from damage during short circuits if the paper is brittle due to thermal aging or oxidation. These failures usually are associated with fast transients such as lightning.
- Extensive treeing in areas of high oil velocity, such as the oil entrance to the windings in forced-cooled designs. This can be associated with static electrification and usually occurs when the oil temperature is less than 40°C and all pumps are in operation.
- Discharges or failure originating from joints in leads. This type of failure usually results from the paper not being tight at the joint in the tape. Discharges start in the oil space at the surface of the cable and propagate out through the joint.

IEEE C57.125-1991, IEEE Guide for Failure Investigation, Documentation, and Analysis for Power Transformers and Shunt Reactors, contains a comprehensive treatment of insulation system failure and analysis of the relative voltage stresses that can lead to discharges.

3.12.5.1.2.2 Thermal or Oxidation Failure Modes — Deteriorated insulation at the end turns of core-form transformers or on the outer turns of line coils in shell-form designs. Such deterioration is caused by local hot spots. The eddy losses are higher in these regions, and the designer may have used added insulation in some regions that have high electrical stress.

- Overheated lap leads. This usually occurs because the designer has used added insulation on the leads. The leads may have added eddy loss because they are in a high leakage flux field.
- Leads with brown or black paper at the surface of the conductor. This results from excessive paper insulation on the lead.
- Joints with deteriorated paper. The resistance of the joint may be too high, or there may be leakage flux heating if the connector is wide.
- Damaged paper or pressboard adjacent to the core or core supports. This type of heating is usually the result of leakage flux heating in the laminations or core joints.
- Paper has lost much of its strength, but there have been no thermal stresses. This is the result of excessive oxygen in the oil. In the initial stages of the process, the outer layers of paper will have more damage than the inner layers.

3.12.5.1.2.3 Magnetic-Circuit Heating — Large gaps at joints can result in localized saturation of the core. The gap area will be black, and there may be low levels of methane, ethane, and some hydrogen.

- Local heating on the surface or at joints. Such heating is caused by excessive burrs on the edges or at end cuts of the laminations.
- Burning at the joint of outer laminations. This can he caused by circulating currents in the outer laminations of cores. It results from an imbalance in leakage flux.

3.12.5.1.2.4 Short-Circuit Failure Modes — Failure at the ends of shell-form coils. This type of failure results from inadequate support of the outer layers of the coils.

- Beam bending between spacers in either shell- or core-form designs, which is caused by stresses higher than the beam strength of the conductor. Missing spacers can also be involved.
- Buckling of inner winding in core-form construction. This results from inadequate strength of the conductor or inadequate spacer support. The evidence is radial distortion of the winding.
- Telescoping of conductors. This occurs in windings with a thin conductor where there is not adequate support by an inner cylinder. The layers slip over adjacent layers.
- Lead distortion.
- Turns or disks telescoping over the end insulation or supports. This results from high axial forces in combination with insulation that may not have been properly dried and compressed.

Short-circuit failures are mechanical failures where the forces *internal* to the transformer generated by the current feeding the *external* fault through the windings. This results in movement of insulation or conductor, which can lead to collapse, breakage, or stretching of insulation and/or conductor. When winding movement or an internal short occurs, magnetic fields and force vectors become abnormal. Stresses may be placed on materials that the designer could not anticipate. The resulting damage can make it difficult to determine the first cause.

3.12.5.1.3 Examples of Multiple Failure Modes

It is important to consider that many problems and failures are a combination of events or problems. Some examples will be used to illustrate the technique.

3.12.5.1.3.1 Example of Excessive Water in the Solid Insulation of a Transformer — Measured water in the oil is high, particularly after the transformer has been loaded and the oil is hot. (At elevated temperatures, water comes from the paper to the oil rather quickly.)

- Power factor of the insulation is increasing.
- The hydrogen content of the oil is increasing.
- The dielectric strength of the oil is decreasing when tested in accordance with ASTM D-1816.

This analysis indicates that the insulation has excessive water and it should be dried. This conclusion could have been made without the power-factor measurements.

3.12.5.1.3.2 Example of Excessive Oxygen in the System (Usually Recognized as >3000 ppm of Oxygen in the Oil) — Generation of CO is high.

- There is no history of overloads.
- Design analysis indicated no excessive hot spots.
- Oxygen appears to vary as the CO increases. (The oxygen is being consumed by the process that forms CO.)
- Internal inspection indicated that the outer layers of paper on a taped cable had greater deterioration than the inner layers.

3.12.5.1.3.3 Example of Multiple Causes — The importance of keeping good records of transformer operation and maintenance events and of making a complete analysis of all data involved in problem solving cannot be overemphasized. Many failures and problems result from multiple causes. The following example demonstrates the importance of diligent investigations.

- Transformer experienced a severe short circuit as the result of a through-fault on the system. Transformer did not fail.
- Oxygen in the oil had been high — 4000 ppm or higher for years.
- Transformer had history of high CO generation.
- Failure occurred some months after a switching event.

The failure was at first attributed to the switching event alone. However, the investigation showed that it was initiated by damage to brittle insulation, probably during the short-circuit event. The brittle condition was caused by the high oxygen in the system. The overvoltage involved in the switching event caused the failure at the damaged paper location.

Another important factor in problem and failure analysis is to use two experienced persons when possible. Each can challenge the ideas expressed by the other and offer suggestions for different approaches in the investigation. Experience has shown that a better analysis results when using this approach.

3.12.5.2 Analysis of Current and Voltage Waveforms

It is important to determine which phase (or phases) of a three-phase transformer or bank of single-phase transformers is (are) involved in a short circuit. Sophisticated measuring devices exist and are often part of the protective scheme for modern transformer installations. High-speed recording devices, like oscillographs and digital recorders, can provide records of the current and voltage waveshapes before, during, and after the fault. Some of these devices will provide current magnitude and phase angle (from

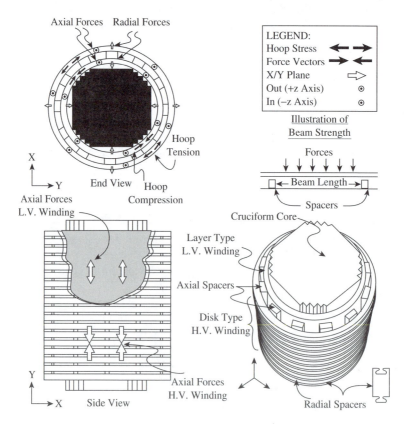

FIGURE 3.12.1 Concentric circular winding. (From IEEE Guide for Failure Investigation, Documentation, and Analysis for Power Transformers and Shunt Reactors, C57.125-1991. ©1991 IEEE. With permission.)

a reference value) of the currents and voltages. Determine, from these devices, the magnitude of the fault and the phase(s) involved by determining which phase voltages distorted because of the IZ drop. Take into account the winding connections when calculating the current flowing in the windings of the transformer. Subsequent investigation should concentrate on this (or these) phase(s).

3.12.5.3 Analysis of Short-Circuit Paths

Observe any evidence of arc initiation and terminus. Signs of partial discharge or streamers across insulation parts may be found. The impurity that initiated the arc may have been destroyed by the ensuing arc. An accurate short-circuit current can lead to analysis of the location of the fault by determining the impedance to the fault location, even if the location is inside the transformer winding. Electrical failures are frequently dielectric system (insulation) failures.

3.12.5.4 Analysis of Mechanical Stresses

Observe any evidence of misalignment of winding components. Coil or insulation that has shifted indicates a high level of force that may have broken or abraded the insulation. This mechanical failure will then manifest itself as an electrical breakdown.

Each type of windings has specific failure modes. Core-form transformers have many different winding arrangements, such as disk windings, layer windings, helical windings, and other variations.Figure 3.12.1 helps explain the forces and stresses in a concentric circular winding. Figure 3.12.2 describes the same force vectors in rectangular windings. Though not shown, a shell-form winding exhibits similar forces.

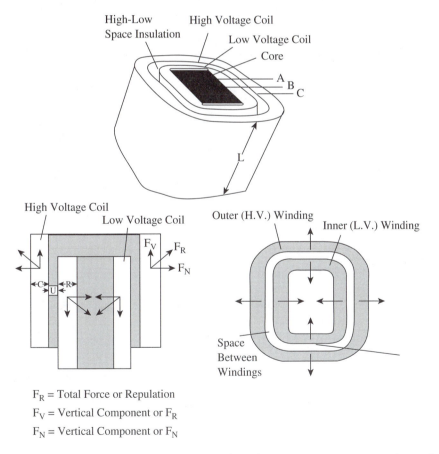

F_R = Total Force or Repulsion

F_V = Vertical Component or F_R

F_N = Vertical Component or F_N

FIGURE 3.12.2 Rectangular winding. (From IEEE Guide for Failure Investigation, Documentation, and Analysis for Power Transformers and Shunt Reactors, C57.125-1991. ©1991 IEEE. With permission.)

3.12.6 Special Considerations

3.12.6.1 Personal Injury

The first person on the scene following an incident involving personal injury must preserve as much evidence as possible. While protection of human life is the first priority, make note as soon as possible of the situation existing at the time of the failure. It may fall to the first technical person on the scene to make the necessary observations and record the data. Interview witnesses and record indicators to preserve as much data as possible.

3.12.6.2 Safety

Every possible safety precaution should be observed when dealing with power transformers. The power system has lethal voltages, and even when a unit is de-energized by automatic action of the protective scheme, the possibility of hazardous voltages remains. All OSHA and electric utility safety requirements must be followed to ensure that workers, investigators, and the public are protected from harm. This includes, but is not limited to, isolation and grounding of devices, following applicable tagging procedures, use of personal protective gear, and barricading the area to prevent ingress by unauthorized or unqualified individuals.

3.13 On-Line Monitoring of Liquid-Immersed Transformers

Andre Lux

On-line monitoring of transformers and associated accessories (measuring certain parameters or conditions while energized) is an important consideration in their operation and maintenance. The justification for on-line monitoring is driven by the need to increase the availability of transformers, to facilitate the transition from time-based and/or operational-based maintenance to condition-based maintenance, to improve asset and life management, and to enhance failure-cause analysis.

This discussion covers most of the on-line monitoring methods that are currently in common practice, including their benefits, system configurations, and application to the various operational parameters that can be monitored. For the purposes of this section, the term *transformer* refers, but is not limited, to: step-down power transformers; generator step-up transformers; autotransformers; phase-shifting transformers; regulating transformers; intertie transmission transformers; dc converter transformers; high-voltage instrument transformers; and shunt, series, and saturable reactors.

3.13.1 Benefits

Various issues must be considered when determining whether or not the installation of an on-line monitoring system is appropriate. Prior to the installation of on-line monitoring equipment, cost-benefit and risk-benefit analyses are typically performed in order to determine the value of the monitoring system as applied to a particular transformer. For example, for an aging transformer, especially with critical functions, on-line monitoring of certain key parameters is appropriate and valuable. Monitoring equipment can also be justified for transformers with certain types of load tap changers that have a history of coking or other types of problems, or for transformers with symptoms of certain types of problems such as overheating, partial discharge, excessive aging, bushing problems, etc. However, for transformers that are operated normally without any overloading and have acceptable routine maintenance and dissolved gas analysis (DGA) test results, monitoring can probably not be justified economically.

3.13.1.1 Categories

Both direct and strategic benefits can arise from the installation of on-line monitoring equipment. Direct benefits are cost-savings benefits obtained strictly from changing maintenance activities. They include reducing expenses by reducing the frequency of equipment inspections and by reducing or delaying active interventions (repair, replacement, etc.) on the equipment. Strategic benefits are based on the ability to prevent (or mitigate) failures or to avoid catastrophe. These benefits can be substantial, since failures can be very damaging and costly. Benefits in this category include better safety (preventing injuries to workers or the public in the event of catastrophic failure), protection of the equipment, and avoiding the potentially large impact caused by system instability, loss of load, environmental cleanup, etc.

3.13.1.2 Direct Benefits

3.13.1.2.1 Maintenance Benefits

Maintenance benefits represent resources saved in maintenance activities by the application of on-line monitoring as a predictive maintenance technique. On-line monitoring can mitigate or eliminate the need for manual time-based or operation-based inspections by identifying problems early and allowing corrective actions to be implemented.

3.13.1.2.2 Equipment Usage Benefits

Equipment usage benefits arise because additional reinforcement capacity may be deferred because on-line monitoring and diagnostics allow more effective utilization of existing equipment. On-line monitoring equipment can continuously provide real-time capability limits, both operationally and in terms of equipment life.

3.13.1.3 Strategic Benefits

Strategic benefits are those that accrue when the results of system failures can be mitigated, reduced, or eliminated. A key feature of on-line monitoring technology is its ability to anticipate and forestall catastrophic failures. The value of the technology is its ability to lessen the frequency of such failures.

3.13.1.3.1 Service Reestablishment Benefits

Service reestablishment benefits represent the reduced need for repair and/or replacement of damaged equipment because on-line monitoring has been able to identify a component failure in time for planned corrective action. Unscheduled repairs can be very costly in terms of equipment damage and its potential impact on worker safety and public relations.

3.13.1.3.2 System Operations Benefits

System operations benefits represent the avoidance of operational adjustments to the power system as a result of having identified the component failure prior to a general failure. System adjustments, in the face of a delivery-system breakdown, can range from negligible to significant. An example of a negligible adjustment is when the failure is in a noncritical part of the network and adequate redundancy exists. Significant adjustments are necessary if the failure causes large, baseload generation to experience a forced outage, or if contractual obligations to independent generators cannot be met. These benefits are driven in part by the duration of the resulting circuit outage.

3.13.1.3.3 Outage Benefits

Outage benefits represent the impact of component failure and resulting system breakdown on end-use customers. A utility incurs direct revenue losses as a result of a system or component failure. A utility's customers, in turn, may also experience losses during failures. The magnitude and/or frequency of such losses may result in the customer's loss of significant revenues.

3.13.2 On-Line Monitoring Systems

The characteristics of transformer on-line monitoring equipment can vary, depending on the number of parameters that are monitored and the desired accessibility of the data. An on-line monitoring system typically records data at regular intervals and initiates alarms and reports when preset limits are exceeded. The equipment required for an on-line transformer monitoring system consists of sensors, data-acquisition units (DAU), and a computer connected with a communications link.

3.13.2.1   Sensors

Sensors measure electrical, chemical, and physical signals. Individual sensor types and monitoring methods are discussed in Section 3.13.3, On-Line Monitoring Applications. Standard sensor output signal levels are 4 to 20 mA, 0 to 1 mA, and 0 to 10 V when an analog representation is used. The sensors can be directly connected to the data acquisition unit(s). Another category of sensors communicates in serial format, as is characteristic of those implemented within intelligent electronic devices (IED).

Information/data about a function or status that is being monitored is captured by a sensor that can be attached directly to the transformer or within the control house. Once captured, the data are transferred to a data-acquisition unit (DAU) that can also be attached to the transformer or located elsewhere in the substation. The transfer is triggered by a predefined event such as a motor operation, a signal reaching a threshold, or the changing state of a contact. The transfer can also be initiated by a time-based schedule such as an hourly measurement of the power factor of a bushing, or any other such quantity.

The method of data collection depends on the characteristics of the on-line monitoring system. A common characteristic of all systems is the need to move information/data from the sensor level to the user. The following represent examples of possible components in a data-collection system.

3.13.2.2   Data-Acquisition Units

A data-acquisition unit collects signals from one or more sensors and performs signal conditioning and analog-to-digital conversions. The DAU also provides electrical isolation and insulation between the measured output signals and the DAU electronics. For example, a trigger could cause the DAU to start recording, store information about the event, and send it to a substation computer.

3.13.2.3   DAU-to-Computer Communications Line

The data-collection process usually involves transferring the data to a computer. The computer could be located within the DAU, elsewhere in the substation, or off-site. The data can be transferred via a variety of communications networks such as permanent direct connection, manual direct connection, local-area networks (LAN), or wide-area networks (WAN).

3.13.2.4   Computer

At the computer, information is held resident for additional analysis. The computer may be an integral part of the DAU, or it may be located separately in the station. The computer is based on standard technology. From a platform point of view, software functions of the substation computer program include support of the computer, the users, communications systems, storage of data, and communications with users or other systems, such as supervisory control and data acquisition (SCADA). The computer manages the DAUs and acts as the data and communications server to the user-interface software. The computer facilitates expert-system diagnostics and contains the basic platform for data acquisition and storage.

3.13.2.5   Data Processing

The first step in data processing is the extraction of sensor data. Some types of data can be used in the form in which it is acquired, while other types of data need to be processed further. For example, a transformer's top-oil temperature can be directly used, while a bushing's sum current waveform requires additional processing to calculate the fundamental frequency (50 or 60 Hz) phasor. The data are then compared with various reference values such as limits, nameplate values, and other measurements, depending on the user's application.

In situations where reference data are not available, a learning period may be used to generate a baseline for comparison. Data are accumulated during a specified period of time, and statistical evaluation is used to either accept or reject the data. In some applications, the rejected data are still saved, but they are not used in the calculation of the initial benchmark. In other applications, the initial benchmark is determined using only the accepted data.

The next data-processing step is to determine if variations suggest actual apparatus problems or if they are due to ambient fluctuations (such as weather effects), power-system variables, or other effects. A combination of signal-processing techniques and/or the correlation of the information obtained from measurements from locations on the same bus can be used to eliminate both the power-system effects and temperature influences.

The next step in processing depends on the sophistication of the monitoring system. However, the data generally need to be interpreted, with the resulting information communicated to the user. One common approach is to compare the measured parameter with the previous measurement. If the value has not changed significantly, then no data are recorded, saved, or transmitted.

3.13.3   On-Line Monitoring Applications

Various basic parameters of power transformers, load tap changers, instrument transformers, and bushings can be monitored with available sensor technologies.

3.13.3.1   Power Transformers

Transformer problems can be characterized as those that arise from defects and develop into incipient faults, those that derive from deterioration processes, and those induced by operating conditions that exceed the capability of the transformer. These problems may take many years to gestate before developing into a problem or failure. However, in some cases, undesirable consequences can be created quite precipitously.

Deterioration processes relating to aging are accelerated by thermal and voltage stresses. Increasing levels of temperature, oxygen, moisture, and other contaminants significantly contribute to insulation degradation. The deterioration is particularly exaggerated in the presence of catalysts and/or through-faults and by mechanical or electromechanical wear. Characteristics of the deterioration processes include sludge accumulation, weakened mechanical strength of insulation materials such as paper-wrapped conductor, shrinkage of materials that provide mechanical support, and improper alignment of tap-changer mechanisms. Excessive moisture accelerates the aging of insulation materials over many years of operation. During extreme thermal transients that can occur during some loading cycles, high moisture content can result in water vapor bubbles. The bubbles can cause serious reduction in dielectric strength of the insulating liquid, resulting in a dielectric failure.

The processes causing eventual problems (e.g., shrinkage of the insulation material or excessive moisture) may take many years to develop, but the consequences can appear suddenly. Continuous monitoring permits timely remedial action: not too early (thus saving valuable maintenance resources) and not too late (which would have costly consequences). Higher loading can be tolerated, as continuous automated evaluation will alert users of conditions that could result in failure or excessive aging of critical insulation structures (Griffin, 1999).

Table 3.13.1 lists the major transformer components along with their associated problems and the parameters that can be monitored on-line to detect them.

3.13.3.1.1   Dissolved-Gas-in-Oil Analysis

3.13.3.1.1.1   Monitored Parameters — Dissolved gas-in-oil analysis (DGA) has proved to be a valuable and reliable diagnostic technique for the detection of incipient fault conditions within liquid-immersed transformers by detecting certain key gases. DGA has been widely used throughout the industry as the primary diagnostic tool for transformer maintenance, and it is of major importance in a transformer owner's loss-prevention program.

Data have been acquired from the analysis of samples from electrical equipment in the factory, laboratory, and field installations over the years. A large body of information relating certain fault conditions to the various gases that can be detected and easily quantified by gas chromatography has been developed. The gases that are generally measured and their significance are shown in Table 3.13.2 (Griffin, 1999). Griffin provides methods for interpreting fault conditions associated with various gas concentration levels and combinations of these gases (Griffin, 1999).

TABLE 3.13.1 Main Tank Transformer Components, Failure Mechanisms, and Measured Signals

Component		Phenomenon	Measured Signals
General	Specific		
Noncurrent-carrying metal components	Core	Overheating of laminations	Top and bottom temperatures Ambient temperature Line currents Voltage Hydrogen (minor overheating) Multigas, particularly ethane, ethylene, and methane (moderate or severe overheating)
	Frames Clamping Cleats Shielding Tank walls	Overheating due to circulating currents, leakage flux	Top and bottom temperatures Ambient temperature Line currents Voltage Multigas, particularly ethane, ethylene, and methane
	Core ground Magnetic shield	Floating core and shield grounds create discharge	Hydrogen or multigas Acoustic and electric PD
Winding insulation	Cellulose: Paper, pressboard, wood products	Local and general overheating and excessive aging	Top and bottom temperatures Ambient temperature Line currents RS moisture in oil Multigas, particularly carbon monoxide, carbon dioxide, and oxygen
		Severe hot spot Overheating	Top and bottom temperatures Ambient temperature Line currents Moisture in oil Multigas, particularly carbon monoxide, carbon dioxide, ethane, hydrogen, and oxygen
		Moisture contamination	Top and bottom temperatures Ambient temperature Relative saturation of moisture in oil
		Bubble generation	Top and bottom temperatures Ambient temperature Total percent dissolved gas-in-oil Line currents Relative saturation of moisture in oil Hydrogen Acoustic and electric PD
		Partial discharge	Hydrogen or multigas Acoustic and electric PD
Liquid insulation		Moisture contamination	Top and bottom temperatures Ambient temperature Relative saturation of moisture in oil
		Partial discharge	Hydrogen Acoustic and electric PD
		Arcing	Hydrogen and acetylene
		Local overheating	Ethylene, ethane, methane
Cooling system	Fans Pumps Temperature-measurement devices	Electrical failures of pumps and fans	Motor (fan, pump) currents Top-oil temperature Line currents

— continued

TABLE 3.13.1 (continued) Main Tank Transformer Components, Failure Mechanisms, and Measured Signals

Component			
General	Specific	Phenomenon	Measured Signals
		Failure or inaccuracy of top liquid or winding temperature indicators or alarms	Ambient temperature Top and bottom temperatures Line currents
	Internal cooling path	Defects or physical damage in the directed flow system	Top and bottom temperatures Ambient temperature Line currents
		Localized hot spots	Carbon monoxide and carbon dioxide
	Radiators and coolers	Internal or external blocking of radiators resulting in poor heat exchange	Top and bottom temperatures Ambient temperature Line currents
Oil and winding temperature forecasting		Overloading of transformer	Top and bottom temperatures Ambient temperature Line currents Moisture in oil Multigas, particularly carbon monoxide, carbon dioxide, and oxygen

[a] Denotes combustible gas. Overheating can be caused both by high temperatures and by unusual or abnormal electrical stress.

Source: Based on IEEE Guide C57.104. With permission.

TABLE 3.13.2 Gases Typically Found in Transformer Insulating Liquid under Fault Conditions

Gas	Chemical Formula	Predominant Source
Nitrogen	N_2	Inert gas blanket, atmosphere
Oxygen	O_2	Atmosphere
Hydrogen [a]	H_2	Partial discharge
Carbon dioxide	CO_2	Overheated cellulose, atmosphere
Carbon monoxide [a]	CO	Overheated cellulose, air pollution
Methane [a]	CH_4	Overheated oil (hot metal gas)
Ethane [a]	C_2H_6	Overheated oil
Ethylene [a]	C_2H_4	Very overheated oil (may have trace of C_2H_2)
Acetylene [a]	C_2H_2	Arcing in oil

Laboratory-based DGA programs are typically conducted on a periodic basis dictated by the application or transformer type. Some problems with short gestation times may go undetected between normal laboratory test intervals. Installation of continuous gas-in-oil monitors may detect the start of incipient failure conditions, thus allowing the user to confirm the presence of a suspected fault through laboratory DGA testing. Such an early warning might enable the user to plan necessary steps required to identify the fault and implement corrective actions. Existing technology can determine gas type, concentration, trending, and production rates of generated gases. The rate of change of gases dissolved in oil is a valuable diagnostic that is useful in determining the severity of a developing fault. A conventional unscheduled gas-in-oil analysis is typically performed after an alarm condition has been reported. The application of on-line dissolved-gas monitoring considerably reduces the risk of detection failure or of a delay in detecting fault initialization because of the typically long intervals between on-site oil sampling.

Laboratory-based sampling and analysis with a frequency sufficient to obtain real-time feedback becomes impractical and too expensive. For critical transformers, on-line gas-in-oil monitors can provide timely and continuous information in a manner that permits load adjustments to prevent excessive gassing from initiating thermal-type faults. This can keep a transformer operating for many months while ensuring that safety limits are observed.

3.13.3.1.1.2 Gas Sensor Development — Early attempts to identify and document the gases found in energized transformers date from 1919. This analysis was conducted by liquid column chromatography (Myers et al., 1981). An early type of gas monitor, still in use in many locations, is a device similar to the Buchholz relay, which was developed in the late 1920s. This type of relay detects and measures the pressure of free gas generated in the transformer and indicates an alarm signal.

The gas chromatograph was first applied to this area in the early 1960s. Its ability to differentiate and quantify the various gases that are generated and found in the insulating oil of transformers and other electrical equipment has proven quite useful (Myers et al., 1981). Beginning in the late 1970s and continuing to the present, efforts have been made to develop a gas chromatograph for on-line applications. These efforts have been focused on analyzing gases in the gas space of transformers and on extracting the gases from transformer oil and injecting the gases into the gas chromatograph. Recently, on-site laboratory-quality analyses have become available utilizing a portable gas chromatograph that is not permanently connected to the transformer.

In the 1980s and early 1990s, an alternative method to using gas chromatography was developed. Sensors based on fuel-cell technology and thermal conductivity detection (TCD) were developed. Both methods use membrane technologies to separate dissolved gases from the transformer oil and produce voltage signals proportional to the amount of dissolved gases. The fuel-cell sensor senses hydrogen and carbon monoxide together with small amounts of other hydrocarbon gases. This method has been successful in providing an early warning of detecting incipient faults initiated by the dielectric breakdown of the insulating fluid and the cellulose found in the solid insulation.

Subsequent efforts have been targeted toward measuring the other gases that can be produced inside the transformer that are detectable by gas chromatography. These efforts are designed to provide on-line access to data that can then be used to indicate the need for further sampling of the insulating oil. The oil is then analyzed in the laboratory to confirm the monitoring data.

During the mid-1990s, a multigas on-line DGA monitor that could detect and quantify the gas concentrations in parts per million (ppm) was developed (Chu et al., 1993; 1994). This monitor samples all seven key gases and was designed to provide sufficient dissolved-gas data, ensuring that analysis and interpretation of faults could take place on-line using the criteria provided in IEEE standards (IEEE C57.104). The sampling approach is noninvasive, with both the extraction and sensor systems external to the transformer (Glodjo, 1998). The system takes multiple oil samples per day and senses changes in the absolute values of gas concentrations and in the ratios of the concentrations of particular selected gases. This information is analyzed along with the transformer load and temperature levels, environmental conditions, and known fault conditions from repair records and diagnostic software programs. A second system developed uses membrane extraction technology combined with infrared spectroscopy (FTIR) sensing for all gases except hydrogen. For hydrogen, this system uses fuel-cell technology. It can also detect all seven key gases (per IEEE Std. C57.104).

3.13.3.1.2 Moisture in Oil

The measurement of moisture in oil is a routine test (in addition to other physical characteristics of the oil) performed in the laboratory on a sample taken from the transformer. The moisture level of the sample is evaluated with the sample temperature and the winding temperature of the transformer. This combination of data is vital in determining the relative saturation of moisture in the cellulose/liquid insulation complex that establishes the dielectric integrity of the transformer. Moisture in the transformer reduces the insulation strength by decreasing the dielectric strength of the transformer's insulation system. As the transformer warms up, moisture migrates from the solid insulation into the fluid. The rate of migration is dependent on the conductor temperature and the rate of change of the conductor temperature. As the transformer cools, the moisture returns to the solid insulation at a slower rate. The time constants for these migrations depend on the design of the transformer and the solid and liquid components in use. The combination of moisture, heat, and oxygen are the key conditions that indicate accelerated degradation of the cellulose. Excessive amounts of moisture can accelerate the degradation process of the cellulose and prematurely age the transformer's insulation system. The existence of a particular type of furanic compound in the oil is also an indication of moisture in the cellulose insulation.

Moisture-in-oil sensors were first successfully tested and used in the early 1990s (Oommen, 1991; 1993). The sensors measure the relative saturation of the water in oil, which is a more meaningful measure than the more familiar units of parts per million (ppm). Continuous measurements allow for detection of the true moisture content of the transformer insulation system and of the hazardous conditions that may occur during temperature cycling, thereby helping to prevent transformer failures.

3.13.3.1.3 *Partial Discharge*

One cause of transformer failures is dielectric breakdown. Failure of the dielectrics inside transformers is often preceded by partial-discharge activity. A significant increase either in the partial-discharge (PD) level or in the rate of increase of partial-discharge level can provide an early indication that changes are evolving inside the transformer. Since partial discharge can lead to complete breakdown, it is desirable to monitor this parameter on-line. Partial discharges in oil will produce hydrogen dissolved in the oil. However, the dissolved hydrogen may or may not be detected, depending on the location of the PD source and the time necessary for the oil to carry or transport the dissolved hydrogen to the location of the sensor. The PD sources most commonly encountered are moisture in the insulation, cavities in solid insulation, metallic particles, and gas bubbles generated due to some fault condition.

The interpretation of detected PD activity is not straightforward. No general rules exist that correlate the remaining life of a transformer to PD activity. As part of the routine factory acceptance tests, most transformers are tested to have a PD level below a specified value. From a monitoring and diagnostic view, detection of PD above this level is therefore cause for an alarm, but it is not generally cause for a tripping action. These realities illustrate one of the many difficulties encountered in PD diagnosis. The results need to be interpreted with knowledge of the studied equipment. Two methods are used for measuring partial discharges: electrical and acoustic. Both of these have attracted considerable attention, but neither is able to yield an unambiguous PD measurement without additional procedures.

3.13.3.1.3.1 Electrical Method — The electrical signals from PD are in the form of a unipolar pulse with a rise time that can be as short as nanoseconds (Morshuis, 1995). Two electrical procedures for partial-discharge measurement exist. These give results in microvolts or picocoulombs. There is no fundamental conversion between the procedures applicable to all cases. The signals exhibit a very wide frequency content. The high frequencies are attenuated when the signal propagates through the equipment and the network. The detected signal frequency is dependent both on the original signal and the measurement method.

Electrical PD detection methods are generally hampered by electrical interference signals from surrounding equipment and the network, as illustrated in Figure 3.13.1. Any on-line PD sensing method has to find a way to minimize the influence of such signals. One way is to use a directional high-frequency field sensor (Lemke, 1987). The high detection frequency limits the disturbance from PD sources at a distance, and the directionality simplifies a remote scan of many objects. Therefore, this type of sensor seems most appropriate for periodic surveillance. It is not known whether this principle has been tried in a continuous monitoring system.

A popular method to interpret PD signals is to study their occurrence and amplitude as a function of the power-phase position; this is called phase-resolved PD analysis (PRPDA). This method can give valuable insight into the type of PD problem present. It is suggested that by identifying typical problem patterns in a PRPDA, one could minimize external influences (Fruth and Fuhr, 1990). The conceptual difficulty with this method is that the problem type must be known beforehand, which is not always the case. Second, the relevant signals may be corrupted by an external disturbance.

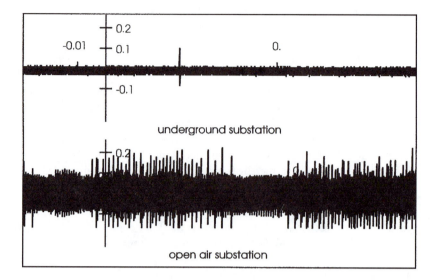

FIGURE 3.13.1 Electric PD measurements on transformers in underground and open-air substations. The overhead transmission lines cause a multitude of signals, making a PD measurement very insensitive. Underground stations are generally fed by cables that attenuate the high-frequency signals from the network, and PD measurements are quite sensitive. Horizontal scale in seconds, vertical scale in mV.

There have been many attempts to use neural networks or adaptive digital filters (Wenzel et al., 1995a), but it is not clear if this has led to a standard method. The problem with this approach is that the measured and the background signals are very similar, and the variation within each of the groups may be much larger than the difference between them. Adaptive filters and neural networks have been used to diminish other background sources such as medium-wave radio and rectifier pulses.

These methods employ a single sensor for the PD measurement. If several sensors of different types or at different locations are employed, the possibilities of reducing external influences are greatly enhanced. Generally, the multisensor approach can be split into two branches: separate detection of external signals and energy flow measurements.

When there is a clear source for the disturbing signals, it is tempting to use a separate sensor as a pickup for those and simply turn off the PD measurements when the external level is too large. Methods like this have the disadvantage of being insensitive during some portion of the measurement time. In addition, a very large signal from the equipment under study may be detected by the external pickup as well, and thus be rejected.

Energy-flow measurements use both an inductive and a capacitive sensor to measure current and voltage in the PD pulse (Eriksson et al., 1995; Wenzel et al., 1995b). By careful tuning of the signals from the two sensors, they can be reliably multiplied, and the polarity of the resulting energy pulse determines whether the signal originated inside or outside the apparatus. This approach seems to be the most promising for on-line electric PD detection.

3.13.3.1.3.2 Acoustic Method — Like electrical methods, acoustic methods have a long history of use for PD detection. The sensitivity can be shown to be comparable with electric sensing. Acoustic signals are generated from bubble formation and collapse during the PD event, and these signals have frequencies of approximately 100 kHz (Bengtsson et al., 1993). Like the electric signals, the high frequencies are generally attenuated during propagation. Due to the limited propagation velocity, acoustic signals are commonly used for location of PD sources.

The main advantage of acoustic detection is that disturbing signals from the electric network do not interfere with the measurement. As the acoustic signal propagates from the PD source to the sensor, it generally encounters different materials. Some of these materials can attenuate the signal considerably; furthermore, each material interface further attenuates the propagated signal. Therefore, acoustic signals can only be detected within a limited distance from the source. Consequently, the sensitivity for PD inside transformer windings, for example, may be quite low. In typical applications, many acoustic sensors are carefully distributed around the tested equipment (Eleftherion, 1995; Bengtsson et al., 1997).

Though not disturbed by the electric network, external influences in the form of rain or wind and non-PD vibration sources, like loose parts and cooling fans, will generate acoustic signals that interfere with the PD detection. One way to decrease the external influence is to use acoustic waveguides (Harrold, 1983) that detect signals from inside the transformer tank. This solution is typically only considered for permanent monitoring of important transformers. As an alternative, phase-position analysis can be used to reject these disturbances (Bengtsson et al., 1997).

A transformer generates disturbing acoustic signals in the form of core noise, which can extend up to the 50 to 100-kHz region. To diminish this disturbance, acoustic sensors with sensitivity in the 150-kHz range are usually employed (Eleftherion, 1995). Such sensors may, however, have less sensitivity to PD signals as well (Bengtsson et al., 1997). The properties of these signals are such that it is relatively easy to distinguish them from PD signals; thus, their main effect is to limit sensitivity.

Regarding the electric multisensor systems discussed here, there are a few descriptions of combined electric and acoustic PD monitoring systems for transformers in the literature (Wang et al., 1997). Rather elaborate software must be employed to utilize the potential sensitivity of these systems. If both the acoustic and the electric parts are designed with the considerations above in mind and an effective software constructed, systems like this will become effective yet costly.

3.13.3.1.4 Oil Temperatures
Overheating or overloading can cause transformer failures. Continuous measurement of the top-oil temperature is an important factor in maximizing the service life. Top-oil temperature, ambient temperature, load (current), fan/pump operations, and direct readings of winding temperatures (if available) can be combined in algorithms to determine hottest-spot temperature and manage the overall temperature conditions of the transformer.

3.13.3.1.5 Winding Temperatures
There is a direct correlation between winding temperature and normally expected service life of a transformer. The hottest-spot temperature of the winding is one of a number of limiting factors for the load capability of transformers. Insulation materials lose their mechanical strength with prolonged exposure to excessive heat. This can result in tearing and displacement of the paper and dielectric breakdown, resulting in premature failures. Conventional winding temperature measurements are not typically direct; the hot spot is indirectly calculated from oil temperature and load current measurements using a widely recommended and described test method (Domun, 1994; Duval and Lamarre, 1977; Feser et al., 1993; Fox, 1983; IEEE, 1995). Fiber-optic temperature sensors can be installed in the winding only when the transformer is manufactured, rebuilt, or refurbished. Two sensor types are available: optical fibers that measure the temperature at one point, and distributed optical fibers that measure the temperature along the length of the winding. Since a distributed fiber-optic temperature sensor is capable of measuring the temperature along the fiber as a function of distance, it can replace a large number of discrete sensors and allow a real-time measurement of the temperature distribution.

3.13.3.1.6 Load Current and Voltage
Maximum loading of transformers is restricted by the temperature to which the transformer and its accessories can be exposed without excessive loss of life. Continuous on-line monitoring of current and voltage coupled with temperature measurements can provide a means to gauge thermal performance. Load current and voltage monitoring can also automatically track the loading peaks of the transformer, increase the accuracy of simulated computer load-flow programs, provide individual load profiles to assist in distribution-system planning, and aid in dynamically loading the transformer.

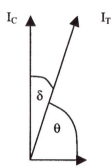

FIGURE 3.13.2 Power-factor graphical representation.

3.13.3.1.7   Insulation Power Factor

The dielectric loss in any insulation system is the power dissipated by the insulation when an ac voltage is applied. All electrical insulation has a measurable quantity of dielectric loss, regardless of condition. Good insulation usually has a very low loss. Normal aging of an insulating material causes the dielectric loss to increase. Contamination of insulation by moisture or chemical substances can cause losses to be higher than normal. Physical damage from electrical stress or other outside forces also affects the level of losses.

When an ac voltage is applied to insulation, the leakage current flowing through the insulation has two components, one resistive and the other capacitive. This is depicted in Figure 3.13.2. The power factor is a dimensionless ratio of the resistive current (I_R) to total current (I_T) flowing through the insulation and is given by the cosine of the angle θ depicted in Figure 3.13.2. The dissipation factor, also known as tan delta, is a dimensionless ratio of the resistive current to the reactive current flowing through the insulation and is the tangent of the angle δ in Figure 3.13.2. By convention, these factors are usually expressed in percent. Due to the fact that theta is expected to be large, usually approaching 90 degrees, and delta is commensurately small, the power factor and dissipation factor are often considered to be essentially equal.

3.13.3.1.8   Pump/Fan Operation

The most frequent failure mode of the cooling system is the failure of pumps and fans. The objective of continuous on-line analysis of pumps and fans is to determine if they are on when they are supposed to be on and are off when they are supposed to be off. This is accomplished by measuring the currents drawn by pumps and fans and correlating them with the measurement of the temperature that controls the cooling system. This can also be accomplished by measuring pump/fan current and top-oil temperature. Mode of operation is verified based on current level. Normal operational modes indicate rotation of fan blades and correct rotation of pump impeller. Abnormal operational modes are usually the result of improper control wiring to those devices.

Pump failures due to malfunctioning bearings could be a source of metallic particles, and such particles could be a potential dielectric hazard. Sensors that detect bearing wear are available. The ultrasonic sensors are embedded in the pump bearings and measure the bearing length, thus determining whether metal loss is occurring.

Furthermore, continuous on-line analysis should take into account that:

- The temperature that controls the cooling system can differ from the temperature measured by the diagnostic system.
- The initial monitoring parameters are set for the cooling stages based on the original transformer design. Any modifications to the cooling sequences or upgrades must be noted, since this will change the monitoring system output.
- The sensitivity of the diagnostic system is influenced by the number of motors that are measured by each current sensor.

3.13.3.2 Instrument Transformers

The techniques available to monitor instrument transformers on-line can be focused on fewer possible degradation mechanisms than those for monitoring power transformers. However, the mechanisms by which instrument transformers fail are among the most difficult to detect on-line and are not easy to simulate or accelerate in the laboratory.

3.13.3.2.1 Failure Mechanisms Associated with Instrument Transformers

While the failure rates of instrument transformers around the world are generally low, the large numbers of installed instrument transformers has led to the development of a database of failures and failure statistics. One problem associated with compiling a database of failures of porcelain-housed instrument transformers is that such failures are often catastrophic, leaving little evidence to determine the cause of the fault. Nevertheless, the following mechanisms have been observed and identified as probable causes of failure.

3.13.3.2.1.1 Moisture Ingress — Moisture ingress is commonly identified as a cause of failure of instrument transformers. The ingress of moisture into the instrument transformer can occur through loss of integrity of a mechanical seal, e.g., gaskets. The moisture penetrates the oil and oil/paper insulation (which increases the losses in the insulating materials) and failure then follows. This would appear to be a particular problem if the moisture penetrates to certain high-stress regions within the instrument transformer. The increase in the dielectric losses will be detected as a change in the power factor of the material and will also appear as increased moisture levels in oil quality tests.

3.13.3.2.1.2 Partial Discharge — The insulation of instrument transformers may have voids within it. Such voids will undergo partial discharge if subjected to a high enough electric field. Such discharges may produce aggressive chemical by-products, which then enlarge the size of the void, causing an increase in the energy of the discharge within the void. Eventually, these small partial discharges can degrade individual insulation layers, resulting in the short-circuiting of stress grading layers. Such a developing fault can be detected in two ways. One is the observance of a change in the capacitance of the device (through the shorting of one stress grading layer), and which may reflect as a change in tan delta. The second is an increase in the partial discharge levels (in pC) associated with the failing item.

3.13.3.2.1.3 Overvoltages — Overvoltages produced by induced lightning surges are also a failure mechanism, particularly where thunderstorms occurred in the vicinity of the failure. More recently, the observance of fast rise-time transients ($T_{rise} \approx 100 \, \eta s$) in substations during disconnect switch operations has led to concerns that these transients may cause damage to the insulation of instrument transformers. There is significant speculation that instrument transformers do not perform well when exposed to a number of disconnect switch operations in quick succession. These disconnector-generated fast transients will remain a suspected cause of failure until more is understood about the stress distribution within the instrument transformer under these conditions. Switching overvoltages are an additional source of overstressing that may lead to insulation failure.

3.13.3.2.1.4 Through-Faults — In order to prevent failures due to the mechanisms outlined above, experience seems to indicate that slower-forming faults are probably detectable and preventable, while fast-forming faults due to damage caused by lightning strikes will be difficult to detect quickly enough to prevent consequential failure of the transformer.

Another possible mechanism may relate to mechanical damage to the insulation after a current transformer (CT) has been subjected to fault current through its primary winding. After current transformer failures, it is often observed in retrospect that one to two weeks prior to the failure the CT had been subjected to a through-fault. Again, it is difficult to state that damage is caused to the CT under these conditions, and additional information would be required before this mechanism can be considered a probable cause of failure.

3.13.3.2.2   *Instrument Transformer On-Line Monitoring Methods*

On-line techniques for the measurement of relative tan delta and relative capacitance (by comparing individual units against a larger population of similar units) have been installed by a number of utilities, with reports of some success in identifying suspect units. On-line partial-discharge measurement techniques may provide important additional information as to the condition of the insulation within the instrument transformer, but research and development work is still under way in order to address issues related to noise rejection vs. required sensitivity and on-site calibration. Other possible future developments may include on-line dissolved-gas analyzers that will be able to detect all gases associated with the partial-discharge degradation of oil/paper insulation. The following subsections review applicable methods for on-line monitoring of instrument transformers.

3.13.3.2.2.1   Relative Tan Delta and Relative Capacitance Measurements — Off-line partial discharge and tan delta monitoring are well-established techniques. These can be supplemented by taking small samples of mineral oil from the instrument transformer for DGA. The development of on-line monitoring techniques is ongoing, but significant progress has been made, particularly with respect to on-line tan delta and capacitance measurements. Laboratory-type tan delta and capacitance measurements usually require a standard low-loss capacitor at the voltage rating of the equipment under test, such that a sensitive bridge technique can be used to determine the capacitance and the tan delta (also know as the insulation power factor) of the insulation. This is not practical for on-line measurements.

This problem is overcome by relying on relative measurements, in which the insulation of one instrument transformer is compared with the insulation of the other instrument transformers that are installed in the same substation. By comparing sufficient numbers of instrument transformers with other similar units, changes in one unit (not explained as normal statistical fluctuations due to changes in loading and ambient temperatures) can be identified. There are two commercially available units that monitor tan delta on-line. In the first, the ground current from each of the three single-phase instrument transformers is detected. This is done by isolating the base of the instrument transformer from its base except at one connection point, which then forms the only current path to ground. This current can then be measured using a suitable sensor. The current consists of two components: a capacitive component (the capacitance of a typical CT to earth being on the order of 0.5 to 1 ηF) and a resistive component dependent upon the insulation loss factor or tan delta of the insulation within the instrument transformer. If each of the three instrument transformers is in similar condition and of similar design, then the phasor sum of the three-phase currents to earth is essentially zero. Any resistive component of current to earth causes slight phase and magnitude shifts in these currents. If all three units on each phase have a low tan delta, then changes in one unit with respect to the other two can be readily detected. As the insulation deteriorates, and possibly as a grading layer is shorted out, a change in the capacitance of the unit will be reflected as a change in the capacitive current to earth. As the measurements are made with respect to other similar units, such measurements are referred to as relative tan delta and relative capacitance change measurements. Figure 3.13.3 shows this arrangement schematically.

Another technique involves comparing each instrument transformer with a number of different units, possibly on the same busbar or on each of three phases. The capacitive and resistive current flows to earth are monitored, and the results for each instrument transformer can then be compared with those values measured on other units. Relative changes in tan delta and capacitance can then be determined, and an alarm is raised if these exceed norms established from software algorithms.

These two techniques are currently in service and have achieved success in detecting instrument transformers behaving in a manner that is markedly different from other similar units. Both measurement tools are trending instruments by detecting changes of certain parameters for a large sample of units over a period of time. Consequently, they can identify an individual unit or units performing outside the parameter variations seen for other units.

3.13.3.2.2.2   On-Line Gas Analysis — The fuel-cell sensor-membrane technology that has been applied widely to power transformers with circulating oil can be applied to instrument transformers. However, in instrument transformers, the oil is confined, and this confinement can affect sensor operation.

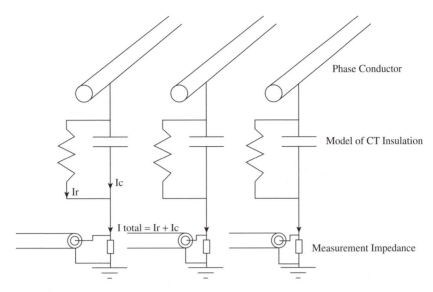

FIGURE 3.13.3 Schematic representation of relative tan delta measurements.

Installation can require factory modifications, depending on the type of sensor that is installed (Boisseau and Tantin, 1993). Typically, the hydrogen sensor is located in an area where the oil is stagnant, especially during periods of low ambient temperatures. This arrangement results in poor accuracy for low hydrogen-concentration levels. For significant hydrogen concentrations (above 300 ppm) in stagnant oil, the accuracy has been determined to be acceptable (Cummings et al., 1988). These constraints may not apply to the thermal conductivity detection (TCD) technology. In this case, the sensor is located externally to the apparatus and utilizes active oil circulation through the monitor while also providing continuous moisture-level monitoring.

3.13.3.2.2.3 On-Line Partial Discharge Measurements — On-line partial-discharge measurement techniques that were discussed in the section on power transformers (Section 3.13.3.1) are also applicable to instrument transformers.

3.13.3.2.2.4 Pressure — Due to partial-discharge activity inside the tank, gases can be formed, which increases the pressure after the gases saturate the oil. A threshold-pressure switch can be used to perform this measurement. The operation of this sensor is possible with an inflatable bellows that is placed between the expansion device and the enclosure. The installation of the device typically requires factory modification. In some applications, pressure sensors take a considerable amount of time (on the order of months) to detect any significant pressure change. The sensitivity of this type of measurement is less than that of hydrogen and partial-discharge sensors (Boisseau and Tantin, 1993). Pressure sensors are also available that mount on the drain valve (Cummings et al., 1988).

3.13.3.3 Bushings

Bushings are subjected to high dielectric and thermal stresses, and bushing failures are one of the leading causes of forced outages and transformer failures. The methods of detecting deterioration of the bushing insulation have been well understood for decades, and conventional off-line diagnostics are very effective at discovering problems. The challenge facing a maintenance engineer is that some problems have gestation time (i.e., going from good condition to failure) that is shorter than typical routine test intervals. Since on-line monitoring of power-factor and capacitance can be performed continuously, and with the same sensitivity as the off-line measurement, deciding whether to apply an on-line system is reduced to an economic exercise of weighing the direct and strategic benefits with the cost.

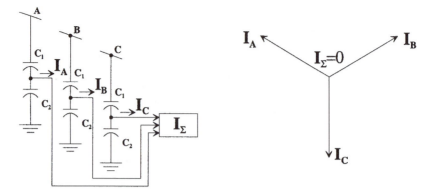

FIGURE 3.13.4 Bushing sum-current measurements.

3.13.3.3.1 Failure Mechanisms Associated with Bushings

The two most common bushing failure mechanisms are moisture contamination and partial discharge. Moisture usually enters the bushing via deterioration of gasket material or cracks in terminal connections, resulting in an increase in the dielectric loss and insulation power factor. The presence of tracking over the surface or burn-through of the condenser core is typically associated with partial discharge. The first indication of this type of problem is an initial increase in power factor. As the deterioration progresses, increases in capacitance will be observed.

3.13.3.3.2 On-Line Bushing Power-Factor and Capacitance Measurements

Measurement of power factor and capacitance is a useful and reliable diagnostic indicator. The sum-current method is a very sensitive method for obtaining these parameters on-line. The basic principle of the sum-current method is based on the fact that the sums of the voltage and current phasors are zero in a symmetrical three-phase system. Therefore, analysis of bushing condition can be performed by adding the current phasors from the capacitance or power-factor taps, as depicted in Figure 3.13.4. If the bushings are identical and system voltages are perfectly balanced, the sum current, I_S, will equal zero.

In reality, bushings are never identical, and system voltages are never perfectly balanced. As a result, the sum current is a nonzero value and is unique for each set of bushings. The initial sum current can be learned, and the condition of the bushings can be determined by evaluating changes in the sum-current phasor. By using software techniques and an expert system to analyze changes to the sum current, changes in either the capacitance or power factor of any of the bushings being monitored can be detected, as shown in Figure 3.13.5.

Figure 3.13.5a depicts a change that is purely resistive, i.e., only the in-phase component of current is changing. It is due to a change in C_1 insulation power factor, and it results in the current phasor change $\Delta I'_A$ from I^0_A to I'_A. The change in current is in phase with A-phase line voltage, V_A, and it is equal to I'_Σ. This is then evidence of a power factor increase for the A-phase bushing.

Figure 3.13.5b depicts a change that is purely capacitive, i.e., only a quadrature component of current is changing. In this case, the change is due to a change in C_1 insulation capacitance, and it results in the current phasor change $\Delta I''_A$ from I^0_A to I''_A. The change in current leads the voltage V_A by 90°, and it is equal to I''_Σ.

Expert systems are also used to determine whether the sum-current change is related to actual bushing deterioration or changes in environmental conditions such as fluctuations in system voltages, changes in bushing or ambient temperature, and changes in surface conditions (Lachman, 1999).

3.13.3.4 Load Tap Changers

High maintenance costs for load tap changers (LTC) result from several causes. The main reasons include: misalignment of contacts, poor design of the contacts, high loads, excessive number of tap changes,

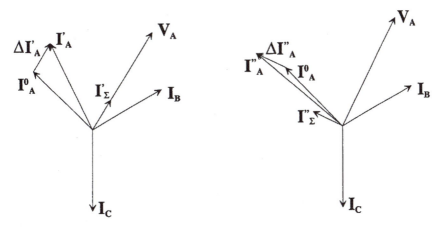

FIGURE 3.13.5 Analysis of bushing sum currents: (a) change in current phasor due to change in power factor of bushing A; (b) change in current phasor due to change in capacitance of bushing A.

mechanical failures, and coking caused by contact heating. Load-tap-changer failures account for approximately 41% of substation transformer failures (Bengtsson, 1996; CIGRE, 1983).

LTC contact wear occurs as the LTC operates to maintain a constant voltage with varying loads. This mechanical erosion is a normal operating characteristic, but the rate can be accelerated by improper design, faulty installation, and high loads. If an excessive-wear situation is undetected, the contacts can burn open or weld together. Monitoring a combination of parameters suitable for a particular LTC design can help avoid such failures.

LTC failures are either mechanical or electrical in nature. Mechanical faults include failures of springs, bearings, shafts, and drive mechanisms. Electrical faults can be attributed to coking of contacts, burning of transition resistors, and insulation problems (Bengtsson, 1996). This section discusses the various parameters that can be monitored on-line that will give an indication of tap-changer condition.

3.13.3.4.1 *Mechanical Diagnostics for On-Load Tap Changers*
A variety of diagnostic algorithms for on-load tap changers can be implemented using drive-motor torque or motor-current information. Mechanical and control problems can be detected because additional friction, contact binding, extended time for tap-changer position transition, and other anomalies significantly impact torque and current.

A signature, or event record, is captured each time the tap changer moves to a different tap. This event can be recorded either as motor torque or as a vibro-acoustic pattern and motor current as a function of time. The signature can then be examined by several methods to detect mechanical and, in the case of vibro-acoustic patterns, electrical (arcing) problems. The following five mechanical parameters can be monitored on-line:

3.13.3.4.1.1 Initial Peak Torque or Current — Initial current inrush and starting torque are related to mechanical static friction and backlash in the linkages. Monitoring this peak value during the first 50 msec of the event provides a useful diagnostic. Increasing values are cause for concern.

3.13.3.4.1.2 Average Torque or Motor Current — Running current or torque provides a measure of dynamic friction and also helps detect binding. Monitoring the average value after initial inrush/startup is a useful diagnostic measure. Motor-current measurement is most effective when the motor directly drives the mechanical linkages. Several common tap-changer designs employ a motor to charge a spring. It is the spring that supplies energy to move the linkages during a tap change. In this case, motor-current measurement is not very effective at detecting mechanical trouble. Torque or force sensors measuring drive force will yield the desired information.

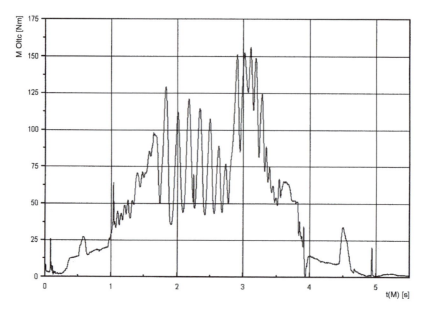

FIGURE 3.13.6 Sample torque curve.

A monitoring system is available that determines the torque curve by measuring the active power of the motor. Anomalies in the torque curve are detected by using an expert system that performs a separate assessment of the individual functions of a switching operation (Leibfried et al., 1998). Figure 3.13.6 is a sample torque curve for a resistance-type tap changer.

3.13.3.4.1.3 Motor-Current Index — The area under the motor-current curve is called the motor index and is usually given in ampere-cycles, based on the power frequency. A similar parameter based on torque can be used. This parameter characterizes the initial inrush, average running conditions, and total running time. Not all types of tap-changer operations have similar index values. An operation through neutral can have a significantly higher index as the reversing switch is exercised. Similarly, tap-changer raise operations can have different index values, depending on whether the previous operation was also a "raise" or a "lower." This is related primarily to linkage backlash. Figure 3.13.7 shows an example of the motor-current curve for a load tap changer, and Figure 3.13.8 shows an example of the motor-current index.

Sequential controls and other operational issues must also be considered. For example, the index will be very large if the tap changer moves more than one step during an operation. The index will be very small if the control calls for a tap change and then rescinds the request before seal-in. All of these situations must be considered when performing diagnostics based on motor-current or torque measurements.

3.13.3.4.1.4 Contact-Wear Model — Monitoring systems are available in which an expert system is used to calculate the total wear on the tap-changer switch contacts. The system issues a recommendation concerning when the contacts should be replaced. The model used by the expert system is based on tap-changer switch-life tests and field experience.

3.13.3.4.1.5 Position Determination — Monitoring systems are available that determine the exact position of the fine tap selector during a switching operation. The system uses this information to correlate tap position with the motor torque. In this manner, the position of the tap changer after a completed switching operation is determined, and the end position of the tap-changer range of operation is monitored.

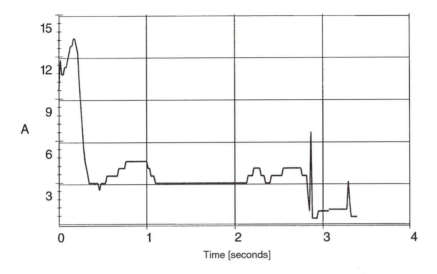

FIGURE 3.13.7 Load-tap-changer motor current during a tap-changing event.

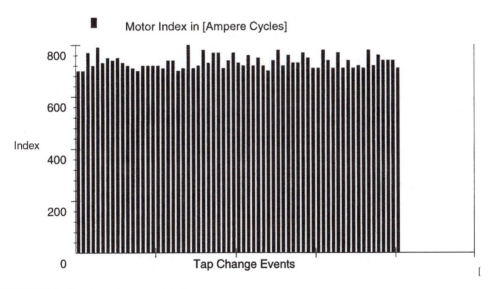

FIGURE 3.13.8 Sample motor-current-index curve.

3.13.3.4.2 *Thermal Diagnostics for On-Load Tap Changers*

A variety of diagnostic algorithms for on-load tap changers can be implemented using temperature data. The heat-transfer pattern resulting from energy losses results in a temperature profile that is easily measured with external temperature sensors. Temperature profiles are normally influenced by weather conditions, cooling-bank status, and electrical load. However, abnormal sources of energy (losses) also impact the temperature profile, thus providing a method of detection. The following four electrical/thermal parameters can be monitored on-line.

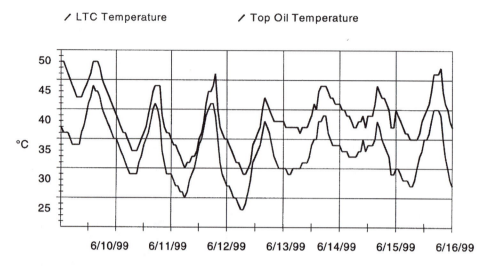

FIGURE 3.13.9 Sample differential-temperature measurement. The top trace is the main-tank top-oil temperature, and the bottom trace is the LTC compartment temperature.

3.13.3.4.2.1 Temperature — The simplest temperature-related diagnostic involves monitoring the temperature level. Load-tap-changer temperature in excess of a certain level may be an indication of equipment trouble. However, there are also many factors that normally influence temperature level. One LTC-monitoring system measures the temperature of the diverter-switch oil and the main-tank oil temperature as a way to estimate the overload capacity of the tap changer.

3.13.3.4.2.2 Simple Differential Temperature — Another simple algorithm involves monitoring the temperature difference between the main tank and load-tap-changer compartment for those tap-changer designs in which the tap changer is in a compartment separate from the main tank. Under normal operating conditions, the main-tank temperature is higher than the tap-changer compartment temperature. This result is expected, given the energy losses in the main tank and general flow of thermal energy from that point to other regions of the equipment. Differential temperature is most effective on external tap-changer designs because this arrangement naturally results in larger temperature differences. Smaller differences are expected on tap changers that are physically located inside the main tank.

Many factors influence differential temperature. Excessive losses caused by bad contacts in the tap changer are detectable. However, load-tap-changer temperature can exceed main-tank temperature periodically under normal conditions. Short-term (hourly) variations in electrical load, weather conditions, and cooling-bank activation can result in main-tank temperatures below the tap changer. Reliable diagnostic algorithms must account for these normal variations in some way. Figure 3.13.9 is a graphical representation of the top-oil temperature in the main tank and of the LTC compartment temperature.

3.13.3.4.2.3 Differential Temperature with Trending — Trending is one method used to distinguish between normal and abnormal differential temperature. When the load-tap-changer temperature exceeds the main-tank temperature, the temperature trends are examined. If the tap-changer temperature is decreasing, this is deemed a normal condition. However, if the tap changer temperature exceeds the main-tank temperature and is increasing, an equipment problem may be indicated.

3.13.3.4.2.4 Temperature Index — Another method used to examine temperature differential involves computing the area between the two temperature curves over a rolling window of time (usually one week). This quantity is called the temperature index and is usually expressed in units of degree-hours. Normal temperature difference (main tank above tap changer) is counted as "negative" area, and the reverse is "positive" area. Therefore, over a period of seven days, the index reflects the general relationship between the two measurements without changing significantly due to normal daily variations in temperature. Under abnormal conditions, the index will exhibit an increasing trend as the load tap changer tends to run hotter relative to the main tank. This method eliminates false alarms associated with simple differential monitoring, but it responds slowly to abnormal conditions. A change in tap-changer temperature characteristics that takes place over the course of several hours will require several days to be reflected in the index. This response time is usually adequate, as the problem developing within the LTC normally requires an extended period to progress to the point where maintenance is required.

3.13.3.4.3 Vibro-Acoustic Monitoring

The vibrations caused by various mechanical movements during a tap-changing operation can be recorded and analyzed for signs of deterioration. This provides continuous control of the transition time as well as an indication of contact wear and detection of sudden mechanical-rupture faults (Bengtsson et al., 1998).

Acoustic monitoring of on-load tap changers has been under development. The LTC operation can be analyzed by recording the acoustic signature and comparing it with the running average representative of recent operations. The signal is analyzed in distinct frequency bands, which facilitates the distinction between problems with electrical causes and those with mechanical causes.

Every operation of the tap changer produces a characteristic acoustic wave, which propagates through the oil and structure of the transformer. Field measurements show that in the case of a properly functioning tap changer, this vibration pattern proves to be very repeatable over time for a given operation.

The acoustic signal is split into two frequency bands. Experience has shown that electrical problems (arcing when there should not be any, notably as for the case of a vacuum-switch-assist LTC) are detected in a higher frequency band than those mechanical in nature (excessive wear or ruptured springs). This system has the intelligence to distinguish imminent failure conditions and normal wear of the LTC to allow for just-in-time maintenance (Foata et al., 1999).

3.13.3.4.4 Dissolved-Gas Analysis

Analysis of gases dissolved in the oil in the load-tap-changer compartment is proving to be a useful diagnostic. Key gases for this analysis include acetylene and ethylene. However, any conclusions to be drawn from a correlation of measured dissolved-gas concentrations with certain types of faults are not yet well documented. The study is complicated by the fact that the basic design and the materials used in the particular tap changer are found to significantly affect the DGA results.

References

Bengtsson, C., Status and trends in transformer monitoring, *IEEE Trans. Power Delivery*, 11, 1379–1384, 1996.

Bengtsson, T., Kols, H., Foata, M., and Leonard, F., Monitoring Tap Changer Operations, Paper 12.209, presented at CIGRE Int. Conf. Large High Voltage Electric Syst., CIGRE, Paris, 1998.

Bengtsson, T., Kols, H., and Jönsson, B., Transformer PD Diagnosis Using Acoustic Emission Technique, in Proc. 10th ISH, Montréal, 1997.

Bengtsson, T., Leijon, M., and Ming L., Acoustic Frequencies Emitted by Partial Discharges in Oil, Paper No. 63.10, in Proc. 7th ISH, Dresden, 1993.

Boisseau, C. and Tantin, P., Evaluation of Monitoring Methods Applied to Instrument Transformers, presented at Doble Conference, 1993.

Boisseau, C., Tantin, P., Despiney, P., and Hasler, M., Instrument Transformers Monitoring, Paper 110-13, presented at CIGRE Diagnostics and Maintenance Techniques Symposium, Berlin, 1993.

Canadian Electricity Association, On-Line Condition Monitoring of Substation Power Equipment and Utility Needs, CEA No. 485 T 1049, Canadian Electricity Association, 1996.

Chu, D., El Badaly, H., and Slemon, C., Development of an Automated Transformer Oil Monitor, presented at EPRI 2nd Conf. Substation Diagnostics, 1993.

Chu, D., El Badaly, H., and Slemon, C., Status Report on the Automated Transformer Oil Monitor, EPRI 3rd Conf. Substation Diagnostics, 1994.

CIGRE Working Group 05, An international survey on failures in large power transformers in service, *Electra*, 88, 1983.

Cummings, H.B. et al., Continuous, on-line monitoring of freestanding, oil-filled current transformers to predict an imminent failure, *IEEE Trans. Power Delivery*, 3, 1776–1783, 1988.

Domun, M.K., Condition Monitoring of Power Transformers by Oil Analysis Techniques, presented at Science, Education and Technology Division Colloquium on Condition Monitoring and Remanent Life Assessment in Power Transformers, IEE Colloquium (digest), no. 075, March 22, 1994.

Duval, M. and Lamarre, C., The characterization of electrical insulating oils by high performance liquid chromatography, *IEEE Trans. Electrical Insulation*, 12, 1977.

Eleftherion, P., Partial discharge XXI: acoustic emission-based PD source location in transformers, *IEEE Electrical Insulation Mag.*, 11, 22, 1995.

Eriksson, T., Leijon, M., and Bengtsson, C., PD On-Line Monitoring of Power Transformers, Paper SPT HV 03-08-0682, presented at Stockholm Power Tech 1995, p. 101.0.

Feser, K., Maier, H.A., Freund, H., Rosenow, U., Baur, A., and Mieske, H., On-Line Diagnostic System for Monitoring the Thermal Behaviour of Transformers, Paper 110-08, presented at CIGRE Diagnostics and Maintenance Techniques Symposium, Berlin, 1993.

Foata, M., Aubin, J., and Rajotte, C., Field Experience with Acoustic Monitoring of On Load Tap Changers, in 1999 Proc. Sixty Sixth Annu. Int. Conf. Doble Clients, 1999.

Fox, R.J., Measurement of peak temperatures along an optical fiber, *Appl. Opt.*, 22, 1983.

Fruth, B. and Fuhr, J., Partial Discharge Pattern Recognition — A Tool for Diagnostics and Monitoring of Aging, Paper 15/33-12, presented at CIGRE International Conference on Large High Voltage Electric Systems, 1990.

Glodjo, A., A Field Experience with Multi-Gas On-Line Monitors, in 1998 Proc. Sixty Fifth Annu. Int. Conf. Doble Clients, 1998.

Griffin, P., Continuous Condition Assessment and Rating of Transformers, in 1999 Proc. Sixty Sixth Annu. Int. Conf. Doble Clients, 1999, p. 8-8.1.

Harrold, R.T., Acoustic waveguides for sensing and locating electric discharges within high voltage power transformers and other apparatus, *IEEE Trans. Power Appar. Syst.*, 102, 1983.

IEEE, Guide for the Interpretation of Gases Generated in Oil-Immersed Transformers, IEEE Std. C57.104, Institute of Electrical and Electronics Engineers, Piscataway, NJ.

IEEE, Guide for Loading Mineral-Oil-Immersed Transformers, IEEE Std. C57.91-1995, Institute of Electrical and Electronics Engineers, Piscataway, NJ, 1995.

Lachman, M.F., On-line diagnostics of high-voltage bushings and current transformers using the sum current method, PE-471-PWRD-0-02-1999, *IEEE Trans. Power Delivery*, 1999.

Leibfried, T., Knorr, W., Viereck, D., Dohnal, D., Kosmata, A., Sundermann, U., and Breitenbauch, B., On-Line Monitoring of Power Transformers — Trends, New Developments, and First Experiences, Paper 12.211, presented at CIGRE Int. Conf. Large High Voltage Electric Syst., 1998.

Lemke, E., A New Procedure for Partial Discharge Measurements on the Basis of an Electromagnetic Sensor, Paper 41.02, in Proc. 5th ISH, Braunschweig, 1987.

Morshuis, P.H.F., Partial discharge mechanisms in voids related to dielectric degradation, *IEE Proc.-Sci. Meas. Technol.*, 142, 62, 1995.

Myers, S.D., Kelly, J.J., and Parrish, R.H., *A Guide to Transformer Maintenance*, Transformer Maintenance Institute, Akron, OH, 1981.

Oommen, T.V., On-Line Moisture Sensing in Transformers, in Proc. 20th Electrical/Electronics Insulation Conf., Boston, 1991, pp. 236–241.

Oommen, T.V., Further Experimentation on Bubble Generation during Transformer Overload, Report EL-7291, Electric Power Research Institute, Palo Alto, CA, 1992.

Oommen, T.V., On-Line Moisture Monitoring in Transformers and Oil Processing Systems, Paper 110-03, presented at CIGRE Diagnostics and Maintenance Techniques Symposium, Berlin, 1993.

Sokolov, V.V. and Vanin, B.V., In-Service Assessment of Water Content in Power Transformers, presented at Doble Conference, 1995.

Wang, C., Dong, X., Wang, Z., Jing, W., Jin, X., and Cheng, T.C., On-line Partial Discharge Monitoring System for Power Transformers, in Proc. 10th ISH, Montréal, 1997, p. 379.

Wenzel, D., Borsi, H., and Glockenbach, E., Pulse Shaped Noise Reduction and Partial Discharge Localisation on Transformers Using the Karhunen-Loéve-Transform, Paper 5627, in Proc. 9th ISH, Graz, 1995.

Wenzel, D., Schichler, U., Borsi, H., and Glockenbach, E., Recognition of Partial Discharges on Power Units by Directional Coupling, Paper 5626, in Proc. 9th ISH, Graz, 1995.

Zaretsky, M.C. et al., Moisture sensing in transformer oil using thin-film microdielectrometry, *IEEE Trans. Electrical Insulation*, 24, 1989.

3.14   U.S. Power Transformer Equipment Standards and Processes

Philip J. Hopkinson

This section[4] describes the power transformer equipment standards approval processes and lists the standards that are in place in the U.S. in 2002. The subsection on accredited standards approval processes (Section 3.14.1) provides an abbreviated description of the methods that are employed in the U.S. for gaining American National Standards Institute (ANSI) approval and recognition. Similarly, an approval process is also shown for International Electrotechnical Commission (IEC) documents.

The U.S. uses a voluntary process for the development of nationally recognized power transformer equipment standards. This chapter describes the U.S. standards accreditation process and provides flow charts to show how accredited standards are approved. With the International Electrotechnical Committee (IEC) taking on increased importance, the interaction of U.S. technical experts with IEC is also described. Finally, relevant power transformer documents are listed for the key power transformer equipment standards that guide the U.S. industry.

3.14.1   Processes for Acceptance of American National Standards

The acceptance of a standard as an American national standard (ANS) requires that it be processed through one of three methods:

1. Canvass list
2. Action of accredited standards committee
3. Action of accredited standards organization

Table 3.14.1 lists the major standards organizations.

All three methods share the common requirement that the process used has been accredited by the American National Standards Institute (ANSI), that the methodology incorporates due process, and that consensus among the interests is achieved. Inherent in that approval is presentation of accepted operational procedures and/or a balloting group that is balanced among users, manufacturers, and general-interest groups. Other considerations include the following:

- The document must be within the scope previously registered.
- Identified conflicts must be resolved.
- Known national standards must have been examined to
    Avoid duplication or conflict
    Verify that any appeal has been completed
    Verify that the ANSI patent policy is met

---

[4] Many thanks are offered by the author for assistance rendered by representatives of the National Electrical Manufacturer's Association, by American National Standards Institute, and by individuals from IEEE, IEC, EEI, EL&P, UL, and especially by Ms. Purefoy, who arranged and documented these contents.

TABLE 3.14.1 Major Standards Organizations

ANSI	American National Standards Institute
ASC C57	Accredited Standards Committee C57 for Power and Distribution Transformers (deactivated December 31, 2002, by mutual decision of IEEE, NEMA, and C57 Committee)
IEEE	Institute of Electrical and Electronics Engineers
NEMA	National Electrical Manufacturers Association
EL&P	Electric Light and Power
EEI	Edison Electric Institute
AEIC	Association of Edison Illuminating Companies
UL	Underwriter's Laboratory
IEC	International Electrotechnical Commission

While the three processes differ in their methods of gaining consensus, all three use common methods for document submittal to ANSI. An explanation of the ANSI submittal process, which involves the use of Board for Standard Revision form 8 (BSR-8) and Board for Standard Review form 9 (BSR-9), is available on-line at http://web.ansi.org/rooms/room_16/public/ans.html.

The BSR-8 form (request to initiate public review of a proposed ANS) is used to submit draft candidate American national standards for public review in ANSI's standards action. This form can be submitted multiple times for the same standard if multiple public reviews are required due to substantive changes in text. If the BSR-8 form is a resubmittal, this should be clearly marked on the form. This form is available via the ANSI reference library or via e-mail to psa@ansi.org.

The BSR-9 form (request for formal approval of a standard as an ANS) is used to submit candidate American national standards for final approval. All of the information requested on the form must be provided. The form itself serves as a checklist for the evidence of consensus that the BSR and the ANSI procedures require. The certification section on the form is the developer's acknowledgement that all items listed as part of the certification statement are true, e.g., all appeals have concluded. This form is available via the ANSI reference library or via e-mail to psa@ansi.org.

An explanation of the accreditation process (accreditation of American National Standards Developers) is available at http://web.ansi.org/rooms/room_16/public/accredit.html.

3.14.1.1 Canvass List

The canvass-list method provides procedures for seeking approval/acceptance of a document without the structure of a committee or an organization. Figure 3.14.1 shows a flow diagram of the ANSI canvass-list method. Under the canvass list, the originator of a standard seeking its acceptance as an American national standard (ANS) must assemble, to the extent possible, those who are directly and materially affected by the activity in question. The standards developer conducts a letter ballot or "canvass" of those interested to determine consensus on a document. Additional interest in participating on a canvass is sought through an announcement in ANSI's publication entitled, "Standards Action." Although canvass developers provide ANSI with internal procedures used in the development of the draft American national standard, the due process used to determine consensus begins after the draft standard has been developed. Standards developers using the canvass method must use the procedures provided in Annex B of the ANSI procedures.

The balloting group, by the above methods, consists of a balance of interests and affected parties. To summarize, once the canvass list is approved and finalized, the document is circulated for voting. Simultaneously, the manager of the canvass list completes and submits the Board for Standard Revision 8 (BSR-8) form for public notification of the undertaking and providing opportunity for comment from persons outside the balloted group. Once the balloting period ends, a minimum of 45 days, and usually after 60 days for transformers, the manager completes Board for Standard Review 9 (BSR-9) to provide (1) validation that the proposal received consensus approval in the balloting and (2) a report on how each of the participants voted. The BSR-9 is forwarded to ANSI for action by the Board for Standard Review. Of particular concern is that the document review be completed under an open and fair procedure and that a consensus of the voting group approve its acceptance.

Underwriter's Laboratories (UL) uses the canvass-list method to obtain ANSI recognition of UL documents.

3.14.1.2 Accredited Standards Committee

The accredited standards committee (ASC) is a second method for gaining "national" acceptance of a standard. Figure 3.14.2 shows a flow diagram of the ANSI committee method. The accredited standards committees are standing committees of directly and materially affected interests created for the purpose of developing a document and establishing consensus in support of this document for submittal to ANSI. The committee method is most often used when a standard affects a broad range of diverse interests or where multiple associations or societies with similar interests exist. The committee serves as a forum where many different interests, without a common membership in an organization or society, can be represented. Accredited standards committees are administered by a secretariat, an organization that takes the responsibility for providing administrative oversight of the committee's activities and ensuring compliance with the pertinent operating procedures.

An accredited standards committee may adopt the procedures provided in Annex A of the ANSI procedures, or it may develop its own operating procedures consistent with the requirements of Section 2.2 of these procedures.

Under current procedures, ASCs are entities established through the coalescence of a balance of interest groups focused on a particular product area. The product area and the committee's organization and organizational procedures are approved by ANSI. The ASC charter is subject to periodic review and reaffirmation but, generally, is unobstructed. The ASC has the option to develop and submit standards for acceptance as ANSs or to process documents that fall within their operational scope that originate in other bodies — trade associations, business groups, and the like. Documents submitted to ASCs are subjected to the same procedures for consideration as in the canvass-list method. A BSR-8 is issued upon receipt of a document for acceptance to initiate committee review and vote. The BSR-8 provides for public notification of the undertaking and for public comment. Once the balloting period is ended, the BSR-9 report is sent to ANSI confirming the voting and the consensus. ANSI reviews the report and provides appropriate approval.

Until recently, IEEE and NEMA used the Accredited Standards Committee to gain ANSI C57 document approvals. ASC C57 was deactivated December 31, 2002, by actions of IEEE, NEMA, and the C57 Committee.

The Electric Light and Power Delegation (EL & P) represents the Edison Electric Institute (EEI) and the Association of Edison Illuminating Companies (AEIC). EL & P, predominantly through EEI, is well represented in the IEEE Transformers Committee. EL & P is not currently a standards development organization, but it votes as a delegation on documents submitted to ANSI C57 for approval.

3.14.1.3 Accredited Standards Organization

The third method for acceptance is the accredited standards organization. The organization method is often used by associations and societies that have, among other activities, an interest in developing standards. This is the method used by IEEE starting January 1, 2003. Figure 3.14.3 shows a flow diagram of the ANSI standards-organization method.

Although participation on the consensus body is open to all interested parties, members of the consensus body often participate as members in the association or society. The organization method is the only method of consensus development in which the standards developer must develop its own operating procedures. These procedures must meet the general requirements of Section 2.2 of the ANSI procedures. This method provides flexibility, allowing the standards developer to utilize a system that accommodates its particular structure and practices.

Under these procedures, an organization demonstrates the openness and balance of its voting groups and its operating procedures. The ASO's purview or authority for processing documents for acceptance as an ANS may be restricted to particular products or a group of products, depending upon organizational interests and goals. The documents developed by the ASO and conforming to the interest-balance and openness procedures are submitted to ANSI utilizing the BSR-8 and BSR-9 reports, in appropriate sequence. The ANSI BSR evaluates the documentation and makes its decision using the criteria as in the other methodologies.

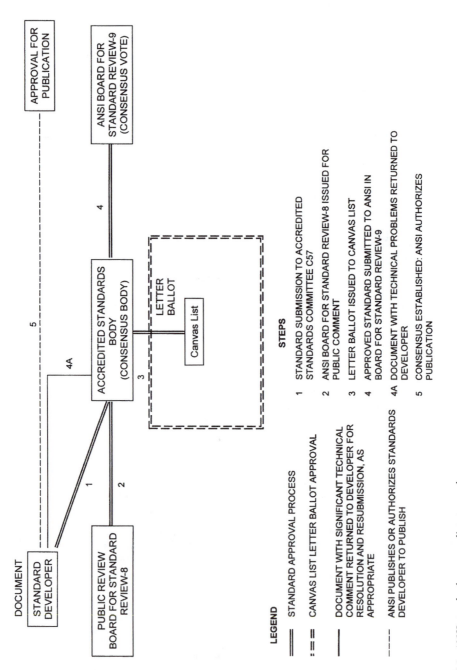

FIGURE 3.14.1 ANSI standards canvass-list approval process.

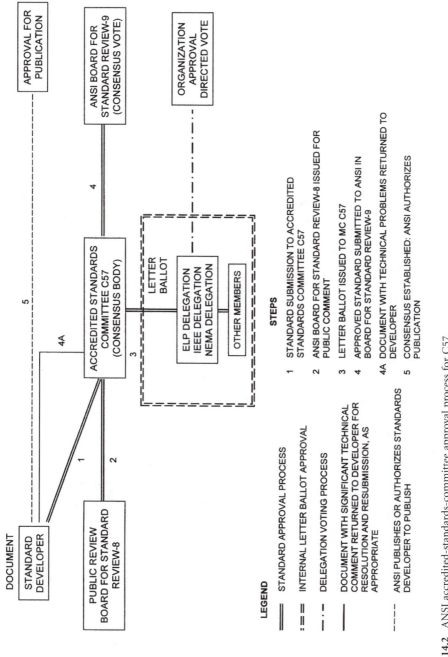

FIGURE 3.14.2 ANSI accredited-standards-committee approval process for C57.

The developer of a standard is presented with a range of options in pursuing the document's acceptance as an American national standard. All methods require prior, or standing, approval until organization scope is changed from ANSI. The canvass method provides the greatest flexibility for the developer but places a greater involvement in assembling the necessary balloting group and, therefore, latitude in determining voting participants. For a developer outside an organization, the canvass list and the ASC methods provide the greatest and quickest access. The ASO route is not a normal venue for outsiders, or nonmembers, without special arrangements or agreements from the sponsoring ASO, particularly, if the document falls outside the scope of the organization's accreditation.

3.14.2 The International Electrotechnical Commission (IEC)

The International Electrotechnical Commission (IEC) is composed of a central office and approximately 100 technical committees and subcommittees. The IEC central office is located in Geneva, Switzerland. All balloting of technical documents is conducted through the central office.

The national committees of each country are responsible for establishing participation status on the various technical committees and subcommittees as well as casting votes on the respective ballots. Participation status by a national committee of a country can be in one of three categories:

- P-member: participates actively in the work, with an obligation to vote on all questions formally submitted for voting within the technical committee or subcommittee, to vote on enquiry drafts and final draft international standards, and whenever possible, to participate in meetings.
- O-member: follows the work as an observer and, therefore, receives committee documents and has the right to submit comments and to attend meetings.
- Nonmember: has neither the rights nor obligations described above for the work of a particular committee. Nevertheless, all national bodies — irrespective of their status within a technical committee or subcommittee — have the right to vote on enquiry drafts and on final-draft international standards. All ballots are cast by the national committees of the respective countries, with one vote per country.

The U.S. national committee is a committee of the American National Standards Institute (ANSI), with headquarters in New York City. U.S. technical interface with IEC TCS/SCS is via a technical advisory group (TAG) and a technical advisor (TA). TAGs are administered by an organizational TAG administrator. TAs are appointed by the U.S. National Committee Executive Committee (EXCO) to represent U.S. interests on the various technical committees and subcommittees. TA appointments are based on technical experience and capability. TAs are responsible for establishing TAGs to develop consensus positions of technical issues and to assist the TA in technical representation. This includes the nomination of working-group experts.

As stated above, all balloting of technical documents is conducted through the central office of the IEC. The ballots are first distributed to the national committees. The national committees next send the ballots to the appropriate technical experts for input. Within the U.S., ballots are submitted to the TA/TAG. The TA/TAG administrator has the responsibility to distribute the ballot to the TAG for direction and/or comment. A consensus process is used to be certain that votes are truly representative of the U.S. position.

The TA/TAG administrator sends all recommended actions to the secretary of the USNC. The secretary then sends the official U.S. vote to the IEC central office. Figure 3.14.4 shows a flow diagram of the IEC ballot process.

3.14.3 Relevant Power Transformer Standards Documents

There are numerous issued documents that apply to the specifications and performance requirements for the various power transformers in the industry today. This section organizes them in ascending order of power ratings, in the following categories:

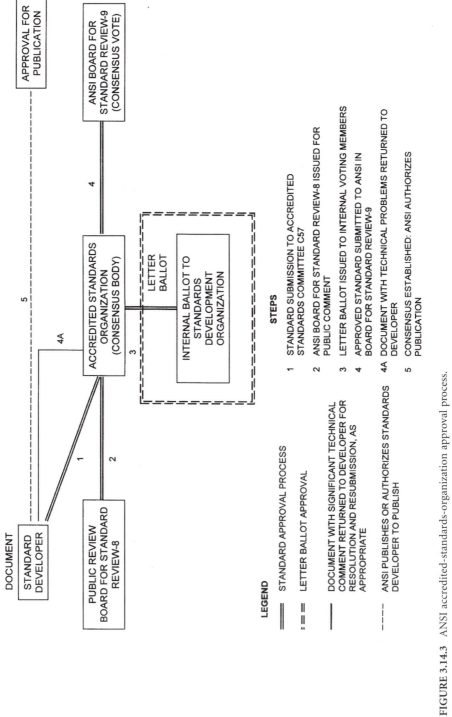

FIGURE 3.14.3 ANSI accredited-standards-organization approval process.

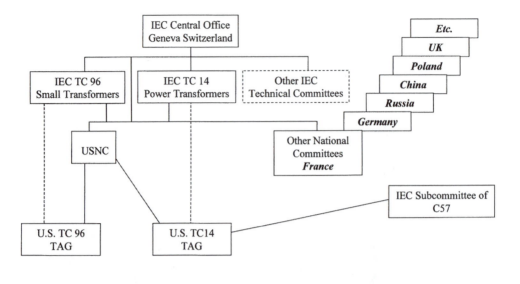

Participation on IEC Technical Committees is limited based on experience and by appointment

FIGURE 3.14.4 IEC technical-committee document approval process.

1. Small dry-type transformers
   NEMA
   UL
   IEC

2. Electronics power transformers
   IEEE

3. Low-voltage medium-power dry-type transformers
   NEMA
   UL
   IEC

4. Medium-voltage and large-power dry-type transformers
   IEEE
   NEMA
   UL
   IEC

5. Liquid-filled transformers
   IEEE
   NEMA
   IEC

3.14.3.1 Small Dry-Type Transformers

3.14.3.1.1 *NEMA ST-1, Specialty Transformers (except General-Purpose Type)*

This standards publication covers control transformers, industrial control transformers, Class 2 transformers, signaling transformers, ignition transformers, and luminous-tube transformers. The publication contains service conditions, tests, classifications, performance characteristics, and construction data for the transformers.

Class 2 Transformers: These transformers are dry-type, step-down, insulating specialty transformers suitable for use in National Electrical Code Class 2 circuits. They are generally used in remote-control, low-energy power, and signal circuits for the operation of bells, chimes, furnace controls, valves, relays, solenoids, and the like. Their secondary voltage is limited to 30 V.

kVA range: 0 to 5 kVA, single phase
Voltage
    Control transformers through 4800 V
    Ignition and luminous tube transformers through 15,000 V

3.14.3.1.2 ANSI/UL 506, Standard for Safety for Specialty Transformers

These requirements cover air-cooled transformers and reactors for general use, and ignition transformers for use with gas burners and oil burners. Transformers incorporating overcurrent or over-temperature protective devices, transient-voltage surge protectors, or power-factor-correction capacitors are also covered by these requirements. These transformers are intended to be used in accordance with the National Electrical Code, NFPA 70.

These requirements do not cover liquid-immersed transformers, variable-voltage autotransformers, transformers having a nominal primary rating of more than 600 V, transformers having overvoltage taps rated over 660 V, cord- and plug-connected transformers (other than gas-tube-sign transformers), garden-light transformers, voltage regulators, swimming pool and spa transformers, or other special types of transformers covered in requirements for other electrical devices or appliances.

These requirements also do not cover:

Autotransformers used in industrial control equipment, which are evaluated in accordance with the requirements for industrial control equipment, UL 508.

Class 2 or Class 3 transformers, which are evaluated in accordance with the Standard for Class 2 and Class 3 Transformers, UL 1585.
    Class 2 transformers: These transformers are dry-type, step-down, insulating specialty transformers suitable for use in National Electrical Code Class 2 circuits. They are generally used in remote-control, low-energy power, and signal circuits for the operation of bells, chimes, furnace controls, valves, relays, solenoids, and the like. Their secondary voltage is limited to 30 V.
    Class 3 transformers: These transformers are similar to Class 2 transformers, but their output voltage is greater than 30 V and less than 100 V.

Toy transformers, which are evaluated in accordance with the Standard for Toy Transformers, UL 697.

Transformers for use with radio- and television-type appliances, which are evaluated in accordance with the requirements for transformers and motor transformers for use in audio-, radio-, and television-type appliances, UL 1411.

Transformers for use with high-intensity discharge lamps, which are evaluated in accordance with the Standard for High-Intensity-Discharge Lamp Ballasts, UL 1029.

Transformers for use with fluorescent lamps, which are evaluated in accordance with the Standard for Fluorescent-Lamp Ballasts, UL 935.

Ventilated or nonventilated transformers for general use (other than compound-filled or exposed-core types), which are evaluated in accordance with the requirements for dry-type general purpose and power transformers, UL 1561.

Dry-type distribution transformers rated over 600 V, which are evaluated in accordance with the requirements for transformers, distribution, dry-type — over 600 V, UL 1562.

Transformers incorporating rectifying or waveshaping circuitry, which are evaluated in accordance with the requirements for power units other than Class 2, UL 1012.

Transformers of the direct plug-in type, which are evaluated in accordance with the requirements for Class 2 power units, UL 1310.

Transformers for use with electric discharge and neon tubing, which are evaluated in accordance with the Standard for Neon Transformers and Power Supplies, UL 2161.

A product that contains features, characteristics, components, materials, or systems new or different from those in use when the standard was developed — and that involves a risk of fire, electric shock, or injury to persons — shall be evaluated using the appropriate additional component and end-product requirements as determined necessary to maintain the level of safety for the user of the product as originally anticipated by the intent of this standard.

3.14.3.1.3 ANSI/UL 1446, Standard for Safety for Systems of Insulating Materials — General

These requirements cover test procedures to be used in the evaluation of Class 120(E) or higher electrical insulation systems intended for connection to branch circuits rated 600 V or less. These requirements also cover (1) the investigation of the substitution of minor components of insulation in a previously evaluated insulation system and (2) the test procedures to be used in the evaluation of magnet wire coatings, magnet wires, and varnishes.

These requirements do not cover a single insulating material or a simple combination of materials, such as a laminate or a varnished cloth.

These requirements do not cover insulation systems exposed to radiation or operating in oils, refrigerants, soaps, or other media that potentially degrade insulating materials.

These requirements shall be modified or supplemented as determined by the applicable requirements in the end-product standard covering the device, appliance, or equipment in which the insulation system is used.

Additional consideration shall be given to conducting tests for an insulating material, such as a coil encapsulant, that is used as the ultimate electrical enclosure.

Additional consideration shall be given to conducting tests for an insulating material or component that is a functional support of, or in direct contact with, a live part.

A product that contains features, characteristics, components, materials, or systems new or different from those in use when the standard was developed — and that involves a risk of fire, electric shock, or injury to persons — shall be evaluated using the appropriate additional component and end-product requirements as determined necessary to maintain the level of safety for the user of the product as originally anticipated by the intent of this standard.

3.14.3.1.4 IEC TC 96

This document covers standardization in the field of safety of transformers, power-supply units, and reactors with a rated voltage not exceeding 1000 V and a rated frequency not exceeding 1 MHz of the following kinds:

Power transformers and power supply units with a rated power less than 1-kVA single phase and 5-kVA polyphase

Special transformers and power supply units other than those intended to supply distribution networks, in particular transformers and power-supply units intended to allow the application of protective measures against electric shock as defined by IEC TC 64

Electrical installations in buildings, with no limitation of rated power, but in certain cases including limitation of voltage

Reactors with a rated power less than 2-kVAR single phase and 10-kVAR polyphase

Special reactors other than those covered by IEC 289

Note: Excluded are switch-mode power supplies (dealt with by SC 22E)

Safety group function:

Special transformers and power supply units other than those intended to supply distribution networks, in particular transformers and power supply units intended to allow the application of protective measures against electric shock as defined by TC 64, with no limitation of rated power but in certain cases including limitation of voltage.

Table 3.13.2 provides a list of relevant documents.

TABLE 3.14.2 Relevant Documents for IEC TC96

IEC TC 96	Description
61558-1	Safety of power transformers, power supply units and similar
61558-2-1	Particular requirements for separating transformers for general use †
61558-2-2	Particular requirements for separating control transformers
61558-2-3	Particular requirements for ignition transformers for gas or oil burners
61558-2-4	Particular requirements for isolating transformers for general use
61558-2-5	Particular requirements for shaver transformers and supply
61558-2-6	Particular requirements for safety isolating transformers for general use
61558-2-7	Particular requirements for toys
61558-2-8	Particular requirements for bell and chime
61558-2-9	Particular requirements for transformers for Class III handlamps (NP)
61558-2-10	Particular requirements for high insulation level transformers (NP)
61558-2-11	Particular requirements for stray field transformers
61558-2-12	Particular requirements for stabilizing transformers
61558-2-13	Particular requirements for autotransformers
61558-2-14	Particular requirements for variable transformers
61558-2-15	Particular requirements for insulating transformers for the supply of medical rooms (CDV)
61558-2-16	Particular requirements for power supply units and similar (NP)
61558-2-17	Particular requirements for transformers for switch mode power
61558-2-18	Particular requirements for medical appliances
61558-2-19	Particular requirements for mainsborne perturbation attenuation transformers w/earthed midpoint (CDV)
61558-2-20	Particular requirements for small reactors (CDV)
61558-2-21	Particular requirements for transformers with special dielectric (liquid SF_6)
61558-2-22	Particular requirements for transformers with rated maximum temperature for luminaries (NP)
61558-23	Particular requirements for transformers for construction sites (CDV)

3.14.3.2 Electronics Power Transformers

3.14.3.2.1 IEEE 295, Electronics Power Transformers

This standard pertains to power transformers and inductors that are used in electronic equipment and supplied by power lines or generators of essentially sine wave or polyphase voltage. Guides to application and test procedures are included. Appendices contain certain precautions, recommended practices, and guidelines for typical values. Provision is made for relating the characteristics of transformers to the associated rectifiers and circuits.

Certain pertinent definitions, which have not been found elsewhere, are included with appropriate discussion. Attempts are made to alert the industry and profession to factors that are commonly overlooked.

This standard includes, but is not limited to, the following specific transformers and inductors.

Rectifier supply transformers for either high- or low-voltage supplies
Filament- and cathode-heater transformers
Transformers for alternating-current resonant charging circuits
Inductors used in rectifier filters
Autotransformers with fixed taps

kVA Range: 0 to 1000+ kVA
Voltage: 0 to 15 kV

3.14.3.3 Low-Voltage Medium-Power Dry-Type Transformers

3.14.3.3.1 NEMA ST-20, Dry-Type Transformers for General Applications

This standards publication applies to single-phase and polyphase dry-type transformers (including autotransformers and non-current-limiting reactors) for supplying energy to power, heating, and lighting

TABLE 3.14.3 NEMA TP-1 Rating Range for Single-Phase and Three-Phase Dry-Type and Liquid-Filled Distribution Transformers

Voltage Class: Primary Voltage = 34.5 kV and below Secondary Voltage = 600 V and below		
Transformer Type	Number of Phases	Rating Range
Liquid rating	Single phase	10–833 kVA
	Three phase	15–2500 kVA
Dry rating	Single phase	15–833 kVA
	Three phase	15–2500 kVA

Note: Includes all products at 1.2 kV and below.

circuits and designed to be installed and used in accordance with the National Electrical Code. It applies to transformers with or without accessories having the following ratings:

1.2-kV class (600 V nominal and below), 0.25 kVA and up
Above 1.2-kV class, sound-level limits are supplied that are applicable to commercial, institutional, and industrial transformers.

This standards publication applies to transformers, commonly known as general-purpose transformers, for commercial, institutional, and industrial use in nonhazardous locations both indoors and outdoors. The publication includes ratings and information on the application, design, construction, installation, operation, inspection, and maintenance as an aid in obtaining a high level of safe performance. These standards, except for those for ratings, may be applicable to transformers having other-than-standard ratings. These standards, as well as applicable local codes and regulations, should be consulted to secure the safe installation, operation, and maintenance of dry-type transformers.

This publication does not apply to the following types of specialty transformers: control, industrial control, Class 2, signaling, oil- or gas-burner ignition, luminous tube, cold-cathode lighting, incandescent, and mercury lamp. Also excluded are network transformers, unit substation transformers, and transformer distribution centers.

3.14.3.3.2 NEMA TP-1, Guide for Determining Energy Efficiency for Distribution Transformers

This standard is intended for use as a basis for determining the energy-efficiency performance of the equipment covered and to assist in the proper selection of such equipment. This standard covers single-phase and three-phase dry-type and liquid-filled distribution transformers as defined in Table 3.14.3.

Products excepted from this standard include:

Liquid-filled transformers below 10 kVA
Dry-type transformers below 15 kVA
Drives transformers, both ac and dc
All rectifier transformers and transformers designed for high harmonics
Autotransformers
Nondistribution transformers, such as UPS transformers
Special impedance and harmonic transformers
Regulating transformers
Sealed and nonventilated transformers
Retrofit transformers
Machine-tool transformers
Welding transformers
Transformers with tap ranges greater than 15%
Transformers with frequency other than 60 Hz
Grounding transformers
Testing transformers

TABLE 3.14.4 NEMA TP-2 Rating Range for Single-Phase and Three-Phase Dry-Type and Liquid-Immersed Distribution Transformers

Transformer Type	Number of Phases	Rating Range
Liquid immersed	Single phase	10–833 kVA
	Three phase	15–2500 kVA
Dry type	Single phase	15–833 kVA
	Three phase	15–2500 kVA

Note: Includes all products at 1.2 kV and below.

3.14.3.3.3 NEMA TP-2, Standard Test Method for Measuring the Energy Consumption of Distribution Transformers

This standard is intended for use as a basis for determining the energy-efficiency performance of the equipment covered and to assist in the proper selection of such equipment. This standard covers single-phase and three-phase dry-type and liquid-immersed distribution transformers (transformers for transferring electrical energy from a primary distribution circuit to a secondary distribution circuit or within a secondary distribution circuit) as defined in Table 3.14.4.

This standard addresses the test procedures for determining the efficiency performance of the transformers covered in NEMA Publication TP-1. Products excepted from this standard include:

Liquid-filled transformers below 10 kVA
Dry-type transformers below 15 kVA
Transformers connected to converter circuits
All rectifier transformers and transformers designed for high harmonics
Autotransformers
Nondistribution transformers, such as UPS transformers
Special impedance and harmonic transformers
Regulating transformers
Sealed and nonventilated transformers
Retrofit transformers
Machine-tool transformers
Welding transformers
Transformers with tap ranges greater than 15%
Transformers with frequency other than 60 Hz
Grounding transformers
Testing transformers

3.14.3.3.4 ANSI/UL 1561, Standard for Safety for Dry-Type General-Purpose and Power Transformers

These requirements cover:

General-purpose and power transformers of the air-cooled, dry, ventilated, and nonventilated types rated no more than 500-kVA single phase or no more than 1500-kVA three phase to be used in accordance with the National Electrical Code, NFPA 70. Constructions include step-up, step-down, insulating, and autotransformer-type transformers as well as air-cooled and dry-type reactors.

General-purpose and power transformers of the exposed core, air-cooled, dry, and compound-filled types rated more than 10 kVA but no more than 333 kVA single phase or no more than 1000 kVA three phase to be used in accordance with the National Electrical Code, NFPA 70. Constructions include step-up, step-down, insulating, and autotransformer-type transformers as well as air-cooled, dry-type, and compound-filled-type reactors.

These requirements do not cover ballasts for high-intensity-discharge (HID) lamps (metal halide, mercury vapor, and sodium types) or fluorescent lamps, exposed-core transformers, compound-filled transformers, liquid-filled transformers, voltage regulators, general-use or special types of transformers covered in requirements for other electrical equipment, autotransformers forming part of industrial control equipment, motor-starting autotransformers, variable-voltage autotransformers, transformers having a nominal primary or secondary rating of more than 600 V, or overvoltage taps rated greater than 660 V.

These requirements do not cover transformers provided with waveshaping or rectifying circuitry. Waveshaping or rectifying circuits may include components such as diodes and transistors. Components such as capacitors, transient-voltage surge suppressors, and surge arresters are not considered to be waveshaping or rectifying devices.

A product that contains features, characteristics, components, materials, or systems new or different from those in use when the standard was developed — and that involves a risk of fire, electric shock, or injury to persons — shall be evaluated using the appropriate additional component and end-product requirements as determined necessary to maintain the level of safety for the user of the product as originally anticipated by the intent of this standard.

3.14.3.3.5   *ANSI/UL 1446, Standard for Safety for Systems of Insulating Materials — General*

These requirements cover test procedures to be used in the evaluation of Class 120(E) or higher electrical insulation systems intended for connection to branch circuits rated 600 V or less. These requirements also cover the investigation of the substitution of minor components of insulation in a previously evaluated insulation system and also the test procedures to be used in the evaluation of magnet wire coatings, magnet wires, and varnishes.

These requirements do not cover a single insulating material or a simple combination of materials, such as a laminate or a varnished cloth.

These requirements do not cover insulation systems exposed to radiation or operating in oils, refrigerants, soaps, or other media that potentially degrade insulating materials.

These requirements shall be modified or supplemented as determined by the applicable requirements in the end-product standard covering the device, appliance, or equipment in which the insulation system is used.

Additional consideration shall be given to conducting tests for an insulating material, such as a coil encapsulant, that is used as the ultimate electrical enclosure.

Additional consideration shall be given to conducting tests for an insulating material or component that is a functional support of, or in direct contact with, a live part.

A product that contains features, characteristics, components, materials, or systems new or different from those in use when the standard was developed — and that involves a risk of fire, electric shock, or injury to persons — shall be evaluated using the appropriate additional component and end-product requirements as determined necessary to maintain the level of safety for the user of the product as originally anticipated by the intent of this standard.

3.14.3.3.6   *IEC Technical Committee 14 Power Transformers*

The purpose of IEC Technical Committee 14 Power Transformers is to prepare international standards for power transformers, on-load tap changers, and reactors without limitation of voltage or power (not including instrument transformers, testing transformers, traction transformers mounted on rolling stock, and welding transformers).

Relevant documents are listed in Table 3.14.5.

3.14.3.4   Medium-Voltage and Large-Power Dry-Type Transformers

3.14.3.4.1   *IEEE Documents*

These standards are intended as a basis for the establishment of performance, interchangeability, and safety requirements of equipment described and for assistance in the proper selection of such equipment.

Electrical, mechanical, and safety requirements of ventilated, nonventilated, and sealed dry-type distribution and power transformers or autotransformers (single and polyphase, with a voltage of 601 V or higher in the highest-voltage winding) are described. Instrument transformers and rectifier transformers are also included.

TABLE 3.14.5 Relevant Documents for IEC Technical Committee 14 Power Transformers

60076-1	General requirements
60076-2	Temperature rise
60076-3	Insulation-levels, dielectric tests, and external clearances in air
60076-4	Guide for lightning impulse and switching impulse testing
60076-5	Ability to withstand short circuit
60076-6	Reactors (IEC 289)
60076-8	Power transformers (application guide)
60076-9	Terminal and tapping markings (IEC 616)
60076-10	Determination of transformer reactor sound levels
60076-11	Dry-type power transformers (IEC 726)
60076-12	Loading guide for dry-type power transformers (IEC 905)
61378	Converter transformers
61378-1	Transformers for industrial applications
61378-3	Applications guide

The information in these standards apply to all dry-type transformers except as follows:

Arc-furnace transformers
Rectifier transformers
Specialty transformers
Mine transformers

When these standards are used on a mandatory basis, the word *shall* indicates mandatory requirements; the words *should* and *may* refer to matters that are recommended or permissive but not mandatory.

Note: The introduction of this voluntary-consensus standard describes the circumstances under which the standard may be used on a mandatory basis.

Relevant IEEE documents are listed in Table 3.14.6.

3.14.3.4.2 NEMA Documents

3.14.3.4.2.1 NEMA TP-1, Guide for Determining Energy Efficiency for Distribution Transformers — This standard is intended for use as a basis for determining the energy-efficiency performance of the equipment covered and to assist in the proper selection of such equipment. This standard covers single-phase and three-phase dry-type and liquid-filled distribution transformers as defined in Table 3.14.7.

Products excepted from this standard include:

Liquid-filled transformers below 10 kVA
Dry-type transformers below 15 kVA
Drive transformers, both ac and dc
All rectifier transformers and transformers designed for high harmonics
Autotransformers
Nondistribution transformers, such as UPS transformers
Special impedance and harmonic transformers
Regulating transformers
Sealed and nonventilated transformers
Retrofit transformers
Machine-tool transformers
Welding transformers
Transformers with tap ranges greater than 15%
Transformers with frequency other than 60 Hz
Grounding transformers
Testing transformers

TABLE 3.14.6 Relevant IEEE Documents

IEEE 259-1994	Standard Test Procedure for Evaluation of Systems of Insulation for Specialty Transformers (ANSI)
IEEE 638-1992	Standard for Qualification of Class 1E Transformers for Nuclear Generating Stations
ANSI C57.12.00-1993	IEEE Standard General Requirements for Liquid-Immersed Distribution, Power, and Regulating Transformers (ANSI)
ANSI C57.12.01-1998	IEEE Standard General Requirements for Dry-Type Distribution and Power Transformers Including Those with Solid Cast and/or Resin-Encapsulated Windings
IEEE C57.12.35-1996	Standard for Bar Coding for Distribution Transformers
ANSI C57.12.40-1994	American National Standard Requirements for Secondary Network Transformers — Subway and Vault Types (Liquid Immersed)
IEEE C57.12.44-1994	Standard Requirements for Secondary Network Protectors
ANSI C57.12.50-1981 (R1989)	American National Standard Requirements for Ventilated Dry-Type Distribution Transformers, 1 to 500 kVA, Single-Phase, and 15 to 500 kVA, Three-Phase, with High-Voltage 601 to 34 500 Volts, Low-Voltage 120 to 600 Volts
ANSI C57.12.51-1981 (R1989)	American National Standard Requirements for Ventilated Dry-Type Power Transformers, 501 kVA and Larger, Three-Phase, with High-Voltage 601 to 34 500 Volts, Low-Voltage 208Y/120 to 4160 Volts
ANSI C57.12.52-1981 (R1989)	American National Standard Requirements for Sealed Dry-Type Power Transformers, 501 kVA and Larger, Three-Phase, with High-Voltage 601 to 34 500 Volts, Low-Voltage 208Y/120 to 4160 Volts
ANSI C57.12.55-1987	American National Standard for Transformers — Dry-Type Transformers Used in Unit Installations, Including Unit Substations — Conformance Standard
IEEE C57.12.56-1986	Standard Test Procedure for Thermal Evaluation of Insulation Systems for Ventilated Dry-Type Power and Distribution Transformers
ANSI C57.12.57-1987 (R1992)	American National Standard for Transformers — Ventilated Dry-Type Network Transformers 2500 kVA and below, Three-Phase, with High-Voltage 34 500 Volts and below, Low-Voltage 216Y/125 and 480Y/277 Volts — Requirements
IEEE C57.12.58-1991 (R1996)	Guide for Conducting a Transient Voltage Analysis of a Dry-Type Transformer Coil
IEEE C57.12.59-2000	Guide for Dry-Type Transformer Through-Fault Current Duration
Draft C57.12.60-1998	Guide for Test Procedures for Thermal Evaluation of Insulation Systems for Solid-Cast and Resin-Encapsulated Power and Distribution Transformers
ANSI C57.12.70-1978 (R1992)	American National Standard Terminal Markings and Connections for Distribution and Power Transformers
IEEE C57.12.80-1978 (R1992)	Standard Terminology for Power and Distribution Transformers
IEEE C57.12.91-1995	Standard Test Code for Dry-Type Distribution and Power Transformers
IEEE C57.13-1993	Standard Requirements for Instrument Transformers
IEEE C57.13.1-1981 (R1992)	Guide for Field Testing of Relaying Current Transformers
IEEE C57.13.3-1983 (R1991)	Guide for the Grounding of Instrument Transformer Secondary Circuits and Cases
IEEE C57.15-1986 (R1992)	Standard Requirements, Terminology, and Test Code for Step-Voltage and Induction-Voltage Regulators
IEEE C57.16-1996	Standard Requirements, Terminology, and Test Code for Dry-Type Air-Core Series-Connected Reactors
IEEE C57.18.10-1998	Standard Practices and Requirements for Semiconductor Power Rectifier Transformers
IEEE C57.19.00-1991 (R1997)	Standard General Requirements and Test Procedures for Outdoor Power Apparatus Bushings
IEEE C57.19.01-1991 (R1997)	Standard Performance Characteristics and Dimensions for Outdoor Apparatus Bushings
IEEE C57.19.03-1996	Standard Requirements, Terminology, and Test Code for Bushings for dc Applications
IEEE C57.19.100-1995 (R1997)	Guide for Application of Power Apparatus Bushings
IEEE C57.21-1990 (R1995)	Standard Requirements, Terminology, and Test Code for Shunt Reactors Rated over 500 kVA
*IEEE C57.94-1982 (R1987)	Recommended Practice for the Installation, Application, Operation, and Maintenance of Dry-Type General Purpose Distribution and Power Transformers
IEEE C57.96-1989	Guide for Loading Dry-Type Distribution and Power Transformers
IEEE C57.98-1993	Guide for Transformer Impulse Tests (an errata sheet is available in PDF format)
IEEE C57.105-1978 (R1999)	Guide for Application of Transformer Connections in Three-Phase Distribution Systems

— continued

TABLE 3.14.6 (continued) Relevant IEEE Documents

IEEE C57.110-1998	Recommended Practice for Establishing Transformer Capability when Supplying Nonsinusoidal Load Currents
IEEE C57.116-1989 (R1994)	Guide for Transformers Directly Connected to Generators
IEEE C57.117-1986 (R1992)	Guide for Reporting Failure Data for Power Transformers and Shunt Reactors on Electric Utility Power Systems
IEEE C57.124-1991 (R1996)	Recommended Practice for the Detection of Partial Discharge and the Measurement of Apparent Charge in Dry-Type Transformers
IEEE C57.134-2000	IEEE Guide for Determination of Hottest-Spot Temperature in Dry-Type Transformers
IEEE C57.138-1998	Recommended Practice for Routine Impulse Test for Distribution Transformers
IEEE PC57.142	A Guide to Describe the Occurrence and Mitigation of Switching Transients Induced by Transformer and Breaker Interaction

TABLE 3.14.7 NEMA TP-1 Rating Range for Single-Phase and Three-Phase Dry-Type and Liquid-Filled Distribution Transformers

Voltage Class:
Primary Voltage = 34.5 kV and below
Secondary Voltage = 600 V and below

Transformer Type	Number of Phases	Rating Range
Liquid rating	Single phase	10–833 kVA
	Three phase	15–2500 kVA
Dry rating	Single phase	15–833 kVA
	Three phase	15–2500 kVA

Note: Includes all products at 1.2 kV and below.

3.14.3.4.2.2   NEMA TP-2, Standard Test Method for Measuring the Energy Consumption of Distribution Transformers — This standard is intended for use as a basis for determining the energy-efficiency performance of the equipment covered and to assist in the proper selection of such equipment. This standard covers single-phase and three-phase dry-type and liquid-immersed distribution transformers (transformers for transferring electrical energy from a primary distribution circuit to a secondary distribution circuit or within a secondary distribution circuit) as defined in Table 3.14.8.

This standard addresses the test procedures for determining the efficiency performance of the transformers covered in NEMA Publication TP-1.

Products excepted from this standard include:

Liquid-filled transformers below 10 kVA
Dry-type transformers below 15 kVA
Transformers connected to converter circuits
All rectifier transformers and transformers designed for high harmonics
Autotransformers
Nondistribution transformers, such as UPS transformers
Special impedance and harmonic transformers
Regulating transformers
Sealed and nonventilated transformers
Retrofit transformers
Machine-tool transformers
Welding transformers
Transformers with tap ranges greater than 15%
Transformers with frequency other than 60 Hz
Grounding transformers
Testing transformers

TABLE 3.14.8 NEMA TP-2 Rating Range for Single-Phase and Three-Phase Dry-Type and Liquid-Immersed Distribution Transformers

Transformer Type	Number of Phases	Rating Range
Liquid immersed	Single phase	10–833 kVA
	Three phase	15–2500 kVA
Dry type	Single phase	15–833 kVA
	Three phase	15–2500 kVA

Note: Includes all products at 1.2 kV and below.

3.14.3.4.3 ANSI/UL 1562, Standard for Safety for Transformers, Distribution, Dry-Type—over 600 V

These requirements cover single-phase or three-phase, dry-type, distribution transformers. The transformers are provided with either ventilated or nonventilated enclosures and are rated for a primary or secondary voltage from 601 to 35,000 V and from 1 to 5,000 kVA. These transformers are intended for installation in accordance with the National Electrical Code.

These requirements do not cover the following transformers:

Instrument transformers
Step-voltage and induction-voltage regulators
Current regulators
Arc-furnace transformers
Rectifier transformers
Specialty transformers (such as rectifier, ignition, gas-tube-sign transformers, and the like)
Mining transformers
Motor-starting reactors and transformers

These requirements do not cover transformers under the exclusive control of electrical utilities utilized for communication, metering, generation, control, transformation, transmission, and distribution of electric energy, regardless of whether such transformers are located indoors (in buildings and rooms used exclusively by utilities for such purposes) or outdoors (on property owned, leased, established rights on private property, or on public rights of way [highways, streets, roads, and the like]).

A product that contains features, characteristics, components, materials, or systems new or different from those in use when the standard was developed — and that involves a risk of fire, electric shock, or injury to persons — shall be evaluated using the appropriate additional component and end-product requirements as determined necessary to maintain the level of safety for the user of the product as originally anticipated by the intent of this standard.

3.14.3.4.4 IEC Technical Committee 14 Power Transformers

The purpose of IEC Technical Committee 14 Power Transformers is to prepare international standards for power transformers, on-load tap changers, and reactors without limitation of voltage or power (not including instrument transformers, testing transformers, traction transformers mounted on rolling stock, and welding transformers).

Relevant documents are listed in Table 3.14.9.

3.14.3.5 Liquid-Filled Transformers

3.14.3.5.1 IEEE Documents

These standards provide a basis for establishing the performance, limited electrical and mechanical interchangeability, and safety requirements of the equipment described. They are also a basis for assistance in the proper selection of such equipment.

These standards describe electrical, mechanical, and safety requirements of liquid-immersed distribution and power transformers as well as autotransformers and regulating transformers, single and polyphase, with voltages of 601 V or higher in the highest-voltage winding. These standards also cover instrument transformers and rectifier transformers.

TABLE 3.14.9 Relevant Documents for IEC Technical Committee 14 Power Transformers

60076-1	General requirements
60076-2	Temperature rise
60076-3	Insulation levels, dielectric tests, and external clearances in air
60076-4	Guide for lightning-impulse and switching-impulse testing
60076-5	Ability to withstand short circuit
60076-6	Reactors (IEC 289)
60076-8	Power transformers — application guide
60076-9	Terminal and tapping markings (IEC 616)
60076-10	Determination of transformer reactor sound levels
60076-11	Dry-type power transformers (IEC 726)
60076-12	Loading guide for dry-type power transformers (IEC 905)
60214-1	Tap changers for power transformers
60214-2	Application guide for on-load tap changers (IEC 542)
61378	Converter transformers
61378-1	Transformers for industrial applications
61378-3	Applications guide

These standards apply to all liquid-immersed distribution, power, regulating, instrument, and rectifier transformers except as indicated below:

Arc-furnace transformers
Specialty transformers
Grounding transformers
Mobile transformers
Mine transformers

Relevant IEEE documents are listed in Table 3.14.10.

3.14.3.5.2   NEMA Documents

3.14.3.5.2.1   NEMA TP-1, Guide for Determining Energy Efficiency for Distribution Transformers — This standard is intended for use as a basis for determining the energy-efficiency performance of the equipment covered and to assist in the proper selection of such equipment. This standard covers single-phase and three-phase dry-type and liquid-filled distribution transformers as defined in Table 3.14.11. Products excepted from this standard include:

Liquid-filled transformers below 10 kVA
Dry-type transformers below 15 kVA
Drives transformers, both ac and dc
All rectifier transformers and transformers designed for high harmonics
Autotransformers
Nondistribution transformers, such as UPS transformers
Special impedance and harmonic transformers
Regulating transformers
Sealed and nonventilated transformers
Retrofit transformers
Machine-tool transformers
Welding transformers
Transformers with tap ranges greater than 15%
Transformers with frequency other than 60 Hz
Grounding transformers
Testing transformers

TABLE 3.14.10 Relevant IEEE Documents

IEEE 62-1995	Guide for Diagnostic Field Testing of Electric Power Apparatus — Part 1: Oil Filled Power Transformers, Regulators, and Reactors
IEEE 259-1994	Standard Test Procedure for Evaluation of Systems of Insulation for Specialty Transformers (ANSI)
IEEE 637-1985	Guide for the Reclamation of Insulating Oil and Criteria for Its Use (ANSI)
IEEE 638-1992	Standard for Qualification of Class 1E Transformers for Nuclear Generating Stations
IEEE 799-1987	Guide for Handling and Disposal of Transformer Grade Insulating Liquids Containing PCBs (ANSI)
IEEE 1276-1997	Trial-Use Guide for the Application of High Temperature Insulation Materials in Liquid-Immersed Power Transformers
ANSI C57.12.00-1993	Standard General Requirements for Liquid-Immersed Distribution, Power, and Regulating Transformers (ANSI)
ANSI C57.12.10-1988	American National Standard for Transformers — 230 kV and below 833/958 through 8333/10 417 kVA, Single-Phase, and 750/862 through 60 000/80 000/100 000 kVA, Three-Phase without Load Tap Changing; and 3750/4687 through 60 000/80 000/100 000 kVA with Load Tap Changing — Safety Requirements
ANSI C57.12.20-1997	American National Standard for Overhead Distribution Transformers, 500 kVA and Smaller: High Voltage, 34 500 V and below: Low Voltage 7970/13 800 Y V and below
ANSI C57.12.22-1989	American National Standard for Transformers — Pad-Mounted, Compartmental-Type, Self-Cooled, Three-Phase Distribution Transformers with High-Voltage Bushings, 2500 kVA and Smaller: High Voltage, 34 500 GrdY/19 920 Volts and below; Low Voltage, 480 Volts and below
IEEE C57.12.23-1992	Standard for Transformers — Underground-Type, Self-Cooled, Single-Phase Distribution Transformers With Separable, Insulated, High-Voltage Connectors; High Voltage (24 940 GrdY/14 400 V and below) and Low Voltage (240/120 V, 167 kVA and Smaller)
ANSI C57.12.24-1994	American National Standard Requirements for Transformers — Underground-Type, Three-Phase Distribution Transformers, 2500 kVA and Smaller; High Voltage, 34 500 GrdY/19 920 Volts and below; Low Voltage, 480 Volts and below — Requirements
ANSI C57.12.25-1990	American National Standard for Transformers — Pad-Mounted, Compartmental-Type, Self-Cooled, Single-Phase Distribution Transformers with Separable Insulated High-Voltage Connectors; High Voltage, 34 500 GrdY/19 920 Volts and below; Low Voltage, 240/120 Volts; 167 kVA and Smaller
IEEE C57.12.26-1992	Standard for Pad-Mounted, Compartmental-Type, Self-Cooled, Three-Phase Distribution Transformers for Use with Separable Insulated High-Voltage Connectors (34 500 GrdY/19 920 Volts and below; 2500 kVA and Smaller)
ANSI C57.12.29-1991	American National Standard Switchgear and Transformers-Pad-Mounted Equipment-Enclosure Integrity for Coastal Environments
ANSI C57.12.31-1996	American National Standard for Pole-Mounted Equipment — Enclosure Integrity
ANSI C57.12.32-1994	American National Standard for Submersible Equipment — Enclosure Integrity
IEEE C57.12.35-1996	Standard for Bar Coding for Distribution Transformers
ANSI C57.12.40-1994	American National Standard Requirements for Secondary Network Transformers — Subway and Vault Types (Liquid Immersed)
IEEE C57.12.44-1994	Standard Requirements for Secondary Network Protectors
ANSI C57.12.70-1978 (R1992)	American National Standard Terminal Markings and Connections for Distribution and Power Transformers
IEEE C57.12.80-1978 (R1992)	Standard Terminology for Power and Distribution Transformers
IEEE C57.12.90-1993	Standard Test Code for Liquid-Immersed Distribution, Power, and Regulating Transformers and Guide for Short Circuit Testing of Distribution and Power Transformers
IEEE C57.13-1993	Standard Requirements for Instrument Transformers
IEEE C57.13.1-1981 (R1992)	Guide for Field Testing of Relaying Current Transformers
IEEE C57.13.3-1983 (R1991)	Guide for the Grounding of Instrument Transformer Secondary Circuits and Cases
IEEE C57.15-1986 (R1992)	Standard Requirements, Terminology, and Test Code for Step-Voltage and Induction-Voltage Regulators
IEEE C57.16-1996	Standard Requirements, Terminology, and Test Code for Dry-Type Air-Core Series-Connected Reactors
IEEE C57.18.10-1998	Standard Practices and Requirements for Semiconductor Power Rectifier Transformers

—continued

TABLE 3.14.10 (continued) Relevant IEEE Documents

IEEE C57.19.00-1991 (R1997)	Standard General Requirements and Test Procedures for Outdoor Power Apparatus Bushings
IEEE C57.19.01-1991 (R1997)	Standard Performance Characteristics and Dimensions for Outdoor Apparatus Bushings
IEEE C57.19.03-1996	Standard Requirements, Terminology, and Test Code for Bushings for DC Applications
IEEE C57.19.100-1995 (R1997)	Guide for Application of Power Apparatus Bushings
IEEE C57.21-1990 (R1995)	Standard Requirements, Terminology, and Test Code for Shunt Reactors Rated over 500 kVA
IEEE C57.91-1995	Guide for Loading Mineral-Oil-Immersed Transformers
IEEE C57.93-1995	Guide for Installation of Liquid-Immersed Power Transformers
IEEE C57.98-1993	Guide for Transformer Impulse Tests (An errata sheet is available in PDF format)
IEEE C57.100-1986 (R1992)	Standard Test Procedures for Thermal Evaluation of Oil-Immersed Distribution Transformers
IEEE C57.104-1991	Guide for the Interpretation of Gases Generated in Oil-Immersed Transformers
IEEE C57.105-1978 (R1999)	Guide for Application of Transformer Connections in Three-Phase Distribution Systems
IEEE C57.109-1993	Guide for Liquid-Immersed Transformer Through-Fault-Current Duration
IEEE C57.110-1998	Recommended Practice for Establishing Transformer Capability when Supplying Nonsinusoidal Load Currents
IEEE C57.111-1989 (R1995)	Guide for Acceptance of Silicone Insulating Fluid and Its Maintenance in Transformers
IEEE C57.113-1991	Guide for Partial Discharge Measurement in Liquid-Filled Power Transformers and Shunt Reactors
IEEE C57.116-1989 (R1994)	Guide for Transformers Directly Connected to Generators
IEEE C57.117-1986 (R1992)	Guide for Reporting Failure Data for Power Transformers and Shunt Reactors on Electric Utility Power Systems
IEEE C57.121-1998	Guide for Acceptance and Maintenance of Less Flammable Hydrocarbon Fluid in Transformers
IEEE C57.131-1995	Standard Requirements for Load Tap Changers
IEEE C57.138-1998	Recommended Practice for Routine Impulse Test for Distribution Transformers
IEEE C57.120-1991	IEEE Loss Evaluation Guide for Power Transformers and Reactors
IEEE PC57.123	Draft Guide for Transformer Loss Measurement
IEEE C57.125-1991	IEEE Guide for Failure Investigation, Documentation, and Analysis for Power Transformers and Shunt Reactors
IEEE PC57.127-2000	Trial Use Guide for the Detection of Acoustic Emissions from Partial Discharges in Oil-Immersed Power Transformers
IEEE PC57.129-1999	Trial-Use General Requirements and Test Code for Oil-Immersed HVDC Converter Transformers
IEEE PC57.130-1998	Trial Use Guide for the Use of Dissolved Gas Analysis during Factory Thermal Tests for the Evaluation of Oil Immersed Transformers and Reactors
IEEE C57.131-1995	IEEE Requirements for Load Tap Changers
IEEE PC57.133-2001	IEEE Guide for Short-Circuit Testing of Distribution and Power Transformers
IEEE PC57.135	Draft Guide for the Application, Specification and Testing of Phase-Shifting Transformers
IEEE PC57.136	Draft Guide for Sound Abatement and Determination for Liquid-Immersed Power Transformers and Shunt Reactors Rated over 500 kVA
IEEE PC57.142	A Guide To Describe the Occurrence and Mitigation of Switching Transients Induced by Transformer and Breaker Interaction

TABLE 3.14.11 NEMA TP-1 Rating Range for Single-Phase and Three-Phase Dry-Type and Liquid-Filled Distribution Transformers

Voltage Class
Primary Voltage = 34.5 kV and below
Secondary Voltage = 600 V and below

Transformer Type	Number of Phases	Rating Range
Liquid rating	Single phase	10–833 kVA
	Three phase	15–2500 kVA
Dry rating	Single phase	15–833 kVA
	Three phase	15–2500 kVA

3.14.3.5.2.2   NEMA TP-2, Standard Test Method for Measuring the Energy Consumption of Distribution
        Transformers — This standard is intended for use as a basis for determining the energy-
efficiency performance of the equipment covered and to assist in the proper selection of such equipment.
This standard covers single-phase and three-phase dry-type and liquid-immersed distribution transformers (transformers for transferring electrical energy from a primary distribution circuit to a secondary
distribution circuit or within a secondary distribution circuit) as defined in Table 3.14.12.
This standard addresses the test procedures for determining the efficiency performance of the transformers covered in NEMA Publication TP-1.

Products excepted from this standard include:

Liquid-filled transformers below 10 kVA
Dry-type transformers below 15 kVA
Transformers connected to converter circuits
All rectifier transformers and transformers designed for high harmonics
Autotransformers
Nondistribution transformers, such as UPS transformers
Special impedance and harmonic transformers
Regulating transformers
Sealed and nonventilated transformers
Retrofit transformers
Machine-tool transformers
Welding transformers
Transformers with tap ranges greater than 15%
Transformers with frequency other than 60 Hz
Grounding transformers
Testing transformers

3.14.3.5.2.3   NEMA TR-1, 1993 (R-1999) Transformers Regulators and Reactors — This publication
provides a list of all ANSI C57 standards that have been approved by NEMA. In addition, it includes
certain NEMA standard test methods, test codes, properties, etc., of liquid-immersed transformers,
regulators, and reactors that are not American national standards.

3.14.3.5.3   IEC 76-1

IEC 76-1 is intended to prepare international standards for power transformers, on-load tap changers,
and reactors without limitation of voltage or power (not including instrument transformers, testing
transformers, traction transformers mounted on rolling stock, and welding transformers). The relevant
documents are listed in Table 3.14.13.

Note that the industry standards-making organizations and participants are constantly in a state of
change. In order to see the most recent standards and guides, please consult the catalogs of the respective
standards-writing organizations.

TABLE 3.14.12   NEMA TP-2 Rating Range for Single-Phase and Three-Phase Dry-Type and Liquid-
Immersed Distribution Transformers

Transformer Type	Number of Phases	Rating Range
Liquid immersed	Single phase	10–833 kVA
	Three phase	15–2500 kVA
Dry type	Single phase	15–833 kVA
	Three phase	15–2500 kVA

Note: Includes all products at 1.2 kV and below.

TABLE 3.14.13 Relevant Documents for IEC 76-1

60076-1	General requirements
60076-2	Temperature rise
60076-3	Insulation levels, dielectric tests, and external clearances in air
60076-4	Guide for lightning impulse and switching impulse testing
60076-5	Ability to withstand short circuit
60076-6	Reactors (IEC 289)
60076-7	Loading guide for oil-immersed power transformer (IEC 354)
60076-8	Power transformers — application guide
60076-9	Terminal and tapping markings (IEC 616)
60076-10	Determination of transformer reactor sound levels
60076-11	Dry-type transformers
60076-12	Loading guide for dry-type transformers
60076-13	Self-protected liquid-filled transformers
60076-14	Guide for the design and application of liquid-immersed power transformers, using high-temperature insulating materials
60076-15	Gas-filled-type power transformers
60214-1	Tap changers for power transformers
60214-2	Application guide for on-load tap changers (IEC 542)
61378	Converter transformers
61378-1	Transformers for industrial applications
61378-2	Transformers for HVDC applications
61378-3	Applications guide

Index

S